Project Management

Annotated Instructor's Guide

Second Edition

AF709549

PEARSON

nccer

Upper Saddle River,
New Jersey
Columbus, Ohio

National Center for Construction Education and Research

President: Don Whyte
Director of Curriculum Revision and Development: Daniele Stacey
Project Management Project Manager: Carla Sly
Production Manager: Tim Davis
Quality Assurance Coordinator: Debie Ness
Editors: Rob Richardson, Matt Tischler, Brendan Coote
Desktop Publishing Coordinator: James McKay

NCCER would like to acknowledge the contract service provider for this curriculum:
MetaMedia Training International, Germantown, Maryland.

This information is general in nature and intended for training purposes only. Actual performance of activities described in this manual requires compliance with all applicable operating, service, maintenance, and safety procedures under the direction of qualified personnel. References in this manual to patented or proprietary devices do not constitute a recommendation of their use.

Copyright © 2008, 1997 National Center for Construction Education and Research, Gainesville, FL 32606. No part of this work may be reproduced in any form or by any means, including photocopying, without written permission of the publisher. Developed by the National Center for Construction Education and Research. Published by Pearson Education, Inc., Upper Saddle River, NJ 07458. All rights reserved. Printed in the United States of America. This publication is protected by Copyright and permission should be obtained from the NCCER prior to any prohibited reproduction, storage in a retrieval system, or transmission in any form or by any means, electronic, mechanical, photocopying, recording, or likewise. For information regarding permission(s), write to: NCCER Product Development, 3600 NW 43rd St, Bldg G, Gainesville, FL 32606.

10 9 8 7 6 5 4 3 2

PEARSON

Contren® Learning Series

ISBN 0-13-604487-5
ISBN 13 978-0-13-604487-1

PREFACE

TO THE INSTRUCTOR

As a new or soon-to-be construction project manager, you're embarking upon a career that requires more brains than brawn. You may spend less time at the job site and more time in meetings. However, the experience you gained as a craftworker will serve you well as you negotiate this new career path.

Construction project managers plan and direct the building and maintenance of everything from bridges to high-rises to wastewater systems. They usually have the satisfaction of seeing a job through from start to finish. In a typical day, project managers might meet with owners, examine a work breakdown structure (WBS), negotiate with subcontractors, and directly supervise crews. Their qualifications include both formal education and informal on-the-job training, and their knowledge encompasses construction materials and methods, mathematics, communications, safety, human resources, scheduling, and customer service, among other areas.

According to the Bureau of Labor Statistics, the outlook for this profession is bright. In 2004, 431,000 persons worked as construction managers, and the need is for 123,000 more by 2014. A typical construction manager earns over $70,000 annually.

This updated edition of *Project Management* introduces you to the many areas of your profession. Since you'll spend a considerable amount of time meeting with people, communications will be important. Behind the scenes, you'll work to make sure that costs and resources are controlled, quality is achieved, and schedules are maintained. While this book gives you an overview of all these areas, your employer will likely provide you with more detailed training.

NEW WITH THIS UPDATE

NCCER and Pearson are pleased to present a companion DVD with this update of *Project Management*. Entitled: *Minor Decisions: Major Impact – How to Deal with Real Issues in Project Management*, the DVD provides instructors with situational scenarios that stimulate discussion about the roles and actions of a project manager. Originally offered as a stand-alone resource, this video is now available to all users of this second edition. Suggested activities and discussion are included in each module should you choose to integrate this instruction into your lesson plans. Refer to the section entitled *Using The DVD With This Annotated Instructor's Guide* for more details.

We invite you to visit the NCCER website at www.nccer.org for the latest releases, training information, newsletter, and much more. You can also reference the Contren® product catalog online at www. crafttraining.com. Your feedback is welcome. You may email your comments to curriculum@nccer.org or send general comments and inquiries to info@nccer.org.

CONTREN® LEARNING SERIES

The National Center for Construction Education and Research (NCCER) is a not-for-profit 501(c)(3) education foundation established in 1995 by the world's largest and most progressive construction companies and national construction associations. It was founded to address the severe workforce shortage facing the industry and to develop a standardized training process and curricula. Today, NCCER is supported by hundreds of leading construction and maintenance companies, manufacturers, and national associations. The Contren® Learning Series was developed by NCCER in partnership with Pearson Education, Inc., the world's largest educational publisher.

Some features of NCCER's Contren® Learning Series are as follows:

- An industry-proven record of success
- Curricula developed by the industry for the industry
- National standardization providing portability of learned job skills and educational credits
- Compliance with the Office of Apprenticeship requirements for related classroom training (CFR 29:29)
- Well-illustrated, up-to-date, and practical information

NCCER also maintains a National Registry that provides transcripts, certificates, and wallet cards to individuals who have successfully completed modules of NCCER's Contren® Learning Series. *Training programs must be delivered by an NCCER Accredited Training Sponsor in order to receive these credentials.*

Contents

Minor Decisions: Major Impact –
How to Deal with Real Issues in Project Management
Using The DVD With This Annotated Instructor's Guide

Contren® Curricula

NCCER's training programs comprise over 50 construction, maintenance, and pipeline areas and include skills assessments, safety training, and management education.

Boilermaking
Cabinetmaking
Carpentry
Concrete Finishing
Construction Craft Laborer
Construction Technology
Core Curriculum:
  Introductory Craft Skills
Drywall
Electrical
Electronic Systems Technician
Heating, Ventilating, and
  Air Conditioning
Heavy Equipment Operations
Highway/Heavy Construction
Hydroblasting
Industrial Maintenance
  Electrical and Instrumentation
  Technician
Industrial Maintenance Mechanic
Instrumentation
Insulating
Ironworking
Masonry
Millwright
Mobile Crane Operations
Painting
Painting, Industrial
Pipefitting
Pipelayer
Plumbing
Reinforcing Ironwork
Rigging
Scaffolding
Sheet Metal
Site Layout
Sprinkler Fitting
Welding

Pipeline
Control Center Operations,
  Liquid
Corrosion Control
Electrical and Instrumentation
Field Operations, Liquid
Field Operations, Gas
Maintenance
Mechanical

Safety
Field Safety
Safety Orientation
Safety Technology

Management
Introductory Skills for the
  Crew Leader
Project Management
Project Supervision

Spanish Translations
Albañilería Nivel Uno
Andamios
Currículo Básico
  Habilidades Introductorias
  del Oficio
Instalación de Rociadores
  Nivel Uno
Orientación de Seguridad
Principios Básicos de
  Maniobras
Seguridad de Campo

Supplemental Titles
Applied Construction Math
Careers in Construction
Your Role in the Green
  Environment

Acknowledgments

This curriculum was revised as a result of the farsightedness and leadership of the following sponsors:

Austin Industrial
Clemson University
Construction Services Enterprise
FMI Corporation
Gaylor, Inc.
Linc Network, LLC
Lincoln Educational Services
M.C. Dean, Inc.
Professional Safety Associates
Sundt Construction
Tansey & Associates, Inc.
Zachry Construction Corporation

This curriculum would not exist were it not for the dedication and unselfish energy of those volunteers who served on the Authoring Team. A sincere thanks is extended to the following:

Dan Barrow
Dennis Bausman
Rick Bubier
Kelley Chisholm
Stephen Clare
Philip Copare
Gregg Corley
G. Leroy Ehlers
Richard J. Goetz
Chris Harrington
Roger Liska
Mike Mosley
Steven P. Pereira
Clint Sundt
Tom Tansey
Gene Tibbs
Douglas Wagner

NCCER PARTNERING ASSOCIATIONS

American Fire Sprinkler Association
Associated Builders and Contractors, Inc.
Associated General Contractors of America
Association for Career and Technical Education
Association for Skilled and Technical Sciences
Carolinas AGC, Inc.
Carolinas Electrical Contractors Association
Center for the Improvement of Construction Management and Processes
Construction Industry Institute
Construction Users Roundtable
Design Build Institute of America
Green Advantage
Merit Contractors Association of Canada
Metal Building Manufacturers Association
NACE International
National Association of Minority Contractors
National Association of Women in Construction
National Insulation Association
National Ready Mixed Concrete Association
National Systems Contractors Association
National Technical Honor Society
National Utility Contractors Association
NAWIC Education Foundation
North American Crane Bureau
North American Technician Excellence
Painting & Decorating Contractors of America
Portland Cement Association
SkillsUSA
Steel Erectors Association of America
Texas Gulf Coast Chapter, ABC
U.S. Army Corps of Engineers
University of Florida
Women Construction Owners & Executives, USA

Product Supplements

Windows/Macintosh-Based

TestGen

ISBN 13 978-0-13-605592-1
ISBN 0-13-605592-3

Ensure test security with NCCER's computerized testing software. This software allows instructors to scramble the module exam questions and answer keys in order to print multiple versions of the same test, customize tests to suit the needs of their training units*, add questions, or easily create a final exam.

**Due to NCCER's Accreditation Guidelines, instructors may not delete existing questions from exams. Doing so may seriously jeopardize either the accreditation status or the training program sponsor or any recognition of trainees, instructors, and trainers through the NCCER National Registry.*

Transparency

Masters

ISBN 13 978-0-13-605612-6
ISBN 0-13-605612-1

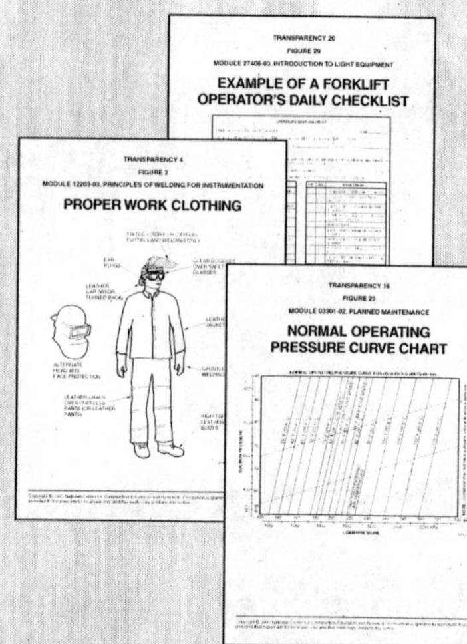

Spend more time training and less time at the copier. In response to instructor feedback NCCER offers loose, reproducible copies of the overhead transparencies referenced in the Instructor's Guides. The transparency masters package includes most of the Trainee Guide graphics, enlarged for projection and printed on loose sheets* for easy copying onto transparency film using your photocopier.

* *Transparency masters are provided on regular, loose sheets of paper, not acetates.*

Microsoft®

PowerPoint® Presentation Slides

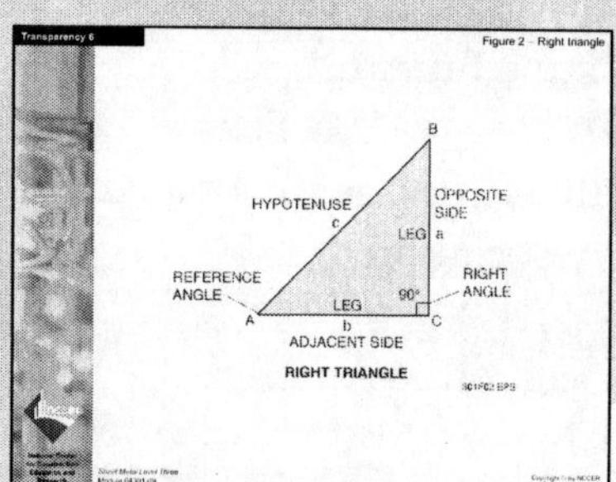

This slide show presentation will be a powerful addition to your classroom discussions! NCCER's Microsoft® PowerPoint® CD features the drawings and photos from the Traineee Guide, sequenced as they are called out in the Annotated Instructor's Guide annotations. Please contact your Pearson Sales Representative for availability.

ISBN 13 978-0-13-212939-8
ISBN 0-13-212939-6

To order these supplements, contact Pearson Education, Inc. at

800-922-0579

Using The DVD With This Annotated Instructor's Guide

Minor Decisions: Major Impact

How to Deal with Real Issues in Project Management

USING THE DVD WITH THIS ANNOTATED INSTRUCTOR'S GUIDE

The DVD *Minor Decisions: Major Impact* included with this Annotated Instructor's Guide is intended to provide instructors with situational scenarios that stimulate discussion about the roles and actions of a project manager, and bring project management issues to life in the classroom. Designed to complement NCCER's *Project Management* curriculum, the topics and situations in this video can apply to many of today's real-world problems in the construction industry.

This AIG contains discussion suggestions and tips to guide your use of this video in the classroom. Adult education theory shows that adults learn best when they can listen, see, discuss, and manipulate the issues or tasks they are learning about. Engage the class in substantive discussions and encourage students to share their experiences and lessons learned. By eliciting the combined experience and knowledge of the students in your class, you will be using solid adult learning theory to help them become effective project managers. Your students will benefit from hearing your own real-world stories and personal experiences in relation to the issues and situations that arise throughout the video scenario.

Lesson Plan

The choice of topics to discuss, the length of time to discuss them, and the sharing of actual experiences all influence every lesson. There are six scenarios, each less than two minutes in length, except for the final module. The recommended time for class discussion is 15 to 20 minutes per scenario. However, the time may vary according to your preferences and the topics covered in the segment.

To plan how to embed the video in your instruction, play and study it in its entirety several times before using the video in the classroom. The story evolves and the segments will cover issues that relate to key topics in project management. An effective way to engage the class is to pause the video after an event is depicted and ask the students what they think about the event that was portrayed. What points were made by the parties involved and are they important to the resolution of the problem? Have any of the class members experienced similar problems or concerns? What might they expect to happen next? Resume play, let the scenario unfold, and follow up with discussion points about the incidents portrayed.

This Annotated Instructor's Guide provides suggestions, tips, discussion questions, and exercises that you can use to engage the students.

Note: A User's Guide insert is included with the enclosed DVD. This insert is intended for those using the DVD as stand-alone instruction (that is, without the Annotated Instructor's Guide).

Introduction

Minor Decisions: Major Impact is a case study illustrating a realistic situation that a project manager is likely to encounter on the job. It brings to life the challenges of managing under pressure, resolving team conflicts, focusing on best practices, and mitigating impacts on schedules and costs.

In this story, Samantha, the project manager, has to referee arguments between foremen; juggle schedules and delays; and resolve resource, safety, and housekeeping issues. Her solutions to her problems are presented before the story ends, although participants are encouraged to provide alternative resolutions.

The scenario is presented in six segments with the following running times in minutes. Each segment is accessed from the main menu on the DVD.

The Missing Tool	1:29
Samantha's Response	1:57
Keith's Defense	1:11
Anthony's Explanation	1:27
Craig's Observation	1:59
The Resolution	3:55

General Themes and Issues

The six video segments cover a number of themes and underlying issues that project managers face and that are covered in the *Project Management* curriculum. These include:

- Management and leadership styles
- Teamwork and attitudes
- Conflict resolution
- Disciplinary action
- Problem identification, fact finding, and root cause analysis
- Planning
- Problem resolution
- Resource control
- Housekeeping
- Safety
- Schedule and cost impacts
- Quality and best practices

WORKING WITH THE PROJECT MANAGEMENT DVD

Equipment Needed

The video can be played using a stand-alone consumer or professional DVD player with a remote control and a display device. It can also be played on a DVD-capable computer connected to a display device and speakers. Using a computer enables you to control the menu from the keyboard.

Consumer DVD Player

Navigate using a standard DVD remote control packaged with your player.

DVD remote controls enable you to navigate using a set of buttons (up, down, left and right arrows, and an ENTER button).

Use the navigation buttons on the remote control to highlight the scenario menu item that you wish to play. Click the ENTER or OK button to select the highlighted scenario. The video will start playing.

Professional DVD Player

Navigate using the player's remote control or a mouse.

If using the remote to access the menu, follow the directions that accompany your professional DVD player. An industrial player will have Audio Out and Video Out connectors. USB connections enable you to connect a mouse to guide navigation.

Computer

Navigate using a computer and a mouse.

Your computer must have a DVD-ROM drive and software loaded onto it to play DVDs. If your computer has an embedded DVD drive, the application software may be included on your system. Many DVD software players can be downloaded via the Internet. Once the software is installed, you can easily navigate through the video using a mouse.

In the classroom, you will need to connect your computer to a projection system for class viewing.

Setup and Practice

Before using the video for the first time in a classroom setting, make sure you are prepared and that the system is properly set up. Play the video to test that the projection and audio system are correctly connected. Practice using the remote control or mouse to guide the segments you want to play, and to pause and resume.

Viewing and Printing Documents

For your convenience, this user's guide is provided as a PDF document on the DVD. You will need Adobe® Acrobat Reader™ software to view and print the manual. If you do not have Adobe Acrobat Reader, you can download it at no cost at www.adobe.com.

Use a computer with a DVD-ROM drive to view and print the .pdf files. To access the files:

- Double-click on *My Computer*.
- Right-click the DVD-ROM drive (typically D:).
- Select *Explore*.
- Double-click the *Instructor Resources folder*.
- Select the documents to open the file.
- Print from your computer.

Additional copies of

Minor Decisions: Major Impact

may be obtained through Pearson Education, Inc. for a cost of $100.00.

Please ask your Sales Representative for ISBN 0-13-609093-1 or 978-0-13-609093-9

Module 44101-08

Introduction to Project Management

NCCER STANDARDIZED CRAFT TRAINING PROGRAM

The National Center for Construction Education and Research (NCCER) provides a standardized national program of accredited craft training. Key features of the program include instructor certification, competency-based training, and performance testing. The program provides trainees, instructors, and companies with a standard form of recognition through a National Craft Training Registry. The program is described in full in the *Guidelines for Accreditation*, published by NCCER. For more information on standardized craft training, contact the NCCER by writing us at 3600 NW 43rd St., Bldg. G, Gainesville, FL 32606; calling 352-334-0911; or emailing info@nccer.org. More information may be found at our website, www.nccer.org.

HOW TO USE THIS ANNOTATED INSTRUCTOR'S GUIDE

Each page presents two sections of information. The larger section displays each page exactly as it appears in the Trainee Module. The narrow column ties suggested trainee and instructor actions to each page and provides icons (detailed below) to call your attention to material, safety, audiovisual, or testing requirements. The bottom of each page includes space for your notes.

The **Audiovisual** icon indicates an appropriate time to show a transparency or other audiovisual aid.

The **DVD** icon indicates an appropriate time to discuss or show segments of the video.

The **Classroom** icon prompts you to define a term, stress a point, ask trainees to explain a concept, or give examples.

The **Demonstration** icon directs you to show trainees how to perform tasks.

The **Examination** icon tells you to administer the written module examination.

The **Homework** icon is placed where you may wish to assign reading for the next class, assign a project, or advise trainees to prepare for an examination.

The **Laboratory** icon is used when trainees are to practice performing tasks.

The **Materials** icon is a reminder for you to gather materials needed for classes, labs, and testing.

The **Performance Testing** icon tells you to administer a performance test or a portion thereof.

The **Safety** icon is used to emphasize safety issues. It is often keyed to *Caution* and *Warning!* statements in the Trainee Module.

The **Teaching Tip** icon indicates additional guidance is available, such as how to conduct an exercise, get the most educational value from a field trip, or encourage class participation. Teaching Tips may expand on a feature (*Think About It, Did You Know?*) or provide *Quick Quizzes* or similar exercises. You will be referred to the Teaching Tips section at the back of the module if there is additional material.

The **Combination** icon indicates that the laboratory listed corresponds with a performance task. If desired, you can note the proficiency of the trainees during the laboratory, and use it to satisfy performance testing requirements.

PREPARATION

Before teaching this module, you should review the Objectives, Performance Tasks, Materials and Equipment List, and Module Outline. Be sure to allow ample time to prepare your own training or lesson plan and gather all required materials and equipment.

MODULE OVERVIEW

This module introduces the role and responsibilities of project management, including technical and management skills. It provides an overview of the phases in a construction project and describes alternate types of project delivery methods.

PREREQUISITES

There are no prerequisites for this module.

OBJECTIVES

Upon completion of this module, the participant will be able to:

1. Define project.
2. Describe the characteristics of a project manager.
3. Describe the basic functions of project management.
4. Cite the importance of ethical approaches to project management.
5. Discuss the flow and phases of a construction project.
6. Describe the four common construction delivery systems.

MATERIALS AND EQUIPMENT LIST

Desktop or laptop computer or DVD player

Multimedia projector and screen

Minor Decisions: Major Impact – How to Deal with Real Issues in Project Management DVD

Transparencies

Blank acetate sheets

Overhead projector and screen

Transparency pens

Whiteboard/chalkboard

Markers/chalk

Pencils and scratch paper

Module Examination (located in the Test Booklet)

ADDITIONAL RESOURCES

This module is intended to present thorough resources for task training. The following reference work is suggested for both instructors and motivated participants interested in further study. This is optional material for continued education rather than for task training.

PMI Standards Committee, *A Guide to the Project Management Body of Knowledge*. PMI Publications, Newton Square, Pa. (2004).

TEACHING TIME FOR THIS MODULE

An outline for use in developing your lesson plan is presented below. Note that each Roman numeral in the outline equates to one session of instruction. Each session has a suggested time period of 2½ hours. This includes 10 minutes at the beginning of each session for administrative tasks and one 10-minute break during the session. Approximately 2½ hours are suggested to cover *Introduction to Project Management*.

Topic	Planned Time
Session I. Introduction to Project Management; Review and Testing	
A. Introduction to Project Management	________
B. What is a Project?	________
C. Characteristics and Functions of a Project Manager	________
D. Ethical Approaches to Project Management	________
E. Phases of a Construction Project	________
F. Construction Project Flow	________
G. Construction Delivery Systems	________
H. Review	________
I. Module Examination	________

1. Participants must score 70% or higher to receive recognition from NCCER.
2. Record the testing results on Craft Training Report Form 200, and submit the results to the Training Program Sponsor.

USING THE PM DVD WITH *INTRODUCTION TO PROJECT MANAGEMENT*

Inform the students that, during the course of this curriculum, they will have an opportunity to engage with a simulated scenario that recreates the real problems that project managers face on the job. Introduce the video story line and let the students know that there will be stimulating class discussions and exercises for each of the succeeding modules in the curriculum.

Before proceeding to the next module, plan to play the video. Depending upon your preferences in conjunction with your lesson plan, either play the video in its entirety or play Segments 1 and 2 to familiarize the students with the characters and the situation. It is recommended that you allow an additional 15 to 20 minutes of discussion time per scenario, so be sure to factor this into your overall lesson plan. You should plan to replay segments as you progress through the curriculum to aid the discussions and exercises, and to emphasize particular points that are relevant to the topics in the modules.

Note: Section 3.0.0, Characteristics and Functions of a Project Manager, presents discussion opportunities for students to evaluate Samantha's performance and uncover how deficiencies in any of these areas may contribute to "The Missing Tool".

Project Management

44101-08

Introduction to Project Management

Assign reading of Module 44101-08.

44101-08
Introduction to Project Management

Topics to be presented in this module include:

Overview

Today's construction industry provides more than just buildings where people live, work, shop, worship, and learn. The construction industry also builds highways, bridges, airports, and tunnels that enable goods and people to move freely about the country. It builds reservoirs, dams, power stations, irrigation systems, sewer systems, and flood control networks that provide water and power and protect public health. Our lives would be significantly different without the construction industry, and considerably less comfortable.

The construction industry is expected to continue to grow, and the need for more trained construction personnel is anticipated. The skills a project manager needs are acquired and refined through experience and education. The National Center for Construction Education and Research (NCCER) has developed this Project Management training curriculum to provide that education and to give the industry the highly qualified, highly trained managers it needs to survive, grow, and prosper.

A construction project is a short-term endeavor based on specifications and requirements that are driven by functional, budgetary, customer, and time constraints. The construction project manager needs to be sensitive to the project itself as well as the customer's desires and company's constraints that can appear between pre-construction and final completion of the project. This module provides an overview of the responsibilities and characteristics of a project manager along with phases and flow of a project.

Instructor's Notes:

Objectives

When you have completed this module, you will be able to do the following:

1. Define project.
2. Describe the characteristics of a project manager.
3. Describe the basic functions of project management.
4. Cite the importance of ethical approaches to project management.
5. Discuss the flow and phases of a construction project.
6. Describe the four common construction delivery systems.

Trade Terms

- Accountable
- Agency construction management (CM)
- Bidding
- Close-out
- Construction
- Construction management (CM) at risk
- Construction manager
- Delegation
- Design-build
- Engineering-procurement-construction (EPC)
- Grapevine
- Project manager
- Span of control
- Traditional construction delivery
- Pre-construction
- Work breakdown structure (WBS)

Prerequisites

There are no prerequisites for this module.

This course map shows all of the modules in the *Project Management* curriculum. The suggested training order begins at the bottom and proceeds up. Skill levels increase as you advance on the course map. The local Training Program Sponsor may adjust the training order.

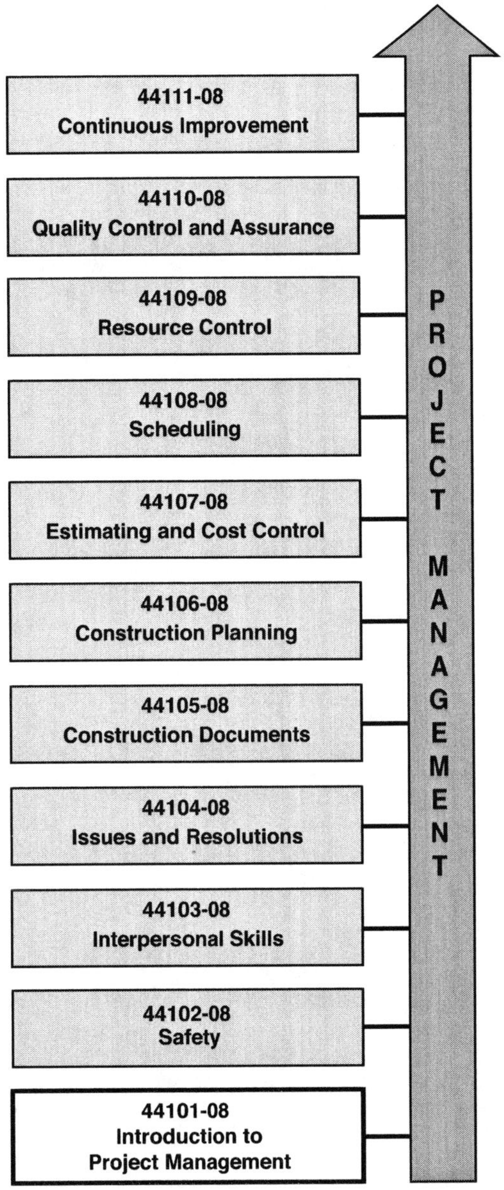

101CMAP.EPS

Ensure that you have everything required to teach the course. Check the Materials and Equipment list at the front of this module.

See the general Teaching Tip at the end of this module.

Explain that terms shown in bold are defined in the Glossary at the back of the module.

Show Transparency 1, Objectives. Review the goals of the module, and explain what will be expected of the participant.

Show Transparency 2, Introduction to Project Management. Discuss today's construction industry in broad terms using the points in Transparency 2.

Show Transparency 3, Modules Overview. Briefly review each module in the curriculum.

See the Teaching Tip for Section 1.0.0 at the end of this module.

MODULE 44101-08 ◆ INTRODUCTION TO PROJECT MANAGEMENT 1.1

Show Transparency 4, Project Definition. Define the term project.

1.0.0 ◆ INTRODUCTION TO PROJECT MANAGEMENT

In most industries, the **project manager** is someone who has achieved a level of competency and experience. Often this person is selected to manage one or more projects. Sometimes this transition from skilled contributor to a manager is not an easy one, since different skills are required in each role. This Project Management training curriculum will identify key concepts as well as best practices to help you become an effective project manager.

See the Teaching Tip for Section 2.0.0 at the end of this module.

See the Teaching Tip for Sections 3.0.0–3.6.0 at the end of this module.

2.0.0 ◆ WHAT IS A PROJECT?

While the term project can apply to almost any activity that requires time, people, and expenditures, this module focuses on projects in today's **construction** industry, where the range of construction projects varies in size and scope. The one constant of every construction project is a clearly specified start and completion. In between these two points is a wide range of synchronized and controlled activities that must meet project specifications.

Show Transparency 5, Project Management Areas of Responsibility and Authority.

Show Transparency 6, Project Management Skills. Briefly discuss each skill area and provide examples of each.

To help project managers direct a project's activities systematically, the Project Management Institute (PMI) conducts extensive research on project management. In 1986, PMI released *A Guide to Body of Knowledge*. It became known as the *PMBOK® Guide*. Subsequent revisions were made, most recently in 2004. Today, the *PMBOK* is a recognized source for establishing project management standards.

Because there are so many ways to organize and run projects, using common definitions accompanied by repeatable steps is very important. The current *PMBOK® Guide* provides the following description: "A project is a temporary endeavor undertaken to create a unique product or service." In the construction industry, the word temporary could mean several months or even years—but ultimately the construction project is completed and the building is occupied.

Describe the three broad areas of responsibility of a project manager. Provide examples of each area.

Stress the importance of planning and provide examples that relate to the three broad areas.

Introduce the concept of the work breakdown structure (WBS) and CPM diagrams.

3.0.0 ◆ CHARACTERISTICS AND FUNCTIONS OF A PROJECT MANAGER

In this course, project manager refers to anyone who directly manages one or more construction projects. The project manager is the person with authority and responsibility for:

- Planning, organizing, staffing, directing, and controlling one or more projects
- Reporting directly to the owner or the owner's designated representative on the status of those projects
- Completing each project on time, within budget and in full compliance with all job specifications, code requirements, drawings, and contractual agreements

There are many subsets included in the above list. The ability to plan is one of the most important characteristics a project manager needs to possess. But also consider other factors. The project manager is an important part of the project team. Through both technical experience and building relationships, the project manager can make the difference between a successful project and one with overruns, personnel attrition, and ultimately failure. What other characteristics should the project manager possess? Today, every project manager must use both technical and management skills to meet the requirements of the construction industry. These management skills include:

- Planning
- Initiating
- Directing
- Organizing
- Communicating
- Controlling

3.1.0 Planning

Planning is an important skill. In the early phases of project startup, the project manager needs to develop a **work breakdown structure (WBS)**. Using a WBS helps the project manager set up project tasks in a hierarchical arrangement that is similar to an organizational chart. Part of the planning process also involves CPM (critical path method), a technique that diagrams tasks based on their longest path through the network. At the onset of the project, either before or after the kick-off meeting, it may be necessary to brainstorm with the project team. Used effectively, brainstorming involves all project members in problem-solving and helps to prioritize tasks. Establishing a risk management plan will help to identify and analyze potential risks during construction.

Brainstorming and risk identification may help to identify resources, safety issues, budgeting constraints, staffing, and regulatory matters such as permits and codes. Most importantly, having enough information will help the project manager assemble the right staff.

Instructor's Notes:

Staffing involves acquiring competent employees, placing them into specific positions, and keeping those positions filled. In many companies, the project manager may be responsible for hiring supervisors, field engineers, and other field personnel. The skills of finding and keeping a high-quality staff are discussed in other modules of this training curriculum.

3.2.0 Initiating

Initiating requires building relationships and being keenly aware of customer needs. The project manager needs to know how to identify and budget resources and also how to effectively evaluate the drawings and specifications to provide accurate estimating. Another skill is the ability to effectively communicate with secondary audiences. For example, the project manager is not always involved with the team, but may have to elicit information from stakeholders and consider their input during the planning process.

3.3.0 Directing

Directing involves guiding and supervising employees in order to accomplish company objectives. This includes assigning tasks and following up to be sure they were completed. Employees cannot be forced to do their best work; instead, they must be motivated. Therefore, it is the project manager's role to develop techniques to motivate employees. The most important characteristics of a project manager are solid leadership and good communication skills.

3.4.0 Organizing

A company's staffing structure is usually presented in a diagram called an organizational chart. An organizational chart shows the horizontal segments, the vertical segments, and the lines of authority between them. By clearly showing each individual's duties and area of authority, a well-prepared organizational chart also illustrates areas where individuals' authority and responsibility overlap and areas in which no one has been assigned authority or responsibility. A typical organizational chart is presented in *Figure 1*.

In addition to the formal lines of authority and responsibility shown in the organizational chart, there are informal channels of communication and contact within every company. Informal channels among company personnel are the result of happenstance, necessity, and social relationships, and they can exist at all levels of management, supervision, and the workforce.

Like the formal company organization, the informal organization grows out of a need to accomplish company goals. For example, a project

Show Transparency 7 (Figure 1). Introduce formal organization structures.

See the Teaching Tip for Section 3.4.0 at the end of this module.

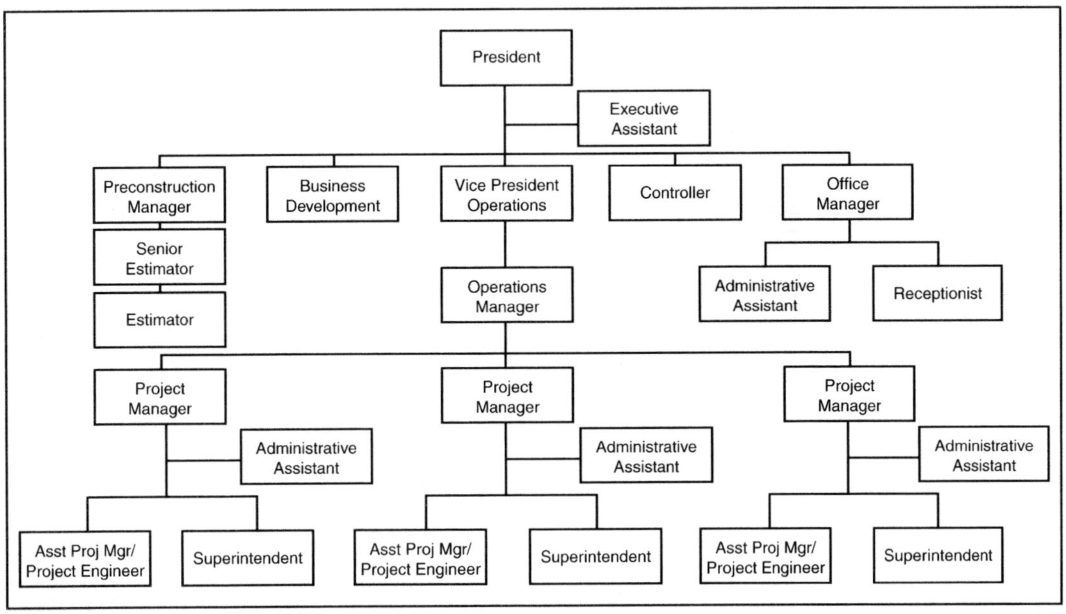

101F01.EPS

Figure 1 ◆ Typical organizational chart.

MODULE 44101-08 ◆ INTRODUCTION TO PROJECT MANAGEMENT 1.3

Discuss formal and informal lines of communication, such as the grapevine.

Emphasize that the project manager must be accountable for outcomes.

Discuss the impact of span of control on the project manager.

Emphasize the need for a communication plan before project kick-off.

Show Scenarios 1 and 2. Have students rate Samantha on the six management skills of a Project Manager discussed in Section 3.0.0.

manager may require immediate verification that an invoice or bill has been paid. Instead of going through the formal channels outlined on the organizational chart, the project manager may simply pick up the phone and call the accounts payable department directly. The accounting staff may provide the answer immediately, and so a new channel of communication is established.

Another example of informal company organization is the **grapevine**. Based on friendship and casual contact, in most companies this network is able to originate and spread information through an office or a project faster than any company memo. Many times, damaging rumors can be stopped and incorrect information can be corrected more quickly through the grapevine than through the formal channels of communication. A company's informal organization is an important management tool, and each project manager must recognize its existence.

As an organization grows, the manager must assign duties to others in order to have time to devote to more important duties and tasks. The act of assigning duties to others is called **delegation**. In delegating activities, the project manager should assign not just the task but the authority and responsibility to complete that task.

Having authority means having the power to act or make decisions. Having responsibility means being **accountable** for the outcome of an assignment. In every act of delegation, authority and responsibility must be carefully defined and properly balanced. The project manager must keep in mind that the ultimate authority and responsibility for delegation remains with the project manager. The project manager is the one accountable to the company for completion of tasks under his or her supervision.

The maximum number of employees that a project manager can control in a given situation is called the project manager's **span of control**. Span of control is also the term used to describe the number of people who report to a project manager. Adding more employees to the manager's team results in a wider span of control. A project manager with fewer people to manage has a narrow span of control. There are advantages to both wide and narrow spans of control. In a wide span of control, there are fewer layers of management, which can result in better communication. On certain types of construction projects, a wider span of control can be more cost-effective because one project manager can direct the project.

A narrow span of control allows the project manager to be able to communicate quickly with supervisors and help direct their tasks. For example, a supervisor can provide feedback and suggestions to handle on-site tasks in a more efficient manner. However, a narrow span of control may have the potential for micro-managing.

Therefore, several factors determine a project manager's span of control. They are:

- The manager's ability and personality
- The quality of communication between the manager and the manager's supervisors
- The number and types of other duties required of the manager
- The skills and training level of the project team
- The types of projects under the manager's command

A project manager should not be responsible for managing more projects or supervising more employees than the individual can effectively monitor and control. A general rule for the number of employees a manager can effectively oversee is between four and eight.

A number of factors can contribute to a project manager losing effective control of the project. Often the reason for poor management is caused by the project manager trying to supervise more work and more employees than can be managed properly. In other words, failure comes from a project manager trying to work beyond a personal span of control.

3.5.0 Communicating

Every project manager should have a communication plan in place before project kick-off. Rather than crafting a plan from scratch, there may already be a communication plan within the company. The amount of project correspondence from requests for information (RFIs), coordinating meeting minutes, change orders, drawings submittals, daily job logs, and emails can be overwhelming. The *PMBOK® Guide* outlines four major steps for implementing a communication plan. The *PMBOK* specifically defines these as:

- *Communications planning* – determining the information and communications needs of the stakeholders, including who needs what information, when will they need it, and how will it be given to them.
- *Information distribution* – making needed information available to project stakeholders in a timely manner.
- *Performance reporting* – collecting and disseminating performance information, including status reporting, progress measurement, and forecasting.
- *Administrative closure* – generating, gathering, and disseminating information to formalize phase or project completion.

Instructor's Notes:

3.6.0 Controlling

Control is the means of measuring performance and correcting any apparent deviation from the project plan. The most effective control techniques involve establishing standard units for measuring progress, identifying deviations as soon as they occur, and correcting them. The elements and process of control will be discussed in more detail in modules covering cost control, resource control, and quality control in this curriculum.

4.0.0 ◆ ETHICAL APPROACHES TO PROJECT MANAGEMENT

Ethics in its simplest definition means doing the moral and right thing. A culture and practice of ethics is essential for today's corporations. Corporate boards of directors are increasingly focusing on ethics. In addition to the legal department, some firms are appointing ethics officers or hiring outside experts to assist in developing a code of ethics and practices for the organization. Practicing professional ethics serves the interests of the firm and the customer and drives business performance.

The major business focus for all construction companies is to make a profit. The business process to attain profit must be based on ethical considerations. Project managers within such environments maintain a high standard. However, there may be times when potential unethical situations arise that are somewhat in the gray area. Consider the following example:

Phil, the project manager, is reviewing estimates for copper tubing and mentions to one of his foremen that the prices are getting too high to meet the specs. The foreman lets Phil know that his cousin has connections and thinks he can get it for three cents off per foot. The foreman also mentions that he has done this before and tells Phil that he will give him all the details so Phil can prepare a product substitution. The foreman mentions that it will be a nice profit on the job.

Ask yourself – what you think your action would be? Project managers need to think through all situations that appear to stray from the plans and specifications and contractual agreements. On a larger scale, what about an owner who pressures the contractor to complete the project by a certain date because of external pressures or public perceptions? Will the construction company cut corners to save time and money? Both of these situations present problems because avoiding ethical responsibilities by the company or individuals within the company may have far-reaching implications.

The project manager must ensure that the project team understands the ethics and values of the company and puts these into practice on the job. Honesty and integrity are the foremost values that govern ethical situations. Practices that the project manager can implement include:

- Preparing ethical bids and requiring the same from suppliers
- Negotiating fairly with the customer and suppliers
- Being candid and honest with customers
- Providing the professional services consistent with best practices and the competencies of the organization
- Complying with regulations and legal procedures
- Adhering to confidentiality and proprietary information provisions and avoiding conflicts of interest

During a construction project, the project manager will encounter many situations that will require good business decisions to resolve conflict. *Table 1* lists some common conflicts that can potentially have ethics implications.

The whole idea of what is ethical and what is not ethical is difficult, and project managers are not perfect. However, the company that integrates ethics with their business model is less likely to avoid turn-over, litigation and client dissatisfaction.

5.0.0 ◆ PHASES OF A CONSTRUCTION PROJECT

A construction project can be divided into three phases or sequences that represent the project flow:

- Development
- Design
- Construction

5.1.0 Development Phase

The development phase is the first step in all construction projects. Beginning with an idea or concept, the owner prepares the required feasibility studies, locates suitable land through research, and obtains the necessary finances. The architect/engineer develops conceptual drawings and models, conceptual estimates, and begins preliminary discussions with various government agencies to examine regulating and permitting issues.

Using the conceptual estimate, the owner and architect/engineer develop the project budget, taking into account all anticipated costs and

Show Transparency 8 (Table 1).

Discuss the importance of ethics and cite a job-related example.

Ask participants if they have encountered any instances of questionable ethics, and discuss their responses with the class.

See the Teaching Tip for Section 4.0.0 at the end of this module.

See the Teaching Tip for Section 5.0.0 at the end of this module.

Show Transparencies 9, 10, and 11, Development Phase, Design Phase, and Construction Phase.

Discuss the subsets of each phase and provide examples as an option.

Table 1 Conflicts with Ethical Implications

Situation	Ethical Concern(s)
Schedule conflicts resulting from lack of support, both internal and external	May impact service to the client
Restrictive administrative procedures that may constrain the project manager's authority and accountability	May impact service to the client
Priorities regarding concurrent projects within the company	The project manager is ethically bound to deliver services contracted
Problems with personnel resources	The project manager is responsible to assign qualified staff to projects
Technical issues and conflicts on how to expedite and reduce costs to obtain solutions	Possible conflicts of interest
Personality conflicts (team members, subcontractors)	Possible issues with confidentiality and conflicts of interest

101T01.EPS

expenses. In addition to analyzing the cost of the project, the owner analyzes the projected return on the investment. If the project is considered financially feasible, then the owner generally seeks financial assistance from lending institutions.

The architect/engineer and owner begin the preliminary review of the project with various government agencies. These reviews are intended to ensure that the project meets all applicable zoning laws, building restrictions, environmental concerns, road access, landscape requirements, and other considerations.

5.2.0 Design Phase

During the design phase, cost estimates are refined, project financing is secured, government regulations are met, and the marketing program is developed.

When the architect/engineer starts the preliminary drawings and specifications, other design professionals—including structural, mechanical, electrical, civil, and other specialized engineers—perform calculations, analyze technical data, and develop project details. This work is translated into drawings and specifications that illustrate all the information that the general contractors, the subcontractors, and the suppliers need to install the components that make up a project.

In addition to developing the drawings and specifications, the architect/engineer will produce all the documents necessary for the owner to obtain bids from interested contractors. With the contract documents complete, the owner selects the method that will be used to choose contractors. The most common method for selecting a contractor is through competitive **bidding**. Other methods an owner may employ to choose a contractor are discussed in a later section of this module.

5.3.0 Construction Phase

The construction phase is usually organized by a general contractor, which may perform work with its own workforce. In addition, the general contractor may seek the services of specialty contractors or subcontractors to perform certain portions of work on the project.

As construction nears completion, the architect/engineer, the owner, and government agencies start their final inspections and acceptance of the nearly completed project. If the project has been managed properly, the work has been performed satisfactorily, the architect/engineer has inspected the project regularly, and local codes have been met throughout, then final inspection and acceptance will proceed quickly. The result is a satisfied client and a profitable project for the contractors. On the other hand, if the inspection reveals faulty work, inferior materials, and code violations, the inspection and acceptance procedures can generate disputes that will undoubtedly result in a dissatisfied client and a loss of profits for the contractor.

Instructor's Notes:

1.6 PROJECT MANAGEMENT

6.0.0 ◆ CONSTRUCTION PROJECT FLOW

Understanding the project flow is an important tool for management of any project. Each of the four phases of project flow is illustrated in *Figure 2*.

- Bidding
- Pre-construction
- Construction
- Closeout

Just as the total project can be divided into phases or sequences, the construction phase can be further divided into several phases. *Figure 3* illustrates the construction flow phase of a typical project.

6.1.0 Bidding Phase

The bidding phase starts when an owner or architect/engineer requests that a general contractor or specialized contractor submit a bid or proposal on all or part of a construction project. The contractor reviews the documents to determine the extent of the work and the requirements of the project.

Usually, the contractor assigns an estimator to develop the estimate. The estimator studies the project documents; prepares a quantity survey (material take-off); and prepares a list of the various materials, components, and subcontractors necessary to complete the project. The total of all the items is summarized. The company's overhead cost and a reasonable profit are added to the cost summary. The final sum becomes the bid amount the company will submit to the owner or architect/engineer.

In a competitive bid situation, the company that submits the lowest bid usually is awarded the contract to do the work. If the project is being negotiated, the owner or architect/engineer typically may not award the project solely on the basis of the lowest bid, but will consider other factors including safety performance, contractor qualifications, financial strength, personnel, and schedule. In any case, the bidding phase ends when the project contract is awarded.

6.2.0 Pre-Construction Phase

The pre-construction phase is often the most important part of the construction flow process. During this phase, the project manager and the project team are involved in several major areas:

- Issuing subcontracts and purchase orders
- Scheduling development
- Reviewing the project estimate
- Identifying manpower and equipment requirements
- Establishing cost control
- Preparing job site mobilization plans

6.2.1 Subcontracts and Purchase Orders

The successful contractor must select subcontractors after carefully reviewing their bids for scope of work, cost, and exceptions to the contract documents. *Figure 4* lists the topics that should be discussed before any contract is awarded.

The contractor must verify that suppliers can meet the requirements of the contract documents before issuing purchase orders for materials and equipment.

6.2.2 Schedule

The project schedule is developed with cooperation and input from all members of the company's project team, the major subcontractors, the key suppliers, and others who can contribute to a realistic schedule. While developing the schedule, the highest priority is assigned to details regarding long-lead items, critical operations, sequencing, duration of activities, equipment deliveries, and labor requirements. Scheduling is discussed in greater detail in another module.

6.2.3 Estimate

The pre-construction review of the estimate is a key function of the project manager. A thorough project team review of the estimate ensures that everyone on the project has a clear understanding of the project costs. The subject of estimating will be discussed in more detail in the *Estimating and Cost Control* module.

6.2.4 Labor and Equipment

Determining the number of craftworkers for the various trades required at each stage of work gives the project team a basis for determining the availability of personnel, identifying possible labor shortage or surpluses, and controlling labor costs and cash flow. Reviewing equipment requirements ensures that construction equipment is available when required, used efficiently on the site, and removed from the job site when it is no longer needed.

Show Transparency 12 (Figure 2). Review the four phases of the traditional construction project in Figure 2. Discuss the subparts of each phase.

Explain the responsibilities and actions the project manager must direct in each phase.

Show Transparency 13 (Figure 3). Discuss how the construction phase of a typical project can be further broken down into subphases.

Show Transparency 14 (Figure 4). Explain how to use the contract award checklist. Emphasize that it can be an effective tool when selecting subcontractors.

MODULE 44101-08 ◆ INTRODUCTION TO PROJECT MANAGEMENT 1.7

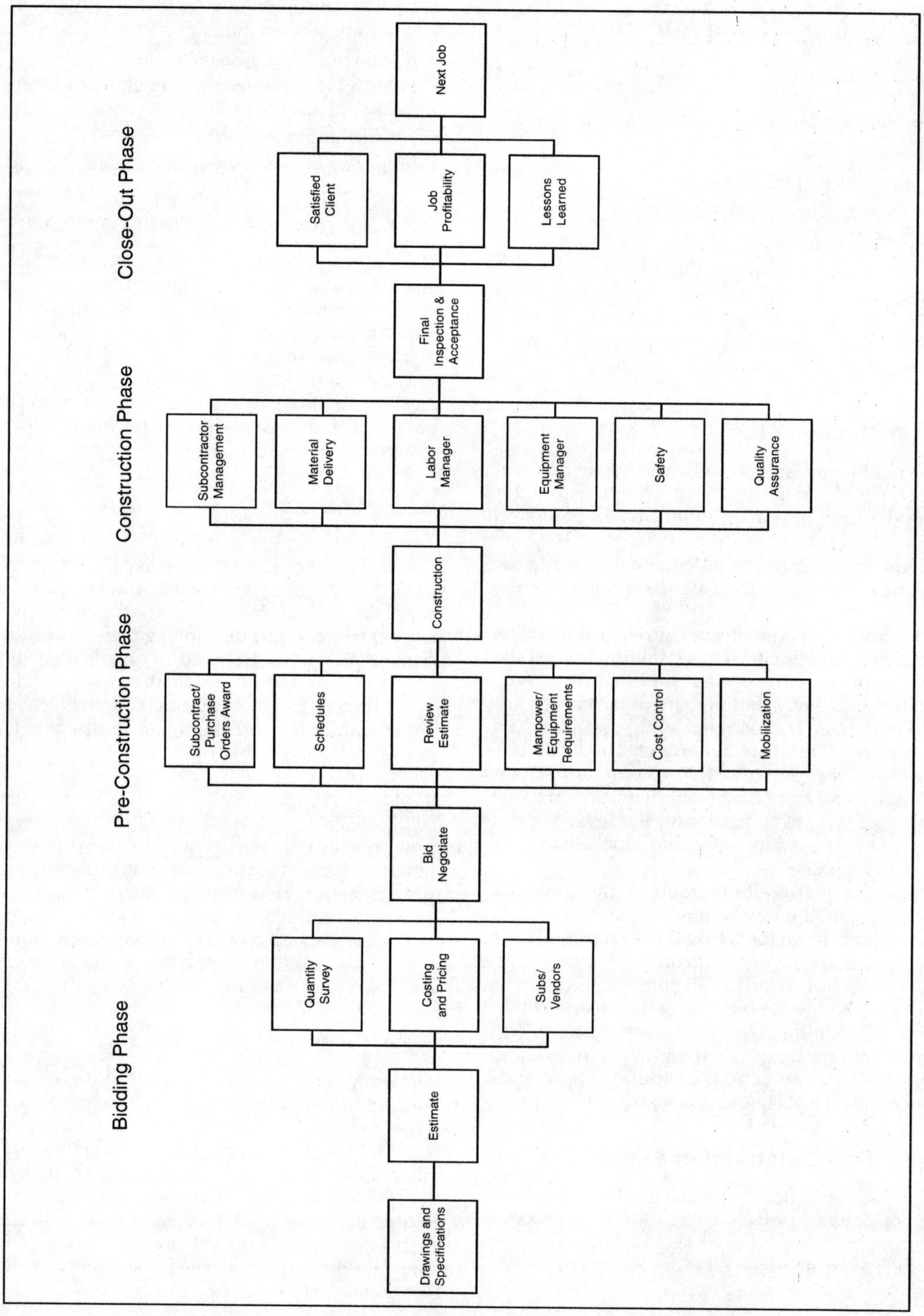

Figure 2 ◆ Flow of a traditional project.

Instructor's Notes:

1.8 PROJECT MANAGEMENT

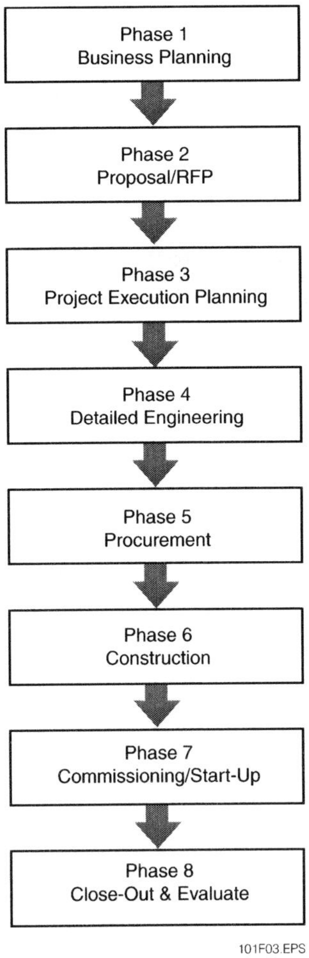

Figure 3 ◆ Construction flow of a typical project.

6.2.5 Cost and Other Controls

Every step of the project, from inception to completion, is based on cost management. The project manager should review the cost control system with every member of the project team, emphasizing the need for accurate and timely cost reporting. The project manager's role and involvement in cost control will be reviewed in detail in the *Estimating and Cost Control* module.

Other controls that need to be implemented and managed are covered in the modules on *Scheduling* and *Resource Control* (including materials, tools, and equipment).

6.2.6 Mobilization

Mobilization includes setting up field site offices and storage trailers, locating temporary facilities, establishing storage layout areas, and other tasks necessary to start the project. *Figure 5* illustrates a typical checklist that may be used to review mobilization item requirements.

6.3.0 Construction Phase

The construction phase, or the actual building of the project, involves subcontractors, equipment, labor coordination, scheduling of material delivery, safety requirements, and quality control and assurance of the project. Before construction begins, plan a pre-construction conference to discuss and distribute project-related material and information to the project team including subcontractors and major vendors and suppliers.

6.4.0 Close-Out Phase

Close-out of the project involves the final inspections and acceptance of the work by the owner, architect/engineer, and government agencies. Record drawings or as-built drawings are also submitted in accordance with the contract requirements. The closeout documents include:

- Punch list
- Guarantees and warranties
- Operation and maintenance manuals
- Release of liens
- Record or as-built drawings
- Other documents as required by the contract

7.0.0 ◆ CONSTRUCTION DELIVERY SYSTEMS

Project managers must be aware of the different construction delivery systems in today's market. The selection of a construction system depends on the project, the type of company, and the owner's preference. The most common construction project delivery systems are:

- The traditional contract delivery system
- The **design-build** system
- The construction management delivery system (CM)
- The engineering-procurement-construction (E-P-C) system

Each of these systems has its advantages and disadvantages which the owner must consider before making any decisions.

Show Transparency 15 (Figure 5), Mobilization Checklist / Status Report.

Discuss the importance of mobilization and the placement of resources on the job site to maintain efficiency and safety.

Show Transparency 16, Construction Delivery Systems. Identify the four types of construction delivery systems.

MODULE 44101-08 ◆ INTRODUCTION TO PROJECT MANAGEMENT 1.9

CONTRACT AWARD CHECKLIST

Project Name	Project Manager	Project Superintendent
Trade Contractor	Trade Contractor Superintendent	Specification Section

	Yes	N/A
Performance and payment	______	______
Bond, if required, on file	______	______
Insurance requirements on file	______	______
General conditions reviewed	______	______
Contract requirements reviewed	______	______
Drawings and specifications reviewed	______	______
Schedule reviewed	______	______
Material status reports reviewed	______	______
Permit requirements reviewed	______	______
Payment request/invoice procedure reviewed	______	______
Safety requirements reviewed	______	______
Safety program received	______	______
City, county, or state licenses on file	______	______
Change order procedures reviewed	______	______
Back charge procedures reviewed	______	______

Comments: ______________________________

101F04.EPS

Figure 4 ◆ Contract award checklist.

Instructor's Notes:

Mobilization Items	PM Responsibility	Superintendent Responsibility	Other (Identify)	Date Required	Date Completed
Permits					
Insurance					
Recruiting and Staffing					
Office Complex					
Temporary Utilities					
Site Security					
Fenced Storage					
Telecommunications					
Temporary Toilet					
Survey/Base Lines					
Disposal Service					
First Aid Facilities					
First Aid Training					
Survey Equipment					
Notice of Commencement					
Urgent Care Facilities					
Emergency Response Services					
Local Post Office Box					
Local Banking Services					
Job Camera/Videos					
Webcam					
Set up Local Accounts					
Verify Burning Regulations					
Copier or Copier Service					
Office Supplies					
Files					
Furniture					
Computers and Networking					
Project Signage					
Project Bulletin Board					

101F05.EPS

Figure 5 ◆ Mobilization checklist/status report.

MODULE 44101-08 ◆ INTRODUCTION TO PROJECT MANAGEMENT 1.11

Explain how the traditional contract delivery system works and provide a situation where it would be used.

Discuss how this system can provide checks and balances.

Explain the advantages and disadvantages of this system.

Discuss why single source procurement is popular with owners.

Provide a job-related example of how design-build would be a good approach and an example of when it would not.

Explain how design-build can provide more management control during the construction process.

Discuss areas where communication complexity may occur.

Explain the differences between Agency CM and CM at risk.

Discuss the roles of the owner, architect, construction manager, and contractor in Agency CM and CM at Risk arrangements.

7.1.0 Traditional Contract Delivery System

The **traditional contract delivery system** has been the standard approach for many years and is still used. This approach involves the owner, contractor and architect/engineer. The traditional contract delivery system reduces risk because the owner has control during the design and construction phases.

Using this approach, the owner contracts separately with the architect/engineer for the preparation of plans and specifications and assistance in the bidding stage. The architect/engineer may also be requested by the owner to oversee certain aspects of the project during construction. For example, a renovation may have requirements to maintain historical aesthetics, so the design firm would coordinate with the contractor to ensure those types of requirements are met.

Once the architect/engineer is under contract, all the design will be completed before a general contractor is selected, usually through a competitive bid process. This allows the owner to carefully analyze the bids to obtain the best price for construction. When a contractor is selected based on the firm fixed price, the owner has a high level of confidence that the estimate is accurate because several bids were reviewed during the selection process.

Once selected, the owner will negotiate a contract with the general contractor who will provide services for the construction of the project. The owner retains responsibility for overall project management. All transactions and contractual matters are handled by the owner. This approach provides checks and balances between the design and construction. Most companies are very familiar with this approach and generally see it as a good way to do business. Additionally, the owner is able to provide influential input throughout the project. Some disadvantages are that the owner must make a significant investment up front to develop the final drawings and specifications. The project design and specifications could also potentially lead to higher bids than anticipated, requiring redesign.

7.2.0 Design-Build

The **design-build** project delivery system is a single source procurement that allows the owner to select and enter into a contract with a design-build firm that will assume the obligation of furnishing design, supervision and construction services during the project. The design-build project delivery system is an attractive arrangement for the owner because it provides a single point of responsibility for design and construction. This arrangement keeps the owner from being in the middle of any disputes between the construction and design. The downside is that it takes away the checks and balances which occur when the design and construction phase are contracted separately. The owner also loses some of the control that it has on the project, and loses the direct advisory relationship with the architect/ engineer that the owner has in the traditional contract delivery system.

7.3.0 Construction Management (CM)

The **construction management system (CM)** may be a good solution for projects with unique security, requirements, regulations, codes and specifications such as medical facilities, power plants, or federal government facilities. The two common controls of CM are **agency CM** and **CM at risk.**

Agency CM involves providing contractor-based management at the early stages of a project to ensure that the project stays within scope during design and throughout construction. The assigned program manager reports directly to the owner and works on fee basis, but also interacts with the architect/engineer and the general contractor along with other parties assigned to mediate and solve problems. The CM program manager's primary focus is to advise the owner and to assist in providing resources from the CM firm that can support different project requirements and logistics. The following are a few examples:

- Maintain project relationships with the trade contractors
- Establish timetables and resolving conflicts
- Plan and coordinate equipment, materials, and suppliers
- Coordinate with local, state, and federal agencies regarding safety and code compliance

Since the CM program manager is involved early in the entire process, monitoring the concept, design, execution, and completion can be effectively managed.

CM at risk involves hiring a construction management firm that meets specific qualifications and often performs the work at a guaranteed maximum price. The **construction manager** participates with the owner in the concept and design of the project and becomes the general contractor once the price is agreed. As a part of the design team, the CM contributes expertise and recommendations early in the process. The CM assumes the risk of subcontracting, manag-

Instructor's Notes:

1.12 PROJECT MANAGEMENT

ing schedules, adhering to contract specifications, and cost overruns.

The CM at risk approach is often more efficient than the traditional contract delivery method. The CM has the flexibility to fast track development. Mobilization can start during the design process. Subcontracts can be let for portions of the work while other portions are being designed.

Many owners find this arrangement less risky since the construction manager is responsible for all costs and overruns and any cost underrun savings remain with the owner.

7.4.0 Engineering-Procurement Construction (E-P-C)

Engineering, procurement, and construction (E-P-C) is a spin-off of the design-build approach. This solution is sometimes known as a turnkey solution because one company will provide design engineering, management and construction services. E-P-C is usually the model that large companies use on complex construction projects such as power plants, bridges, and refineries.

Explain from a management perspective how E-P-C is somewhat similar to design-build (single point of contact with owners).

Cite some projects where E-P-C is used.

Ask participants if they have any questions about the different delivery systems.

Have participants complete the Review Questions, and go over the answers prior to administering the Module Examination.

Review Questions

1. What is the definition of a project?

 ______

 ______

2. List three common functions of a project manager.
 1. ______
 2. ______
 3. ______

3. During the planning phase, a project manager needs to develop a WBS in order to ______.

4. List the four phases of a construction project.
 1. ______
 2. ______
 3. ______
 4. ______

5. When does the bidding phase begin?

 ______

6. Schedule development and reviewing the project estimate are part of the ______ phase.

7. Briefly describe a situation you would consider unethical (do not use the one described in this module).

 ______

 ______

 ______

 ______

8. When developing the project schedule, the project manager or estimator should ______.
 a. consult with the owner's capital projects officer
 b. modify an existing schedule that is similar to the scope
 c. consult with members of the project team

9. Select two documents that are submitted for project closeout.
 a. cost control summary report
 b. punch lists
 c. documentation of all change orders
 d. as-built drawings

10. List five factors that determine a project manager's span of control.
 1. ______
 2. ______
 3. ______
 4. ______
 5. ______

MODULE 44101-08 ◆ INTRODUCTION TO PROJECT MANAGEMENT 1.13

Summarize the major concepts presented in this module.

Summary

A project is a finite endeavor that requires a project manager to systematically direct its activities and the team performing the work. The project manager has a responsibility to plan, organize, and control the project and complete it within the budget and time agreed and in compliance with the specifications, codes, and contract. The project manager and the team must practice ethical standards of conduct in performing the job and in their relations with stakeholders.

There are well-defined phases to a construction project and to its flow. The major phases include development, design, and construction. Project flow includes bidding, pre-construction, construction, and closeout. There are also different approaches to a construction project that include the traditional contract delivery system, design-build, construction management, and engineering-procurement-construction.

The project manager must be involved on a daily basis in controlling the use of resources and reducing costs by such means as:

- Implementing more efficient work methods
- Using more effective tools and equipment
- Providing training and re-training of personnel
- Monitoring productivity on the job site and working to improve it
- Reviewing and following up on production records

The project manager's responsibility is to control production so that the project is constructed for the estimated cost, within the scheduled time, and to the quality standards designated in the specifications and plans.

Notes

Instructor's Notes:

1.14 PROJECT MANAGEMENT

Trade Terms Quiz

1. Making an offer to perform work described in contract documents is known as ________.
2. ______________ is a building project that is typically divided into three phases; development, design, and construction.
3. A ______________ is the person responsible for the planning, coordination, and controlling of a project from start to finish.
4. Being ______________ means you are subject to giving an account or being answerable.
5. A hierarchical definition of the tasks and activities of a project is called the __________.
6. ______________ is the act of empowering or the assignment of duties to others.
7. Performing final inspections and preparing documents for handing over the project is known as ______________.
8. An informal person-to-person means of circulating information or gossip is referred to as the ______________.
9. The ______________ consists of the maximum number of projects and employees that a person can control in a given situation.
10. ______________ is the planning phase that takes place before on-site work can begin.
11. ______________ is the standard approach to construction development in which the owner contracts separately with the architect/engineer and the contractor.
12. ______________ is single source procurement for an owner.
13. This delivery system is a form of design-build known as ______________.
14. A firm contracted to provide design expertise and deliver the construction project at a firm price at no risk to the owner is known as ______________.
15. ______________ involves the use of a manager as the owner's agent and advisor during a construction project.

Trade Term List

Accountable
Agency construction management (CM)
Bidding
Closeout
Construction
Construction management (CM) at risk
Construction manager
Delegation
Design-build
Engineering-procurement-construction (EPC)
Grapevine
Project manager
Span of control
Traditional construction delivery
Pre-construction
Work breakdown structure (WBS)

Have participants complete the Trade Terms Quiz, and go over the answers.

Administer the Module Examination. Record the results on Craft Training Report Form 200, and submit the results to the Training Program Sponsor.

MODULE 44101-08 ◆ INTRODUCTION TO PROJECT MANAGEMENT 1.15

Trade Terms Introduced in This Module

Accountable: Subject to giving an account; answerable.

Agency construction management (CM): A firm that provides various services to owners in the course of a construction project. These services can vary from program management to design and construction.

Bidding: Making an offer to perform work described in contract documents at a specified cost.

Close-out: Performing final inspections and documents to prepare for handing the project over for the owners' acceptance and possession.

Construction: A building project that is typically divided into three phases: development, design, and construction.

Construction management (CM) at risk: The form of construction management and delivery in which the CM firm assumes the burdens and risk for the project; the CM is involved in the design phase and performs the construction delivery at a firm agreed-upon price.

Construction manager: The supervisor responsible to the project manager for day-to-day activities on the construction site.

Delegation: The act of empowering to act for another; assignment of duties to others.

Design-build: A construction delivery system whereby the contractor provides both design and construction services.

Engineering-Procurement-Construction (EPC): A variation of the design-build approach wherein there is a single point of contact with the owner throughout the project.

Grapevine: An informal person-to-person means of circulating information or gossip.

Project manager: The person responsible for the planning, coordination, and controlling of a project from inception to completion, meeting the project's requirements and ensuring completion on time, within cost, and to quality standards specified in contract documents.

Span of control: The maximum number of employees that a person can control in a given situation; also the supervisory ratio of from three-to-seven individuals, with five-to-one being established as optimum.

Traditional contract delivery: The standard approach to construction development in which the owner contracts separately with the architect/engineer and the contractor.

Pre-construction: The planning phase that takes place before on-site work begins, involving issuing contracts, schedule development, identifying labor and equipment requirements, and other preparatory work.

Work breakdown structure (WBS): A hierarchical decomposition of deliverables which breaks down the scope of work into unique and distinguishable activities and tasks.

Instructor's Notes:

Resources & Acknowledgments

References

PMI Standards Committee, *A Guide to the Project Management Body of Knowledge*. PMI Publications, Newton Square, Pa. (2004).

MODULE 44101-08 ◆ INTRODUCTION TO PROJECT MANAGEMENT 1.17

MODULE 44101-08 — TEACHING TIPS

The following are suggested activities or instructional methods to help you teach the material in this module.

General

When you call on someone to answer a question, the rest of the class relaxes or even tunes out because they expect that the question and answer will take place only between you and the participant you called on. Instead, use this technique to involve more participants in answering questions and to keep them on their toes.

1. Ask participants to define a term or explain a concept.
2. After one participant has answered, ask a participant seated nearby if the answer is right. Then ask whether a participant in the back of the room agrees.
3. Ask participants to explain why they think an answer is right or wrong.
4. Use the session to clear up incorrect ideas and encourage participants to learn from their mistakes.

Section 1.0.0 *Introduction*

1. Provide examples related to construction in your area for the following talking points:
   - The nature of today's construction industry is diverse.
   - Today's construction industry provides many solutions.
   - Construction is a growth industry.
   - Project management skills are essential.
2. Inform the participants that this Project Management Training Curriculum will identify key concepts as well as best practices to help them become effective project managers. Briefly review each module in the curriculum.

Section 2.0.0 *What is a Project?*

Explain that an important characteristic of every project is its finite time frame and schedule.

Explain the importance of how projects generate profit or loss to a company. Emphasize there are many aspects to a project, and provide some examples.

Remind participants that the *PMBOK® Guide* provides global standards that can be customized or integrated into a company's standards.

Ask for and address any questions from the participants.

Sections 3.0.0 – 3.6.0 *Characteristics and Functions of a Project Manager*

1. Ask participants to define the primary areas of project management and authority.
2. Review and briefly discuss their responses.
3. Provide an overview of project management areas of authority and responsibility. Elaborate on the following points:
   - All projects have unique characteristics
   - The importance of time, budget, and quality as they relate to controlling and directing the project team
   - Logistics, challenges, and management factors when managing several projects
   - Why the project manager is responsible to upper management
   - The need for effective communication to facilitate project tasks
   - The importance of understanding contractual requirement
   - Compliance with plans and specifications and all applicable codes
4. Identify the six broad skill areas project managers must possess.
   - Describe the importance of each
   - Provide examples of the consequences for each

Section 3.4.0 *Organizing*

When discussing Section 3.4.0, use a section on the organization chart and walk through the lines of authority and communication to illustrate how a construction issue can be resolved.

Request that participants draw and present the organization chart for their company and project. Then ask what changes, if any, they would recommend to the organizational structure of their company or project. Have them briefly discuss the rationale for those changes.

Section 4.0.0 *Ethical Approaches to Project Management*

Review the ethics example in Section 4.0.0. Think of a similar situation relating to an ethical issue. The purpose of the exercise is to get participants to think about ethics and identify potential gray areas that could lead to problems. Use the following guidelines for the exercise:

1. Divide the participants into groups.
2. Present the situation either verbally or use a handout.
3. Have each group analyze the situation.
4. Instruct each group to present its findings and recommendations.
5. Have each group critique the other responses.
6. Summarize the findings and provide feedback.
7. Conduct a brief question-and-answer period.

Estimated time for this exercise is 20 minutes.

Section 5.0.0 *Phases of a Construction Project*

Review the three phases of a construction project (development, design, and construction) and the components of each phase.

1. Development Phase
   a. Project Idea or Concept
   b. Land Research
   c. Feasibility Study
   d. Conceptual Drawings
   e. Preliminary Government Review
   f. Finance Review
   g. Conceptual Estimate
   h. Preliminary Plans and Specifications
2. Design Phase
   a. Government Review
   b. Marketing
   c. Finance
   d. Estimating
   e. Working Drawings
   f. Bid Negotiation
3. Construction Phase
   a. Build
   b. Final Inspection and Acceptance

MODULE 44101-08 — ANSWERS TO REVIEW QUESTIONS

	Answer	Section
1.	A project is a temporary endeavor undertaken to create a unique product or service. A project also has a predetermined start and completion.	(2.0.0)
2.	1. Planning, organizing, staffing, directing, and controlling one or more projects. 2. Reporting directly to the owner or the owner's designated representative on the status of those projects. 3. Completing each project on time, within budget and in full compliance with all job specifications, code requirements, drawings, and contractual agreements.	(3.0.0)
3.	A WBS is used to help the project manager set up project tasks in a hierarchical arrangement that is similar to an organizational chart.	(3.1.0)
4.	1. Bidding 2. Pre-construction 3. Construction 4. Closeout	(6.0.0)
5.	The bidding phase starts when an owner or architect/engineer requests that a general contractor or specialized contractor submit a bid or proposal on all or part of a construction project.	(6.1.0)
6.	Pre-construction phase.	(6.2.0)
7.	Answers will vary. Look for responses that present situations where potential conflicts may arise. For example, hiring, releasing information, or cutting safety regulations. Answers may also include examples listed in Table 1, Conflicts with Ethical Implications.	(4.0.0)
8.	c.	(6.2.2)
9.	b and d.	(6.4.0)
10.	1. The manager's ability and personality 2. The quality of communication between the manager and the manager's job supervisors 3. The number and type of other duties required of the manager 4. The skills and training level of the project team 5. The types of projects under the manager's command	(3.4.0)

MODULE 44101-08 — ANSWERS TO TRADE TERM QUIZ

1. bidding
2. Construction
3. project manager
4. accountable
5. work breakdown structure (WBS)
6. Delegation
7. closeout
8. grapevine
9. span of control
10. Pre-construction
11. Traditional contract delivery
12. Design-build
13. engineering-procurement-construction (EPC)
14. Construction management (CM) at risk
15. Agency CM

CONTREN® LEARNING SERIES — USER UPDATE

NCCER makes every effort to keep these textbooks up-to-date and free of technical errors. We appreciate your help in this process. If you have an idea for improving this textbook, or if you find an error, a typographical mistake, or an inaccuracy in NCCER's Contren® textbooks, please write us, using this form or a photocopy. Be sure to include the exact module number, page number, a detailed description, and the correction, if applicable. Your input will be brought to the attention of the Technical Review Committee. Thank you for your assistance.

Instructors – If you found that additional materials were necessary in order to teach this module effectively, please let us know so that we may include them in the Equipment/Materials list in the Annotated Instructor's Guide.

Write: Product Development and Revision
National Center for Construction Education and Research
3600 NW 43rd St, Bldg G, Gainesville, FL 32606

Fax: 352-334-0932

E-mail: curriculum@nccer.org

Craft ______ Module Name ______

Copyright Date ______ Module Number ______ Page Number(s) ______

Description ______

(Optional) Correction ______

(Optional) Your Name and Address ______

Module 44102-08

Safety

NCCER STANDARDIZED CRAFT TRAINING PROGRAM

The National Center for Construction Education and Research (NCCER) provides a standardized national program of accredited craft training. Key features of the program include instructor certification, competency-based training, and performance testing. The program provides trainees, instructors, and companies with a standard form of recognition through a National Craft Training Registry. The program is described in full in the *Guidelines for Accreditation*, published by NCCER. For more information on standardized craft training, contact the NCCER by writing us at 3600 NW 43rd St., Bldg. G, Gainesville, FL 32606; calling 352-334-0911; or emailing info@nccer.org. More information may be found at our website, www.nccer.org.

HOW TO USE THIS ANNOTATED INSTRUCTOR'S GUIDE

Each page presents two sections of information. The larger section displays each page exactly as it appears in the Trainee Module. The narrow column ties suggested trainee and instructor actions to each page and provides icons (detailed below) to call your attention to material, safety, audiovisual, or testing requirements. The bottom of each page includes space for your notes.

The **Audiovisual** icon indicates an appropriate time to show a transparency or other audiovisual aid.

The **DVD** icon indicates an appropriate time to discuss or show segments of the video.

The **Classroom** icon prompts you to define a term, stress a point, ask trainees to explain a concept, or give examples.

The **Demonstration** icon directs you to show trainees how to perform tasks.

The **Examination** icon tells you to administer the written module examination.

The **Homework** icon is placed where you may wish to assign reading for the next class, assign a project, or advise trainees to prepare for an examination.

The **Laboratory** icon is used when trainees are to practice performing tasks.

The **Materials** icon is a reminder for you to gather materials needed for classes, labs, and testing.

The **Performance Testing** icon tells you to administer a performance test or a portion thereof.

The **Safety** icon is used to emphasize safety issues. It is often keyed to *Caution* and *Warning!* statements in the Trainee Module.

The **Teaching Tip** icon indicates additional guidance is available, such as how to conduct an exercise, get the most educational value from a field trip, or encourage class participation. Teaching Tips may expand on a feature (*Think About It, Did You Know?*) or provide *Quick Quizzes* or similar exercises. You will be referred to the Teaching Tips section at the back of the module if there is additional material.

The **Combination** icon indicates that the laboratory listed corresponds with a performance task. If desired, you can note the proficiency of the trainees during the laboratory, and use it to satisfy performance testing requirements.

PREPARATION

Before teaching this module, you should review the Objectives, Performance Tasks, Materials and Equipment List, and Module Outline. Be sure to allow ample time to prepare your own training or lesson plan and gather all required materials and equipment.

MODULE OVERVIEW

This module stresses the importance of job site safety and identifies the project manager's duties and responsibilities regarding safety. It covers loss prevention and creating a zero-accident work environment. Several checklists are presented as references.

PREREQUISITES

Please refer to the Course Map in the Participant Module. Prior to training with this module, it is recommended that the participant shall have successfully completed *Project Management,* Module 44101-08.

OBJECTIVES

Upon completion of this module, the participant will be able to do the following:

1. Recognize the need for an effective job site safety and loss prevention program.
2. Identify the project manager's duties and responsibilities with respect to safety and loss prevention.
3. Identify the direct and indirect costs of accidents.
4. Identify potential areas for loss and evaluate the risks.
5. Identify methods of risk control.
6. Understand OSHA's Focused Inspection Program.
7. Evaluate subcontractors on the basis of past safety experience.
8. Identify the need for and types of employee participation in safety programs.
9. List things to be considered when dealing with the press.
10. Plan, implement, and evaluate a job site safety program with assistance from staff safety professionals or outside consultants.

MATERIALS AND EQUIPMENT LIST

Desktop or laptop computer or DVD player
Multimedia projector and screen
Minor Decisions: Major Impact – How to Deal with Real Issues in Project Management DVD
Transparencies
Blank acetate sheets
Overhead projector and screen
Transparency pens
Whiteboard/chalkboard
Markers/chalk
Pencils and scratch paper
OSHA Forms 300 and 301
OSHA Poster (Form 3165)
Module Examinations (located in the Test Booklet)

ADDITIONAL RESOURCES

This module is intended to present thorough resources for task training. The following reference works are suggested for both instructors and motivated participants interested in further study. These are optional materials for continued education rather than for task training.

Contren® Learning Series, *Field Safety*, 2003

Contren® Learning Series, *Safety Technology*, 2003

The *OSHA Recordkeeping Handbook* (OSHA Publication #3245) is available from the OSHA website at: www.osha.gov/pls/publications /pubindex.list

Other information relating to OSHA standards and regulations may be obtained from the OSHA website at www.osha.gov, local OSHA offices, or from the U.S. Department of Labor – OSHA Publications Office, Room N3101, 200 Constitution Avenue, N.W., Washington, D.C. 20210

Construction Industry Institute (2001). *Making Zero Accidents a Reality*. Retrieved from: www.construction-institute.org/scriptcontent/cpi2001slides/mathis_cpi.ppt

Field Inspection Reference Manual (FIRM). OSHA. Retrieved from www.osha.gov

Heinrich, H.W., Petersen, D., Roos, N.R., Brown, J., and Hazlett, S. *Industrial Accident Prevention: A Safety Management Approach* (1980)

Meyer, S.W. and Pegula, S.M. *Injuries, Illnesses, and Fatalities in Construction, 2004*. Bureau of Labor Statistics, posted 24 May 2006. Retrieved from: www.bls.gov/iif (1 October 2007)

TEACHING TIME FOR THIS MODULE

An outline for use in developing your lesson plan is presented below. Note that each Roman numeral in the outline equates to one session of instruction. Each session has a suggested time period of 2½ hours. This includes 10 minutes at the beginning of each session for administrative tasks and one 10-minute break during the session. Approximately 15 hours are suggested to cover *Safety*.

Topic	Planned Time
Session I. Project Manager Duties, Accident Costs, and Hazard Control	
A. Project Manager's Duties and Responsibilities	____________
B. Accident Costs	____________
C. Cost of Administering an Effective Safety Program	____________
D. Causes of Accidents	____________
Session II. Areas for Loss and Hazard Control	
A. Potential Areas for Loss	____________
B. Process of Hazard Control	____________
C. Planning the Job	____________
Session III. Complying with Regulations	
A. Occupational Safety and Health Act	____________
B. Recommended Employer's Safety and Health Program	____________
C. Multi-Employer Work Sites	____________
Session IV. Pre-Qualification of Contractors and Subcontractors and Pre-Task Job Planning	
A. Pre-Qualification of Contractors and Subcontractors	____________
B. Pre-Task Safety Planning	____________
C. Job Safety Analysis	____________

Session V. Key Elements of a Prevention Safety Program, Part One

A. Management Support and Policy Statement
B. Policy on Alcohol and Drug Abuse
C. Assignment of Responsibilities
D. Employee Screening, Selection, and Placement
E. Safety Rules

Session VI. Key Elements of a Prevention Program, Part Two; Review and Testing

A. Orientation and Training
B. Safety Meetings and Employee Involvement
C. Crisis Management
D. Emergency Reporting and Response
E. Post-Incident Claims Management
F. Review
G. Module Examination
   1. Participants must score 70% or higher to receive recognition from NCCER.
   2. Record the testing results on Craft Training Report Form 200, and submit the results to the Training Program Sponsor.

USING THE PM DVD WITH *SAFETY*

Instruct the students to look at the main characters and the background workers on the site as each scenario is played. Have them point out whether there are any failures to adhere to proper PPE criteria.

As each scenario unfolds, ask the students to indicate how the actions of the individuals and the condition of the job site may jeopardize safety. What safety problems were brought to light?

Note: Section 7.3.0 Assignment of Responsibilities, presents discussion opportunities for students to observe the project manager, foremen, and superintendent's attitudes towards safety and accountability.

Project Management

44102-08

Safety

Assign reading of Module 44102-08.

44102-08
Safety

Topics to be presented in this module include:

Overview

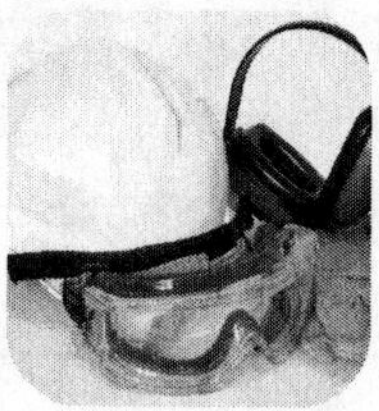

As a management tool, a safety and loss prevention program has a twofold purpose:

- To provide project managers with a basic understanding as to how a company's safety performance affects the profit and loss picture today and its ability to remain competitive in the future.
- To serve as a guide and resource for project managers in carrying out their duties with respect to safety and loss prevention.

Your safety and loss prevention program should provide the basic framework for meeting the company's obligation for operating in a safe, efficient, and profitable manner. It should be incorporated into all phases of your operation and involve all employees at every level.

This module contains suggestions, checklists, and audit forms that may be useful to you. Keep in mind that these are resources and references that you may modify, expand, or condense to suit your specific needs.

Instructor's Notes:

Objectives

When you have completed this module, you will be able to do the following:

1. Recognize the need for an effective job site safety and loss prevention program.
2. Identify the project manager's duties and responsibilities with respect to safety and loss prevention.
3. Identify the direct and indirect costs of accidents.
4. Identify potential areas for loss and evaluate the risks.
5. Identify methods of risk control.
6. Understand OSHA's Focused Inspection Program.
7. Evaluate subcontractors on the basis of past safety experience.
8. Identify the need for and types of employee participation in safety programs.
9. List things to be considered when dealing with the press.
10. Plan, implement, and evaluate a job site safety program with assistance from staff safety professionals or outside consultants.

Trade Terms

Accident
Action levels
Experience modification rate (EMR)
Focused inspection program
Incidence rates
Occupational Safety and Health Administration (OSHA)
OSHA recordable incidence rate
Personal protective equipment (PPE)
Risk assessment codes
Workers' compensation
Worker exposure

Prerequisites

Before you begin this module, it is recommended that you successfully complete *Project Management,* Module 44101-08.

This course map shows all of the modules in the *Project Management* training curriculum. The suggested training order begins at the bottom and proceeds up. Skill levels increase as you advance on the course map. The local training program sponsor may adjust the training order.

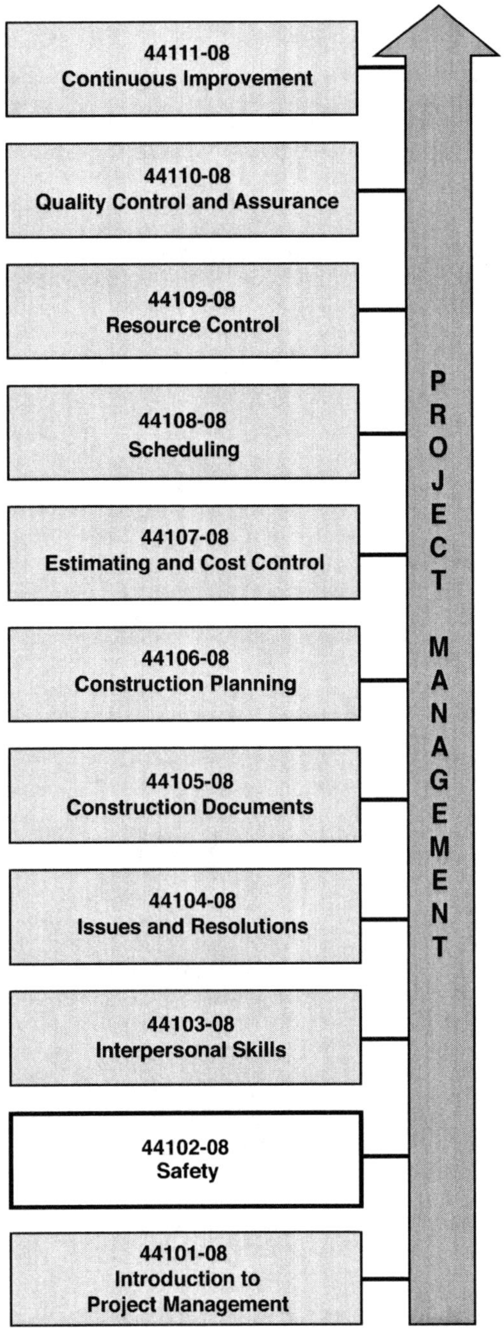

102CMAP.EPS

Ensure that you have everything required to teach the module. Check the Materials and Equipment list at the front of this module.

See the general Teaching Tip at the end of this module.

Show Transparency 1, Objectives. Review the goals of the module, and explain what will be expected of the participant.

Explain that terms shown in bold are defined in the Glossary at the back of this module.

MODULE 44102-08 ◆ SAFETY 2.1

Explain that OSHA, the Occupational Safety and Health Administration, promotes safety at work sites.

Emphasize that skilled employees are construction contractors' most valuable assets and keeping them safe must be a priority.

Review the consequences of not having a Safety and Loss Prevention Program.

Show Transparency 2, Project Manager's Duties.

1.0.0 ◆ INTRODUCTION

Construction work, by its nature, can be hazardous. According to figures from the Bureau of Labor Statistics, the construction industry had a total **Occupational Safety and Health Administration (OSHA)** recordable injury/illness rate of 6.3 in 2005. This means that more than six workers out of every 100 experienced a job-related injury or illness that required some form of medical treatment. This incidence rate was relatively unchanged from 2004. The mining industry, often thought of as hazardous, had a total recordable incidence rate of 3.6 in 2005.

Skilled employees are a contractor's most valuable resource. The company cannot afford to lose them because of preventable job-related injuries and illnesses.

Owners and users of construction services have recognized that they have a direct economic stake in the safety performance of their contractors. **Accident** costs are an expense to the contractor, which are passed on in one way or another to the client.

As a direct result, many owners and purchasers of construction services are pre-qualifying contractors based on their past safety performances. It is becoming commonplace for a client to ask a contractor for his total **OSHA recordable incidence rate** and **workers' compensation** (WC) **experience modification rate (EMR)**. Contractors with high **incidence rates** and EMRs are sometimes being excluded from bidding.

There are several principal reasons for a contractor to have an effective safety program: The contractor has a moral and legal obligation to provide employees with a safe place to work that is free from recognized hazards. There are economic reasons for having an effective safety program. Increased insurance costs due to past accident experience can affect future bids. Likewise, the hidden or uninsured costs of accidents or injuries can rob a company of its profits.

Owners and general contractors are increasingly becoming aware of the impact that a contractor's safety performance has on overall job costs. A major accident can adversely affect a contractor's reputation and also give the owner some bad press. Failure to have an effective safety and loss prevention program can result in:

- Needless pain and suffering for injured workers and their families
- Economic hardship for workers and the company
- Missed opportunities to bid on future work
- Loss of the competitive edge
- Increased insurance and operating costs
- Citations and fines for violations of safety and health regulations
- Damage to the company's name and reputation

Successful contractors have recognized that effective safety management is a money saver. Accidents have high direct and indirect costs, costs which management can limit. Contractors who fail to recognize this economic fact of life will find themselves frequently underbid and may be forced out of the market.

To make progress against the human and economic losses that result from construction accidents, you, as project manager, must recognize the connection between loss prevention measures and the bottom line.

A company's goal in terms of securing jobs, doing quality work and making a profit cannot be achieved without planning. The same degree of planning is necessary to achieve safety and loss prevention objectives. An effective safety and loss prevention program is a carefully thought-out action plan, an integral part of the overall project execution plan. It is not, and cannot be, an afterthought or add-on that results from an accident.

1.1.0 Project Manager's Duties and Responsibilities

The project manager has the responsibility to complete the job safely, on time, within specifications, and within budget. To fulfill his responsibilities the project manager, sometimes with the help of others, has the following duties:

- Identify hazards and potential risks
- Evaluate the risk
- Develop a plan of action
- Set the proper attitude through leadership
- Execute the plan
- Audit compliance and evaluate performance
- Make necessary adjustments
- Investigate accidents and take corrective action
- Manage claims
- Close out the job and evaluate overall performance

When carrying out these duties, the ideals of safety, quality, and production must be inseparable. Management gets not what it *expects* but what it *inspects*. Supervisors and employees who are praised or rewarded for quality and production (or chastised for the lack of it) without a cor-

Instructor's Notes:

2.2 PROJECT MANAGEMENT

responding emphasis on safety will tend to place more importance on these areas than on safety. It is your job to see that supervisory emphasis on quality, safety, and production is the same.

2.0.0 ◆ ACCIDENT COSTS AND IMPACT

An accident is defined as an unplanned event that may or may not result in personal injury or property damage. Accidents are often categorized by their severity and impact:

- *Near-miss* – An unplanned event or occurrence in which no one was injured and no damage to property occurred, but during which either could have happened. Near-miss incidents are warnings which should not be overlooked or taken lightly.
- *Property damage* – An unplanned event that resulted in damage to tools, materials or equipment but no one was injured.
- *Minor injuries* – Personnel may have received minor cuts, bruises, or strains but the injured workers returned to full duty on their next regularly scheduled work shift.
- *Serious or disabling injuries* – Personnel received injuries that resulted in temporary or permanent disability. Included in this category would be lost time accidents, restricted duty or motion cases and those which resulted in permanent partial or permanent total disability.
- *Fatalities.*

Studies have shown that for every serious or disabling injury, there were 10 injuries of a less serious nature and 30 property damage accidents (*Figure 1*). A further study showed that 600 near-miss incidents occurred for every serious or disabling injury.

In many cases, the elements of chance, timing, or luck are the only reasons that the near-miss incidents are not more serious. Safety and loss prevention efforts should be focused on preventing the frequent occurrences, not necessarily concentrated on the one or two serious injuries. The key to an effective safety program is controlling the frequency of incidents; frequency breeds severity.

Review and discuss the five types of accidents. Ask participants to provide examples of each type of accident.

Show Transparency 3, (Figure 1).

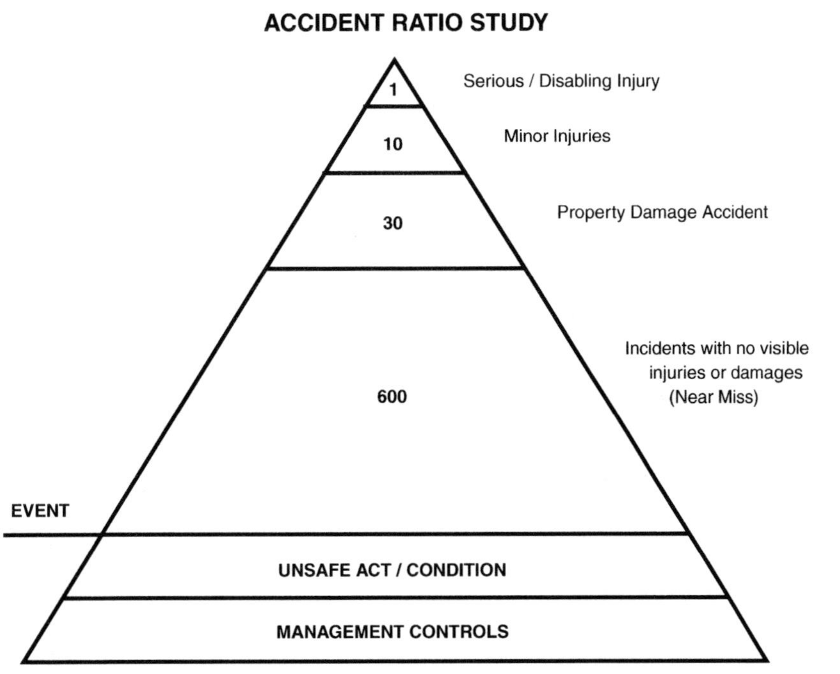

Figure 1 ◆ Accident ratio study.

MODULE 44102-08 ◆ SAFETY 2.3

Show Transparency 4 (Figure 2). Emphasize that the indirect costs of accidents far outweigh the direct costs.

Point out that indirect costs may include costs associated with accident investigation, corrective measures, lower employee morale, absenteeism, and poor customer/community relations.

Show Transparency 5 (Figure 3).

2.1.0 Accident Costs

When an accident happens, everyone loses—the injured worker, the employer, and the insurance company. The only one who usually benefits from an accident is the plaintiff's attorney. Accident costs are often classified as direct (insured) and indirect (uninsured).

Direct, or insured, costs include medical costs and other workers' compensation insurance benefits as well as liability and property damage insurance payments. Of these, claims under workers' compensation, the insurance that covers workers on the job, are the most substantial of direct costs.

For most construction companies, the direct costs of accidents are not fixed. They vary depending upon each company's accident experience. This is reflected in the company's workers' compensation EMR.

Indirect (uninsured) or hidden costs can be compared to the hidden nine-tenths of an iceberg, with the tip of the iceberg representing the direct or insured costs (see *Figure 2*). Studies have shown that the hidden costs of accidents can and usually do exceed the direct costs of accidents from two to seven times.

2.1.1 Calculating Accident Costs

Real dollars are lost when an accident occurs. These dollars have a tremendous effect on the company's profit margin. For example, consider a company that operates on a profit margin of three percent of its gross income. The company suffers a loss of $50,000 due to accidents. In order to make up for that loss, the company must increase its gross income by $1,666,667! The accident cost worksheet (*Figure 3*) shows how direct and indirect costs are recorded and calculated.

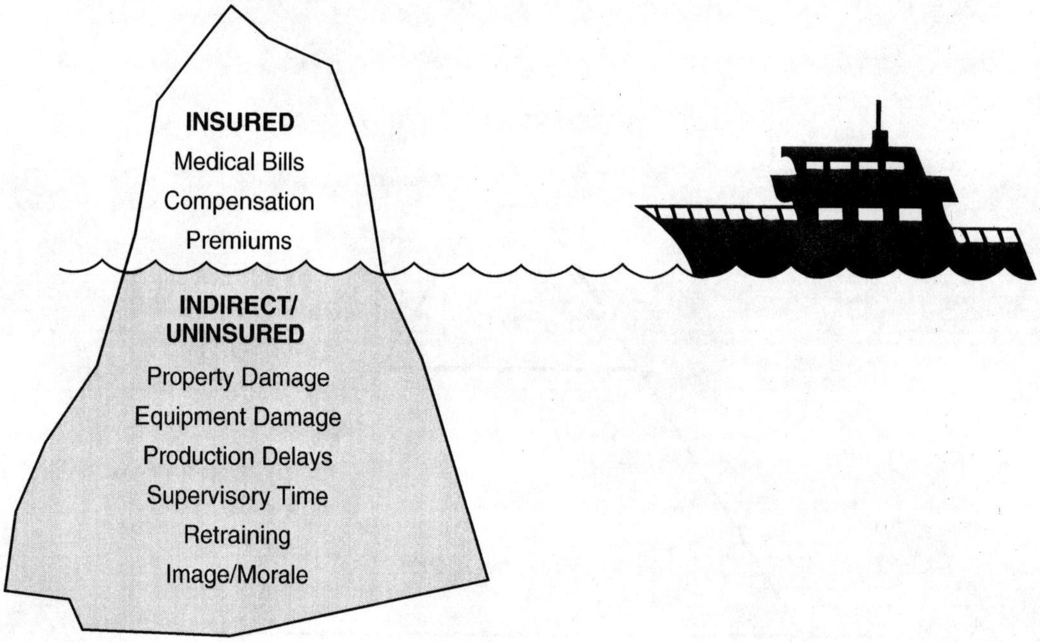

102F02.EPS

Figure 2 ◆ Hidden costs of accidents.

Instructor's Notes:

2.4 PROJECT MANAGEMENT

Person Injured: ____________________

Type of accident: _________________

Nature of Injury: ____________________

Direct (Insured) Costs		Indirect (Uninsured) Costs	
List Items	Cost in Dollars	List Items	Cost in Dollars
Subtotal A		Subtotal B	

Total A + B = ________________

102F03.EPS

Figure 3 ◆ Accident cost worksheet.

MODULE 44102-08 ◆ SAFETY 2.5

Explain the correlation between accident experience ratings and future insurance costs.

Show Transparency 6 (Figure 4). Explain that in this example, the insurance costs equal the payroll in dollars times the workers' compensation rates divided by 100.

Show Transparency 7 (Figure 5).

See the Teaching Tip for Sections 2.1.0–2.1.3 at the end of the module.

2.1.2 Accident Experience vs. Future Insurance Costs

What effect, if any, does a contractor's accident experience have on future insurance costs? It has significant impact. Contractors with poor accident experience will generally pay more in workers' compensation insurance than those with a good record. The difference is in their workers' compensation experience modifier.

Experience rating is a method of modifying future workers' compensation (WC) insurance premiums by comparing a particular company's actual losses to the losses normally expected for that company's type of work. The average rate for a particular class of work is called the book rate or manual rate. Contractors with better than average loss experience have a modifier less than 100-percent, a credit factor. Those with worse than average experience have a modifier (multiplier) over 100-percent, a penalty factor. The experience modifier rate is multiplied by total workers' compensation cost to discover what the company will pay for workers' compensation insurance. *Figure 4* shows the manual workers' compensation premium rates for a typical job in Louisiana. The insurance costs equal the payroll (in dollars), times the workers' compensation rates, divided by 100. As you can see, the WC rate can have quite an impact on job cost.

Figure 5 illustrates the effect of an EMR of 1.1.

2.1.3 Case Study: Stan's Accident

On Monday at 8:30 a.m., Stan fell while working. His leg was injured and he struck his head, causing a wound and considerable bleeding. Co-workers Dave, Sam, Roy and construction supervisor, Mack, gave Stan assistance and first aid. Mack drove Stan to the clinic for treatment, arriving at the clinic at 9:30 a.m.

Stan had a concussion and it required 18 stitches to close the head wound. His leg was badly bruised with a slight fracture of the tibia, or shin-bone. Mack drove Stan home from the clinic and arrived back at work at 12:50 p.m. He was back to work by 1:00 p.m. Production had stopped due to problems that only Mack could resolve.

Typical Workers' Compensation Premium Rates for an Established Contractor in Louisiana (Stock Insurance Companies)

Assigned Code	Classification	Payroll	WC Rate (Preferred)	Insurance Cost
5403	Carpenters	$120,000.00	$30.28	$36,336.00
5213	Cement Finishers	55,000.00	14.18	$7,799.00
5190	Electricians	62,000.00	5.75	$3,565.00
5183	Pipefitters	61,000.00	4.59	$2,799.90
5040	Iron/Steel Erection (Metal Bridges/Exteriors)	72,000.00	36.25	$26,100.00
Totals		**$370,000.00**		**$76,599.00**

As of 9/05

102F04.EPS

Figure 4 ◆ Manual workers' compensation premium rates.

Instructor's Notes:

2.6 PROJECT MANAGEMENT

Total Workers' Compensation Costs (from *Figure 4*)	$76,599.00
	× 1.1
EMR	$84,258.90

102F05.EPS

Figure 5 ◆ The effect of an EMR of 1.1.

Pete, a subcontractor, had a three-person crew who could not do their scheduled work because Mack's crew had not completed their work. This resulted in an extra four hours charged to the job by the contractor plus an extra $450 for rental on their equipment.

Mack spent an additional two hours reviewing, discussing, and preparing the accident report. The insurance clerk took an hour to process the necessary forms.

The medical clinic charges were:

Doctor $180
X-rays $250
Treatment and supplies $156
Medication $45
Drug test
(required by company for accidents) $110
Follow-up visit $100

Stan returned to work the following Friday but was only about 75 percent as productive for the first week back (five workdays). He was given full pay for the day of the accident but no additional pay during the recovery time. He was not off long enough to collect workers' compensation for his lost wages.

Mack, Dave, and Roy worked a total of 60 hours overtime to get production back on schedule. Mack was not paid any overtime. Dave and Roy were each paid for 20 hours overtime. Fringe benefit cost for all employees and contractors was 38 percent of base pay.

Rates of pay:

Stan Ward $28/hr
Mack Solis $5,500/month
based on a 173-hr. month (no overtime)
Sam $22/hr
Roy $22/hr
Dave $22/hr
Insurance Clerk $14/hr
Sub contract sheet metal crew
– 2 mechanics each earning $22/hr
– 1 helper $15/hr

Most of the accident costs have been stated in this story. Medical costs may have been paid by the workers' compensation fund, to which the employer contributed. In reality, the employer bears all costs directly or indirectly and, therefore, so does most of the construction industry in the United States.

A company's EMR is based on the first three of the last four years' accident experience. In other words, your EMR for the 2008 policy year would be based on your losses in 2004, 2005, and 2006. What you do today will have a significant impact on your operations for years to come.

2.2.0 Cost of Administering an Effective Safety Program

The Business Roundtable's A-3 report estimated that the cost of maintaining an effective safety program is approximately 2.5 percent of the direct labor costs. This includes salaries for safety, medical and clerical personnel, cost of safety meetings, inspections of tools and equipment, orientation meetings, **personal protective equipment** and miscellaneous supplies and equipment. However, most of these items are required by state and federal laws and the law-abiding contractor would have to pay for them anyway. The likely added cost of a safety program beyond what is legally required is probably closer to 1 percent of the direct labor costs as opposed to 2.5 percent. These same studies concluded that for every dollar invested in a safety program, the contractor could save as much as $3.20.

2.3.0 Causes of Accidents

Accidents and injuries result in needless pain and suffering as well as financial hardship for the employee, family, and contractor. Employers have moral, legal, and financial obligations to prevent accidents and injuries. To do this you must understand what causes accidents. Accidents are caused by:

- Unsafe acts
- Unsafe conditions
- Combinations of unsafe acts and conditions
- Acts of God

William Heinrich, a noted psychologist, suggested that accidents resulted from unsafe acts 88 percent of the time, unsafe conditions 10 percent of the time, and Acts of God 2 percent of the time. Later studies suggest that most accidents involve both unsafe acts and unsafe conditions. Most safety professionals today agree that we should

Emphasize that the cost of administering an effective safety program is much less expensive than the cost of accidents.

See the Teaching Tip for Section 2.3.0 at the end of the module.

MODULE 44102-08 ◆ SAFETY 2.7

Explain that identifying and correcting root causes of safety problems is important in maintaining a safe work site.

Describe the three levels of accident causation. Ask participants to provide examples for each.

Show Transparency 8 (Figure 6). Use this transparency to describe the flow of the three accident causes.

Assign reading of Sections 3.0.0–3.3.0.

Discuss potential areas for loss. List the people, property, and other factors that accidents can involve or affect.

not ignore unsafe conditions but a greater focus should be on the actions of people.

While it is important to focus on unsafe acts and unsafe conditions, you must identify and correct the root cause(s) if you are to effect permanent change. Unsafe acts and unsafe conditions are merely symptoms of the problem. The management systems that failed to detect, failed to correct, or failed to anticipate the unsafe acts and unsafe conditions are the root causes on which you should focus your attention.

2.3.1 The Three Levels of Accident Causation

The drawing in *Figure 6* indicates the flow of the three accident causes. An accident that is considered a Level I direct cause is an event which may or may not cause an injury or damage. In accident investigation, look for the energy sources. These energy sources generally fall into one of three categories: by accident type, by energy source, and/or by hazardous materials.

Level II, or indirect causes, of accidents are the unsafe acts or unsafe conditions that precede the accident event. In the past, many investigators have looked only for these unsafe acts or conditions. When found, the act or condition was corrected. It was later found that these were just symptoms of a greater problem; thus they are classified as indirect causes.

The basic, or root causes, of an accident are Level III causes. These include system failures that permitted, encouraged, allowed, failed to identify and correct, failed to follow-up on, failed to anticipate or predict the unsafe act or condition. These causes are those which would effect permanent results when corrected. They are weaknesses which not only affect the single accident being investigated, but also might affect many other future accidents and operational problems.

3.0.0 ◆ AREAS FOR LOSS AND HAZARD CONTROL

The planning guidelines (*Appendix A*) and checklists (*Appendixes B* through *F*) included in this module may be helpful in identifying and evaluating hazards associated with a prospective job. Other resources that may be helpful include:

- Closeout reports or critiques from similar jobs
- Owner or client safety manuals, rules, or procedures
- Local, state, and federal safety, health and environmental regulations

3.1.0 Potential Areas for Loss

Before you can develop and implement an effective safety program, you must identify the areas where you might reasonably expect a loss. Construction accidents can involve or affect:

- Employees
- Subcontractors
- The general public
- The building under construction
- Building materials and equipment
- Adjacent structures/exposures
- The environment

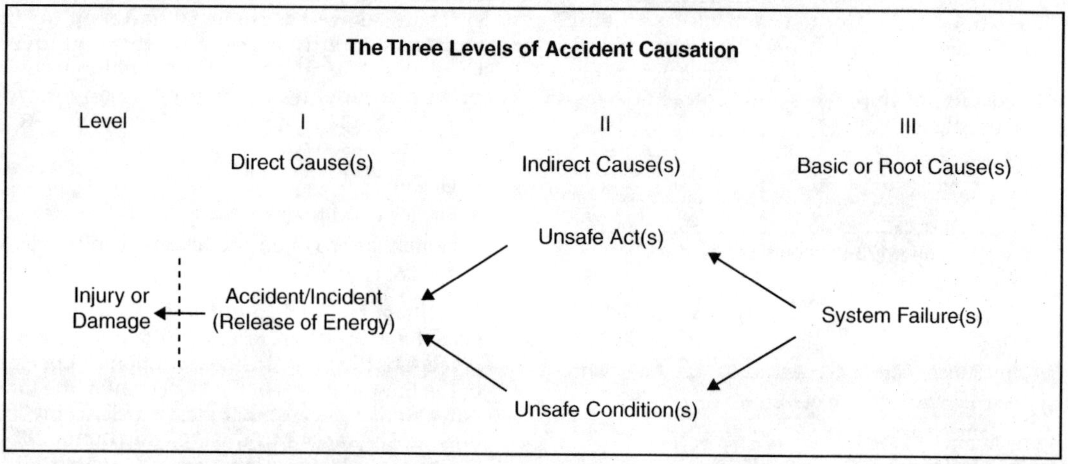

Figure 6 ◆ The three levels of accident causation.

Instructor's Notes:

2.8 PROJECT MANAGEMENT

3.2.0 Process of Hazard Control

Small contractors who do not have a permanent safety staff should strongly consider using outside consultants or insurance company safety representatives to help develop and implement a hazard control program. Most insurance companies provide safety and loss prevention services, the cost of which is usually included in your premium. Proper planning in the early phases of the job can reduce or eliminate costly delays and interruptions due to accidents and injuries. Find out what services are available from your insurance carrier and use them to your advantage.

3.2.1 Hazard Identification and Evaluation

Identifying and evaluating hazards is the first step in the process of hazard control. It must start in the pre-bidding stage of the job and continue forward. In many cases, the identification of potential hazards associated with the job may be done as a team effort and involve one or more of the following:

- Project manager/estimator
- Job superintendent
- Safety manager, staff safety specialist, insurance company loss control representative, or outside consultant
- Risk/insurance manager or insurance agent
- Personnel manager

3.2.2 Ranking the Hazards by Risk

Ranking the hazards by risk is the second step in the hazard control program. This involves assigning a rank of the hazards by consequence (severity) and probability (frequency). The relative consequences of hazards may be ranked as follows:

- *Catastrophic* – may cause death or loss of the structure
- *Critical* – may cause severe injury, illness or property damage
- *Marginal* – may cause minor injury, illness or property damage
- *Negligible* – probably would not affect personnel safety or health, but nevertheless would result in some violation of specific criteria

The hazard probability category may be ranked as follows:

- Likely to occur immediately or within a short time
- Probably will occur in time
- Possibly will occur in time
- Unlikely to occur

After estimating the consequence and probability you should estimate **worker exposure**, the number of persons who will be regularly exposed to the hazard. The following scheme can be used to estimate exposure:

- Greater than 50 people regularly exposed
- From 10 to 49 people regularly exposed
- From 5 to 9 people regularly exposed
- Less than 5 people regularly exposed

When the hazards have been ranked according to all three criteria (consequence, probability, and exposure), the next step is to assign a single **risk assessment code** considering the results of your analysis (see *Table 1*). Risk assessment codes can be assigned as follows:

- Critical
- Serious
- Moderate
- Minor
- Negligible

A risk that has catastrophic consequences, is highly probable, and involves the exposure of many people would obviously earn a risk assessment code of 1. A risk with negligible consequences is not very likely, involves the exposure of a few people, and gets a code of 5. Risks falling between these extremes are judgment calls, but examination of the three criteria should help you place them relatively easily.

Table 1 Risk Assessment Matrix

Matrix for determining degree of risk (Risk Assessment Code – RAC)

Probabilities \ Consequence	I	II	III	IV
A	1	1	2	3
B	1	1	2	3
C	2	3	3	4
D	3	4	4	5

102T01.EPS

Emphasize the importance of proper safety planning in the early phases of a job.

Ask participants to list persons who may be involved in identifying potential hazards on a job site.

Explain that hazards are ranked by consequence, probability, and exposure.

Show Transparency 9, Hazard Consequence Categories. Provide examples of each.

Show Transparency 10, Hazard Probability Categories. Provide examples of each.

Show Transparency 11, Exposure Categories.

Explain that after hazards are ranked, they are assigned a Risk Assessment Code.

Show Transparency 12, Risk Assessment Codes.

Show Transparency 13 (Table 1). Using examples, explain how the Risk Assessment Matrix is used.

Explain that risks falling between the extremes of 1 and 5 are judgment calls made by reviewing the three risk categories.

MODULE 44102-08 ◆ SAFETY 2.9

Review and discuss risk management techniques.

Ask participants to list methods that a contractor may use to deal directly with a hazard.

Discuss hazard control techniques and provide examples of each.

Assign reading of Sections 4.0.0–4.4.1.

Review the four major sets of federal regulations that may affect construction operations.

Of course, many small contractors will not need to go into this much detail, but they must use some method to identify the potential hazards and evaluate their exposure. This must be done before a course of action can be planned.

Management decision-making is the next step in the process of hazard control. Risk management techniques can be used to eliminate the contractor's exposure to the risk or the hazard itself can be dealt with. Risk management techniques include, but are not limited to, the following:

- *Risk avoidance* – Don't bid the job.
- *Risk sharing* – A joint venture arrangement may be employed where a contractor more experienced in this type of work handles the more hazardous operations.
- *Risk transfers* – The more hazardous work is completely subcontracted to contractors specializing in this type of job.

If the contractor elects to deal directly with the hazard, the following are methods that can be employed:

- Engineer out the hazard or use alternate construction methods that eliminate or minimize the risk and exposure.
- Protect the workers from the hazard by barricading, guarding, or shielding.
- Use of management controls such as limiting worker exposure, special training, and education.
- Use of personal protective equipment.

Although the use of personal protective equipment is sometimes the only practical method, particularly in construction, it is the least desirable method. Personal protective equipment must be properly selected, used, and maintained. If you can eliminate the hazard and thereby eliminate the need for personal protective equipment, you are much better off. Failure to use the proper personal protective equipment is frequently listed on accident reports as the cause of the accident. This is an unsafe act that possibly could be prevented by eliminating the hazard.

There are three major areas where hazard control techniques can be applied:

- At the source, by eliminating the hazard
- Along the path, by barricading or guarding
- At the receiver (worker), through the use of personal protective equipment

3.3.0 Planning the Job

Once the hazards have been identified and evaluated and methods for eliminating or controlling the hazards have been developed, these hazard control techniques must be included in every phase of the job from the bidding process to the job closeout. The checklist shown in *Appendix B* of this chapter can be useful in planning the safety aspects of the job.

Many progressive contractors conduct a design and constructability review when planning the job. Examples of items that may be reviewed include the following:

- What size cranes will be required and can they access the required area?
- Would pre-assembly of structural sections at ground level reduce the elevated work hazards?
- Where should electrical power lines be located to minimize problems with crane operations?

4.0.0 ◆ COMPLYING WITH REGULATIONS

As a project manager, you should be aware of the various safety, health, and environmental regulations that affect your job. In general, there are four major sets of federal regulations which may affect your operations. In some cases, these federal regulations may be enforced by state agencies under an agreement with the federal government.

- *Occupational Safety and Health Administration (OSHA)* – Conducts inspections of workplaces to determine compliance with the Occupational Safety and Health (OSH) Act and with specific OSHA standards. When violations of the OSH Act or OSHA standards are found, OSHA is authorized to issue citations to employers, propose penalties, and require abatement of hazards.
- *Department of Transportation (DOT)* – Regulations primarily cover vehicles with a gross weight of more than 10,000 pounds. These regulations govern driver selection and training; the operation, maintenance, and inspection of the vehicles; handling of cargo, and vehicle placarding.
- *Environmental Protection Agency (EPA)* – Regulations deal with the prevention of ground, air, and water pollution. There are specific regulations concerning the handling and storage of hazardous materials, the reporting of inventories and spills or releases of hazardous materi-

Instructor's Notes:

2.10 PROJECT MANAGEMENT

als, and the generation, storage, treatment, and disposal of hazardous wastes.

- *Chemical Safety and Hazard Investigation Board (CSB)* – The CSB is an independent investigative agency authorized under the Clean Air Act to investigate chemical incidents to determine the conditions and circumstances which have led to an incident and to identify the cause or causes so that similar incidents might be prevented. Its mission is to enhance the safety of workers and the public by uncovering the underlying causes of accidental chemical releases at fixed facilities and motivating remedial action by both the private and public sectors. The chemical incidents which the CSB investigates are those which result from the production, processing, handling or storage of chemical substances (not limited to extremely hazardous substances) causing death, serious injury, substantial property damage (including damage to natural resources), or evacuations of the public.
  - Consistent with each agency's statutory responsibilities, OSHA and CSB will coordinate their fact-finding efforts. The CSB onsite representatives may discuss factual data pertaining to an incident with other on-site agencies.
  - However, the CSB is an independent agency tasked with certain oversight responsibilities; it is not an enforcement agency. To ensure that during the conduct of an investigation the CSB is not to be perceived as an extension of a state or federal enforcement investigation, its investigative activities will be separate and distinct from those of other on-site agencies. Interviews of witnesses and requests for documents will be conducted or requested separately by the CSB unless the company or person(s) involved request otherwise. While the CSB will cooperate with other onsite entities, its focus is different and its interaction with all parties must and will be distinct from the activities of enforcement agencies.

In addition to these federal regulations, there are numerous state and local regulations. Before bidding a job you should find out what safety, health, and environmental regulations will apply. It must be clearly understood that federal, state, and local regulations are usually minimum standards and mere compliance with these minimum standards does not guarantee that you will have an effective safety and loss prevention program.

4.1.0 Occupational Safety and Health Act

Under OSHA, employers are required to provide their employees with a safe and healthful workplace, free from recognized hazards. Failure to comply with these regulations can result in fines, and in some cases, jail terms. OSHA compliance officers can and will make job site inspections to check compliance with these laws. These inspections are most often unannounced, and may be the result of a random job site selection process, a complaint, or in response to an accident. Although you can request that OSHA compliance officers obtain a search warrant, most companies do not require them to do so. You should be familiar with your company's policies and procedures concerning inspections by regulatory agencies.

Employees must be informed of their rights under the OSH Act. To do this, the OSHA Poster (OSHA Form No. 3165) must be prominently displayed in a conspicuous place on the job site where notices to employees are customarily posted.

4.1.1 Regulated Substances and Access to Medical and Exposure Records

In addition to the hazard communication standard, OSHA also has individual standards, which regulate employee exposure to certain hazardous substances. One such substance of concern to the construction industry is asbestos. If employee exposures to these regulated substances exceed certain levels (referred to as **action levels**), employers must begin compliance activities such as air monitoring, employee training, and medical surveillance.

In most cases, employers will be required to maintain employee medical and exposures records related to the regulated substances for the duration of employment plus 30 years. Under OSHA's access to employee exposure records, employees or their designated representatives have the right to examine and copy their medical and exposure records.

When you bid a job, you must determine if your employees will or could be exposed to hazardous chemicals and plan accordingly. When evaluating potential employee exposure, you must look at the materials your employees will be handling as well as those used by other contractors who might be working around or near your employees.

Remind participants that there are also state and local regulations that may apply.

Give examples of your state and local regulations for safety, health, and the environment.

Emphasize that federal, state, and local regulations are usually minimum standards, and that compliance with minimum standards does not guarantee an effective safety and loss prevention program.

See the Teaching Tips for Sections 4.0.0 and 4.1.0 at the end of the module.

Point out that OSHA also has individual standards that regulate employee exposure to certain hazardous substances, such as asbestos.

Emphasize that when bidding on a job, it is important to determine whether employees might be exposed to such hazards, and plan accordingly.

MODULE 44102-08 ◆ SAFETY 2.11

Review and discuss recordkeeping requirements.

Review and discuss instances in which OSHA Form 300 must be completed.

Emphasize that all entries on Form 300 must be made no later than six working days after a recordable injury or illness has occurred.

Explain why not every injury is recordable.

Define occupational injury, occupational illness, and medical treatment when used to determine whether an injury is recordable.

Obtain copies of OSHA Forms 300 and 301. Explain the correlation between the two forms.

4.1.2 Recordkeeping Requirements

All contractors who employ 11 or more people must make, keep, and preserve records of occupational injuries and illnesses and make these records available to representatives of the Secretary of Labor upon request. The recordkeeping requirements of the OSH Act apply to almost all private sector employers. Government entities are exempt from the Act but may be subject to the same regulations in those states that have their own occupational safety and health plans.

If an on-the-job accident occurs that results in the death of an employee or in the hospitalization of three or more employees, all employers, regardless of number of employees, must report the accident, in detail, to the nearest OSHA office within eight hours. In states with approved plans, employers report such accidents to the state agency responsible for safety and health programs.

The two forms that OSHA provides for the purpose of keeping injury and illness records are Form 300 and Form 301 (these forms can be found at www.osha.gov). OSHA Form 300, log and summary of occupational injuries and illness, must be completed in all of the following cases:

- Occupational death
- Nonfatal occupational illness
- Nonfatal occupational injuries that involve one or more of the following:
  - Loss of consciousness
  - Restrictions of work or motion
  - Transfer to another job
  - Medical treatment other than first aid

Entries on the form must be made no later than six working days after a recordable injury or illness has occurred. Not every injury is recordable. The following definitions are used to determine whether or not an injury is recordable:

- *Occupational injury* – Any injury (e.g., cut, fracture, sprain, amputation) that results from a work accident or from exposure to a single incident in the work environment. Work environment comprises the physical location, equipment, materials processed or used, and the kinds of operations performed in the course of an employee's work, whether on or not on the employer's premises. Conditions resulting from bite wounds (e.g., insects, snakes, animals) or from one-time exposure to chemicals are considered injuries.
- *Occupational illness* – An illness is any abnormal condition or disorder, other than one resulting from an occupational injury, caused by exposure to environmental factors associated with employment. It includes acute and chronic illnesses or diseases that may be caused by inhalation, ingestion, or direct contact.
- *Medical treatment* – This includes treatment administered by a physician or a registered professional under the standing orders of a physician. Medical treatment does not include first-aid treatment (one-time treatment of minor scratches, cuts, burns, splinters, and so forth, which do not ordinarily require medical care) even though provided by a physician or registered professional.

Data entered on the Form 300 must be summarized annually and the summary posted in a location customarily used for posting public information. The information must be posted no later than February 1 and remain in place for three months, regardless of whether any injuries or illness were recorded during the year.

OSHA Form 301 is for recording accident investigation information. For every recordable injury or illness entered on the Form 300, additional information must be recorded on the OSHA Form 301 or a similar form, such as a workers' compensation form, that contains all the items found on Form 301. Supplementary records must be completed and filed at the work establishment within a specific number of workdays after the employer has received information that an injury or illness has occurred.

Ordinarily, records must be maintained at each establishment, the term establishment being defined on the reverse side of OSHA Form 300. If an employer has more than one establishment, a separate set of records must be maintained at each one. Some employees are subject to common supervision, but do not report to or work at a fixed establishment on a regular basis. Records for employees engaged in physically dispersed activities that occur in construction, installation, repair, or service operations may be maintained at the field office or mobile base of operations. Records may also be maintained at an established central location, with the address and telephone number of the location made available at the work site.

The distinction between fixed and non-fixed establishments for the purpose of determining where OSHA records should be kept rests on the nature and duration of the operation and not on the type of structure in which the business is located. Generally, any operation at a given site for more than one year is considered a fixed estab-

Instructor's Notes:

2.12 PROJECT MANAGEMENT

lishment. Also, fixed establishments are often where clerical, administrative, or other business records are kept. For example, a construction crew repairing a bridge for two months is considered working in a non-fixed establishment, while a crew repairing a bridge for a year and a half is considered working at a fixed establishment.

All records must remain in the establishment for five years after the date of the accident or illness. If an establishment changes ownership, the new employer must preserve the records for the remainder of the five-year period. The new employer is not responsible for updating records of the former owner.

Records may be inspected and copied at any reasonable time by authorized federal or state government representatives. Also, the log and summary shall be available to employees, former employees, and their representatives for examination and copying in a reasonable manner and at a reasonable time.

4.1.3 OSHA's Focused Inspection Program

An analysis of the causes of construction industry fatalities for the years 2003-2004, conducted by OSHA and the National Institute for Occupational Safety and Health (NIOSH) in 2004, is shown in *Table 2*.

All other fatalities accounted for only 13 percent of the total. By attacking these four high hazard areas, using its new **focused inspection program,** OSHA hopes to reduce accidents, injuries, and fatalities in the construction industry.

The focused inspection program applies strictly to general safety inspections in the construction industry and is *not* applicable to health inspections. If the practices on the job site demonstrated that the contractor's safety and health program was being followed, the job site qualifies for a focused inspection, whereby the compliance officers focus their attention on the four leading causes of construction fatalities: struck-by, caught-in-between, electrocutions, and falls.

- *Struck-by* – Accidents involving material handling, equipment and vehicle operation.
- *Caught-in-between* – Accidents involving trench cave-ins, vehicle operation, and material handling.
- *Electrocutions* – Accidents involving contact with overhead wires, use of defective tools, failure to disconnect power source before repairs, improper ground fault protection, etc.
- *Falls* – Accidents involving failure to provide and/or use appropriate fall protection.

Table 2 Construction Fatalities 2003-2004

Construction Industry Fatalities 2003-2004	
Falls	36%
Transportation Incidents	23%
Struck by / Caught in	18%
Electric Current	10%

102T02.EPS

Representatives from the Construction Industry Trade Association have commended OSHA for focusing its efforts on the leading causes of fatal accidents in construction. Rather than wasting time in search of nuisance citations on a job site with good safety practices, the focused inspection program aims at reducing the root causes of accidents, injuries, and fatalities. The program encourages OSHA compliance officers to be safety officials, rather than compliance police and allows them to move on quickly to their next inspection if no violations of the four hazard areas are found. OSHA's intent with this program is to find the job sites and persons with serious disregard for safety; however, to qualify for a focused inspection, the controlling contractor must have an effective written safety and health program with a designated person responsible for implementing the program.

If the job site does not reflect practices outlined in the program, or in absence of a program and person to implement it, the compliance officer will conduct the traditional wall-to-wall or comprehensive general schedule inspection of the job site. The focused inspection program provides an incentive for contractors to have and follow written safety and health programs, and it rewards the many contractors who have incorporated a strong emphasis on job site safety and health in their daily operations.

Implementation of an effective safety program at the work site is, therefore, the key to OSHA's goals of eliminating fatalities and serious injuries from as many work sites as possible, recognizing the efforts of conscientious good actor contractors, and concentrating comprehensive inspections on bad actor work sites. Although the current OSHA requirement that contractors have a safety program at the work site does not outline what is required in a program, the proper elements and the benefits of such programs have been known for some time: there must be man-

Explain the distinction between fixed and non-fixed establishments.

Emphasize that all records must remain in the establishment for five years after the date of the accident or illness.

Point out that if an establishment changes ownership, the new employer must preserve the records for the remainder of the five-year period.

Ask participants to name the four leading causes of construction fatalities.

Show Transparency 14 (Table 2). Point out that all other fatalities combined account for only 13 percent of the total.

Review and discuss OSHA's focused inspection program.

MODULE 44102-08 ◆ SAFETY 2.13

Ask participants to list the key elements of an effective safety and health program.

Review and discuss these key elements, citing components of each.

agement commitment, employee participation, work site hazard analysis, hazard prevention, and control, and safety training.

At construction sites, assignment of ultimate safety and health responsibility is often difficult to determine because there may be several contractors at the site with varying degrees of control. Therefore, a fundamental principle of occupational safety and health is that employers be held responsible for ensuring that their own employees are not exposed to hazards. OSHA's current policy recognizes that it is not sufficient to hold responsible only those employers who have employees exposed to hazards. Employers who have contractual responsibility for site safety and those who cause unsafe conditions are also responsible. The general contractor is often in the best position to assure safe employment for all workers at a construction site, from the beginning of a job to its completion. Consequently, while all contractors should have programs, OSHA will be looking to see if the controlling contractor, sometimes called the prime or general contractor, has a program. Such top-level responsibility is necessary if the maximum benefit of a program is to be realized.

In an effective program, management demonstrates that worker safety is an essential value of the organization and it can carry out their commitment to safety protection with as much vigor as it does other organizational aims, such as production and sales. Management commitment must be reflected in the safety program and on the work site in a way that is visible to all employees.

In addition, there must be program evaluation or quality control to make sure that the commitment to an effective safety program is actually carried out. In other words, what happens when a problem or an accident occurs? Is there an investigation, are corrective measures taken, and are employees consulted? Does a safety official check the work site on a frequent basis to monitor ongoing conditions?

Management also must provide for and encourage employee involvement in the development and operation of the safety program and in all decisions that affect the employees' safety. This will enable the employees to commit their insight and energy to achieving the goals and objectives of the program.

Why place so much reliance on safety programs? Because they demonstrate pre-planning, awareness, and concern. Properly developed and implemented plans will address all hazards at the work site, not just the four areas OSHA will focus on. Their successful implementation allows OSHA to more efficiently use its limited resources to check only for the significant hazards at a work site and then rapidly move on to another work site where OSHA's limited resources can be better used to correct deficiencies. Effective safety programs do more than protect; they serve as examples and solutions to other contractors as to how problem areas can be handled.

4.2.0 Recommended Employer's Safety and Health Program

The following guidelines for project managers include the tasks and behaviors that should be included in a safety and health program.

- Provide leadership
  - Establish, issue, and communicate policy statement to employees
  - Revise program annually
  - Participate in safety meetings and inspections
  - Commit adequate resources
  - Incorporate safety rules and procedures into site operations
  - Observe safety rules
- Assign responsibility
  - Make safety designee on site knowledgeable and accountable
  - Ensure supervisors (including foremen) understand safety and health responsibilities
  - Require employees to adhere to safety rules
- Identify and control hazards
  - Involve supervisors in periodic site safety inspections
  - Put preventative controls in place (PPE, maintenance, engineering controls)
  - Take action to address hazards
  - Establish a safety committee, where appropriate
  - Make technical resources available
  - Enforce safety procedures
- Train and educate
  - Provide basic safety training to supervisors
  - Require specialized training when needed
  - Ensure employee training program is ongoing and effective
- Keep records and analyze hazards
  - Maintain records of employee illnesses and injuries; post regularly
  - Require supervisors to perform accident investigations, determine causes, and propose corrective action

Instructor's Notes:

2.14 PROJECT MANAGEMENT

- Evaluate injuries, near misses, and illnesses for trends; initiate corrective action

- Provid first aid and medical assistance
  - Keep first aid supplies and medical service available
  - Inform employees of medical results
  - Provide emergency procedures and training, where necessary

4.3.0 Multi-Employer Work Sites

Multi-employer work sites pose particular risks for employers. The following information was extracted from the OSHA Compliance Officer's Field Inspection Reference Manual (F.I.R.M.). All references in this section are to the F.I.R.M.

4.3.1 Issuance of Citation

On multi-employer work sites, citations are typically issued to employers whose employees are exposed to hazards. Additionally, the employers will be cited, whether or not their own employees are exposed. Citations may be written for:

- The employer who actually creates the hazard (the creating employer)
- The employer who is responsible, by contract or actual practice, for safety and health conditions on the work site; employer who has the authority for ensuring that the hazardous condition is corrected (the controlling employer)
- The employer who has the responsibility for actually correcting the hazard (the correcting employer)

4.3.2 Legitimate Defense

Prior to issuing citations to an exposing employer, it must first be determined whether the available facts indicate that employer has a legitimate defense to the citation. Legitimate defenses may include:

- The employer did not create the hazard.
- The employer did not have the responsibility or the authority to have the hazard corrected.
- The employer did not have the ability to correct or remove the hazard.
- The employer can demonstrate that the creating, the controlling and/or the correcting employers, as appropriate, have been specifically notified of the hazards to which his/her employees are exposed.
- The employer has instructed his/her employees to recognize the hazard and informed them how to avoid the dangers associated with it.
  - An exposing employer must have taken appropriate alternative means of protecting employees from the hazard.
  - When extreme circumstances justify it, the exposing employer shall have removed his/her employees from the job to avoid citation.

If an exposing employer meets all of these defenses, that employer shall not be cited. If employers on a work site with employees exposed to a hazard meet these conditions, then the citation shall only be issued to the employers who are responsible for creating the hazard and/or who are in the best position to correct the hazard. In such circumstances the controlling employer and/or the hazard-creating employer shall be cited even though no employees of those employers are exposed to the violative condition. Penalties for such citations are calculated based on all exposed employees of all employers as the number of employees for probability assessment.

4.3.3 Employer/Employee Responsibilities

The Act states: "Each employee shall comply with occupational safety and health standards and all rules, regulations, and orders issued pursuant to the Act which are applicable to his own actions and conduct." The Act does not provide for the issuance of citations or the proposal of penalties against employees. Employers are responsible for employee compliance with the standards. Concerted refusals to comply with the Act will not bar the issuance of an appropriate citation where the employer has failed to exercise full authority to the maximum extent reasonable, including discipline and discharge.

4.3.4 Affirmative Defenses

An affirmative defense is any matter that, if established by the employer, will excuse the employer from a violation that has otherwise been proved by the OSHA representative. Although affirmative defenses must be proved by the hearing, OSHA must be prepared to respond whenever the employer is likely to raise, or actually does raise, an argument supporting a defense. Some of the more common affirmative defenses include:

- Unpreventable employee misconduct or isolated event which includes a violative condition that was unknown to the employer; and in violation of an adequate work rule, which was effectively communicated and uniformly enforced.

Review and discuss which citations may be issued on a multi-employer work site.

Point out that citations are typically given to employers whose employees are exposed to hazards.

Review and discuss possible legitimate defenses for work citations.

Point out that employers are responsible for employee compliance with the standards.

Emphasize that even if an employee refuses to comply, employers may be cited if they fail to exercise maximum authority, including discipline or discharge.

Review and discuss common affirmative defenses.

MODULE 44102-08 ◆ SAFETY 2.15

Explain that OSHA regulations are broken down into parts, and Part 1926 refers to construction industry standards.

Review the major subparts under 1926.

- Compliance with the requirements of a standard is functionally impossible or would prevent performances of required work and there are no alternative means of employee protection.
- Compliance with a standard would result in greater hazards to employees than non-compliance, there are no alternative means of employee protection; and an application of a variance would be inappropriate.

4.3.5 General Duty Clause Violations

In the case of general duty clause violations, only employer(s) whose own employees are exposed to the violation may be cited.

Show Transparency 16 (Table 3). Note that each subpart is broken down further into sections and subsections.

4.4.0 OSHA Part 1926

Under Title 29, Chapter XVII is set aside for the Occupational Safety and Health Administration. Under Chapter XVII, the regulations are broken down into Parts. Part 1926, for example, is the standard dealing with occupational safety and health standards, commonly known as the construction industry standards. Under each part, such as Part 1926, major blocks of information are broken down into subparts. The major subparts in the 1926 standards include:

Suggest that participants may wish to color-code frequently used sections of their standards books.

Assign reading of Sections 5.0.0–6.2.10.

- Subpart C – General Safety and Health Provisions
- Subpart D – Occupational Health and Environmental Controls
- Subpart E – Personal Protective and Life Saving Equipment
- Subpart F – Fire Protection and Prevention
- Subpart G – Signs, Signals and Barricades
- Subpart H – Materials Handling, Storage, Use and Disposal
- Subpart I – Tools—Hand and Power
- Subpart J – Welding and Cutting
- Subpart K – Electrical
- Subpart L – Scaffolds
- Subpart M – Fall Protection
- Subpart N – Cranes, Derricks, Hoists, Elevators and Conveyors
- Subpart O – Motor Vehicles, Mechanized Equipment and Marine Operations
- Subpart P – Excavations
- Subpart Q – Concrete and Masonry Construction
- Subpart R – Steel Erection
- Subpart S – Underground Construction, Caissons, Cofferdams and Compressed Air
- Subpart T – Demolition
- Subpart U – Blasting and the Use of Explosives
- Subpart V – Power Transmission and Distribution
- Subpart W – Rollover Protective Structures; Overhead Protection
- Subpart X – Stairways and Ladders
- Subpart Y – Diving
- Subpart Z – Toxic and Hazardous Substances

Each Subpart is further broken down into sections. Look at one subpart in detail: Subpart D – Occupational Health and Environmental Controls. The index of Subpart D is shown in *Table 3*.

4.4.1 Color Coding

To simplify identification of the standards, you can color-code your standards book. Generally speaking, color coding two levels below the section heading will be adequate. Color-code those

Table 3 Subpart D Index

SUBPART D – OCCUPATIONAL HEALTH AND ENVIRONMENTAL CONTROLS	
1926.50	Medical services and first aid
1926.51	Sanitation
1926.52	Occupational noise exposure
1926.53	Ionizing radiation
1926.54	Non-ionizing radiation
1926.55	Gases, vapors, fumes, dusts and mists
1926.56	Illumination
1926.57	Ventilation
1926.58	[Reserved]
1926.59	Hazard communication
1926.60	Methylenedianiline
1926.61	Retention of DOT markings, placards and labels
1926.62	Lead
1926.64	Process safety management of highly hazardous chemicals
1926.65	Hazardous waste operations and emergency response
1926.66	Criteria for design and construction of spray booths

102T03.EPS

Instructor's Notes:

2.16 PROJECT MANAGEMENT

particular sections of the standards you use frequently. Note that there is a subject index in the back of the standards book. This index can be very helpful in locating specific standards by picking out a keyword from any given hazard description.

5.0.0 ◆ PRE-QUALIFICATION OF CONTRACTORS AND SUBCONTRACTORS

Many of the progressive users of construction services and successful general contractors are pre-qualifying prospective contractors and subcontractors based in part on past safety performance. Checking the statistics could save you from production delays due to injuries and accidents, and possibly even third-party lawsuits.

As a minimum, companies are asking for a contractor's workers' compensation EMR for the previous three years and their OSHA incidence rate for the same period. Some are also asking contractors to complete a questionnaire based on the Business Roundtable's A-3 Report. Incidence rates are the number of injuries and/or illnesses or lost workdays per 100 full-time employees and calculated as:

$$IR = (N/EH) \times 200{,}000$$

Where:			
	IR	=	Incidence Rate
	N	=	Number of injuries and/or illnesses or lost workdays
	EH	=	Total Hours worked by all employees during the reference period
	200,000	=	Base for 100 full-time equivalent workers (working 40 hours per week, 50 weeks per year)

The injury/illness statistics are taken directly from your OSHA Form 300 log. With respect to man-hours, you can use the actual hours worked if you have the information available or you can use 166.7 hours each month or 2000 hours each year per person.

You can use this same formula to compute rates for occupational injuries, illnesses, lost workday cases, nonfatal cases without lost workdays, or the number of nonfatal cases with lost workdays. Simply replace the number of injuries in the formula with the number for which you want rates computed.

Incidence rates may also be interpreted as the percentage of employees that will suffer the degree of injury for which the rate was calculated. That is, if the incidence rate of lost workday cases is 5.1 per 100 per year full-time employees, then about 5 percent of the contractor's employees incurred a lost workday injury.

6.0.0 ◆ PRE-TASK SAFETY PLANNING

The pre-project planning guidelines focused on the project as a whole. Safety planning must also be applied to specific jobs and tasks associated with the project. Two techniques that may be used are job safety analysis and pre-task hazard analysis.

6.1.0 Job Safety and Pre-Task Hazard Analysis

Job safety analysis (JSA) and pre-task hazard analysis are both tools for identifying and controlling potential hazards. The three major differences between the two methods are:

- Timing
- Level of detail
- Use

Pre-task hazard analyses are typically done just before starting a task and do not include a high level of detail. JSAs are typically done in advance and contain significantly more detail. JSAs are more formal and include key job steps, whereas the pre-task hazard analyses are more general and usually do not include each job step.

JSAs are typically done for recurring or repetitive tasks and are usually retained and referenced. The completed JSA often becomes the basis for standard operating procedures or maintenance procedures. JSAs are often used as training tools, whereas pre-task hazard analyses are primarily used to raise awareness and maintain focus. However, when performing a pre-task hazard analysis, one should ask, "Do we have any JSAs or standard operating procedures that would cover the job at hand?" If so, the applicable JSA or procedure may need to be reviewed as part of the pre-task hazard analysis.

Both processes can be helpful in accident prevention, but the intent and use of each must be addressed in company procedures and clearly understood by all. Both processes require training on hazard recognition, evaluation, and control, as well as how to perform and document the specific analysis.

Note that pre-qualification of contractors and subcontractors is based in part on past safety performance, including the experience modification rate (EMR) and OSHA incidence rate.

Show Transparency 16, OSHA Incidence Rate. Explain each factor used to determine incidence rates, and perform a sample problem.

Point out that pre-project planning guidelines help project managers focus on the project as a whole, but that safety must also be planned into specific jobs and tasks associated with the project.

Review and discuss Appendix A, Pre-Project Planning Safety Checklist.

Explain that a job safety analysis (JSA) and pre-task hazard analysis are tools used to identify and control potential hazards.

Ask participants to list three major differences between job safety analysis and pre-task hazard analysis. Review and discuss these differences.

Note that while processes may be different, job safety analysis and pre-task hazard analysis can both be helpful in accident prevention.

Explain that once hazards are identified, methods of elimination or control should be developed and incorporated into the job plan.

Show Transparency 17 (Figure 7). Note that many companies use a similar form for pre- and post-job safety meetings.

Review and discuss the purpose of JSAs and the four basic steps in the JSA process.

Ask participants to list the three methods of conducting JSAs. Review and discuss the advantages and disadvantages of each.

Explain that when selecting jobs to be analyzed, most people start with the worst job first.

See the Teaching Tips for Section 6.2.3 and 6.2.0–6.2.10 at the end of the module.

6.1.1 Pre-Task Safety Planning

Pre-task safety planning is a process for identifying and evaluating potential hazards associated with a given task or work assignment. Once the hazards have been identified, methods of eliminating or controlling the hazards should be developed and incorporated into the job plan.

6.1.2 Process of Pre-Task Analysis

Pre-task safety plans or pre-task hazard analyses are typically performed just prior to starting the task. The evaluation may be performed individually or collectively. The ideal situation is to have the entire work group involved in the process. Some companies mandate that the supervisor coordinate the process while others encourage individual crew members to lead the process. When employees are working alone, they are encouraged to conduct a personal pre-task hazard analysis.

Most companies provide a pocket-sized form with a list of items to consider. Potential hazards identified are checked off or noted on the form and personal protective equipment requirements are usually indicated. In most cases there are spaces for the workers to sign indicating they have reviewed, understand, and agree with the precautions listed. Some pre-task analysis forms incorporate the JSA process that requires listing each step, the associated hazards, and the appropriate safeguards.

Completed pre-task hazard analysis forms are typically reviewed by the supervisor at the start of the workday or sometime during the day. The forms are also used to audit safe work practices during the day. Upon completion of the task, completed forms are usually collected and retained for a pre-determined length of time. Pre-task hazard analyses are conducted each time a new task is assigned. It is possible for a craftworker to complete or review two or more pre-task hazard analysis forms on any given day.

Some firms use the pre-task hazard analysis form for pre- and post-job safety briefings (*Figure 7*). In some cases, workers initial the completed form at the end of the day indicating the job was completed safely and without incident.

6.2.0 Job Safety Analysis

Job safety analysis is a procedure used to review job methods and uncover hazards that may have been overlooked in the layout or design of the equipment, tools, processes or work area; developed after production started; or resulted from changes in work procedures or personnel. It is also a powerful planning and training tool.

6.2.1 Performing a Job Safety Analysis

The four basic steps in performing a job safety analysis are as follows:

Step 1 Select the job to be analyzed.

Step 2 Break the job down into successive steps or activities and observe how these actions are performed.

Step 3 Identify the hazards and potential accidents; only an identified problem can be corrected or eliminated.

Step 4 Develop safe job procedures to eliminate the hazards and prevent potential accidents.

6.2.2 Methods of Conducting JSAs

The three basic methods for conducting a job safety analysis are direct observation, group discussion, and group discussion using a videotape of a job. A fast and efficient method of conducting a JSA is through direct observations of job performance. In many cases, however, this method may not be practical or desirable. For example, new jobs and those that are done infrequently do not lend themselves to direct observation. When this is the case, the JSA can be made through discussions with persons familiar with the job. Individuals often involved in the process include, but are not limited to, frontline supervisors, safety specialists, engineers, experienced employees, and outside contractors.

6.2.3 Selecting Jobs to be Analyzed

When selecting jobs to be analyzed, most people start with the worst first. You should be guided by the following factors:

- *Frequency of accidents* (including near misses) – A job that repeatedly produces accidents is a candidate for a JSA. The greater the number of incidents associated with a job, the greater its priority claim for a JSA.
- *Production of disabling injuries* – Every job that resulted in a serious or disabling injury should be given a JSA.
- *Severity potential* – Some jobs may not have a history of accidents but may have the potential for severe injury.
- *New or revised jobs* – Jobs created by changes in equipment or in processes have no history of accidents, but their accident potential may not be fully appreciated. Analysis should not be delayed until accidents or near-misses occur.

Instructor's Notes:

2.18 PROJECT MANAGEMENT

PRE-JOB SAFETY BRIEFING / PRE-TASK HAZARD ANALYSIS

Task/work to be performed

____

____

Tools/equipment/materials involved

____

____

Physical Hazards

- Falls on same elevation
- Falls from elevation
- Pinch points
- Rotating/moving equipment
- Electrical hazards
- Hot/cold substances/surfaces
- Strains/sprains/repetitive motion
- Struck by falling/flying objects
- Sharp objects

Hazardous Chemicals/Substances

- Flammable materials
- Reactive materials
- Corrosive chemicals
- Toxic chemicals
- Oxidizers
- Hazardous wastes
- Biohazards
- Radiation hazards
- Other

____

____

Energy Sources

- Where is the energy?
- What is the magnitude of the energy?
- What could happen or go wrong to release the energy?
- How can it be eliminated or controlled?
- What am I going to do to avoid contact?

Permits Required

- Welding & Burning
- Lockout/Tagout
- Excavation
- Line Entry
- Electrical Hot Work
- Confined Space Entry
- Critical Lift
- Vehicle Entry
- Other

____

____

PPE Requirements

- Safety Glasses
- Goggles
- Face Shield
- SCBA
- Respirator
- Gloves
- Chemical Suit
- Head Protection
- Safety Shoes/Boots
- Full Body Harness
- Lanyard
- Lifeline
- Other

____

____

Special Precautions

Employee Signatures

____

____

____

____

102F07.EPS

Figure 7 ◆ Pre-task hazard analysis form.

MODULE 44102-08 ◆ SAFETY 2.19

Review the steps for job observation.

Point out that the employee being observed should be experienced, capable, cooperative, and willing to share ideas.

Review and discuss the five common errors made when performing a job analysis.

Review and discuss the questions for identifying hazards and potential accidents.

Ask participants to list the most common accident types and provide examples of each type of accident.

- *Multiple employee exposure* – Jobs that expose more than one individual to potential hazards should also be analyzed.

6.2.4 Job Observation

It is important to select an experienced, capable, and cooperative person who is willing to share ideas. If the employees have never participated in a job safety analysis, explain to them the purpose, which is to make a job safe by identifying and eliminating or controlling hazards. Show the workers a completed job safety analysis.

Step 1 Select the right person to observe.

Step 2 Brief the employee on the purpose of the job safety analysis.

Step 3 Observe the person as the job is performed and try to break it down into basic steps.

Step 4 Record each step.

Step 5 Check the breakdown with the person involved.

6.2.5 Common Errors

There are five common errors that are often made when performing a job analysis, as follows:

- Making the breakdown so detailed that an unnecessarily large number of steps are listed.
- Making the job so general that basic steps are not recorded.
- Failure to identify the education and experience level of the target audience.
- Failure to identify end uses (training, actual procedure, basis for procedure).
- Conducting JSAs for all jobs instead of identifying jobs that require JSAs.

6.2.6 Identifying the Hazards and Potential Accidents

The purpose of a JSA is to identify all hazards, both physical and environmental. To do this, ask yourself these questions about each step:

- Is there a danger of striking against, being struck by, or otherwise making harmful contact with an object?
- Can the employee be caught in, on, by, or between objects?
- Is there a potential for a slip, trip, or fall? If so, will it be on the same elevation or to a different elevation?
- Can the worker suffer a strain by pushing, pulling, lifting, bending, or twisting?
- Is the environment hazardous to workers' safety or health?

6.2.7 Accident Types

This section lists the most common types of accidents by category.

- Struck by
  - Moving or flying object
  - Falling material
- Struck against
  - Stationary or moving object
  - Protruding object
  - Sharp or jagged edge
- Contact with
  - Acid
  - Electricity
  - Heat
  - Cold
  - Radiation
  - Toxic and noxious substances
- Caught
  - In
  - On
  - Between
- Fall to
  - Same level
  - Lower level
- Overexertion
  - Lifting
  - Pulling
  - Pushing
- Rubbed or abraded by
  - Friction
  - Pressure
  - Vibration
- Bodily reaction from
  - Voluntary motion
  - Involuntary motion

Instructor's Notes:

6.2.8 Writing a JSA

Follow these recommendations for writing JSAs:

- Put any qualifying statements first, not last.
- Start each instruction with an action word.
- Each instruction should be observable.
- Each instruction should be measurable.

6.2.9 Danger vs. Caution Signal Words

Potential hazards and dangers exist in the workplace that can cause accidents and injuries. Adding safety labels and stickers in key areas of the workplace, or tags to specific tools, materials, and equipment can serve as constant reminders to workers about the hazards and dangers that exist. They also can serve as reminders of what to do in an emergency.

OSHA 29 CFR 1910.145 (f) (5): "Danger tags shall be used in major hazard situations where an immediate hazard presents a threat of death or serious injury to employees. Danger tags shall be used only in these situations." An example of an OSHA danger tag is shown in *Figure 8*.

ANSI Z 535.3-2002 (4.11.1): "Danger indicates an imminently hazardous situation which, if not avoided, will result in death or serious injury. This signal word is to be limited to the most extreme situations. This signal word should not be used for property damage hazards unless personal injury risk appropriate to this level is also involved." An example of an ANSI danger tag is shown in *Figure 9*.

OSHA 29 CFR 1910.145 (f) (6): "Caution tags shall be used in minor hazard situations where a non-immediate or potential hazard or unsafe practice presents a lesser threat of employee injury. Caution tags shall be used only in these situations." An example of an OSHA caution tag is shown in *Figure 10*.

ANSI Z 535.3-2002 (4.11.3): "Caution indicates a potentially hazardous situation which, if not avoided, may result in minor or moderate injury. It may also be used to alert against unsafe practices that may cause property damage." An example of an ANSI caution tag is shown in *Figure 11*.

When evaluating a given procedure, ask, "What should the employee do, or not do, to eliminate this particular hazard or to prevent this potential accident?" The answer must be specific and concrete to be beneficial. General precautions such as be careful, use caution, or be alert are useless. Answers should state what to do and how to do it.

102F08.EPS

Figure 8 ◆ OSHA danger tag.

102F09.EPS

Figure 9 ◆ ANSI danger tag.

102F10.EPS

Figure 10 ◆ OSHA caution tag.

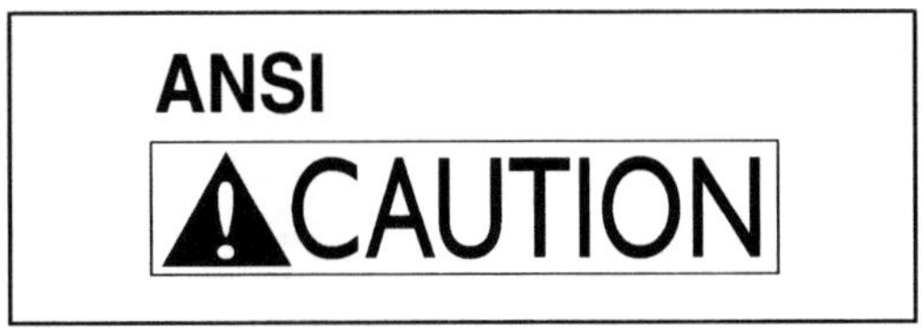

102F11.EPS

Figure 11 ◆ ANSI caution tag.

For example, the recommendation, "Make certain that wrench does not slip or cause loss of balance," is incomplete. It does not tell how to prevent the wrench from slipping. Here is a more complete recommendation: "Set the wrench properly and securely. Test its grip by exerting a slight pressure on it. Brace yourself against something immovable, or take a stance with feet wide apart before exerting full pressure. This prevents loss of balance if the wrench slips."

Job safety analyses can be very beneficial if they are performed correctly. They not only result in a safer job, but also increase productivity and eliminate waste. Take the time to do them correctly; more importantly, use them in your daily work.

Review and discuss how to write a JSA. Give examples of qualifying statements, action words, and observable and measurable instructions.

Emphasize the importance of using safety labels and stickers to warn of hazards and dangers.

Show Transparency 18 (Figures 8 and 9). Discuss the difference between OSHA and ANSI danger tags, and give examples when each would be used.

Show Transparency 19 (Figures 10 and 11). Discuss the difference between OSHA and ANSI caution tags, and give examples when each would be used.

Emphasize that when writing procedures, participants need to be sure the instructions are specific and concrete, stating what to do and how to do it.

MODULE 44102-08 ◆ SAFETY 2.21

Note that the final step of a JSA is to develop safe job procedures.

Review and discuss the recommended safe job procedures.

Show Transparency 20 (Figure 12). Review the steps discussed in Sections 6.2.0–6.2.10.

Show Transparency 21 (Figure 13).

See the Teaching Tip for Sections 6.2.0-6.2.10 at the end of this module.

Assign reading of Sections 7.0.0–7.5.0.

Show Transparency 22 (Figure 14). Review and discuss the key points of a safety policy statement.

Discuss the impact of drug and alcohol abuse in the workplace.

6.2.10 Recommended Safe Job Procedures

The final step in conducting a JSA is to develop a recommended safe job procedure to prevent the occurrence of potential accidents. The principle solutions are:

- Find a new way to do the job.
- Change the physical conditions that create the hazard.
- Eliminate remaining hazards by changing work methods or procedures.
- Reduce the necessity of doing a risky job, or at least the frequency with which it must be performed.

The use of a job hazard analysis form (*Figure 12*) is helpful to organize these steps and document results of JSAs. The form shown in *Figure 13* is convenient to use in your daily work to streamline your job hazard identification process. Job hazard analysis can be used to train personnel how to perform a job safely.

7.0.0 ◆ KEY ELEMENTS OF A PREVENTION PROGRAM

There are certain basic elements common to most effective safety and loss prevention programs. The degree and extent to which these elements need to be included in your program will vary depending on the size and nature of your operations.

7.1.0 Management Support and Policy Statement

One such element is the management support and policy statement (*Figure 14*). This is an expression of management's philosophies and goals toward safety and loss prevention. It should be signed by the president or chief executive officer and communicated to all employees who are affected by it.

A good policy statement will recognize management's responsibility for safety and loss prevention without diminishing each employee's role. It should be clear, concise and understandable. It should express a strong, positive commitment to safety, yet it must be realistic. Statements that appear to elevate safety above every other management concern will lack credibility.

7.2.0 Policy on Alcohol and Drug Abuse

Most experts agree that at least one out of five construction workers, or 20 percent of a company's workforce, has an alcohol or drug abuse problem. The negative impact of this issue on the work environment is reflected in these results:

- Increased workers' compensation and health care costs
- Decreased productivity
- Increased tool and materials costs
- Increased absenteeism

Many owners and contractors have recognized this problem and have implemented plans to combat the situation. If other contractors in your area have drug and alcohol abuse programs and you don't, you may be getting more than your 20 percent share of the employees who have a problem.

7.3.0 Assignment of Responsibilities

Effective management must create an assignment of responsibilities, which includes the duties and responsibilities of each employee, from the president of the company down to the hourly employees, and must be clearly defined and communicated. Along with these duties and responsibilities must come accountability and authority. Employees cannot be held accountable for items over which they have no control.

The following sections contain a basic description of the safety responsibilities typically assigned to various positions within an organization.

7.3.1 Executive and Operating Management

The corporate executive or operating management has the overall responsibility for safety and holds the project manager, superintendents, and foremen responsible for the safety of their personnel. An executive also has the authority to authorize necessary expenditures. They are the ones who develop and/or implement new or revised policies and procedures. They actively participate in the safety program by:

- Conducting safety audits and observations
- Reviewing and responding to subordinates' reports and safety activities
- Evaluating management's overall effectiveness
- Making safety performance part of job appraisal

7.3.2 Safety Supervisor/Coordinator

A safety supervisor serves in a staff capacity without line authority. Typical duties include coordinating safety activities, maintaining and analyzing accident records, conducting safety educational activities for supervisory personnel as well as activities to stimulate and maintain employee interest. The safety supervisor also serves on the

Instructor's Notes:

Job Title: Sharpening and Replacing a Rotary Mower Blade

Date of Analysis: ____________

Job Location: ______________________________

Required or Recommended PPE: Gloves and Safety Glasses

Step	Hazard	New Procedure or Protection
1. Disconnect spark plug wire.	1. Striking against housing (SA). Burn hand.	1. Do not use excessive force. Allow mower to cool.
2. Remove gasoline.	2. Spillage – fire, inhalation.	2. Ventilation, no smoking, proper container. Flush away with water (if necessary).
3. Invert mower.	3. Caught between (CB). Spilling gasoline, overexertion.	3. Tip properly (grass catcher chute up). Be sure cap is tight. Lift properly, use leg muscles.
4. Remove dull blade.	4. Knuckles SA blade.	4. Securely block blade – wooden block. Use proper size **socket wrench with extender.**
5. Check for bent blade.	5. None.	5. None.
6. Sharpen and balance dull blade.	6. Cutting hand; SA vise.	6. Wear gloves. **Avoid contact with sharp blade.**
7. Reassemble blade to mower.	7. SA blade.	7. Block blade. **Wear gloves, avoid contact with sharp blade.**
8. Return mower to cutting position.	8. Overexertion. Caught between.	8. Use leg muscles, not back. Wear gloves.
9. Reconnect spark plug wire.	9. None.	9. None.
10. Add gasoline.	10. Fire.	10. Ventilate, no smoking, proper container.
11. Operate mower.	11. Normal operating hazards.	11. **Check for excessive vibration or unusual noise.**

102F12.EPS

Figure 12 ◆ Job hazard analysis form case study.

MODULE 44102-08 ◆ SAFETY 2.23

JOB HAZARD ANALYSIS FORM

Job Title: Date of Analysis:
Job Location: Conducted by:
PPE: Staffing:
Tools, Materials and Equipment: Duration:

Step	Hazards	Quality Concern	Environmental Concern	New Procedure or Protection

102F13.EPS

Figure 13 ◆ Job hazard analysis form.

XYZ Construction, Inc.
Safety Policy Statement

XYZ Construction is firmly committed to operating all of its projects in a safe and efficient manner. In order to achieve and maintain this goal, the ideals of safety, quality and production must be inseparable. Safety, loss prevention and quality control procedures must be integrated into all phases of our operation.

Management has the ultimate responsibility for developing and implementing the necessary policies and procedures that will result in job sites which are free of recognized hazards and in compliance with applicable safety, health and environmental regulations. Recognizing this important responsibility, Management will commit the necessary time, manpower and financial resources to carry out these activities.

Each and every employee has the responsibility to follow these policies and procedures and incorporate them into their daily activities. Working together as a team we can construct high quality buildings in a safe and cost-effective manner. Your cooperation is sincerely appreciated and fully expected. Anything less is unacceptable.

Charles Lewis, President

102F14.EPS

Figure 14 ◆ A policy statement issued by an executive officer.

Instructor's Notes:

2.24 PROJECT MANAGEMENT

safety committee and reviews and evaluates accident investigations. This person also advises management on ways to correct deficiencies on current safety protocols and issues regular reports showing safety performance and accident trends with information gained through regular audits.

7.3.3 Project Manager/Superintendent

The project manager and/or superintendent are responsible for the development and implementation of the job site safety plan. They regularly audit compliance with this plan, making changes as needed. It is their job to inform any subcontractors of job safety requirements and audits compliance. The project manager is responsible for reviewing all accident investigations and taking the appropriate corrective action. Other duties include the following:

- Evaluating and appraising the foremen's safety performance
- Updating safety items on agendas for planning and scheduling meetings
- Making employee safety observations/contacts
- Holding safety meetings with job site supervisory personnel

7.3.4 Supervisor/Foreman

The supervisor/foreman is responsible for the safety of the crew. They train employees in safe work practices, hold crew safety meetings, observe employees for unsafe acts, and take corrective action as needed. A supervisor/foreman is also responsible for obtaining prompt first aid for an injured worker. The supervisor/foreman may also:

- Inspect tools, equipment and area for unsafe conditions
- Investigate and report all accidents
- Discuss safety with individual employees
- Audit subcontractors' compliance with job safety requirements
- Report to management items/practices beyond his control

7.3.5 Employees

Employees work in accordance with accepted safety practices, may make safety suggestions, or serve on the safety committee. They are responsible for reporting any unsafe conditions or practices they observe at the work site. In order to work safely, employees should not undertake jobs they do not understand.

7.4.0 Employee Screening, Selection and Placement

Selecting the right employee for the job is critical. The employee must have the necessary skills and physical capabilities to perform the work safely.

Likewise, the worker must have a good attitude toward safety. Although they are in a minority, some people make a living out of getting hurt on the job and suing the employer. Abusers of alcohol and drugs can also cause major problems. You must establish practical, cost-effective methods of screening and hiring workers, but you must do so in a non-discriminatory manner. This can be done through:

- Written applications
- Reference checks
- Division of Motor Vehicles (DMV or MVR) record checks
- Pre-employment drug screening
- Pre-job assignment physical exams
- Use of outside firms to review public court records for the names of individuals who have sued their employer (Note: in some states this is not permitted)
- Skill testing and/or strength testing

7.5.0 Safety Rules

Safety rules are a necessary part of any program. They should be logical and enforceable, and they should be prepared and presented in terms that are easily understood. Safety rules should specify employee safety responsibility, including to whom the rules apply, where they apply, and how they apply. Rules are of no value unless they are communicated, understood, and enforced. Penalties for the violation of safety rules must also be understood by all parties concerned. Enforcement of safety rules must be fair, consistent, and uniform.

Safety rules are generally divided into the following categories:

- General/company safety rules that apply to all employees
- Client, job, or site-specific safety rules
- Craft or special safety rules that apply to a specific type of work operation

7.6.0 Orientation and Training

Training can be broken down into three major categories: new employee orientation, job-specific safety training, and supervisory training.

Explain the safety responsibilities of each position within an organization.

Emphasize that workers must be selected using practical, cost-effective methods in a non-discriminatory manner.

Review methods used to hire employees.

Review and discuss the key points of safety rules.

Ask participants to list safety rule categories and provide an example of each.

Assign reading of Sections 7.6.0–7.14.0.

Explain the three major categories of training and when training should be conducted.

Before each scenario, ask students to observe each manager's attitudes towards safety and accountability.

MODULE 44102-08 ◆ SAFETY 2.25

Point out that the project manager should ensure that training is performed and is adequate.

Review the minimum training requirements for supervisory personnel.

Ask participants to suggest other training that may be beneficial.

Review and discuss the key points for planning safety meetings.

Review the project manager's roles in safety meetings.

Review and discuss techniques to motivate employees towards taking an interest in safe work practices.

Review Appendix F, Guidelines for Developing a Crisis Management Plan.

Statistics show that employees who have been on the job 30 days or less account for 25 percent of all construction injuries. This clearly illustrates the need for an effective orientation program. Training should be conducted at the following times:

- When a new employee is hired
- When an existing employee is assigned to a new job
- When a new job or work is initiated

Training should cover the following issues:

- Correct work procedures
- Care, use, maintenance, and limitations of any required personal protective equipment
- The hazards associated with any harmful materials used, including warning properties, symptoms of overexposure, first aid, and clean-up procedures for spills
- Where to go for help

New employee orientation and job-specific safety training are usually done by the employee's foreman or supervisor. The project manager should see that the training is being performed and that it is adequate.

One area often overlooked is the need for supervisory training. Just because an employee is a skilled craftworker doesn't guarantee that they will be an effective supervisor. Supervisory personnel should, as a minimum, be trained in the following areas:

- Leadership skills
- Job planning
- Employee training methods
- Job analysis
- Conducting safety meetings
- Conducting employee observation and equipment inspections
- Accident investigation

7.7.0 Safety Meetings and Employee Involvement

Maintenance of employee interest and involvement is a key element in any safety program. Safety meetings are often used for this purpose. They can vary from a formal presentation to a five minute tailgate or toolbox talk.

Safety meetings, when properly conducted and held in a timely fashion, can be used to exchange information regarding specific safety matters; to diffuse potential job disruptions by providing an outlet for critical issues; to provide a written record of the actions taken; and to establish an effective communications link between management and employees. Both federal and state governmental regulations require employers to train their employees on the hazards associated with the work and the necessary safeguards to be taken. Safety meetings can be used as training sessions.

The key to conducting effective safety meetings is proper planning. Consider the following points when planning your safety meetings:

- Limit your discussion to one or two topics
- Obtain any needed facts or materials in advance
- Plan the presentation
- Set a timetable and stick to it
- Be sincere
- Promote discussion
- Reach a decision or conclusion
- Review status of outstanding items from previous meetings

As a project manager, you should hold periodic safety meetings with your first line foremen and make sure they are holding the necessary meetings with their crews. You should sit in on some of the meetings and constructively critique the supervisor's performance. Let them know that you want the meetings to be useful and productive. To do this, they will need adequate time and resources to plan the meetings.

It is important to use techniques to motivate employees to take an interest in safe work practices. Some of these techniques are as follows:

- Establish employee safety committees
- Assist in safety audits/inspections
- Assist in development of JSAs
- Hold safety contests
- Provide training programs such as first aid/CPR
- Assist in safety meeting preparation

7.8.0 Crisis Management

When a catastrophic event occurs, you must have a well-prepared and rehearsed crisis management plan in place. Guidelines for developing a crisis management plan are included as *Appendix F*.

7.9.0 Emergency Reporting and Response

Provisions should be made in advance for reporting and responding to job-site emergencies. An emergency is a situation that may fall under one of the following categories:

Instructor's Notes:

2.26 PROJECT MANAGEMENT

- Personnel injuries
- Fires, explosions
- Tornadoes
- Cave-ins or building collapse
- Civil disturbances

Advance planning can reduce response time and in many cases minimize the severity of the incident. The names, addresses and phone numbers of the nearest medical, fire, police and emergency response agencies should be posted at each job-site phone. All personnel should be familiar with the site emergency reporting and response procedures. Adequate provisions should be made for prompt access to first aid and follow-up medical care. It is highly desirable to have at least two people on each job site trained in basic first aid and CPR.

One area often overlooked by contractors is dealing with the news media. When serious accidents happen, the press usually responds. How you respond can have a dramatic effect on how your firm is perceived by the public. You and your supervisors should be familiar with your firm's policy on dealing with the news media. If you don't have such a policy, you should see that one is developed and implemented. Tips on dealing with the media during emergencies are included as *Appendix E*.

7.10.0 Post-Incident Claims Management

Most injured personnel genuinely want to get back to work following an incident. To help facilitate this, implement a post-incident claims management program. Guidelines for doing so are included as *Appendix C*.

7.11.0 Inspections, Employee Observations and Audits

A system of inspections, employee observations, and audits is necessary to maintain an effective safety and loss prevention program. Inspections should be made to detect the presence of and correct unsafe conditions. Employee observations should be made to detect unsafe work practices and procedures. Safety audits should be made to monitor the use and effectiveness of the company's safety policies and procedures. Safety inspections and employee observations are normally performed by a foreman. Project managers normally review inspection reports and audit compliance with the firm's policies and procedures. A sample checklist for coordination of site-specific safety activities is included as *Appendix B*. A sample job-site safety inspection checklist is included as *Appendix D*.

7.12.0 Accident Investigation and Analysis

All accidents, injuries, illnesses, and near-miss type incidents should be reported and promptly investigated. There are several reasons why these investigations are important:

- To determine the cause or causes, so action can be taken to prevent a recurrence
- To meet insurance company requirements
- To satisfy government regulations
- To document facts and preserve evidence if lawsuits are filed later
- To detect trends and identify potential problem areas

The primary accident investigation function has always been performed by a foreman. It is generally agreed that the foreman is most knowledgeable of the work area and the employees, and thus best able to determine most of the underlying causes of an accident. Depending on the nature of the incident and other conditions, accidents may also be investigated by the safety coordinator, the safety committee, or management. In any case, the project manager should be made aware of all reported or suspected incidents as soon as possible. Likewise, the project manager should review every report for accuracy and completeness. He/she should be certain the causes of the incident have been properly identified and that the corrective action is appropriate and being implemented.

7.13.0 Records

Good recordkeeping is essential to an effective safety program. Without good records it is virtually impossible to make any kind of analysis or measure the effectiveness of a program. It should also be understood that certain records are required by state and federal safety and health regulations. *Figure 15* provides a partial listing of the types of records which should be kept. The length of time required to keep them varies considerably. It is recommended that you establish your own internal recordkeeping system, noting the type of records, who will keep them, and for how long.

Discuss advance preparation for emergency reporting and response.

Review and discuss Appendix E.

Review and discuss Appendix C.

Explain why inspections, employee observations, and audits are performed.

Review Appendices B and D.

List reasons why accident investigations are important.

Explain the project manager's role in accident investigation.

Show Transparency 23 (Figure 15). Review the types of records that should be kept and how long they should be stored.

MODULE 44102-08 ◆ SAFETY 2.27

Review and discuss the nine best industry practices for achieving higher levels of safety performance.

Many states allow claimants up to one year from the date of an accident to file a lawsuit. By keeping non-required records for two years, you can usually be assured that records will be available to help you in your defense, if they are needed.

7.14.0 Program Evaluation and Follow Up

This is one of the most important but often neglected elements of an effective program. To be assured that your program is meeting the company's goals and objectives, it must be evaluated on a regular basis and modified as needed. The project manager has the responsibility for completing this evaluation of the job-site safety program on a regular basis.

When reviewing and implementing your safety program, keep in mind the nine industry best practices from the Construction Industry Institute (CII). Research indicates that contractors who implement the practices listed below achieve higher levels of safety performance than those who do not.

Nine Industry Best Practices

- Demonstrated management commitment
- Staffing for safety
- Safety planning – pre-project/pre-task
- Safety training and education
- Worker involvement and participation
- Recognition and rewards
- Subcontractor management
- Accident/incident reporting and investigation
- Drug and alcohol testing

Record Retention List

Type of Report	Record Retention
Employee applications	+ 30 yrs.
Employee medical and exposure records	duration of employment
Employee authorizations/ releases	duration of employment
Employee training records	duration of employment
OSHA 300 logs and summaries	last 5 yrs.
First Reports of Injury	last 5 yrs.
Accident Investigations	last 5 yrs.
Job-site safety inspections	2 yrs. (recommended)
Minutes of safety meetings	2 yrs. (recommended)
Equipment inspection certifications	2 yrs. (recommended)

102F15.EPS

Figure 15 ◆ Record retention list.

Instructor's Notes:

2.28 PROJECT MANAGEMENT

Review Questions

1. List four major reasons why it is important for a contractor to have an effective safety and loss prevention program.

   1. ________
   2. ________
   3. ________
   4. ________

2. What are the duties of the project manager with respect to safety and loss prevention?

   ____

   ____

   ____

3. List the three major causes of accidents.

   1. ________
   2. ________
   3. ________

   In terms of accident prevention, where should you focus the majority of your attention?

   ____

4. List two examples each of both direct and indirect accident costs.

   1. ________
   2. ________

5. What is the ratio of near-miss type accidents to those which result in:

   Property damage ____
   Minor injuries ____
   Serious or disabling injuries ____

6. What is the estimated cost of an effective safety program?

   ____

7. When making job-site inspections and observations, where should you focus the majority of your attention – on unsafe conditions or unsafe acts?

   ____

8. List the potential areas for loss within the construction industry.

   ____

9. In a process of hazard control, what is used to determine the exposure category?

   ____

10. How long do you have to complete an incident report on the OSHA 300 Injury-Illness log?

    ____

11. What are the four high hazard areas targeted under OSHA's focused inspection program?

    1. ________
    2. ________
    3. ________
    4. ________

12. Who is held responsible for ensuring employees at a multi-employer work site are not exposed to hazards?

    ____

13. What is the host employer's role at multi-employer work site?

    ____

14. What percentage of the nation's construction workforce is estimated to have an alcohol or drug abuse problem?

    ____

15. When drafting safety rules, what key points should be considered?

    ____

    ____

    ____

16. Workers who have been on the job fewer than 30 days account for what percentage of the total construction injuries?

    ____

Have the participants complete the Review Questions, and go over the answers prior to administering the Module Examination.

MODULE 44102-08 ◆ SAFETY 2.29

Review Questions

17. What methods can be used to evaluate a contractor's past safety performance?

____

18. Who normally conducts accident investigations?

____

What is the project manager's role in accident investigations?

____

____

____

19. If your employees are exposed to OSHA-regulated substances such as asbestos, how long must you keep their medical and exposure work?

____

20. When making a statement to the news media, what are some things you should consider?

____

____

____

____

Instructor's Notes:

____

____

____

Summary

Summarize the major concepts presented in the module.

A safety and loss prevention program is a management tool to help you achieve the company's goals and objectives. To be effective, it should be put in writing, communicated to all involved parties, and above all, incorporated into your daily activities.

Your written program should be detailed enough so that it is clearly understood by everyone, but not so cumbersome that people are intimidated by the size of the document. Your safety program manual should not contain any rules, policies, or procedures that you cannot or do not intend to enforce. It is far better to have a small, concise safety and loss prevention program that is incorporated into your daily operations than to have a large, well-documented manual that covers all the bases but is not completely functional.

Remember, as a project manager it is your responsibility to get the job done safely, on time, and within budget. To do this you must identify the hazards, assess and evaluate the risk, and develop a plan of action. This action plan must start in the pre-bid stage and follow the job through to completion. Regular employee observations, safety inspections, and audits will be needed to evaluate your performance.

Notes

MODULE 44102-08 ◆ SAFETY 2.31

Have participants complete the Trade Terms Quiz, and go over the answers.

Administer the Module Examination. Be sure to record the results of the testing on Craft Training Report Form 200, and submit the results to the Training Program Sponsor.

Trade Terms Quiz

1. The agency of the United States Department of Labor created to prevent work-related injuries and issues is called __________.
2. __________ is the amount of exposure to a substance that requires medical attention.
3. An unplanned event that interrupts the normal progress of an activity is an __________.
4. The number of lost workdays per 100 full-time employees per year is known as an __________.
5. The __________ is a unit of measure used to evaluate a company's safety performance.
6. __________ provides insurance in exchange for the release of the right to sue for negligence.
7. The __________ is a rate computation to determine a surcharge or credit to workers' compensation.
8. The clothing, helmets, goggles, and other gear designed to protect the wearer are called __________.
9. __________ is the number of persons who will be exposed to a toxic substance in the course of their employment.
10. A numerical rating based on probability, consequences, and exposure is the __________.
11. The process of limiting an inspection to the four high hazards is called a __________.

Trade Term List

Accident
Action levels
Experience modification rate (EMR)
Focused inspection program
Incidence rates
Occupational Safety and Health Administration (OSHA)
OSHA recordable incidence rate
Personal protective equipment (PPE)
Risk assessment codes
Workers' compensation
Worker exposure

Instructor's Notes:

2.32 PROJECT MANAGEMENT

Trade Terms Introduced in This Module

Accident: An unplanned event that may or may not result in personal injury or property damage that interrupts the normal progress of an activity. Accidents are categorized by their severity and impact and invariably preceded by an unsafe act, unsafe condition, or both.

Action level: The level of exposure to a substance or physical agent (usually one half of the permissible exposure limit) that requires medical surveillance under a particular OSHA standard.

Experience modification rate (EMR): A rate computation to determine surcharge or credit to workers' compensation premium based on a company's previous accident experience. Accident experience from three previous years, not to include last year, is used to determine the rate.

Focused inspection program: OSHA's partial inspection process focusing on the hazards that cause the most deaths in the construction industry: falls, electrical, caught in/between, and struck by.

Incidence rate: The number of injuries and/or illnesses or lost workdays per 100 full-time employees per year or 200,000 hours of exposure.

Occupational Safety and Health Administration (OSHA): An agency of the United States Department of Labor created by Congress under the Occupational Safety and Health Act. Its mission is to prevent work-related injuries, illnesses, and deaths by issuing and enforcing rules (called standards) for workplace safety and health.

OSHA recordable incidence rate: A computation of the total number of OSHA-defined recordable injuries and illnesses times 200,000 divided by the hours worked for the period in question. It is one unit of measure used to evaluate a company's safety performance.

Personal protective equipment (PPE): The protective clothing, helmets, goggles, or other gear designed to protect the wearer's body or clothing from injury by electrical hazards, heat, chemicals, and infection, for job-related occupational safety and health purposes.

Risk assessment code: A numerical rating of the risk associated with a hazard based on probability, consequence(s), and exposure.

Workers' compensation: Provides insurance to cover medical care and compensation for employees who are injured in the course of employment, in exchange for mandatory relinquishment of the employee's right to sue their employer for negligence.

Worker exposure: The number of persons who will be regularly exposed to a toxic substance or harmful physical agent in the course of employment through any route of entry (inhalation, ingestion, skin contact, absorption, etc.).

MODULE 44102-08 ◆ SAFETY 2.33

Resources & Acknowledgments

Additional Resources

Contren® Learning Series, *Field Safety*, 2003.

Contren® Learning Series, *Safety Technology*, 2003.

The *OSHA Recordkeeping Handbook* (OSHA Publication #3245) is available from the OSHA website at: www.osha.gov/pls/publications/pubindex.list.

Other information relating to OSHA standards and regulations may be obtained from the OSHA website at www.osha.gov, local OSHA offices, or from the U.S. Department of Labor – OSHA Publications Office, Room N3101, 200 Constitution Avenue, N.W., Washington, D.C. 20210

Acknowledgment

This module was developed in cooperation with Steven P. Pereira, CSP, Principal, Professional Safety Associates, Inc., Denham Springs, Louisiana.

References

Construction Industry Institute (2001). *Making Zero Accidents a Reality*. Retrieved from: www.construction-institute.org/scriptcontent/cpi2001slides/mathis_cpi.ppt.

Field Inspection Reference Manual (FIRM). OSHA. Retrieved from www.osha.gov.

Heinrich, H.W., Petersen, D., Roos, N.R., Brown, J., and Hazlett, S. *Industrial Accident Prevention: A Safety Management Approach* (1980).

Meyer, S.W. and Pegula, S.M. *Injuries, Illnesses, and Fatalities in Construction, 2004*. Bureau of Labor Statistics, posted 24 May 2006. Retrieved from: www.bls.gov/pub/cwc/sh20060519ar01p1.htm (1 October 2007).

Instructor's Notes:

2.34 PROJECT MANAGEMENT

Appendix A

Pre-Project Planning Safety Checklist

MODULE 44102-08 ◆ SAFETY 2.35

PRE-PROJECT PLANNING SAFETY CHECKLIST

1. **General Information:**

Job Number: ________________ Client: ____________________________________

Location: ____

Client Contact: ____________________Phone: ____________Fax: ____________

Start Date: ______________________ Completion Date: __________________________

2. **Scope of Work:** Briefly describe the project and your scope of work.
(Type and size of project, materials of construction and construction methods).

____

____

____

____

Peak Employment: Company __________________Subcontractors ______________

Will the job involve any unusual or high risk work? If so, specify the nature of the potential problem(s) and the proposed solutions.

____

____

____

____

Will the job involve any of the following? If so, describe.

_____ Blasting

_____ Pile driving

_____ Tunneling or major excavations

_____ De-watering

_____ Demolition

_____ Work at extreme heights

_____ Work over water

_____ Underpinning

_____ Handling or exposure to asbestos, silica, hexavalent chromium, hazardous wastes, lead or OSHA-regulated substances

_____ Work in, on, or adjacent to equipment handling flammable, toxic, or otherwise hazardous chemicals

____

____

____

____

102A01a.EPS

Instructor's Notes:

____

____

____

2.36 PROJECT MANAGEMENT

3. **Hazardous Processes, Materials or Equipment:**
Identify processes, materials or equipment that may expose employees to hazardous conditions either in routine work or emergency situations. Obtain Material Safety Data Sheets or Hazardous Waste Sheets on all hazardous materials to which employees may reasonably be exposed. Find out how the client is complying with the OSHA Hazard Communications Standard and how this information will be conveyed to contractor employees. Find out if the proposed work falls under the OSHA Process Safety Management Standard 1910.119.

4. **Client Safety Rules and Procedures:**
Obtain copies of all client safety rules, procedures, or manuals that contractors are expected to follow. Do not overlook emergency plans or procedures. List any unusual or special safety requirements.

5. **Medical Surveillance/Industrial Hygiene Monitoring Requirements:**
List any medical surveillance or industrial hygiene monitoring requirements for contractor personnel. Determine who will be expected to provide these services. List any special requirements, such as no-beard policies, drug screening tests, showering facilities, and decontamination facilities.

6. **Regulatory Permit Requirements and Inspection Schedules:**
List all required permits and inspection schedules and who is responsible for obtaining these permits and coordinating the necessary inspections.

7. **Site Location and Layout:**
A. Briefly describe site conditions and layout (rural, residential, commercial, industrial, congested, etc.).
B. Describe adjacent exposures. How might they be affected by the project? Is there any need to take photos of the adjacent structures or have a physical inspection made by an independent third party?
C. Utilities: List all utilities available at or near the site. If possible, obtain the names and phone numbers of area utility representatives.
  1. Water:
  2. Electricity:
  3. Gas:
  4. Sewer:
  5. Other:
  6. Have all existing utilities been located and marked appropriately?
  7. Will an assured grounding program or ground fault circuit interrupters be used to protect temporary electrical circuits? Who is responsible for coordinating this activity?

D. Field office, storage and laydown areas:
  1. Any designated areas?
  2. Accessible by trucks, forklifts and cranes? Will vehicles be able to drive through as opposed to backing out?
  3. Any restricted or congested roads/areas? If so, describe.
  4. Parking facilities for employees, contractors and visitors?
  5. Any special requirements for temporary buildings with respect to location, size, materials of construction, etc.?

102A01b.EPS

MODULE 44102-08 ◆ SAFETY 2.37

8. Personnel Health and Safety:

A. Personal protective equipment: What equipment is required and who will provide it?

_____ Hard hats
_____ Safety shoes
_____ Chemical-resistant boots
_____ Fire-retardant clothing
_____ Plano safety glasses, with or without side shields
_____ Prescription safety glasses, with or without side shields
_____ Gloves, specify: _______________
_____ Safety harnesses and lanyards, specify type: ________________
_____ Chemical-resistant clothing, specify: _______________
_____ Lifelines
_____ Personnel nets
_____ Debris nets
_____ Face shields/goggles
_____ Respiratory protection, specify type: _____________________
_____ Other (specify): ____
____

B. What temporary toilet facilities will be needed? Will any showering facilities be required?

C. Is a drinking water supply available on site?

D. What method(s) will be used to protect wall and floor openings?

E. Any special scaffolding requirements? If so, specify. ______________________________

F. Any materials or personnel hoists or elevators? If so, who will operate, inspect and maintain them?

G. Overhead protection required at building entrances?

9. Emergency Reporting and Response:

A. List all emergency communication systems and onsite phone numbers.

B. How will onsite medical and first aid be handled? Who will provide what type and level of care? What first aid supplies and rescue equipment will be required/provided?

C. Location and telephone number of the nearest offsite medical facilities and ambulance service companies.

D. Name, address, and phone number of company physician(s).

E. Obtain information on the client's emergency assembly areas and evacuation routes.

F. What onsite emergency notification and response procedures will be used? Obtain information on the client's emergency alarm system and signals. If possible, obtain lists of emergency alarm codes. Try to get a tape recording of the alarm signals.

G. What emergency escape equipment (if any) is needed? Who will provide it and where will it be located?

H. What offsite emergency reporting and response procedures will be used?

I. Name, address, and phone number of fire department.

J. Will contractor personnel be expected to serve on a Hazardous Materials Emergency Response Team? If so, specify the level and extent of contractor involvement.

102A01c.EPS

Instructor's Notes:

____
____
____

10. Fire Protection and Prevention:

A. Is there an adequate number of active fire hydrants on site? What provisions will be made to keep them unobstructed and accessible?

B. List the number, type, and size of portable fire extinguishers to be provided and by whom.

C. Will standpipes and/or sprinkler systems be required to follow the building up floor by floor? ___Yes ___No ___N/A. Briefly describe type of systems, maintenance and inspection requirements, and responsibilities during the construction phase.

D. Is there a site fire brigade? Will contractors be expected to serve on this brigade?

E. Will temporary heating be used? ___Yes ___No ___N/A. If so, what type?

F. Requirements for the storage and handling of flammable liquids. Any secondary containment required? Describe.

G. How, where and at what frequency will trash be removed and disposed of?

H. How will hazardous waste be handled?

11. Administration:

A. Who will be responsible for collecting and disseminating Material Safety Data Sheets?

B. Who will be responsible for hiring site personnel?

C. What screening and placement procedures will be used?

_____ Written applications

_____ Reference checks

_____ DMV records check

_____ Medical exams

_____ Drug screening

_____ Other (specify):

D. Who will be responsible for New Employee/Contractor Orientation and what materials and/or handouts will be used?

E. What types of safety inspections are required, at what frequency, and by whom?

F. What types of safety meetings will be held, at what frequency, and with whom?

12. Special Tools or Equipment Requirements:

Will any special tools or equipment be required? If so, specify types and brands.

A. Explosion-proof lighting

B. Non-sparking tools

C. Air-operated tools or equipment

D. Ground fault circuit interrupters (GFCI)

E. Extension cords

F. Air movers or other gas freeing equipment

G. Gas testing equipment

H. Other (specify)

102A01d.EPS

MODULE 44102-08 ◆ SAFETY 2.39

13. Work Permit Requirements:

Obtain as much information as possible concerning the client's work permitting procedures. Are permits required for the following types of work?

A. Hot/hazardous work

B. Cold/safe work

C. Vessel entry (confined space)

D. Gas testing

E. Equipment isolation including lockout/tagout

F. Excavations

G. Vehicle entry

H. Critical lifts

I. Scaffolding

J. Other (specify):

14. Training Requirements:*

List any special safety training requirements such as:

A. Respirator

B. Hazard communication

C. Handling of certain hazardous materials (i.e., asbestos, lead). Specify the material(s).

D. Hazardous waste site

E. Response to spills or releases of hazardous materials. Specify what level of response will be required by contractor personnel.

F. Certification for operators of trucks, cranes, manlifts, etc.

G. Other (specify):

*Important Note: Find out who will perform and document this training, the client or the contractor. What training aids such as video tapes, films, slide/tape presentations are available from the client for contractor use?

15. Reporting Requirements:

List any special reporting requirements the client may have, such as:

A. Accident, injuries, illnesses or near-misses

B. Safety meetings

C. Safety inspections

D. Regulatory agency visits

E. Other (specify)

16. Site Security:

A. Describe the security measures required for the job during normal working hours and after hours.

B. If controlled access is required, how will it be handled and by whom?

C. Will any burglar alarm system, guard dog, or watchman services be used?

___Yes___No___N/A. If so, describe.

D. Is/will site lighting be adequate?

102A01e.EPS

2.40 PROJECT MANAGEMENT

Instructor's Notes:

17. Public Liability:

A. Will any sidewalks or streets have to be blocked?

___Yes ___No ___N/A. If so, describe.

B. Any overhead protection and/or lighting required?

___Yes ___No ___N/A. If so, describe.

C. Briefly describe any barricading, lighting, signs, traffic control devices or flagmen that may be required.

18. Environmental Hazards and Controls:

A. EPA, State, or local Right to Know Requirements? Describe.

B. If asbestos or lead is suspected or anticipated, what special precautions, monitoring and training will be required? Who will coordinate and pay for these activities?

C. Will any wastes generated at or leaving the site be classified as hazardous and require disposal as a hazardous waste? ___Yes ___No ___N/A. If so, will it be necessary to obtain any types of generator numbers and or permits? Who will obtain the necessary permits?

D. Will there be any open burning or on-site landfills?

___Yes ___No___N/A. If so, describe.

E. Will any site activities create noise, air, water, or ground pollution problems?
___Yes ___No ___ N/A. If so, describe potential problems and proposed solutions.

F. Is secondary containment required for flammable/hazardous material storage?

19. Cranes/Hoists and Lifting Devices:

A. Will there be any cranes, hoists, or lifting devices on site?

___Yes ___No. If so describe.

B. Who will operate, maintain, and inspect these devices? How will they be certified?

C. Any heavy, unusual, or critical lifts to be made?

___Yes ___No. If so, describe.

D. Any requirements for inspections, certifications, or load tests?

___Yes ___No. If so, describe.

E. Any requirements for critical lift plans?

___Yes ___No. If so, describe.

F. Any planned use or prohibition on use of crane suspended personnel baskets?

___Yes ___No. Explain.

20. Signs, Posters and Bulletin Boards:

A. List all signs, posters, notices required for the project:

_____ OSHA Employee Rights poster

_____ Workers' Compensation notice

_____ Company/job-site rules

_____ Emergency phone numbers

_____ Location of MSDSs

_____ Employee access to medical and exposure records

_____ Others (specify)

B. Where will these notices and signs be posted?

102A01f.EPS

MODULE 44102-08 ◆ SAFETY 2.41

21. Contract Specifications:

A. Bonding requirements: (specify)

B. Insurance requirements:

1. Types of coverage and policy limits
2. Who will provide what coverage?
3. Any special requirements needing insurance carrier approval?
4. Have certificates of insurance been requested and received from?
   a. Client/owner
   b. Prime or general contractor
   c. Subcontractors
   d. Others
5. Names, addresses and phone numbers of key insurance company representatives:
   a. Agent/broker
   b. Claims representative
   c. Loss control or safety representative
6. Any hold harmless or indemnity agreements? If so, obtain copies.
7. Any unusual safety and loss prevention requirements, such as:
   a. Site specific safety and health plan required?
   b. Site safety representative required?
   c. Drug, alcohol, and substance abuse policy required?
   d. Any pre-employment and/or follow-up medical exams required?
   e. Other (specify)

22. Additional Information or Comments:

Source of Information:

Name: ______________________ Title ______________ Date ________

Name: ______________________ Title ______________ Date ________

Report Prepared by:

Name: ______________________ Title ______________ Date ________

Report Reviewed by:

Name: ______________________ Title ______________ Date ________

102A01g.EPS

Instructor's Notes:

____

____

____

Appendix B

Checklist for Coordination of Site-Specific Safety Activities

This checklist is a suggested guide for firms and individuals involved as owners, construction managers, or contractors who must decide how safety and other project considerations will be properly assigned and/or accomplished. This checklist may be customized with specific questions for individual projects.

CHECKLIST FOR COORDINATION OF SITE-SPECIFIC SAFETY ACTIVITIES

	Owner	Construction Manager	Individual Contractor (Specify)
1. Job site safety program:			
a. First aid facility (or kit)			
b. Safety bulletin board including OSHA poster & emergency phone numbers			
c. Safety meetings			
d. Safety inspections			
e. Accident investigations			
2. Safety rails and covers for openings			
3. Fire protection program			
4. Fire prevention			
5. Street and sidewalk protection and maintenance			
6. Special notices to utilities, adjoining property owners, etc.			
7. Temporary water (installation, cost)			
8. Temporary toilets			
9. Temporary telephones			
10. Temporary heat (prior to enclosure)			
11. Temporary heat (after enclosure)			
12. Temporary power/light including GFCIs			
13. Temporary ladders and temporary stairs including access ramps and runways			
14. Allocation of site storage space			
15. Job site security fence and maintenance			
16. Temporary roads and parking area			
17. Hoisting during construction			
18. Hoisting after structure is complete			
19. Required process clean-up			
20. Final clean-up and window washing			
21. Road and street cleaning			

102A02.EPS

MODULE 44102-08 ◆ SAFETY 2.43

Appendix C

Guidelines for Developing a Medical Surveillance and Post-Accident Claims Management Program

1. Develop detailed job descriptions for each trade or job function. Be sure to identify the essential job functions as well as the non-essential or marginal job functions. The company physician will need this information to tailor a physical examination that meets your needs and does not conflict with the Americans with Disabilities Act.
2. Select a physician or group of physicians to act as your company physician. These individuals should be well-qualified, licensed medical practitioners who are specialists in occupational or industrial medicine. Have them visit a jobsite to see exactly what you do. Don't overlook the small industrial medical clinics. They often cater to the contractors' needs and will perform post-job offer physicals, drug tests and post-accident medical care or referral to qualified specialists. Your workers' compensation insurance carrier or claims administrator may be able to give you the names of physicians to consider.
3. Identify all applicable OSHA regulations that require medical surveillance, such as the Asbestos, Lead and HAZWOPER Standards and provide the company physician with a copy of the standards. Have the physician develop a medical surveillance program that will meet your specific needs. Have the physician provide your firm with any OSHA mandated physicians' opinion letters concerning the employee's fitness for duty or limitations.
4. When setting up a job site, obtain and post the names, physical addresses, telephone numbers, and personal contacts for the following:
   - Nearest trauma center
   - Nearest fully-equipped hospital
   - Nearest burn unit
   - Company physician
   - Industrial clinics
   - Police, fire, and ambulance
   - Insurance company claims representatives
   - Insurance company loss control or safety representatives
5. Draft a form authorizing treatment for an individual who reportedly has been injured on the job. (See the sample form in this Appendix.) Mandate that the employee return the completed form to his/her supervisor immediately after treatment but no later than the start of the employee's next regularly scheduled shift. If possible, have the employee's supervisor accompany the injured worker to the doctor. Try to encourage the employee to use the company physician or industrial medical clinic.
6. Establish a daily mandatory check-in by management (telephone contact) with all injured workers until they return back to work. If you have difficulty locating an injured worker, report your findings to your insurance company claims representative. Check in with the injured worker's doctor(s) on a regular basis to determine the employee's status and recovery progress. If the employee or his physician does not want to talk to you on this matter, have your company physician make the call and report back to you.

Instructor's Notes:

2.44 PROJECT MANAGEMENT

7. Implement a light-duty policy for personnel who are injured on the job, who are capable of performing some portions of their regular work or other productive tasks that will not aggravate their condition. Review this policy with your company physician.
8. As part of your alcohol, substance abuse, and contraband policy, mandate post-accident drug testing for all personnel involved in an accident or injured on the job. Many companies require post-accident drug testing for all incidents requiring medical treatment or those that had the potential to result in personal injury or property damage in excess of $500.00.
9. Evaluate your state's second injury fund and be sure you are taking full advantage of the program if one of your employees re-injures him/herself or aggravates a pre-existing condition. This will necessitate developing a medical and occupational history questionnaire to be completed by new hires.
10. Regularly request a printout from your insurance carrier or administrator of your workers' compensation payments and reserves. Review this information carefully and promptly report any discrepancies or concerns to your claims representative.
11. Set up a system for filing and managing employee medical and exposure records. These records must be kept in locked file cabinets and kept strictly confidential. Employee job related, medical, and exposure records should be kept separate from employee's personal medical records. Establish a mechanism and procedure that allows employees to access and copy their personal medical and exposure records. Review this procedure with all employees when they are hired and annually thereafter.

MODULE 44102-08 ◆ SAFETY 2.45

Treatment Authorization / Release Form

Date: ______________________________

Billing information: ______________________

____

Phone No.: __________________________

Direct inquiries/bills to: ____

Mr./Ms. ______________________________ has reportedly been injured on the job.

Nature of injury and first aid provided: ______________________________________

Signature of person providing first aid ____________________________ Date: __________

Signature of person authorizing treatment ________________________ Date: __________

Consent to drug/alcohol test ______________________________ Date: __________

(Patient's Signature)

All personnel receiving medical treatment must be drug tested using the following methodology:

Release of Information by Employee

I agree and consent to the testing and to the release of the results to authorized management. I understand that management may rely upon the test results in making decisions related to my employment as outlined in the Alcohol and Drug Abuse Policy. I authorize the release of all medical and dental records, reports and tests to my employer and/or its insurance carrier. I understand that I may choose or select any treating physician in any field or specialty. The employer does not make any recommendation, but will assist the employee in securing prompt examination or treatment by a physician decided on by the employee. The employer will assist with a list of doctors for the employee to choose from if requested, but it is not intended to be a recommendation of the employer.

Date: ________________ Signature of Employee __________________________

Physician Information

Diagnosis: ____

Treatment: ____

Medication Administered: ______________________________________

____

Work Status: Employee may return to: ☐ Full Duty ☐ Other (specify any limitations)

____

Any follow-up treatment required? ☐ Yes ☐ No (If yes, specify)

____

Physician's Name: ______________________________________

Address: ____

Telephone No.: ____

102A03.EPS

2.46 PROJECT MANAGEMENT

Instructor's Notes:

Appendix D

Sample Job-Site Safety Inspection Checklist

A check in the "no" column indicates non-conformance to an OSHA general policy or safety standard and user should refer to the published OSHA standard for specific details for compliance. User is cautioned that this checklist does not apply to every type of operation and therefore does not list those items covering every hazard relating to a physical or occupational disease exposure. It contains those items where OSHA citations have been issued most frequently.

	Yes	No	Corrective Action
GENERAL			
Pre-job safety meeting held with subcontractor(s)			
Monthly safety meetings held with supervisor and subcontractor(s)			
Competent contractor personnel assigned responsibility to inspect job site for safety			
U.S. Department of Labor "Safety and Health Protection on the Job" poster posted			
Management safety policy directives posted			
Safety bulletins and accident prevention material posted			
Log of Employee Injuries readily available			
Copies of Supplemental Records of Job Injuries readily available			
Copies of Worker Injury Investigation Reports kept at job site			
Emergency Phone Numbers posted at each phone			
SANITATION, MISCELLANEOUS			
Portable containers used to dispense drinking water (required types)			
HAZARD COMMUNICATION			
Written Hazard Communication Program			
Labels and other forms of warning on containers			
Material Safety Data Sheets maintained			
Employees trained on Hazard Communication			

102A04a.EPS

MODULE 44102-08 ◆ SAFETY 2.47

	Yes	No	Corrective Action
FIRE PREVENTION AND PROTECTION			
Required portable fire-fighting equipment available, properly located and maintained with inspection records			
Approved metal safety cans used for handling and use of flammables			
GENERAL			
In areas where flammables are stored or where operations present a fire hazard, "No Smoking or Open Flame" signs posted			
Indoor and outdoor storage of flammables in approved containers or cabinets with warning signs posted			
Form and scrap lumber and all other debris kept clear from work areas			
Combustible scrap and debris removed from work areas at regular intervals			
Containers provided for collection, separation of waste, trash, oil, and used rags			
Solvent waste, oily rags, and flammable liquids kept in fire-resistant, covered containers until removed from job site			
PERSONAL PROTECTION			
Hearing protective devices provided for and worn by workers where noise levels are excessive			
Hard hats provided for and worn by workers			
Eye and face protection provided for and worn by workers where exposed to potential eye or face injury			
Workers required to wear footwear adequate for their assigned work			
Job-specific respiratory protection programs as required			
Physician's opinion letters and respirator fit test records available			
Breathing air meets CGA – Grade D specifications			
Airless spray guns (operating at 1000 psi) equipped with safety device preventing pulling of trigger until safety device is manually released			

102A04b.EPS

Instructor's Notes:

	Yes	No	Corrective Action
SCAFFOLDING			
All scaffolds and components are capable of supporting four times the maximum intended load			
Scaffolds and components are inspected by a competent person before each work shift			
Scaffolds are erected, dismantled, and altered under the direction of a competent scaffold builder			
All open sides and ends of platforms more than 10' above ground on floor level provided with top rails (42' high), midrails and toeboards (4' high)			
Where workers pass or work under scaffolds, screen or other overhead protection provided			
Platform planks laid together tight preventing tools, etc., from falling through			
Planks secured to prevent movement			
Overhead protection on scaffolds where workers exposed to overhead hazards			
Ladders used to gain access to scaffold work platforms			
ROLLING SCAFFOLDS (Manually Propelled)			
Wheel brakes set while in use			
No riders on work platform while moving			
Work levels 10' or more above ground or floor level have guardrails and toeboards			
All cross and diagonal bracing in place and properly connected			
Height does not exceed four times least base dimension unless outriggers are used			
Ladder access to work platform			

102A04c.EPS

MODULE 44102-08 ◆ SAFETY 2.49

	Yes	No	Corrective Action
SWINGING SCAFFOLDS - 2 Point Suspension Roof			
Hooks or irons securely installed and anchored to a sound structural part of the building			
Sheaves fit size of rope used			
Vertical lifelines, independent support lines, and suspension lines are not attached to each other or attached to the same anchor point			
ELECTRICAL, INCLUDING LOCKOUT/TAGOUT			
For power circuits, exposed or concealed, where accidental contact by tool/equipment may be hazardous, warning signs posted and all workers advised of hazard			
Regular inspections made to assure effective grounding of noncurrent carrying metal parts of portable, and/or plug connected equipment. GFCIs installed on all 110-120V temporary circuits			
Temporary lights equipped with guards to prevent accidental contact with bulb			
Lockout/tagout procedures are followed when servicing or repairing equipment that could start or move, injuring personnel (all forms of energy)			
Blocking or pinning devices used to hold equipment in elevated positions meet OEM specifications			
LADDERS			
Ladders regularly inspected and destroyed when found defective			
Side rails extend 36' above landing or provision of grab rails			
Top of ladders tied-in to prevent displacement			
Double cleat ladders not exceeding 24' in length			
Single cleat ladders not exceeding 30' in length			
Rungs inset into edges of side rails or filler blocks used			
All job-built ladders constructed to conform with standards			
No metal ladders used within 10' of electrical lines			
Step ladders used only in full open position			

102A04d.EPS

Instructor's Notes:

	Yes	No	Corrective Action
Step ladders of sufficient height so that top two steps do not have to be used to perform work			
All manufactured single and extension ladders equipped with ladder shoes			
EXCAVATIONS, TRENCHING, AND SHORING			
Excavated material effectively stored and retained at least 2 feet or more from edge of excavations			
Utility company contacted and advised of proposed excavation work to determine underground utility exposure or when overhead power lines are involved			
Substantial stop logs or barricades installed when mobile equipment working adjacent to excavation			
Trenches over 5' in depth shored to standard, laid back to stable slopes or provided with other equivalent protection where hazard of moving ground exists			
Trenches less than 5' protected where hazardous ground movement exists			
Trenches 4' deep or more provided with ladder located no more than 25' of lateral travel			
Protective systems used in excavations over 20' deep are designed by a registered professional engineer			
Excavations, protective systems, and surrounding areas are inspected on a daily basis by a competent person			
WELDING, CUTTING, AND BURNING			
Proper personal protective equipment provided, in good condition and being used			
All oxygen and acetylene gauges in working condition			
Oxyacetylene torches and hoses fitted with flashback arresters and reverse flow check valves			
All oxygen and acetylene hoses in good condition and free of grease and oil			
Mechanical lighters used for lighting torches—No cigarette lighters or matches			
Oxygen and acetylene cylinders stored in upright position in designated areas with caps in place			

102A04e.EPS

MODULE 44102-08 ◆ SAFETY 2.51

	Yes	No	Corrective Action
Oxygen and acetylene hoses properly located to avoid damage by moving equipment or creating a tripping hazard			
Electric arc welding cables in good condition and properly attached by lugs to welding machine			
Rod holders in good condition			
Shields provided to protect other workers from flash burns			
Welders' shields and helpers' goggles in good condition and equipped with proper lenses			
Fire extinguishers provided within 25' of welding, cutting, or burning operations			
All flammable material removed from welding, cutting, and burning operations area			
Protection provided to prevent slag, etc. from falling on workers below			
All welding cable positioned to eliminate tripping hazards			
FLOOR AND WALL OPENINGS			
Wall openings (30' high, 18' wide or greater) from which there is a drop of 4' or more and bottom of opening less than 3' above working surface provided with guardrails			
Bottom of wall openings less than 4' above work surface provided with standard toeboard (4' high)			
Open-sided floors 6' or more above floor or ground level provided with standard railing and toeboard or other equivalent perimeter protection			
Stairways when used during construction have handrails on all open sides, guardrails at landings			
Floor hole covers are properly sized, secured from movement and marked "hole" or "cover"			
MATERIAL HANDLING EQUIPMENT			
Approved canopy guards on forklift trucks and roll-over protection on all earth-moving equipment			
Rated capacity posted on all lifting, hoisting equipment clearly visible to the operator			
Mobile equipment operators are properly trained			

102A04I.EPS

Instructor's Notes:

	Yes	No	Corrective Action
ROOFING			
Pitched roof over 4:1 pitch At heights of 6' or more, workers are protected from falls by guardrails, nets, or personal fall arrest systems			
Roofing bracket scaffolds secured			
Crawling boards or chicken ladders secured and with evenly spaced cleats			
Ladders extended 36' above eaves and secured top and bottom			
Flat and low pitch roofs less than 4:1 pitch At heights of 6' or more, personnel are protected from falls by guardrails, nets, personal fall arrest systems, safety monitoring systems, warning line systems, or any combination thereof			
Protective equipment when handling hot pitch			
Ladder access same as for pitched roofs			
Materials stored 6' from edge			
All openings covered or protected by guardrails			
CONCRETE FORMING AND POURING			
Vertical re-steel protected by covering when employees working above			
Employees on the face of formwork or re-steel are protected from falls of 6' or more by personal fall arrest system, safety net systems, or positioning device systems			
Walk and standing boards when pouring horizontal surfaces on re-steel			
Employees provided with and using personal protective equipment while pouring concrete			
No riding concrete buckets			
All pump concrete lines secured at all joints			
Properly guarded work platform for wall, columns, beams, etc.			
Power troweling machines equipped with positive on/off switch			

102A04g.EPS

MODULE 44102-08 ◆ SAFETY 2.53

	Yes	No	Corrective Action
HAND AND POWER TOOLS			
Powder-actuated tools provided with safety shield/guard and operator has evidence of special training in their use			
Portable power circular saws provided with proper functioning automatic-return lower guard and fixed upper guard			
All fixed power woodworking tools provided with a disconnect switch that can be locked or tagged in the OFF position			
Defective tools – equipment tagged as unsafe, controls locked in OFF position or physically removed from job site			
DEMOLITION			
Dust controlled by wetting			
Employees provided with approved respirators and goggles			
Chutes properly erected and drop area barricaded off			
Floor openings protected			
Employee access to building provided with overhead protection			
Stairways in building used for access properly lighted and maintained			
BLASTING			
Certified blaster in charge			
Storage and handling of explosives conforms to Federal requirements and all local ordinances			
Excess explosives, caps, det-cord, etc. removed to approved storage area prior to blast			
Standard warning system used			
Two-way radios turned off			
Signs posted			
No other work within 100' of loading area			
Wood or plastic poles used for tamping			
Blasting machines only used in electric blasting			

102A04i.EPS

Instructor's Notes:

	Yes	No	Corrective Action
Waste explosives and containers disposed of as required by regulations			
Record of all explosives kept			
Post-blast inspection completed			
Misfires handled as specified by regulations			
Access to work area controlled			
CRANES			
Operators are physically fit and properly trained			
Rated load capacities, recommended operating speeds, special hazard warnings are conspicuously posted			
Operating manual is available in crane			
Hand signal charts are posted			
Daily inspections are made by a competent person			
Periodic and annual inspections are documented			
Inspections of all rigging equipment are performed and documented			
Overhead power lines are clearly identified and adequate clearance is maintained or lines are de-energized			
Swing radius is barricaded			
Crane suspended platforms are properly designed, rigged, and load-tested prior to each use. Rated load capacity is posted on the basket			
STEEL ERECTION			
Foundations for structural steel have cured to 75 percent of compression strength before work starts			
Permanent floors are installed as erection of structural members progresses but there are no more than 8 stories between the erection floor and the uppermost permanent floor			
A fully-planked deck or net is maintained within 2 stories or 30', whichever is less, directly under any erection work being performed			
Deck and roof openings are adequately protected			

102A04j.EPS

MODULE 44102-08 ◆ SAFETY 2.55

	Yes	No	Corrective Action
At least 2 wrench-tight bolts are in place on each connection before hoist line is released			
All personnel on unprotected walking/working surfaces more than 15' above the next level are protected from fall hazards by guardrail systems, safety net systems, personal fall-arrest systems, positioning device systems, or fall-restraint systems			
Connectors working at unprotected elevations between 15' and 30' are provided with and wear personal fall-arrest or restraint systems capable of being tied off			
Connectors working at unprotected heights of 30' or more are protected by nets or personal fall-arrest or fall-restraint systems			

102A04k.EPS

2.56 PROJECT MANAGEMENT

Instructor's Notes:

Appendix E

Tips on Dealing with the Media During Emergencies

The effort of preparing for the media in an emergency does not come easily. Even a routine call from a local reporter is enough to cause panic to many executives and owners. Construction sites are always in the public eye, and many of them are near public highways, commercial buildings and residential areas. It is a safe bet to say that when something happens to go wrong, the media will be there along with emergency personnel.

For the most part, reporters are good people who are trying to do a very tough job. They are professionals who want to find out enough about something to be able to explain it to everyone. Given their limitations of time and background knowledge, they do a commendable job. If a major construction accident occurs, the media will want to know:

- What fell down or blew up?
- Was anyone hurt?
- Why did it happen?
- How is the community being affected?
- What is the construction company doing to fix it?

The biggest mistake that a contractor can make is to be unprepared when the media calls. In an emergency, you should know what reporters will show up, and that you will have to deal with them. Keep in mind that the reporters are going to get someone's story to tell. Most contractors would prefer that it was theirs. But first, a contractor needs to understand a few basic rules when dealing with the media during emergencies.

Rule One: Know Your Company Policy

- Many contractors, for example, prefer to use certain words in times of trouble, and to avoid others. "Line break" sounds much better than "explosion." In any case, avoid words such as "disaster," and "catastrophe."
- Avoid discussion of any legal questions. All questions of liability should be referred to the company's attorney.
- Be sure of the facts. No guesses or speculation.
- Avoid giving personal opinions. They will be perceived as the views and policy of your company.
- Keep a brief form of the company's emergency procedures close at hand.

Rule Two: Know the Media

- Media representatives are professionals. Treat them with professional courtesy and respect, and they are likely to return the favor.
- Never patronize a reporter. The insult may be answered on the air or in print.
- Never lie.
- Do not make jokes.
- Speak simply and plainly.
- Never push the media away. Reporters may begin to get the idea the company is hiding something.
- Never deal in hypothetical situations. Do what any polished political candidate would do. Point out that the company cannot discuss something which has never happened.

Rule Three: Plan Ahead

- Plan ahead for any encounter with the media, especially in a crisis. Brush up on the facts. Review the company policy, work out a statement, and use notes, if necessary.
- Always assume that the conversation is being recorded, whether on the telephone or in person.
- Never go "off the record" with a reporter.
- Tell what positive steps the company is taking. In an emergency, for example, be prepared to tell exactly what the company's crews are doing to clear up the problem.

MODULE 44102-08 ◆ SAFETY 2.57

- Do not be afraid to state that the company does not know all the answers yet.
- If the answers are not known, do not guess, particularly when the company's representative is being asked about the cause of an accident or about injuries suffered in an accident.

Rule Four: Take Control

- Unless and until someone of higher authority arrives to handle official statements, the supervisors (or superintendents) are the only sources of information for the media. The important thing is to control the interview.
- Be helpful, to the extent possible. The company's helpfulness will reflect well when the story comes out.
- Know how to politely decline an interview. One reason could be that there is no time to talk. Another could be that there are others in the company who are more knowledgeable in the area being discussed.
- Know how to buy time. The contractor is within their rights to ask for a few minutes to collect some thoughts. When a reporter calls on the telephone, the company may explain that they are tied up at the moment. Promise to call back in a few minutes and then use the time to plan. Then be sure that the company calls back.
- Note where the interview is taking place. The company can exercise their right to be interviewed where they choose.
- Know how to end an interview. Determine ahead of time, if possible, how much time the reporter(s) will need.

Contractors need to realize that the press has a job to do and that is to get the story. They can get the company's story or someone else's opinion. Give them the facts as they are known to be correct. State the obvious. Stay within the guidelines of company policy and then take care of the accident scene.

Instructor's Notes:

2.58 PROJECT MANAGEMENT

Appendix F

Guidelines for Developing a Crisis Management Plan

Don't wait for a catastrophic event to occur to develop your crisis management plan. Failure to respond promptly and responsibly to an emergency may result in injuries to personnel, damage to property and the environment, business interruption, loss of customers, damage to the company's image and, in the worst case, failure of the firm.

A good manager knows that a well-managed safety and loss prevention program helps minimize the risk of catastrophic events but also understands that a crisis can occur and failure to respond quickly and correctly could have far-reaching consequences.

When a catastrophic event occurs, the affected business must have a well-prepared and rehearsed crisis management plan in place. The plan should be multi-layered to provide immediate response from onsite personnel and include provisions to expand response activities with off-duty and outside personnel on short notice. The plan should make the best use of all available resources including public and private sectors. It must be a functional plan that recognizes the need to protect the following:

- Health and welfare of people (employees and the public)
- Environment
- Property and equipment
- Continuity of operations (customers and stockholders)
- Company's reputation

The plan must address all types of emergencies that could reasonably be encountered and meet the requirements of appropriate regulatory agencies such as the State Police, OSHA, Mine Safety Health Administration (MSHA), Environmental Protection Agency (EPA), Department Of Transportation (DOT), United States Coast Guard (USCG), and the Department of Environmental Quality (DEQ). It should include an Incident Command Structure that can work well with on- and off-site personnel including the local emergency response planning commission.

A comprehensive plan anticipates and prepares for:

- Onsite response, shelter-in-place and evacuation
- Off-site notification and response
- Response to community needs
- Dealing with the press
- Incident investigation
- Post-accident clean up and restoration of operations
- On- and off-site claims handling including litigation
- Post-accident regulatory inquiries

The plan should include onsite response to:

- Injuries
- Chemical spills or releases
- Fires
- Explosions
- Severe weather
- Bomb threats
- Civil disturbances
- Labor disputes
- Security breaches
- Terrorism
- Transportation emergencies
- Workplace violence

The plan should include off-site notification of:

- Police, fire and ambulance
- Regulatory agencies and the LEPC
- National Emergency Response Center (for chemical spills)
- Company officials/employees

MODULE 44102-08 ◆ SAFETY 2.59

- Neighbors
- Next of kin
- News media
- Insurance companies
- Attorney
- Trauma counselors
- Emergency response contractors

The plan should anticipate response by:

- Police, fire and ambulance
- Local, state and federal agencies
- Media representatives
- Community leaders
- Employees and family members

The plan should anticipate investigations by one or more of the following groups:

- OSHA
- MSHA
- EPA and DEQ
- USCG
- Federal Railroad Administration (FRA)
- National Highway Transportation Safety Board (NHTSB)
- State Police/Law Enforcement
- Chemical Safety Board
- Plaintiff Attorneys
- Local District Attorney

The plan should establish the protocol and logistics for handling investigations with respect to:

- Retention of counsel
- Release of information
- Site access
- Photography
- Collection and preservation of evidence
- Chain of custody
- Duplication of documents
- Access to proprietary information
- Employee interviews
- Employee access to counsel
- Pre-citation settlement agreements
- Claims handling and management

All companies should regularly conduct a hazard evaluation and risk assessment of their operations to assure that controls are in place to adequately protect people, equipment, and the environment, as well as the assets of the company. Hazards should be eliminated or reduced to an acceptable level of risk. A crisis management plan should be in place to address unexpected emergencies.

Instructor's Notes:

2.60 PROJECT MANAGEMENT

MODULE 44102-08 — TEACHING TIPS

The following are suggested activities or instructional methods to help you teach the material in this module.

General

When you call on someone to answer a question, the rest of the class relaxes or even tunes out because they expect that the question and answer will take place only between you and the participant you called on. Instead, use this technique to involve more participants in answering questions and to keep them on their toes.

1. Ask participants to define a term or explain a concept.
2. After one participant has answered, ask a participant seated nearby if the answer is right. Then ask whether a participant in the back of the room agrees.
3. Ask participants to explain why they think an answer is right or wrong.
4. Use the session to clear up incorrect ideas and encourage participants to learn from their mistakes.

Sections 2.1.0 – 2.1.3 *Accident Costs*

This exercise will give participants the opportunity to recognize the direct and indirect costs associated with an accident. Participants will need their Participant Guides and pencil and paper to takes notes for discussion. The Instructor will need an overhead projector and screen, Transparency 5 (Figure 3, Accident Cost Worksheet), and a dry-erase marker so that the calculations for the case study may be presented to the class. Allow 20 to 30 minutes for this exercise.

1. Have participants read through the entire scenario once out loud (Section 2.1.3).
2. Display Transparency 5 (Figure 3, Accident Cost Worksheet).
3. Have participants read through the scenario out loud again, one paragraph at a time.
4. Ask participants to calculate the direct and indirect costs.
5. Fill out the Accident Cost Worksheet on the overhead projector.
6. Once the final costs are calculated, emphasize how one accident affects not only the injured worker, but also his co-workers, time and money for the job, EMRs, and future insurance rates. Also explain that while in reality some of the medical costs may have been paid for by the workers' compensation fund, ultimately the employer bears all costs directly or indirectly; so does most of the construction industry in the United States.

Section 2.3.0 *Causes of Accidents*

This exercise will familiarize participants with the causes of accidents. Participants will need their Participant Guides and pencil and paper to takes notes for discussion. Allow 20 minutes for this exercise.

1. Ask participants to list the four causes of accidents.
2. For each cause, ask participants to provide one example.
3. Ask participants to determine which two causes are involved in most accidents.
4. Review and discuss the participants' answers and examples.

Section 4.0.0 *Federal Regulations*

This exercise will familiarize participants with the four major sets of federal regulations that can affect the construction industry. Participants will need their Participant Guides and pencil and paper to takes notes for discussion. Allow 10 to 15 minutes for this exercise.

1. Ask participants the following questions and discuss their answers:
   - What regulations would apply to vehicles with a gross weight of more than 10,000 pounds?
   - What agency is authorized to investigate chemical incidents to determine the conditions and circumstances that may have lead to an incident, and to identify the causes to prevent similar incidents in the future?
   - What regulations would apply for incidents involving ground, air, and water pollution?
   - What administration conducts inspections of workplaces?
   - What agency may coordinate its fact-finding efforts with OSHA?

Section 4.1.0 *Employee Rights*

This exercise will familiarize participants with employee rights. Participants will need their Participant Guides and pencil and paper to takes notes for discussion. Instructors may obtain a copy of OSHA Poster (OSHA Form No. 3165) to show to class. Allow 20 minutes for this exercise.

1. As per the OSH Act, review and discuss the following employee rights with participants:
   - Employees have the right to notify their employer or OSHA about workplace hazards. Employees may ask OSHA to keep their name(s) confidential.
   - Employees have the right to request an OSHA inspection if they believe that there are unsafe and unhealthy conditions in the workplace. Employees or their representatives may participate in that inspection.
   - Employees can file a complaint with OSHA within 30 days of retaliation or discrimination by their employer for making safety and health complaints or for exercising their rights under the OSH Act.
   - Employees have the right to see OSHA citations issued to their employer. The employer must post the citation at or near the place of the alleged violations.
   - The employer must correct workplace hazards by the date indicated on the citation and must certify that these hazards have been reduced or eliminated.
   - Employees have the right to copies of their medical records and records of exposures to toxic and harmful substances or conditions.
   - The employer must post the Employee Rights Notice in the workplace.
   - Employees must comply with all occupational safety and health standards issued under the OSH Act that apply to their own actions and conduct on the job.
2. Ask participants if they have any questions regarding Employee Rights.
3. You may also review and discuss the following Employer Rights as per the OSH Act:
   - Employers must furnish employees a place of employment free from recognized hazards.
   - Employers must comply with the occupational safety and health standards issued under the OSH Act.

Section 6.2.3 *Job Analysis Selection*

This exercise will familiarize participants with the factors used to select jobs for analysis. Participants will need their Participant Guides and pencil and paper to takes notes for discussion. Allow 20 minutes for this exercise.

1. Ask participants to list the factors used to select jobs for analysis.
2. Review and discuss their answers.
3. Ask participants to identify the factor related to each of the following situations and discuss their answers:
   - Jobs that expose more than one person to potential hazards
   - Jobs that result in serious or disabling injuries
   - Jobs that may not have a history of accidents but may have the potential for severe injury
   - Jobs that repeatedly produce accidents
   - Jobs that were created by changes in equipment or processes and have no history of accidents because the accident potential may not yet be fully appreciated

Sections 6.2.0 – 6.2.10 *Job Safety Analysis*

This exercise will give participants the opportunity to use what they learned in these sections to write a mock job safety analysis. Participants will need their Participant Guides and pencil and paper to takes notes for discussion. The Instructor will need an overhead projector and screen, Transparency 21 (Figure 13, Job Hazard Analysis Form), and a dry-erase marker so that the exercise may be conducted as a class. Allow 40 to 50 minutes for this exercise.

1. Before you begin writing the JSA, choose the task you want to analyze. Select a task that is relatively simple, such as changing a drill bit or proper lifting procedures. Simpler tasks may be performed in the classroom and can be observed by participants. Choosing a more complex task would eliminate this step in the classroom exercise.
2. Once you have chosen a job task, have participants break down the job into successive steps or activities and discuss how the actions would be performed.
3. Correctly perform the task for participants so that they may observe the task being performed. Ask participants to take notes to record each step of the task.
4. Collect feedback from participants and compile a class list of steps.
5. Review the common errors listed in Section 6.2.5 and make sure the collective list does not include these errors.
6. Ask participants to list all the hazards and potential accidents that could occur throughout the job. Point out that only the problems that can be corrected or eliminated should be noted. Use the questions in Section 6.2.6 to help identify hazards.
7. Collectively write instructions for the JSA, reviewing Section 6.2.8.
8. Ask participants to develop recommended safe job procedures to prevent the occurrence of potential accidents.
9. Once the JSA is written, ask a participant to perform the task, step-by-step, using the JSA, noting the instruction, hazard, and procedure or protection for each. Make sure that no steps were missed or overlooked.

MODULE 44102-08 — ANSWERS TO REVIEW QUESTIONS

	Answer	Section
1.	Any four of the following: 1. Moral obligations 2. Legal obligations 3. Economic obligations 4. Direct costs 5. Indirect costs	(1.0.0)
2.	1. Identify hazards and potential risks 2. Evaluate the risk 3. Develop a plan of action 4. Set the proper attitude through leadership 5. Execute the plan 6. Audit compliance and evaluate performance 7. Make necessary adjustments 8. Investigate accidents and take corrective action 9. Manage claims 10. Close out the job and evaluate overall performance	(1.1.0)
3.	Any three of the following: 1. Unsafe acts 2. Unsafe conditions 3. Combinations of unsafe acts and conditions 4. Acts of God In terms of accident prevention, a greater focus of attention should be placed on the unsafe actions of people.	(2.3.0)
4.	Direct, or insured, costs include (list two): 1. Medical costs 2. Other workers' compensation insurance benefits 3. Liability and property damage insurance payments. Indirect (uninsured) or hidden costs can include (list two): 1. Property damage 2. Equipment damage 3. Production delays 4. Supervisory time 5. Retraining	(2.1.0)
5.	property damage = 20:1 minor injuries = 60:1 serious or disabling injuries = 600:1	(2.0.0)
6.	The cost of maintaining an effective safety program is estimated to be between 1% and 2.5 % of the direct labor costs.	(2.2.0)
7.	Unsafe acts	(2.3.0)

	Answer	Section
8.	1. Employees 2. Subcontractors 3. The general public 4. The building under construction 5. Building materials and equipment 6. Adjacent structures/exposures 7. The environment	(3.1.0)
9.	1. Greater than 50 people regularly exposed 2. From 10 to 49 people regularly exposed 3. From 5 to 9 people regularly exposed 4. Less than 5 people regularly exposed	(3.2.2)
10.	Entries on the form must be made no later than six working days after a recordable injury or illness has occurred.	(4.1.2)
11.	1. Struck-by accidents 2. Caught-in-between accidents 3. Electrocutions 4. Falls	(4.1.3)
12.	On multi-employer work sites, citations are typically issued to employers whose employees are exposed to hazards. Additionally, the employers will be cited, whether or not their own employees are exposed.	(4.3.1)
13.	The host employer's role at a multi-employer job site is to keep employees safe and the work site free from hazards.	(4.3.0)
14.	It is estimated that at least one out of five construction workers, or 20% of a company's workforce, has an alcohol or drug abuse problem.	(7.2.0)
15.	1. Be logical and enforceable 2. Be prepared and presented in terms that are easily understood 3. Specify employee safety responsibility, including to whom the rules apply, where they apply, and how they apply 4. Be communicated, understood, and enforced 5. Have penalties for violation that must be understood by all parties concerned 6. Have enforcement procedures that are fair, consistent and uniform	(7.5.0)
16.	It is estimated that employees who have been on the job 30 or fewer days account for 25% of all construction injuries.	(7.6.0)
17.	To evaluate a contractor's past safety performance, companies may ask for a contractor's workers' compensation EMR for the previous three years and their OSHA incidence rate for the same period.	(5.0.0)
18.	The foreman normally conducts accident investigations. The project manager should be made aware of all reported or suspected incidents as soon as possible. The project manager should also review every report for accuracy and completeness. He/she should be certain the causes of the incident have been properly identified and that the corrective action is appropriate and being implemented.	(7.12.0)

Answer	Section
19. Employee medical and exposure records must be kept for the duration of their employment.	(7.13.0—Figure 15)
20. Any of the following:	(Appendix E)

1. Avoid trouble words such as "holocaust," "disaster," and "catastrophe."
2. Avoid questions regarding legal issues.
3. Be sure of the facts—do not guess or speculate.
4. Avoid giving personal opinions, as they may be perceived as the views and policy of your company.
5. Keep a brief form of the company's emergency procedures close at hand.
6. Treat media representatives with professional courtesy and respect, and they are likely to return the favor.
7. Never patronize a reporter. The insult may be answered on the air or in print.
8. Do not lie or make jokes.
9. Speak simply and plainly.
10. Never deal in hypothetical situations. Point out that the company cannot discuss something which has never happened.
11. Plan ahead for any encounter with the media, especially in a crisis. Brush up on the facts.
12. Review the company policy, work out a statement, and use notes, if necessary.
13. Always assume that the conversation is being recorded, whether on the telephone or in person.
14. Never go "off the record" with a reporter.
15. Tell what positive steps the company is taking. In an emergency, for example, be prepared to tell exactly what the company's crews are doing to clear up the problem.
16. Unless and until someone of higher authority arrives to handle official statements, the supervisors (or superintendents) are the only sources of information for the media. The important thing is to control the interview.
17. Be helpful, to the extent possible. The company's helpfulness will reflect well when the story comes out.
18. Know how to politely decline an interview. One reason could be that there is no time to talk. Another could be that there are others in the company who are more knowledgeable in the area being discussed.
19. Know how to buy time. Contractors are within their rights to ask for a few minutes to collect some thoughts. When a reporter calls on the telephone, a company representative may explain that they are tied up at the moment. Promise to call back in a few minutes and then use the time to plan. Then be sure that the company calls back!
20. Note where the interview is taking place. The company representative can exercise his right to be interviewed where he chooses.
21. Know how to end an interview. Determine ahead of time, if possible, how much time the reporter(s) will need.

MODULE 44102-08 — ANSWERS TO TRADE TERMS QUIZ

1. Occupational Safety and Health Administration (OSHA)
2. Action level
3. accident
4. incidence rate
5. OSHA recordable incidence rate
6. Workers' compensation
7. experience modification rate (EMR)
8. personal protective equipment (PPE)
9. Worker exposure
10. risk assessment code
11. focused inspection program

CONTREN® LEARNING SERIES — USER UPDATE

NCCER makes every effort to keep these textbooks up-to-date and free of technical errors. We appreciate your help in this process. If you have an idea for improving this textbook, or if you find an error, a typographical mistake, or an inaccuracy in NCCER's Contren® textbooks, please write us, using this form or a photocopy. Be sure to include the exact module number, page number, a detailed description, and the correction, if applicable. Your input will be brought to the attention of the Technical Review Committee. Thank you for your assistance.

Instructors – If you found that additional materials were necessary in order to teach this module effectively, please let us know so that we may include them in the Equipment/Materials list in the Annotated Instructor's Guide.

Write: Product Development and Revision
National Center for Construction Education and Research
3600 NW 43rd St, Bldg G, Gainesville, FL 32606

Fax: 352-334-0932

E-mail: curriculum@nccer.org

Craft ______ Module Name ______

Copyright Date ______ Module Number ______ Page Number(s) ______

Description ______

______

______

______

(Optional) Correction ______

______

______

(Optional) Your Name and Address ______

______

______

Module 44103-08

Interpersonal Skills

NCCER STANDARDIZED CRAFT TRAINING PROGRAM

The National Center for Construction Education and Research (NCCER) provides a standardized national program of accredited craft training. Key features of the program include instructor certification, competency-based training, and performance testing. The program provides trainees, instructors, and companies with a standard form of recognition through a National Craft Training Registry. The program is described in full in the *Guidelines for Accreditation*, published by NCCER. For more information on standardized craft training, contact the NCCER by writing us at 3600 NW 43rd St., Bldg. G, Gainesville, FL 32606; calling 352-334-0911; or emailing info@nccer.org. More information may be found at our website, www.nccer.org.

HOW TO USE THIS ANNOTATED INSTRUCTOR'S GUIDE

Each page presents two sections of information. The larger section displays each page exactly as it appears in the Trainee Module. The narrow column ties suggested trainee and instructor actions to each page and provides icons (detailed below) to call your attention to material, safety, audiovisual, or testing requirements. The bottom of each page includes space for your notes.

The **Audiovisual** icon indicates an appropriate time to show a transparency or other audiovisual aid.

The **DVD** icon indicates an appropriate time to discuss or show segments of the video.

The **Classroom** icon prompts you to define a term, stress a point, ask trainees to explain a concept, or give examples.

The **Demonstration** icon directs you to show trainees how to perform tasks.

The **Examination** icon tells you to administer the written module examination.

The **Homework** icon is placed where you may wish to assign reading for the next class, assign a project, or advise trainees to prepare for an examination.

The **Laboratory** icon is used when trainees are to practice performing tasks.

The **Materials** icon is a reminder for you to gather materials needed for classes, labs, and testing.

The **Performance Testing** icon tells you to administer a performance test or a portion thereof.

The **Safety** icon is used to emphasize safety issues. It is often keyed to *Caution* and *Warning!* statements in the Trainee Module.

The **Teaching Tip** icon indicates additional guidance is available, such as how to conduct an exercise, get the most educational value from a field trip, or encourage class participation. Teaching Tips may expand on a feature (*Think About It, Did You Know?*) or provide *Quick Quizzes* or similar exercises. You will be referred to the Teaching Tips section at the back of the module if there is additional material.

The **Combination** icon indicates that the laboratory listed corresponds with a performance task. If desired, you can note the proficiency of the trainees during the laboratory, and use it to satisfy performance testing requirements.

PREPARATION

Before teaching this module, you should review the Objectives, Performance Tasks, Materials and Equipment List, and Module Outline. Be sure to allow ample time to prepare your own training or lesson plan and gather all required materials and equipment.

MODULE OVERVIEW

This module discusses the values and expectations of the workforce, building relationships, and satisfying stakeholders. It provides information on the principles of effective communication, how to apply the management grid, and using relationship skills to create a leadership environment. It describes behavioral interviewing and professional development of personnel.

PREREQUISITES

Please refer to the Course Map in the Participant Module. Prior to training with this module, it is recommended that the participant shall have successfully completed *Project Management,* Modules 44101-08 and 44102-08.

OBJECTIVES

Upon completion of this module, the participant will be able to do the following:

1. Briefly describe workforce expectations.
2. Describe how stakeholders are identified.
3. Define effective communication skills.
4. Apply human relations skills to the project management role.
5. Apply the Managerial Grid.
6. Define the leadership environment.
7. Describe mentoring and coaching.
8. Apply behavioral interview techniques.
9. Construct professional development plans.

MATERIALS AND EQUIPMENT LIST

Desktop or laptop computer or DVD player
Multimedia projector and screen
Minor Decisions: Major Impact – How to Deal with Real Issues in Project Management DVD
Transparencies
Blank acetate sheets
Overhead projector and screen
Transparency pens
Whiteboard/chalkboard
Markers/chalk
Pencils and scratch paper
Copies of Worksheets for Teaching Tips
Module Examination (located in the Test Booklet)

ADDITIONAL RESOURCES

This module is intended to present thorough resources for task training. The following reference works are suggested for both instructors and motivated participants interested in further study. These are optional materials for continued education rather than for task training.

Bennett, Jeffrey, "Diversity of Construction Workforce Presents Employee Benefits, Challenge," *The Business Review* (Albany), accessed at www.bizjournals.com/albany/stories/2004/07/26/focus7.html

Blake, R. and J. Mouton, *The Managerial Grid: The Key to Leadership Excellence.* Houston: Gulf Publishing Co., 1964

Construction Managers, *Occupational Outlook Handbook*, U.S. Bureau of Labor Statistics, accessed at stats.bls.gov/oco/ocos005.htm

Construction Managers – Job Description, accessed at www.careerplanner.com/Job-Descriptions/Construction-Managers.cfm

Employee Selection and Hiring Strategies – Get the Right People on Your Bus. Available at www.therainmakergroupinc.com/services/Item.asp?ID=23. Accessed Aug. 28, 2007

FMI Corporation, Raleigh, N.C.

Mochal, Tom, "Manage These Three Aspects of Change in Your Project," TechRepublic, June 4, 2007. Accessed at articles.techrepublic.com/5100-10878_11618851.html

Nesby, Tom, "Diversity: New Realities for Construction Companies," Nesby & Associates, *Seattle Daily Journal of Commerce*, www.djc.com/special/construct99/10050621.htm

Project Management Institute, www.pmi.org

Workforce Central Florida. Accessed at www.workforcecentralflorida.com

TEACHING TIME FOR THIS MODULE

An outline for use in developing your lesson plan is presented below. Note that each Roman numeral in the outline equates to one session of instruction. Each session has a suggested time period of 2½ hours. This includes 10 minutes at the beginning of each session for administrative tasks and one 10-minute break during the session. Approximately 12½ hours are suggested to cover *Interpersonal Skills*.

Topic	Planned Time
Session I. Values and Expectations, Stakeholders, and Communication Skills	
A. Values and Expectations of the Workforce	________
B. Stakeholders	________
C. Effective Communication Skills	________
Session II. Human Relationships, Part One	
A. Technical Knowledge and Skills	________
B. Clear Thinking Skills	________
C. Ethical Behavior	________
D. Management Style	________
Session III. Human Relationships, Part Two	
A. Change Management Skills	________
B. Conducting an Effective Meeting	________
Session IV. Human Relationships, Part Three; Leadership	
A. Team-Building	________
B. The Leadership Environment	________

Session V. Leadership, Building Relationships, Interview Techniques, and Professional Development Plans; Review and Testing

A. Building Relationships ____________

B. Behavioral Interview Techniques ____________

C. Professional Development Plans ____________

D. Review ____________

E. Module Examination ____________

1. Participants must score 70% or higher to receive recognition from NCCER.
2. Record the testing results on Craft Training Report Form 200, and submit the results to the Training Program Sponsor.

USING THE PM DVD WITH *INTERPERSONAL SKILLS*

Section 4.0.0 presents an opportunity to discuss whether or not Samantha is an effective communicator. Ask students to describe which skills made her an effective communicator and which rendered her ineffective.

Suggested response: Overall, Samantha was an effective communicator. She was able to listen to the workers and eventually solve the problem.

However, she yelled at, threatened, and placed blame on Keith, which led to an argument that could have been avoided had she attempted to speak with him in a calm, rational manner.

Samantha is an active listener. Throughout the scenario, she paid close attention to what the workers were saying and listened to all the facts before forming an opinion. She asked questions when necessary to clarify the message and made sure that the message was understood.

Section 5.4.0 presents an opportunity for students to describe the different management styles they think Samantha uses throughout the video. Ask the students where they would place Samantha on the Managerial Grid.

Suggested response: Most often, Samantha demonstrated characteristics of Processor, Juggler and Team Leader management styles.

Ask students to describe the kind of management style Samantha demonstrated during her conversation with Keith. Ask them how she could have handled her conversation differently.

Suggested response: Initially, Samantha demonstrated the Sarge management style when talking with Keith. She started her conversation with Keith by yelling at him because his sense of urgency in getting the job done was not apparent. Perhaps instead of yelling, she should have been more levelheaded when confronting Keith.

In assessing Samantha's management style, ask students which characteristics stand out and which does she use to her advantage. Ask them which characteristics and functions could she improve upon.

Suggested response: Samantha had the ability to see the issues between Keith and Anthony as a sign of legitimate concern. She showed reasonable concern for both tasks and people by pushing for productivity, but also trying to resolve the issues between Keith and Anthony. She was able to assess the situation in terms of immediate actions to be taken.

If Samantha had identified that poor housekeeping and missing tools were an issue prior to the argument between Keith and Anthony, she may have been able to meet her long-term goal of staying on schedule and finishing the project on time without extra labor.

(continued)

Section 6.0.0, The Leadership Environment, presents an opportunity to discuss power as it relates to management. Discuss with students the type of power Samantha uses and whether or not it works to resolve the problems.

Suggested response: Samantha used position power as project manager to tell Keith that production schedules, deadlines, and asset allocations were "her headache." She threatened Keith with "consequences" for his part in production delays.

Samantha also used personal power throughout the scenario by trying to combine the goals of the project with the work goals of the team members.

Section 6.4.0, Leadership and Motivation, provides an opportunity to rate Samantha as a leader and whether or not she knows how to motivate her team.

Suggested response: Samantha was an effective leader in that she was sensitive to the environment, while recognizing data and emotional stresses throughout the scenario. She searched for the root cause of frustration among her workers, but ultimately focused on the most important goal: getting the job done on time and within budget.

Discuss with students the motivational techniques that might have been used in the situation presented in the video and what effect it may have had on completing the work on time, improving relationships, and job safety.

Mentoring and coaching could be used to help Anthony overcome his laid-back attitude and Keith his anger. Team-building techniques could bring this group together and create a more positive, cooperative environment. While time is tight on this job, some activities could be designed to reinforce the necessity for collaboration and to build respect among the team members. Samantha should also reinforce the shared vision and strategies for the team's mission for this job. Going forward, she can devise routine team-building activities that will make future projects work more smoothly to bring the project in on time and on budget.

Project Management

44103-08

Interpersonal Skills

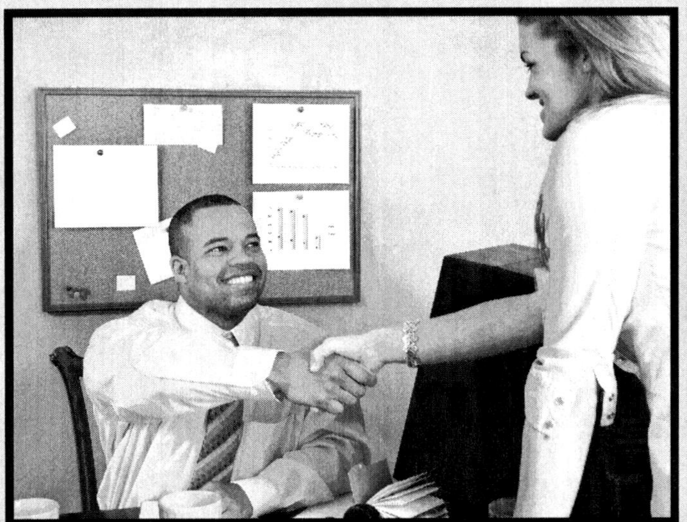

Assign reading of Module 44103-08.

44103-08
Interpersonal Skills

Topics to be presented in this module include:

Overview

The purpose of this module is to help project managers develop interpersonal skills by identifying the abilities and skills that they bring to the job, learning from their experiences, refining their talents for analyzing people and situations, and developing stronger communication skills.

Instructor's Notes:

Objectives

Upon completion of this module, you will be able to do the following:

1. Briefly describe workforce expectations.
2. Describe how stakeholders are identified.
3. Define effective communication skills.
4. Apply human relations skills to the project management role.
5. Apply the Managerial Grid.
6. Define the leadership environment.
7. Describe mentoring and coaching.
8. Apply behavioral interview techniques.
9. Construct professional development plans.

Trade Terms

Active listening
Behavioral interview techniques
Change
Ethics
Management style
Passive listening
Professional development plan (PDP)
Stakeholders
Two-way communication

Prerequisites

Before you begin this module, it is recommended that you successfully complete *Project Management,* Modules 44101-08 through 44102-08.

This course map shows all of the modules in the *Project Management* curriculum. The suggested training order begins at the bottom and proceeds up. Skill levels increase as you advance on the course map. The local Training Program Sponsor may adjust the training order.

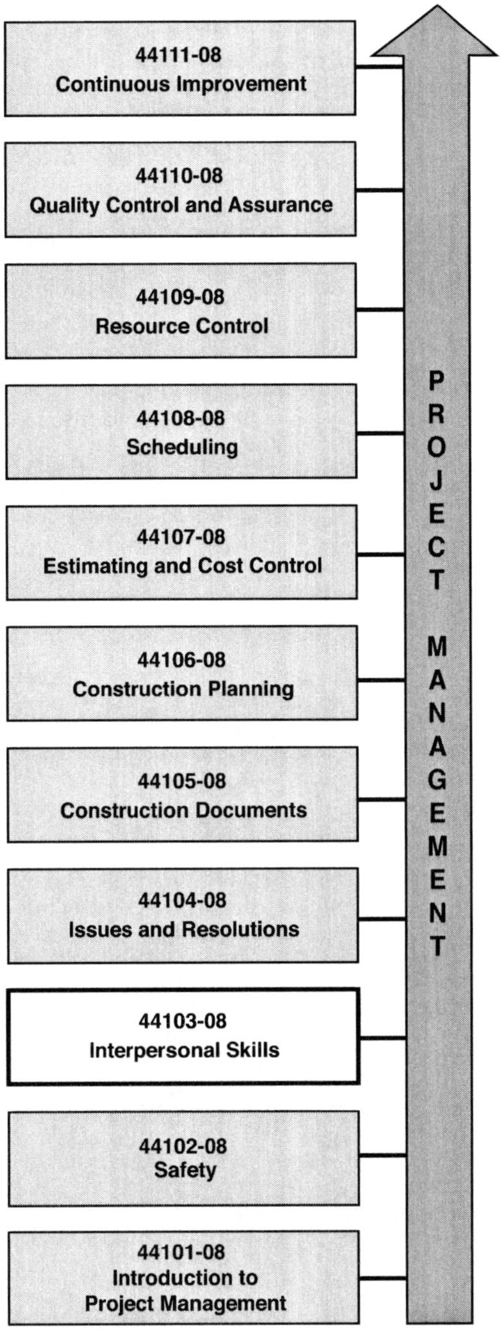

103CMAP.EPS

Ensure that you have everything required to teach the module. Check the Materials and Equipment list at the front of this module.

See the general Teaching Tip at the end of this module.

Show Transparency 1, Objectives. Review the goals of the module, and explain what will be expected of the participant.

Explain that terms shown in bold are defined in the Glossary at the back of this module.

MODULE 44103-08 ◆ INTERPERSONAL SKILLS 3.1

Explain that there are several skills common to successful project managers.

Ask participants to list these skills.

Discuss diversity in the construction industry (age, gender, ethnicity, etc.)

Ask participants to discuss the expectations of today's workers.

1.0.0 ◆ INTRODUCTION

One of the major concerns of the construction and maintenance industry is the selection and training of project managers. There is no one set of established attributes that portrays the ideal project manager. The reason is simple. The qualities a project manager needs for success in one organization might be deemed unnecessary in another. It is difficult to define the perfect project manager.

There are several sets of skills common to successful project managers. Among them are problem-solving skills and the ability to act effectively under a variety of conditions, direct the activities of people, and to achieve specific goals. This module will provide information about communication and human relations skills, the expectations of today's diverse workforce, and relationships with **stakeholders**. Also included are tips about leadership style, interview techniques, and professional development plans.

2.0.0 ◆ VALUES AND EXPECTATIONS OF THE WORKFORCE

During the past several years, the construction industry in the United States has experienced a trend in worker expectations and diversity. These two issues are converging at a rapid pace. At no time has there been such a generational merge in the workforce ranging from The Silent Generation (1925–1945), Baby Boomers (1946–1964), Gen X (1965–1979) and the Millennials (1980–2000). This trend, combined with diversity initiatives by the construction industry, has created a climate that is both exciting and challenging.

Today, many construction companies are aggressively seeking a broad base of candidates for the construction industry. To do this effectively, they are using their own resources as well as relying on associations with the government and other construction trade organizations.

All current research indicates that this industry will be more dependent on the critical skills of a diverse workforce—a workforce that is both culturally and ethnically fused. Across the United States, the construction industry is aggressively seeking to bring new workers into its ranks, specifically women and racial and ethnic minorities. Diversity is no longer solely driven by social and political issues, but by consumers who need hospitals, malls, bridges, power plants, refineries, and many other commercial and residential buildings.

The expectations of both generational and diverse workers pose a challenge in training skilled craftworkers and managers.

What are the expectations of today's workers? Undoubtedly, they include good wages, good working conditions, benefits, opportunities in professional development, and fairness and recognition. The following excerpt from an article by Jeffrey Bennett illustrates a few examples of the values and expectations of today's diverse workforce.

> Diversity in terms of age, education, gender, and ethnic background brings advantages to any business, but it also presents numerous challenges that must be addressed by growing companies. One challenge that immediately comes to mind is employee benefits.
>
> How does a construction company with limited resources develop a comprehensive benefit-plan strategy that is suitable for the entire workforce? Consider the 60-year-old male equipment operator who is nearing retirement and the 35-year-old female controller who is trying to figure out how to pay the daycare cost for her new baby.
>
> The equipment operator is probably not very interested in a cafeteria plan that offers pre-tax payment of dependent-care expenses, but he would like to learn more about post-retirement health care solutions.
>
> How about that 24-year-old laborer who is as healthy as a horse? Do you think he would like to benefit from the upside of his good health by participating in a health care savings account, which would allow him to carry forward cash for the future rather than giving it all to an insurance company?
>
> At the same time, I would expect the 45-year-old baby-boomer engineer with three kids wants the protection of a strong long-term disability plan. He probably also lies awake at night wondering if his investment choices in the 401(k) plan are the right choices to meet his savings goals.

A construction company cannot develop an internal diversity plan in a vacuum. To create and implement such a plan requires a detailed assessment of the people on its team and development of a business strategy that can be managed and implemented from an operational and cost perspective. Part of this strategy includes a means to communicate this vision and strategy to its workforce.

Many contractors are exploring creative approaches to make sure that their workforce and subcontractors represent the diversity within the communities they serve. They do not want to be perceived as a company that devalues the inclusion of minorities and women. Tom Nesby of

Instructor's Notes:

3.2 PROJECT MANAGEMENT

Nesby and Associates identifies how to set up a program using eight steps to increasing diversity in a company, designed to identify minority and women contractors and suppliers. His recommendations are presented here:

1. Have senior management issue a statement of commitment to diversity, including a policy outlining the company's strategic intent, with key goals and responsibilities.
2. Set a budget for building a Minority and Women-Owned Business Enterprise (MWBE) contractor and supplier development program.
3. Publicize the program within your company and in the community.
4. Form alliances and build relationships with community-based organizations that focus on business development for women and people of color.
5. Provide supplier and contractor diversity education to all employees. Include stakeholder teams (general contractors, sub-contractors, and MWBEs) in the education. Hire a professional consultant to assist in the development and to roll out the educational strategy.
6. Network with minority and women associations to find qualified women and people of color as candidates for employment and contract opportunities. Some progressive organizations include Black Contractors Association, National Association of Women in Construction, Black and Hispanic chambers of commerce, and state commissions for minority affairs.
7. Make diversity a criterion in all business decisions.
8. Track and monitor your challenges and successes.

More construction companies are committed to diversity because they realize there are tangible benefits that a broad labor pool can provide to meet the demand of their customers.

3.0.0 ◆ STAKEHOLDERS

Stakeholders are not just people with a financial stake in a construction project. In fact, they can be individuals, organizations, or government agencies that will influence many aspects of the project. This can include money, schedule, design, and other factors.

Today's construction firms need to know who the critical stakeholders are and they often need to factor their opinions and influence into the project design. In a traditional project, the high-level team may involve the owner or owners, the architect, and the general contractor. Historically, most interaction was between the general contractor and the architect along with subcontractors for both. The owners had a vision of the end result and this trickled down to the project team.

Today, the traditional model is still valid but there are many players who go beyond the inner circle of the project team. For example, there may be several owners, and depending on the complexity of the project, owners may bring in non-traditional stakeholders in contractual advisory roles. These could range from the financial officer, investors and analysts, architectural design firms, owners' project representative, and others who have financial or emotional investment.

A simple analogy can offer a vivid illustration. If a project is an archery target, the bull's eye is occupied by the core group, including the project team and influential stakeholders. The next larger ring represents other stakeholders like customers, third-party vendors, or others with a vested interest in the project. The next larger ring after that represents others like local or federal government agencies, pressure groups, and potential customers.

This analogy shows that the project scope will involve many stakeholders with different opinions, motivations, and varied viewpoints. All of them, along with the owners, influence the project design, and ultimately the construction. Project managers need to be aware that they will have to communicate at some time with various stakeholders.

The stakeholders are also those who are affected by—or may affect—the project. Their perspectives need to be respected and taken into account in order for a project to be successful. Stakeholders can have positive or negative views regarding a given project, and often don't agree with one another, making it a challenge to reconcile their varied viewpoints.

3.1.0 Gathering Information from Stakeholders

To ensure that you are responsive to all stakeholder needs, it's important to gather information and consult with them on aspects of the project that affect their interests. If possible, all stakeholders—internal and external—should be included in the process. Information gathering can be formal or informal. Meetings and discussions are informal ways to collect stakeholder information. A formal approach may be required to identify specific information about the project. One efficient approach is to prepare a list, such as the excerpt from the pre-bid interview list in *Figure 1*.

Review and discuss recommendations for increasing diversity in a company.

Define stakeholders.

Explain that with many stakeholders there will be many different opinions, motivations, and viewpoints that influence project design and construction.

Emphasize that project managers will have to communicate with various stakeholders.

Show Transparency 2 (Figure 1). Explain that preparing a list of questions is a good way to gather information from stakeholders.

See the Teaching Tip for Section 4.0.0 at the end of the module. Perform this exercise before beginning Section 4.0.0.

Ask participants to list three common communication strategies.

Ask participants to list different ways that project managers may use two-way communication.

Point out the risk of messages being misinterpreted.

Emphasize that regardless of the communication method, feedback is necessary to ensure the message is understood.

Show Transparencies 3 and 4 (Figures 2 and 3). Explain how communication works when done properly, and how interference can affect a message.

Ask students to determine if Samantha is an effective communicator and have them provide examples.

PRE-BID INTERVIEW QUESTIONS

Safety:

- How does the owner describe the firm's safety culture?
- Discuss site-specific safety issues, such as lockout/tagout and permit procedures.
- Does owner have a behavioral-based safety program?
- What are the recordable injury rate, lost time rate, and E.M.R?
- Contractor intends to fully implement its safety process (discuss details).
- Confirm that all safety supplies will be provided by owner.

Medical Surveillance
(what may present a hazard)

- Asbestos?
- Lead?
- Pathogenic organisms?
- Food waste?
- Hazardous waste?

Welding (do we need certified welders?)
If so:

- Who pays for failed tests and retests?
- Who pays for testing service & coupons?
- Who pays for time (wages)?
- Does owner accept current certification?

Tools/Supplies/Consumables (who provides?)

- E/Os will provide their own craft tool boxes.
- Small tools will be provided by owner (confirm).
- Consumables will be provided by owner (confirm).
- Specialty tools (owner's or contractor's).
- Scaffolding, etc. (owner's or contractor's).

103F01.EPS

Figure 1 ◆ Pre-bid interview checklist.

4.0.0 ◆ EFFECTIVE COMMUNICATION SKILLS

The project manager spends a great deal of time interacting with people during a construction project. Often during these interactions, problems stem from poor communication. It is important to recognize the three common communication strategies that you can use to deliver information: **two-way communication**, identifying distractions and noise, and **active** and **passive listening**. You will need to recognize subtle forms of communication, including verbal and non-verbal cues. You will need to communicate clearly in verbal and written forms and by using body language. And you will need to communicate with and gather information from all stakeholders in the project.

4.1.0 Two-Way Communication

Two-way communication is not just communicating directly to a person or a group. Depending on the circumstances, project managers will communicate using different modes. For example, they will often communicate face-to-face, on the telephone, through email or instant messaging, in written correspondence, or by two-way radio. Most of these communication methods enable people to send information efficiently and effectively, although they also may increase the chances of messages being misinterpreted.

To make sure your message is understood, try to get feedback. For example, request a reply to an email or ask for a return call with questions or concerns. There are always going to be situations in which your communication was not understood, but following up can reduce such problems.

4.2.0 Identifying Distractions And Noise

For someone to understand a message, there cannot be any interference that prevents or distorts the meaning of the message. Communication normally works when a sender (the project manager) delivers a message to a receiver (a subcontractor).

In *Figure 2*, the project manager communicates a message to the subcontractor, who then responds appropriately to the project manager. A successful communication cycle occurs when the subcontractor receives the message, decodes it and then responds.

However, when there is interference, the message coming to the receiver is not clear (*Figure 3*). Therefore, the response going back to the project manager is also not clear.

Interference is created by distractions and noise. For example, the conversation might be taking place on a job site while a crane is being positioned and your cell phone is ringing. Noise refers to anything introduced into the message that is not included in it by the sender, such as sounds in the environment. Noise can also come from ambiguity, cultural differences, and the receiver's psychological state.

Instructor's Notes:

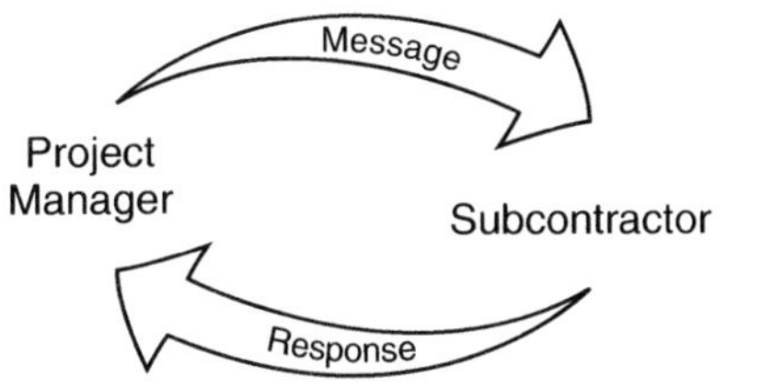

103F02.EPS

Figure 2 ◆ A typical communication cycle.

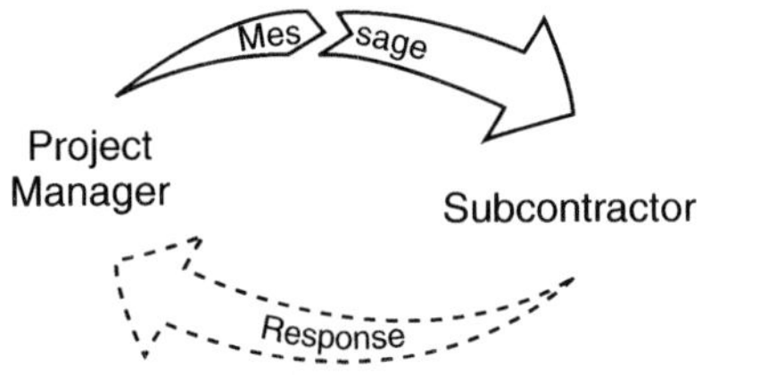

103F03.EPS

Figure 3 ◆ A typical communication cycle with interference.

4.3.0 Active and Passive Listening

Active listening, a vital part of good communication, involves reacting to the speaker using verbal and non-verbal cues to indicate that the message is getting through. Observe your own reactions when you listen to a message. To show you are paying attention, you may nod your head, look at the speaker, repeat, restate, or paraphrase what you heard.

Here are some tips to help you be an active listener.

- Pay close attention to what people are talking about. Devote your full attention to the speaker.
- Don't be too quick to judge. Listen to all the facts before forming an opinion.
- Where appropriate, ask questions or, in your own words, restate the facts or concepts you heard. And ask the speaker if you are correct in how you understood the message.
- Internalize what has been said by making a mental review as well as a few notes about what was presented.

Passive listening is mechanical and effortless. If workers are awake and halfway paying attention, they will choose to hear what is being presented and come away with some understanding of major and minor points. Detecting passive listening can be challenging. Some cues that may indicate a passive audience are facial expressions (frowning, smirking, or even occasional eye-rolling) may indicate that people may be listening and reacting to some but not all parts of the message. Other forms of body language such as tapping, fidgeting, head tilting, and folded arms are nonverbal cues to observe. However, keep in mind that those perceived as passive listeners may not be directly involved with the issue being discussed, so be sure to know the audience being addressed.

5.0.0 ◆ HUMAN RELATIONS

Human relations skills become more essential as a manager moves up in an organization. Effective project managers rely on quality communication on the job site, in the office, and among and between the general contractor, the subcontractors, and the owner. The following areas may need development:

- Technical knowledge and skills
- Clear thinking skills
- Ethical behavior
- **Management style**
- Skills in managing **change**
- Skills for conducting effective meetings
- Team-building skills

5.1.0 Technical Knowledge and Skills

Technical knowledge and skills are important and provide many opportunities for advancement in the construction industry. Generally, however, technical skill becomes less important as a person moves up the management ladder.

5.2.0 Clear Thinking Skills

The ability to think clearly, on the other hand, becomes more important as a person assumes greater management responsibilities. Successful project managers coordinate resources (human, functional, and technical), act on shifting priorities, and find solutions to problems. These abilities demand mental flexibility and sound judgment.

5.3.0 Ethical Behavior

The mark of a successful project manager is a strong sense of **ethics**. Ethics and responsibility are indistinguishable when it involves managing a construction project. As a project manager, you

Review and discuss tips to becoming a better active listener.

Define passive listening.

Ask participants to list cues that indicate a passive listener.

Assign reading of Module Sections 5.0.0–5.6.2.

Point out that human relations skills become more essential as a manager advances in an organization.

Note that technical skills tend to become less important as a person moves up the managerial ladder, and clear thinking skills become more important.

Review and discuss clear thinking skills.

MODULE 44103-08 ◆ INTERPERSONAL SKILLS 3.5

Review and discuss the factors of the PMI code of ethics.

Explain that a successful leader balances concern for tasks with concern for people.

Review and discuss the four factors that shape management style.

Show Transparency 5 (Figure 4). Explain that the grid is the visual framework for understanding the eight behavioral management styles.

Review and discuss the eight managerial styles.

Discuss the issues that a project manager should communicate to the team to help them understand how changes may affect them.

Show Transparency 6 (Figure 5). Explain how these tips help to manage change.

Explain that when change occurs in one part of an organization it will affect other parts as well.

are the person who must plan, implement, and track all activities. The Project Management Institute (PMI) has established a code of ethics built around the following:

- Responsibility and accountability
- Reporting unethical or illegal conduct
- Promoting an environment of respect
- Fairness and an awareness of conflicts of interest
- Not promoting favoritism or discrimination

5.4.0 Management Style

One of the project manager's most important responsibilities is managing people. Project managers need to influence those around them so that their team members can achieve company and personal goals. This responsibility demands exceptional management and leadership skills.

There are several management models that are useful in developing and assessing management and leadership skills. One classic model is the Managerial Grid created by Robert Blake and Jane Mouton. The grid looks at a manager's role and the concern that the manager should demonstrate for tasks and people.

A project manager demonstrates concern for a task by clearly defining a job, setting a critical path, tracking material, and spending labor hours effectively. In other words, concern for a task takes the form of controlling the job from start to finish and directing the efforts of team members to perform specific tasks, in specific ways, and at specific times.

The project manager demonstrates concern for people by devoting time to discuss the job thoroughly with supervisors, providing support and direction throughout the project, and lending an attentive and responsive ear when problems arise. Concern for people means caring about each individual's concerns, both personal and professional.

The secret to successful leadership is to balance these two types of concern. The project manager must develop a management style that is both goal-oriented and people-oriented.

The term management style refers to the manner in which a project manager handles situations, often stressful ones, and is shaped by at least four factors:

- The job or task being managed
- The beliefs and practices of the employees
- The values and personality of the manager
- The culture of the organization

For example, some jobs require more task-oriented behavior than others. Some employees need explicit instructions and follow-up, while others do not. Some managers believe that employees must be tightly controlled, while others believe their mission is to help their employees develop their skills and talents. Still others feel that company policies and procedures are the law and that fairness is best served by managing by the book. The Managerial Grid shown in *Figure 4* is a visual framework for understanding eight behavioral management styles that reflect levels of concern for people and tasks. Refer to the *Did You Know* for a detailed explanation of the management styles illustrated in the grid.

5.5.0 Change Management Skills

Today, construction and maintenance businesses are undergoing massive changes. Customer demands, new legislation, and company and employee needs affect how you do business. Change is constant, so it is essential to manage it effectively in order to provide superior products and services to your customers.

Project managers are more likely to succeed when the people involved or affected by changes understand what to expect. Elements you should communicate to your team are:

- Goals
- How the goals will be met
- Rewards connected with the change

In *Figure 5* you will find a checklist and tips that will help you with the process of managing change.

Project managers need to ensure that those involved or affected by a change have a clear picture of what will be different. Project managers and the participants in the change need to prepare for emotional shifts that may occur in themselves or others during the course of the work. Emotions affect how those involved accept the change. They usually occur in a predictable sequence that is called emotional sequencing shifts.

Many believe negative reactions are only associated with imposed or unavoidable change. Even when change is voluntary or seen as helpful or useful, people will react emotionally. Do not assume that desired change is immune to resistance or implementation problems; any stressful situation, including even the most welcome changes, brings with it some doubt and frustration. It is your job as the leader of the change to overcome these barriers. Change, positive or negative, in one part of an organization will affect other parts of the organization.

Instructor's Notes:

3.6 PROJECT MANAGEMENT

DID YOU KNOW? Assessing a Project Manager's Style

The Managerial Grid is designed to help project managers identify the management style and methods they normally use and to revise their styles as needed. The grid is a two-dimensional graph divided into quadrants by a horizontal scale and a vertical scale. The horizontal scale represents the project manager's task-orientation; the vertical scale represents people-orientation. Both scales are graduated, the numbers run from 1 to 9, with 1 being the lowest value on each scale, and 9 the highest.

Bull of the Woods

Managers who use this style will most likely respond to questions about their decisions or methods with the response: "Do it my way or hit the highway!" Punishment appears to be the only form of motivation they know. The Bull of the Woods approach achieves some degree of success, but the success is usually short-lived. Sabotage, payback, and goofing off when the Bull is not around are the most common ways employees get revenge on this type of individual.

The Sarge

This type is also highly task-oriented and generally will ask, tell, and then yell if a proper sense of urgency in getting a job done is not apparent. The Sarge uses punishment and reward as motivational tools. This style of management is effective with new employees and trainees, but seasoned craftsworkers do not usually respond positively to this style.

The Abandoner

This style demonstrates both a low concern for tasks and people. The Abandoner would rather jump ship than rock the boat. When faced with conflict, Abandoners abdicate responsibility. They often use statements like, "I don't care how you run the pipe, but the architect said..." or "I don't agree with the office either, but we have to follow what they said."

The Bureaucrat

This manager follows the book. If the schedule indicates a specific task this week, then do it that week. Bureaucrats are fair and consistent with all employees and seldom play favorites. Their style is an effective form of management because it is fair, consistent and predictable. However, this inflexibility may lead them into the trap of following a procedure or a schedule blindly without paying proper attention to circumstances.

The Sweetheart

Sweethearts have a high concern for relationships with people but a low concern for tasks. They want their employees to like them but seldom earn employee respect. If asked how they want a job done or why, they try to portray doing the task as something the employees should do as a personal favor to them—"Hey, buddy, do this for me, will ya?"—or worse yet, sidesteps the question with a joke. Employees and peers usually pay back Sweethearts by manipulating them into doing jobs that they do not want or like to do.

The Processor

This is an effective style. Processors generally see a question or a complaint for what it is—a sign of legitimate concern. If they see an apprentice not pulling his weight, they might recognize that the apprentice needs training. If they see a foreman falling behind schedule, they would look for the root cause behind the problem. Although the Processor is effective, this style is not always appropriate. For instance, if the foreman's problem is personal, the foreman might not want to discuss it with the project manager. Probing by the Processor could lead to hard feelings.

The Juggler

The Juggler shows a reasonable concern for both tasks and people. This is the most common style of management in the construction industry. This style pushes for productivity but is also concerned with employee feelings. The Juggler generally assesses a situation in terms of what needs to be done immediately. Managers with this style frequently rely on snap decisions that are effective only if most of those decisions are correct. If they make too many wrong decisions, their employees perceive them as manipulators and withhold their trust. There is a danger, too, that the flexibility they demonstrate will be seen as inconsistency or unfairness; as a result, the Juggler often achieves short-term goals but fails to achieve long-term ones.

The Team Leader

The Team Leader shows an extremely high level of concern for both people and tasks. They are effective when the jobs they manage require continuous communication and interaction, when there is time available to talk things over, and when there is a great deal of trust on the job and within the company.

Ask students to assess Samantha's management styles throughout the video. Then ask them which style she used the most with Keith.

Discuss with students which styles Samantha uses the best and which could be improved upon.

MODULE 44103-08 ◆ INTERPERSONAL SKILLS 3.7

Review and discuss the reaction to change example.

Review and discuss the three statements that are critical to the successful implementation of change.

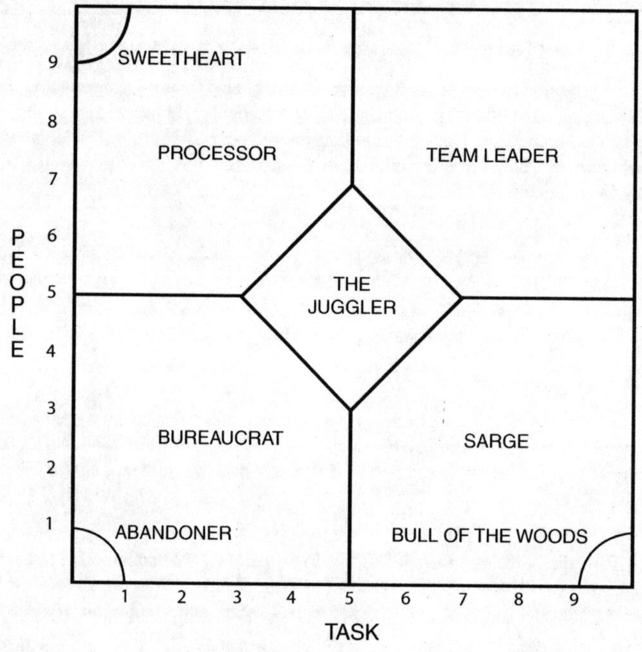

Figure 4 ◆ Managerial grid.

CHECKLIST FOR MANAGING CHANGE PROCESSES

___ Provide participation

___ Provide enough time

___ Start small

___ Avoid surprises

___ Don't carry excess baggage

___ Work with recognized leadership

___ Treat people with dignity

___ Reverse the positions

103F05.EPS

Figure 5 ◆ Change process checklist.

One example of reaction to change might be that of a company purchasing a new telephone system. Before it is implemented, you may hear remarks like, "Great! You mean I'll be able to talk to Joe in Estimating and Mary in Accounting at the same time? That's what we've always needed." During implementation, however, you are likely to hear remarks like, "It's too complicated and I don't understand it. I don't want to use it. In fact, I'm not going to use it, so how do I get my old phone back?" After the change process has taken place and individuals adapt to the new system, you may hear remarks like, "I don't see how we got along without it. It's so easy to use. I'm very satisfied."

In most cases, the participants' old behaviors and ideas were replaced by new ones and they adapted to the new system. Feelings and attitudes of involved personnel are important during the transition from the enthusiasm for success in the early stages of the change to satisfaction upon completion.

The following three statements are critical to the successful implementation of change:

- The level of positive feelings concerning a change is directly related to a person's expectations of what will be involved.

Instructor's Notes:

3.8 PROJECT MANAGEMENT

- The more a person learns about what is involved in the changes, the more pessimistic this person may become about his or her ability or willingness to accomplish the task.
- The level of pessimism or optimism about a change is relative to the information that is available about the new requirements of the individual or the group due to the change.

5.5.1 Change and Emotional Sequencing Shifts

People affected by change usually go through emotional sequencing shifts. Each shift is the result of adaptations made by the individuals affected. Change is an ongoing process and may proceed slowly or rapidly. These effects may vary in the length of time they affect the people involved.

- *Shift 1: Innocence* – People feel good about the changes at hand. This is normal for most tasks, since very few would get off the ground without participant support. This is usually a period of high hopes. The change appears to be perfectly designed, and the individuals involved are ready for anything; they expect the change to proceed smoothly. This is not likely, however, since most of the initial conclusions are made without enough information. It is almost inevitable, in fact, that you will encounter unanticipated difficulties.
- *Shift 2: Skepticism* – After the program is developed, and more and more difficulties surface, morale may drop. Few are prepared for unplanned events, inadequate or slanted data, lack of resources, or political resistance. This is the shift where critical problems are identified. Few solutions exist at this point, and the changes may seem impossible or unrealistic. People involved with the process are now faced with disbelief, as they realize just how hard it is to implement the program. At this stage they are often not sure if they will succeed. This is the critical stage for most change processes—a time when the individuals encounter barriers that often cause increased doubt and pessimism.

  At this point of the process, the change leader needs to watch for signs of lost commitment and a withdrawal of personal investment by the participants. It's normal for negative feelings to surface, and even in some ways desirable, since it allows us to examine the situation more closely. However, this may cause an atmosphere of extreme caution and concern. For change to be successful, the leader must confront the negative responses to the problems occurring during implementation.

  Participants now have a greater level of knowledge about the change and are more aware of the barriers to change. If they perceive these barriers as being too great to overcome, they may not be committed to the process and withdraw. If a change leader is not careful, participants may abandon the task, leaving the leader with no personnel, little support, fewer assets, and ultimately no work. People abandon tasks both overtly and clandestinely. Both types of behavior reflect a high level of pessimism and a low level of commitment.

  Overt behavior is easily observed by other participants in the change. People involved in the change outwardly object and refuse to participate. Clandestine behavior is not easily discerned by other participants in the change, as these attitudes are not expressed as openly. People involved may go through the motions expected, and appear to be participating fully, but are in reality participating in limited fashion.

  Both forms of withdrawal are dangerous to the welfare of change work, but clandestine withdrawal is more likely to seriously damage a project's likelihood for success. Public resistance to change is easily recognizable, and measures can be taken to directly manage it. Clandestine resistance, on the other hand, is rarely acknowledged externally and is difficult to deal with directly. The following examples of a clandestine resister's thinking may convey a better understanding of the factors involved:

  – *Secret:* "In this company, you just don't tell management that their pet idea is causing problems."
  – *Unconcerned:* "I've tried to talk about our problem concerning the changes, but they aren't interested in how we feel."
  – *Projection:* "There's no way this will work, and our project manager knows it, but he's afraid if he says anything, he won't get a promotion."
  – *Apathetic:* "I really want this change to succeed, but I just don't have the time to get involved."
  – *Unable:* "I'm not sure I can do this. Every time I try something it goes wrong. I wish I could leave without anyone thinking I failed."
  – *Herd mentality:* "I wonder if everyone in the company is as confused as I am? This approach just doesn't make sense, but I don't want to make waves or appear uninterested."

Explain that emotional shifts are adaptations made by the individuals affected.

Discuss the first emotional shift, innocence. Explain that initially people usually feel good about change.

Discuss the second emotional shift, skepticism. Explain that morale may drop as implementation of the change progresses and difficulties surface.

Review examples of clandestine resisters. Note that this form of behavior is rarely acknowledged and is difficult to deal with directly.

MODULE 44103-08 ◆ INTERPERSONAL SKILLS 3.9

Review and discuss the third emotional shift, awareness. Explain that if skeptics are confronted successfully, those involved may begin to view change more positively.

Review and discuss the fourth emotional shift, wisdom. Explain that as people become more confident in themselves and the change, they become more knowledgeable.

Review and discuss the fifth emotional shift, mastery. Explain that when the change is completed, the outcome might be different from what is expected during the first stage.

Ask participants to list the three aspects of change that can occur on a project.

Review and discuss scope change, configuration change, and all other changes.

– *Image:* "It's final! I'm not going to do anything else for this system. Everything I do gets put down—I'm sick of it!"

Since these ideas are rarely if ever shared with others, individuals who privately abandon a change will usually appear to be involved.

In some situations, and given certain circumstances, abandonment is one alternative. However, since the aim is for the project to be successfully completed, public and private abandonment must be avoided. If it occurs among key players in the process, the change's success becomes less likely.

- *Shift 3: Awareness* – If the concerns of the skeptics are confronted successfully, the individuals involved with the change will usually begin to view it in a different light. Pessimism declines, and participants begin to understand the situation. Participants realize there are barriers to overcome, but they now fully understand what they face and are aware that it's possible to succeed. The participants move from pushing against the immovable barriers to flowing with the achievements. The perception of the change has moved from being merely possible to probable, if not certain. Problems have not disappeared, but the participants now know what they are facing, and they know they can deal with it. They have survived the past and feel confident that they can handle the future.
- *Shift 4: Wisdom* – As the group becomes more confident in itself and with the change, the individuals involved become knowledgeable. Despite the problems they face, they are certain they will succeed. The light at the end of the tunnel is evident, and new challenges are met with strong efforts.
- *Shift 5: Mastery* – When the change is completed, the outcome might be different from what is expected during the first stage. Each effect of the change sequence has altered the way the participants perceive their work and themselves. This is a good time to review "lessons learned."

5.5.2 Critical Change

Tom Mochal, president of TenStep, Inc., a project management methodology company, writes about critical change and how to manage aspects of change in a project.

You can't predict every change that will happen in your project, but you can manage them as they occur.

It's said that the only constant in the world is change. You can make a great plan, but you can't account for every potential change that may occur. The longer your project is, the more likely you'll be dealing with changes. Since you can't predict every change, the best that you can do is that manage the changes that do happen. There are three aspects of change that can occur on a project.

- *Scope change* – This is the most important change to manage. Scope is defined at two levels—high-level scope describes the boundaries of the project and the major deliverables to be built; low-level scope is defined through your approved requirements.

  The purpose of scope change management is to identify change and manage it effectively. It also protects the project team from changes made after the commitments to schedule and budget are agreed to. In other words, the project team committed to a deadline and budget based on the high-level and detailed scope definition. If the deliverables change during the project (and usually this means that the client wants additional items), the estimates for cost, effort, and duration may no longer be valid. If the sponsor agrees to include the new work into the project scope, the project manager has the right to expect that the current budget and deadline will be modified (usually increased) to reflect this additional work. This process ultimately brings the appropriate information to the project sponsor and allows the sponsor to decide if the modification should be approved based on the business value and the impact to the project in terms of cost and schedule.
- *Configuration change* – Configuration management is the identification, tracking and managing of all the assets of a project, and the characteristics (metadata) of the assets. (In some organizations this process is more narrowly defined to mean only the management of the physical assets.) In most projects, configuration management doesn't come into play. However if your project uses or builds a large number of components, parts, artifacts, equipment, and so on, configuration change management could be very important.
- *All other changes* – Your project may experience changes that don't necessarily fall under scope change management or configuration management. These changes can be grouped into a general change management category. For instance, let's say one of your team members leaves and needs to be replaced. This would not be an example of scope change or configuration change. It's a general change. In this case, you may need to document the fact that a resource change occurred, determine the impact of the change, and put a plan in place to manage the change. In many respects,

Instructor's Notes:

you'll follow a similar process to that of a scope change request.

One of the key differences between general change management and scope change management is that you expect that if a scope change is requested and approved, you will change your budget and schedule to accommodate the change. You should not have that same expectation for non-scope related changes. For example, in the example above when a team member needed to be replaced, there was definitively a change, and there will probably be an impact on the project. However, there is no expectation that this change will result in an approved schedule or budget change.

As a project manager, you should focus on making sure that you manage scope change, which is a primary culprit when projects have problems. However, you should be aware that your project may need to have a configuration management and general change management process as well. Managing all these aspects of change may save you countless hours of trouble.

5.6.0 Conducting Effective Meetings

Thousands of dollars a year are spent on meetings, yet too many meeting hours are unproductive; while most meetings are held to solve problems, few do. The purpose of this section is to help you make the meetings you conduct more productive.

5.6.1 Meeting Preparation

As with other management functions in the construction and maintenance industry, planning is the key to success. Your first responsibility is to plan. Before scheduling a meeting, ask yourself the following questions:

- Is there another method to accomplish my objectives?
- Would a telephone call meet my needs?
- Is there a report that contains the information I need?
- Is this matter really crucial?
- What is the purpose of this meeting?
- What are my intended outcomes?

If the answers to the previous questions indicate a need for a meeting, begin your plan using the following steps:

Step 1 Define your outcome. Your outcome is what you want to achieve by the end of a meeting. Each meeting has its own outcome, such as updating the schedule or closing out change orders.

Step 2 Select the topics to cover. Select topics that will help you achieve your outcome. It is a mistake to include more topics just because everyone is gathered together. By trying to cover too much in one meeting, you may end up missing your outcome altogether. Confine your meeting to topics that will help you achieve your goal. If there are additional issues to cover, then do so at a subsequent meeting or with another method.

Step 3 Identify attendees. Attendees should be invited so they can share the vital information that they have to contribute. Attendees should have the authority to make the necessary commitments, have ownership of the subjects to be covered, or because they have support that you need. Do not force them to come if they don't have a contributing role. Send them copies of the minutes instead.

Step 4 Determine the time and location. Select a time and location to be determined by the attendees' convenience, pressing needs, and your schedule. If Monday morning meetings are usually poorly attended, don't schedule one at that time. Ask the attendees for a convenient time and place. Don't hesitate to schedule a meeting off the job site or out of the office, especially if you are likely to have in-house interruptions.

Step 5 Prepare an agenda. An agenda is a valuable tool. In addition to stating the time and place of the meeting, the topics to be covered, and the time allocated for each topic, an agenda can give participants a preview of the desired outcome of the meeting and remind them about information they should bring with them.

Step 6 Distribute the agenda to the attendees. Let the attendees know the agenda well in advance, so they can come to the meeting prepared. Distributing the agenda will help the attendees gather any information they need, get signed off, and obtain instructions from their superiors. Any feedback may alert you to problems your agenda might create or opportunities for you to explore.

5.6.2 Conducting the Meeting

Set the tone and tempo for the meeting from the start. If you begin late, the meeting will run

Explain why most meetings are unproductive.

Review and discuss the questions project managers should ask before deciding to conduct a meeting.

Review and discuss the steps to planning a meeting.

MODULE 44103-08 ◆ INTERPERSONAL SKILLS 3.11

Emphasize the importance of setting the tone and tempo for the meeting from the start.

Review and discuss the rules for conducting a meeting.

Assign reading of Sections 5.7.0–6.5.0.

Point out the importance of team building.

Explain that employee involvement and empowerment increases loyalty, fosters ownership, and enables work-related decision-making.

Note that managers work through their team, and must apply leadership and interpersonal skills to build a high-endurance team.

behind schedule. If you don't state the ground rules, distracting or disruptive behavior may occur. If you want participation and you don't tell the attendees so, you probably will not receive any. If you do not follow the agenda, the attendees will not either.

Step 1 Start on time. If key players are not there, start anyway. Skip down your agenda and begin with issues and topics that can be handled by the rest of the group. If you don't start the meeting on time, those attendees who arrived on time will probably not be prompt for future meetings.

Step 2 State the ground rules. They will facilitate productive meetings. If you do not have ground rules, create them at your next meeting. Ground rules might include the following:

- Be on time
- Be courteous
- Listen
- Follow agenda
- Participate
- Complete assignments
- Be prepared

Remind participants that these rules apply to all. Furthermore, everyone is responsible for enforcing them. Include them in every agenda.

Step 3 State the outcomes. Focus your meeting by stating the outcomes. There may be several hidden agendas among the participants or other burning issues they want to discuss. If it's not on the agenda, it has no place in the meeting. However, it may be necessary to schedule discussion of your attendees' concerns at another time and place. If there are issues that arise that are not part of the agenda or drift into what appears will be an unresolved issue, a common technique is to "park" the issue, usually on a white board for future discussion or as agenda items for future meetings. Using this technique ensures that all issues will eventually get resolved.

Step 4 Review the agenda. Read the agenda aloud. Make sure you stress the time allocated for each topic, and identify the person responsible for that topic. A flip chart is a good tool for listing the agenda and ground rules. Prepare the flip chart before each meeting. Use one page to list the entire agenda. Include times for each topic. Then, prepare one sheet for each topic. Turn the sheet as you complete each topic and record information and input on the flip chart. Appoint one attendee to compile notes and keep minutes either on the chart or on the computer.

Step 5 Facilitate the meeting by encouraging all points of view. Paraphrase input from others as the meeting moves along. Use brainstorming techniques. Ask open-ended questions and use nonverbal and verbal reinforcement techniques to encourage input. Use outside facilitators as needed.

Step 6 Close the meeting. Well-planned agendas, smoothly-facilitated meetings, and goals accomplished all go for naught if the meeting is not closed properly. The following is a checklist for closing the meeting:

- List action items
- Assign action items
- Set scheduled completion and/or reporting dates for assignments
- Summarize accomplishments
- Thank participants
- Schedule next meeting
- Send out minutes

5.7.0 Team Building

Teamwork is based on collaboration. Employee involvement and employee empowerment increase loyalty, foster ownership, and enable work-related decision-making. A team-based environment helps everyone understand and buy into the reality that thinking, planning, decisions, and actions are better off when done in collaboration—no one person ever has complete ownership.

A manager works through his team. You will need to apply leadership and interpersonal skills to build a high-performance team. In this section, you will learn some of the characteristics of winning teams and some techniques to establish an environment that leads to successful teamwork.

5.7.1 Characteristics of Exceptional Teams

What makes a great team? First, there is a fundamental realization that in a team-oriented environment, you contribute to the overall success of the company. You work with fellow members of the project team to produce results. Even though

Instructor's Notes:

you may have a specific job function, you are teamed up with other members of your company and possibly members from your company's subcontractors.

Teams share a common mission and goal. Every team member must be committed to this aim. Then, the team must share responsibility and collaborate to accomplish the mission and goals.

Team members must communicate with each other. They typically share a common language and terminology, leaving little room for misinterpretation. Issues and problems are discussed openly and directly without confrontation, with the clear purpose of solving problems. Team members support each other and hold each other accountable, but do not disparage anyone for their viewpoints, suggestions, or comments.

All team members participate in decision-making, identifying and solving problems, and implementing solutions. They contribute their ideas and shared knowledge to innovate. They analyze their processes and procedures and find ways to improve them (this is discussed in the *Continuous Improvement* module).

Individuals on the team are focused on and take ownership of the tasks they have to perform. They rise to challenges and obstacles that get in the way and solicit solutions from the team when they need help. Teams are outward-focused. They are responsive to and support other teams. They strive to achieve the results expected by their stakeholders. Successful teams are dynamic and purposeful and know that they are helping the organization move forward.

5.7.2 The Project Manager's Role in Building a Winning Team

Teams are formed, either formally or informally, within an organization. The challenge is to make sure that the teams in your organization excel as a unit and produce results for the stakeholders and satisfaction for the team members.

Your role as the leader is to shape the team and empower its members. Of course, you will hire competent, talented people. But you will also have to hire people who work well with others. This module will help you with techniques for finding the right people to make your team work. After hiring new team members, you have to provide the guidance, environment, and ground rules to let your team flourish. The following key tactics will help you build a winning team:

- *Establish the shared vision and strategies for the team's mission* – It is crucial to develop a shared vision and commitment. Both people and organizations need to establish a strategic framework for significant success. This framework consists of:
  - A vision for your future
  - A mission that defines what you are doing
  - Values that shape your actions
  - Strategies that zero in on your key approaches
  - Goals and action plans to guide your daily, weekly, and monthly actions

  The team must understand the vision and strategy, why it is the best strategy, and what it will take to accomplish the mission. You need to continually reinforce this message and obtain buy-in to the direction. If they don't know, understand or agree with it, they won't support it.
- *Establish the lines of authority* – This will allow everyone to know what to expect and how to get things done. Then, delegate.
- *Set goals and identify tasks* – Involve the team in goal setting and strategies for achieving the goals. Let them identify tasks and approaches to getting them done. Involve them in the decision-making process. Then hold them accountable for the outcomes.
- *Communicate* – First, you must be a good communicator with the team. Then, you must help the team members communicate with you and with each other. Reinforce listening skills. Listen to your team and let them know they must listen respectfully to each other. Help them realize that sharing, not hoarding, of ideas and knowledge is vital for both the team's and individual's success.

  Bring your team together in formal meetings to discuss the higher level issues. This is also a good time to formally reinforce the mission and goal. Encourage positive and informal interactions among group members and let them work on the detailed issues. Keep meetings and discussions interactive; invite the team to challenge assumptions as part of good communication. Get their feedback. If they hold back, resentment and misunderstanding can build up. Vary the venue for meetings. Hold some off-site meetings and arrange informal get-togethers.
- *Create a climate of innovation* – Challenge the team to innovate and achieve breakthroughs. Let them know that they can take risks in certain areas to make things substantially better. Encourage creativity.
- *Address the needs of members* – You need to be concerned about the ambitions and needs of each team member. It is your job to help each

Review and discuss the characteristics of exceptional teams.

Ask participants to list the key tactics a project manager may use to build a winning team.

MODULE 44103-08 ◆ INTERPERSONAL SKILLS 3.13

Emphasize that leadership is a major factor in successful, profitable projects.

Ask participants to list the three factors that enhance a leader's position power. Review and discuss these factors.

Point out that personal power stems from a leader's genuine concern for the trust and confidence of team members, resulting in employee compliance and personal commitment.

Discuss the benefits of a leader's technical knowledge.

Discuss with students the type of power Samantha uses in the video.

one attain his or her goals. Prepare a professional development plan for each member. Do their performance reviews on time and provide feedback on areas that need improvement.

- *Recognize individual and team contributions* – Let the team know that they, too, are responsible for helping each other realize their aspirations; that, collectively, they create winning teams by making each person a winner. Be aware of and reverse jealousy, cynicism, or defensive behavior. Each member must respect and be concerned for the welfare of each person on the team. Send the message that they need to work together without your presence or involvement. Involve the team members in hiring new members and have them contribute to acknowledging and rewarding other members.
- *Reinforce the team as a unit* – Plan activities and exercises that build the team's confidence and reliability on each other. Get away from the workplace. Bring in a facilitator to design and conduct team-building activities.
- *Thank and celebrate* – Let them know, individually and collectively, when they've done a great job and thank them often. They want to know how they are performing. Celebrate success; invite others in the organization to celebrate with the team. This broader recognition can reinforce the unity and shared pride in being a member of the winning team.

6.0.0 ◆ THE LEADERSHIP ENVIRONMENT

Making a profit is a significant goal of every job, and it's the project manager's responsibility to see that projects do not drift away from profitability. Construction and maintenance projects are human endeavors. Humans make them succeed or fail. One key to achieving successful, profitable projects is leadership, which generates pride in craftwork and strong team morale.

Successful leaders realize their job is to positively influence people. They must understand why people act the way they do so they can predict behavior. They also accept their managerial power and use it to direct and change employee behavior when necessary.

6.1.0 Power

Power is the ability to do or act and to get results by influencing the actions of others. It's the tool a leader uses to obtain employee commitment to company goals. Power can be related to a position or a person.

6.1.1 Position Power

Position power originates from the leader's position in the organization. It is the authority to give and withdraw rewards and punishment in the name of the company. Power is given by the organization and can be withdrawn by the organization. It is essential for project managers to realize that, if they do not use their position power wisely and judiciously, they can lose it far more easily than they gained it.

Three factors that enhance a leader's position power are:

- *Fear* – A leader whose power is based on fear gets compliance from others because they believe that if they do not follow the leader they will suffer. For example, workers fear the denial of a pay raise, being assigned to the most troublesome jobs, or outright termination.
- *Politics* – Leaders who base their power on politics gain compliance from others because they believe the leader is on the "inside" and can give or take away their status in the organization.
- *Title* – Leaders who base their power on their titles alone achieve compliance from others because they believe that title or role in an organization give leaders the right to decide on given issues.

6.2.0 Personal Power

Personal power stems from a leader's ability to project genuine concern for the trust and confidence of team members. This type of power provides a manager with employee compliance and personal commitment. Employee commitment grows with the leader's demonstrated commitment to combine the company's goals with the work goals of the team members. Personal power cannot be granted or taken away by the company; it can only be earned or lost by the individual manager.

Sources of personal power include technical knowledge, job information, and personal attributes.

6.2.1 Technical Knowledge

A leader's technical knowledge can earn respect. If you have worked in the trades, for example, your suggestions and opinions will carry more weight than they will from someone who's always had a white-collar job.

Instructor's Notes:

6.2.2 *Job Information*

A leader's knowledge of the job situation and of company developments related to the job are key sources of personal power. A manager who can accurately answer the questions of the project team and give the team insight into what is really going on is usually perceived as a manager with genuine power.

6.2.3 *Personal Attributes*

A leader's willingness to listen, sense of humor, understanding, and sympathy for employee situations and problems earn influence. These qualities demonstrate interest in others and can create a willingness from team members to mesh personal goals and desires with those of the leader.

Personal power and position power feed on one another. A leader perceived to have personal power is frequently given more authority and responsibility. When a company expands a manager's authority and responsibilities, others in the organization perceive formerly undiscovered attributes and abilities in the leader. As successful leaders attain responsibility and authority, they garner a strong following among team members and peers.

6.3.0 Using Power

Power is as power does. When others believe a leader will use power to make decisions, assert authority, and reward and punish, they usually follow. A leader who refuses to exercise power soon earns the title of "paper tiger" and loses influence. A successful project manager exercises power effectively.

Use power with caution, though. The management style and the type of power a project manager may employ in a given situation should depend on the circumstances. For example, position power generally works better when dealing with employees with lower level skills and knowledge. Personal power may work well with every type of employee.

6.4.0 Leadership and Motivation

Leadership is a motivational tool. Motivation is an action-response relationship between two people. Effective project managers develop close working relationships with team members to learn which actions on their part will lead to desired responses. This is not an easy task, as people respond differently because of their individual backgrounds and needs. It is difficult for a leader to find a set of motivational tools that suits everyone on the project team; faced with this difficulty, a poor leader resorts to pigeonholing people, or dealing with every situation with the same heavy-handed authoritarian control. An effective leader looks at each team member and each situation individually and tailors motivational efforts accordingly.

6.5.0 Effective Leaders

Effective leaders are sensitive to their environment while recognizing data and emotional stresses in any given situation. They search for the root causes of frustration among and between superintendents, foremen, teams, the project owner, the general contractor, the subcontractors, and other stakeholders (e.g., vendors). Through it all, successful leaders focus on the most important goal: completing the job on time and within budget.

Few project managers can stay abreast of all of these factors, since conditions change rapidly. Among all the elements in the job environment, one stands out as the key to the project manager's success: learning the probable responses to expect from team members.

Employees respond or do not respond for a variety of reasons.

- *Will* – An employee will do the task assigned.
- *Won't* – An employee will not do the task assigned.
- *Can* – An employee can do the task assigned.
- *Can't* – An employee can't do the task assigned.

These reasons can be combined. For example, one response is "will and can." A leader who assigns a task to an employee who will and can do the assignment seldom needs to be directed. Instead, the leader facilitates the employee's ownership of the task.

Another response is "can but won't." When an employee is capable of doing the task but refuses to do so, the project manager must uncover the reasons and convince the employee of the necessity of the task.

A third possible response is "will but can't." Here the employee is willing to do the job but is unable to do so. The leader's response should be to supply the necessary instruction or training.

Yet another possible response is "will and won't." Here the employee sometimes will cooperate and do the job and other times will not. The leader's task then is to get the employee to commit to consistency and emphasize the importance of compliance to that commitment.

Discuss the benefits of a leader's knowledge of job information.

Discuss the benefits of a leader's personal attributes.

Discuss the importance of using power properly.

Discuss the importance of motivation.

Review and discuss the reasons that employees respond or do not respond.

Discuss with students whether or not they thought Samantha was an effective leader. If not, what kind of motivational techniques could she have used to improve the performance of others?

MODULE 44103-08 ◆ INTERPERSONAL SKILLS 3.15

Review why successful leaders are both goal- and people-oriented.

Emphasize that effective project managers must establish good external and internal relationships.

Ask participants to name two methods used to build external relationships.

Assign reading of Sections 7.0.0–9.1.0.

Show Transparency 7 (Figure 6). Review the components of a job description.

Explain that job descriptions provide a clear picture of the job role and performance standards.

Another possible configuration is "can and can't." A leader who has assigned a task to an employee who sometimes can and sometimes can't do the assigned task must assess the situation. How does the task an employee cannot perform differ from a task that the employee can do? The answer provides guidelines for instruction or training.

The situational leader knows that success depends on personalities. The successful leader focuses on the issue at hand, and applies the management style that fits the people involved. The project manager demonstrates concern for people by devoting time to discuss the job thoroughly with foremen, providing support and direction throughout the project, and lending an attentive and responsive ear when problems arise. Concern for people means caring about each individual's concerns—both personal and professional.

The secret to successful leadership is to balance these two types of concern. The project manager must develop a management style that is both goal-oriented and people-oriented. Effective management styles were discussed earlier, specifically that the team leader must display a high level of concern for both people and tasks (refer again to *Figure 4*, the Management Grid). This is effective when the jobs they manage require continuous communication and interaction. The following is an actual job-related example of how a project manager can build human relations by acknowledging the workforce.

> On my last project, I (as the PM) stood at the project entrance every Friday morning (rain, shine, or snow) and shook hands with each employee and told them how much I appreciated their effort and attendance. This message from the "big guy" was sincerely received by the workforce.
>
> It gave each of them the opportunity to get to know me, and I got to know them. Because of this simple team-building activity, that project had incredibly low absenteeism and turnover and productivity was better than estimated.

The project manager can't concentrate on management staff and leave the field team building to the supervisors. The ability to build and keep positive relations requires honest effort and consistency. Human relations and team building do not require complicated initiatives. Sometimes, as in this project manager's experience, simple actions can make people feel appreciated.

7.0.0 ◆ BUILDING RELATIONSHIPS

Effective project managers must establish both good external and internal relationships. Customers are the focus of the external relationships, while people within one's company, including crew members, compose the internal ones.

7.1.0 External Relationships

Building external relationships usually occurs in one of two ways. First, there is the referral method, in which a previous or current client recommends your company. The second method is through marketing and sales. In both cases, customer service becomes a key component to building and maintaining business relationships.

Aside from being technically competent, the project manager and salesperson need to distinguish your firm and your service from the competition—otherwise you won't meet your customers' real needs.

7.2.0 Internal Relationships

Because project managers work closely with their supervisors and other company personnel to hire, orient, and train workers, it's important to learn how to have the right mix of people assigned to crews and how to manage and motivate them.

7.2.1 Hire the Right People

The project manager needs to rely on the supervisor for managing the day-to-day interaction that occurs at the job site. It is also the project manager's responsibility to ensure that the supervisors have updated and clearly written job descriptions to help them select the right people.

In large companies, hiring is handled by a human resources department. However, if you are working at a small company, hiring may be the responsibility of the project manager and supervisors. If this is the case, then the job description (*Figure 6*) becomes a valuable tool. It provides a clear picture of the job role and performance standards, and should contain the following information:

- Job title
- General position summary
- Supervisor information/reporting relationships
- Principle duties and responsibilities
- Job specifications
- Other requirements that include proper judgment, knowledge, and use of tools

Instructor's Notes:

3.16 PROJECT MANAGEMENT

SAMPLE JOB DESCRIPTION

Position: Pre-Construction Manager

General Summary: Oversee pricing on all projects and delegate estimating responsibilities.
Oversee and coordinate all ROM, schematic, design, and final pricing.

Reports to: Senior Construction Manager

Duties and Responsibilities:

- Identify bid strategies
- Coordinate overall bid schedule and ensure timely preparing of bid
- Review final bid for submission
- Oversee bid packages
- Oversee schematic and design budgets
- Coordinate design-build subcontractors reviews
- Oversee and assign budgets and proposals
- Support marketing with proposal development
- Evaluate estimator performance
- Conduct quarterly and annual performance and salary reviews
- Prepare professional development plans for direct reports

Knowledge, Skills, and Experience Required:

- Minimum 15 years in commercial construction business
- Ability to coordinate all design disciplines on a project
- Ability to conceptually estimate project costs
- Ability to explore methods and means at concept and design stages
- Experience evaluating building systems at concept and design stages
- Exceptional communication skills with owners, architect/engineers, and staff
- Excellent presentation skills
- Willingness to participate in the estimating process when warranted

103F06.EPS

Figure 6 ◆ Example of a job description.

MODULE 44103-08 ◆ INTERPERSONAL SKILLS 3.17

Ask participants to list ways to build strong internal relationships.

Discuss attributes of both mentoring and coaching.

Review and discuss the tips on employee hiring and selection.

Ask participants what the first consideration in the company selection process should be.

Review and discuss the selection process.

7.2.2 Build Great Workers Out of Good Ones

Perhaps the best way to build strong internal relationships is to focus on your employees through coaching, mentoring, and training.

According to most practitioners, both mentoring and coaching need a long-term commitment to be successful. These roles are necessary because they offer a helpful set of tools for career development. It takes time for a new hire to understand how things work in an unfamiliar environment. Coaching, using a thoughtful combination of observation and critique, is an excellent approach to use because it provides tips, feedback, what to and what not to do, as well as other information.

For example, a new craftworker at the site may be very experienced, but because the project has unique OSHA standards, a coach can reduce the learning curve regarding safety issues and job-related matters. Or a new project manager is told that the company has transitioned to the new CSI 2004 format. A coach will be able to help the project manager interpret and use the new numbering system.

Mentoring is a bit different than coaching. The mentor is a person who usually has a lot of experience and is very good at helping a new person understand the company culture, politics, organizational structure, business processes, and client relationships. For example, a new construction manager's mentor may find it important to explain the best way to establish a relationship between the owner, architect, and general contractors on a government construction project. Generally, a mentor leads by example, is a team player and, to a large degree, a role model.

Taking the time to show new hires how your company operates is essential. Effective mentoring takes place when experienced workers are assigned to new hires to teach, guide, and advise. Mentoring and coaching improve performance and build internal relationships. This approach is not just for the problem or less-productive employee. Positive feedback is necessary for even the best workers, not just those who need to improve. As a project manager, take time to make sure a mentoring program is in place. Encourage your supervisors to follow through in building strong internal relationships.

8.0.0 ◆ BEHAVIORAL INTERVIEW TECHNIQUES

Businesses can run into problems when the wrong people are hired. Not only is it a money problem but it also affects productivity. For example, replacing someone restarts the hiring cycle, taking time and energy away from more productive work. Therefore, to avoid costly hiring mistakes, companies have moved from relying on educated guesses to employing better selection processes. The Rainmaker Group has the following advice on employee hiring and selection:

- Traditional hiring methods (resume, interview, and background checks) only provide a 14 percent likelihood of a successful hiring decision.
- The cost of employee turnover can range from ½ to 4 times the employee's annual salary and benefits.
- People are good at convincing hiring managers they are right for the job even when they know they might not be.
- Many times you won't realize you've made a poor hiring decision until the team member has already become an emotional part of your team.
- With the right tools and know-how you can improve the chances of hiring success by as much as 75 percent.

The first consideration in a company selection process is compliance with the laws that govern the hiring of potential candidates—both state and federal laws. These laws provide guidance for establishing a new selection process or revising an existing one.

What are the features of a good selection process? And if a company does not have one, what are the steps in creating one? Today, many companies are augmenting their existing selection process with standardized testing, performance testing, and using outside resources to implement strategies to reduce and prevent litigation. In the construction industry, a company not only hires individual workers but also interviews and selects subcontractors and temporary employees. A successful selection process abides by the law and establishes guidelines.

A construction company needs a mix of individuals that include professions to support the project from an administrative perspective such as project managers, project engineer field engineer, contracts administrator, architect, other design consultants, and so forth. The company also needs to hire individuals and subcontractors for the jobsite. These include foremen, craftworkers, superintendents, and so forth. The selection process for the construction company needs to be consistent. A company cannot afford turnover at the job site or within its office. The selection process for a company will usually include two major areas, the pre-interview process and the interview process.

Instructor's Notes:

3.18 PROJECT MANAGEMENT

8.1.0 Pre-Interview Process

An established and consistent pre-interview process will likely involve the project manager and other assigned individuals. The major responsibility the project manager has in preparing for the interview includes the following steps.

Step 1 Revise or create new job descriptions. The job description needs to focus on the tasks and include acceptable standards as well as education, experience, competencies, and skill sets for the job.

Step 2 Determine the selection criteria and whether a position will need additional testing to measure proficiency. In larger companies, assigned selection committees are usually in place. However, for smaller companies, the project manager may be required to select a committee. If so, the members need to knowledgeable about the specific positions. The type of position will dictate the make-up of the committee members.

Step 3 Check references and education qualifications. This is often the responsibility of human resources; however, the project manager may be required to check for personal references and educational requirements. In this case, the project manager or assigned persons need to ensure that all laws regarding confidentiality are followed.

Step 4 Interviewing and hiring are both time consuming and somewhat stressful for both the candidate and the interviewers. To relieve the pressure, some companies will use software packages to generate interview questions. Although a list of questions can be created in a short time, it needs to be checked by human resources or another administrator familiar with the legal boundaries for interviewing. Be sure to iron out all the logistics for the interview. Focus on the following:

- Review, discuss, and decide if the questions are legal, appropriate and fair.
- Ensure everyone has and understands the questions.
- Determine the time required.
- Determine how many people will be involved.

All the questions need to focus on whether the person can do the job correctly and professionally. Questions also need to be within legal guidelines. Listed below are question topics that should be avoided:

- Race, color, religion, sex (gender), national origin
- Languages spoken at home (if part of the job description, you can ask in what languages the candidate is fluent)
- Family: spouse's employment, child care, marital status, where parents were born, where the candidate was born, if family lives locally, sexual orientation
- Home ownership, car ownership
- Arrest record (you may ask if candidates have ever been convicted of a felony, not if they've ever been arrested)
- Disabilities
- Citizenship (unless required by a federal contract subject to security clearances)

If the candidate volunteers information on any of the above "unsafe" questions, say something like, "That isn't information I need for this interview," and move on to safer territory.

It is advisable to check with the company human resources department to find out what the procedures are if the candidate volunteers inappropriate information.

8.2.0 Interview Process

A popular method of interviewing involves using **behavioral interview techniques**, or questions to help the interviewer get a better picture of a candidate's experience. These types of questions are helpful to discover a person's experience and other job-related skills. This type of questioning can offer some insight as to what the person's feelings are and possibly their personality. The rationale for behavior-based questions is that if hiring managers can detect past behavior and performance, then it's likely they can find a candidate who will behave in a manner that suits the position.

This type of interview process captures information about the candidate in three skill areas: content, functional (transferable), and adaptive (self management). Content skills are specific to what the person does, such as take-off procedures, project planning, or welding. Functional skills are related to information, things, or people, such as organizing, managing, developing, communicating, and so forth. Adaptive or self-management skills are personal characteristics, such as dependable, team player, self-directed, or punctual.

Behavioral interviewing is designed to minimize a hiring manager's personal impressions,

Describe the pre-interview and interview processes.

Review and discuss the steps in the pre-interview process.

Review and discuss the topics that should be avoided during the interview process.

Explain the behavioral interview techniques that help an interviewer get a better picture of the candidate's experience.

Ask the participants to list the three skill areas that behavioral interview techniques can detect.

Discuss content, functional, and adaptive skills.

MODULE 44103-08 ◆ INTERPERSONAL SKILLS 3.19

Review and discuss the acceptable questions that may be asked of a candidate during the interview process.

Review and discuss the STAR formula used to answer behavioral interview questions.

Show Transparency 8 (Figure 7). Review and discuss the sample PDP.

Ask participants to list key points typically found in most professional development plans.

Explain that training keeps employees prepared, flexible, and up-to-date on new techniques, procedures, and safety measures.

which can be a powerful influence clouding the decision to hire. Focusing on how a person really feels, acts, and relates experiences can make hiring decisions easier. During the interview it is acceptable to inquire about a person's work ethic, school attended, how they handled job-related challenges, or even non-job related questions.

It's not necessary to omit traditional interview questions. They can be modified and turned into behavioral questions. For example, instead of asking the candidates how they would behave in a hypothetical situation, rephrase the question and ask, "How would you prepare the staging area at the job site for duct delivery?"

To eliminate general statements, ask for specific details. For example, "What are the factors that determine actual job costs?"

It may be possible to look at existing questions and modify them. Having questions that require specific responses will also help to make the interview a more structured event. This forces the candidate to think how to best answer the question, and they may not get a chance to deliver any prepared stories. Everyone participating in the interview should take detailed notes about the candidate's responses.

Some candidates are savvy when it comes to interviewing and may use the STAR formula to answer behavioral interview questions. STAR is the acronym for situation, task, action, result.

Situation – The candidate listens, then carefully sets the stage for the interviewer by providing an overview of the situation, important background, and any other information. For example, they may discuss their roles and responsibilities, and how they were forced to multitask because of scheduling. They may tend to amplify the situation.

Task – The person describes the tasks involved in the situation, discuss the goals and objectives, and the importance of completing the project (or situation).

Action – They provide details about their actions. "I took this action ... because I anticipated..." The interviewer will be able to get a clear picture about what the person did rather than an "I should have" response.

Results – The candidate provides specific information about the results of their actions and stresses their ability to provide favorable solutions. They will provide information that puts them in a positive light.

Obviously, not every person interviewed will use the STAR method, but the interviewer should be able to detect who is using it. The information obtained may help set the candidate apart from others and make the selection process a bit easier.

9.0.0 ◆ PROFESSIONAL DEVELOPMENT PLANS

A **professional development plan (PDP)** greatly increases the productivity and capabilities of individuals and teams through the use of best practices for specific skill areas. It provides a consistent set of standards to help individuals attain knowledge, technical skills, and in some instances, certification. The PDP is the roadmap that depicts career paths for project managers as well as other skill areas, and also allows for flexibility in identifying individual interests. Using the PDP provides opportunities for you and others to take personal ownership in developing your professional growth.

The structure and organization of a PDP will vary from company to company. Every PDP will likely contain the following:

- Goals and objectives that focus on standards. For example, project managers may use the PMI standards as part of their objectives to maintain standards.
- Seminars, workshops, and courses that support an individual's professional development goals.
- A defined period for achieving professional development goals. Evaluation may occur on a quarterly basis or at an individual's yearly review. Results are measured against the stated goals.

There are no set formulas for developing a PDP, although many generic templates are available online. Most companies will usually have a process in place. Larger companies may rely on human resources to provide guidance, while smaller companies will rely on their staff. In many instances this will be the project manager. A sample professional development plan from an actual company is included as an example in *Figure 7*.

9.1.0 Training

Most companies provide training to their new employees. For example, in a union shop, the apprentice will attend training classes in addition to working on the job. Supervisor-implemented training programs for all employees can be found in many non-union shops. These programs keep the company's employees prepared, flexible, and up-to-date on new techniques, procedures, and safety measures.

Instructor's Notes:

Sample Professional Development Plan (PDP)

NAME		START DATE	
POSITION		MID-YEAR REVIEW DATE	
MANAGER		YEAR END DATE	

	Measure	Actions	Mid-Year Update	Year-End Update	Weight	% of PRF Achieved
	How should this goal be measured?	Describe specific steps or actions to achieve this goal and the timing or dates associated	What is the status of this goal at mid year?	What is the status of this goal at year end?	Total must equal 100%	How much of the PRF is achieved at year end?
Standard goals:	These are goals that are common to all managers in all groups.					
Client Evaluations	No negative					
Profit goals	On target for the job					
				Total:	25%	
Individual goals: (min 3)	These goals should be individualized and specific to the manager. Any goal identified should be aligned with one of the company's corporate goals for success (below).					
Build our people						
Build loyal clients						
Achieve operational excellence						
Achieve consistent profitability						
				Total:	75%	

Training and Development Plan:

Each manager is required to participate in 40 hours of structured training per year.
This structured training must be aligned with one of our strategic goals listed below.

Skill area for training	Source	Location	Date	Complete?	Hours
				Total:	

Core Management Skills: These are the core skill areas that all managers should have well developed.

- Project Management
- Construction Industry Knowledge
- Quality Management
- Leadership
- Financial Understanding
- Verbal and Written Communications
- Client Relationships
- Leveraging Technology

A manager may also pursue training that deepens discipline-specific knowledge or advances the ability to lead within the company.

Strategic Goals:

- Build our people
- Build loyal clients
- Achieve operational excellence
- Achieve consistent profitability

103F07.EPS

Figure 7 ◆ Sample professional development plan

MODULE 44103-08 ◆ INTERPERSONAL SKILLS 3.21

Have participants complete the Review Questions, and go over the answers prior to administering the Module Examination.

Review Questions

1. What is an effective way to make sure your message is understood?

   ____

   ____

2. List four tips for active listening.

   1. ____________________________
   2. ____________________________
   3. ____________________________
   4. ____________________________

3. What four skills must project managers develop as they assume greater management responsibility?

   1. ____________________________
   2. ____________________________
   3. ____________________________
   4. ____________________________

4. The most ineffective style of management listed below is the _____.

   a. Bureaucrat
   b. Sarge
   c. Processor
   d. Abandoner

5. When there are changes to business procedures or projects, what are the three important elements to communicate to a team?

   1. ____________________________
   2. ____________________________
   3. ____________________________

6. The two essential components of effective meeting management are preparation and _____.

   a. closing the meeting
   b. scheduling the next meeting
   c. taking attendance at the meeting
   d. facilitating the meeting

7. List three factors that enhance a leader's position of power.

   1. ____________________________
   2. ____________________________
   3. ____________________________

8. Describe the difference between coaching and mentoring.

   ____

   ____

9. Name the three types of information obtained using the behavioral interview technique.

   1. ____________________________
   2. ____________________________
   3. ____________________________

10. A professional development plan (PDP) greatly increases the ____________ and ____________ of individuals and team members.

Instructor's Notes:

____

____

____

Summary

Summarize the major concepts presented in the module.

Project managers in today's construction industry must possess and execute a blend of both people skills and technical skills. Understanding and using interpersonal skills is essential. As a project manager, you will need to be able to communicate effectively with all people involved in the project. This will range from day-to-day communications with today's diverse workforce to communication and correspondence with architects, general contractors, owners, and other stakeholders.

The project manager must be able to coordinate resources (human, functional, and technical), to meet project priorities and implement solutions in a changing work environment. These abilities demand mental flexibility and sound judgment. Another attribute of a successful project manager is a sense of ethics and responsibility.

With responsibility comes the need to bring people together, to provide leadership, and to develop and maintain teamwork. Because every construction project is fluid, changes inevitably occur. The project manager must be able to help the project team not only cope with change, but help them get involved in the process, so they become part of the solution. You also need to effectively staff the project with qualified craftworkers and subcontractors. Building your staff requires interview skills that you need to use to find the right person for every job opening.

The project manager must build a skilled, satisfied workforce by implementing a personal development plan for his team. Using the skills in this module will enable you to blend managerial, technical, and interpersonal skills for successful project completion.

Notes

MODULE 44103-08 ◆ INTERPERSONAL SKILLS 3.23

Have participants complete the Trade Terms Quiz, and go over the answers.

Administer the Module Examination. Be sure to record the results of the testing on Craft Training Report Form 200, and submit the results to the Training Program Sponsor.

Trade Terms Quiz

1. One of the methods of _________ is granting opportunities to present viewpoints.
2. _________ is a system of moral principles, rules and standards of conduct.
3. _________ is an event that occurs when something passes from one state/phase to another.
4. Giving undivided attention to a speaker in a genuine effort to understand the speaker's point of view is called _________.
5. A _________ is an individualized document used to record an employee's current training needs and short-and long-term career goals.
6. Interviewing based on discovering how the interviewee acted in specific employment-related situations is known as _________.
7. _________ is an approach adopted by managers in exercising authority, encouraging participation in decision-making, motivating staff, communicating information, and maintaining control.
8. _________ are specific persons or groups who have vested interests in a project.
9. When your audience does not respond to your message, this is known as _________.

Trade Terms List

Active listening
Behavioral interview techniques
Change
Ethics
Management style
Passive listening
Professional development plan (PDP)
Stakeholders
Two-way communication

Instructor's Notes:

3.24 PROJECT MANAGEMENT

Trade Terms Introduced in This Module

Active listening: Giving undivided attention to a speaker in a genuine effort to understand the speaker's point of view.

Behavioral interview techniques: Interviewing based on discovering how the interviewee acted in specific employment-related situations.

Change: An event that occurs when something passes from one state or phase to another.

Ethics: A system of moral principles, rules and standards of conduct.

Management style: Approach adopted by managers in exercising authority, encouraging participation in decision-making, motivating staff, delegating authority, communicating information and maintaining control.

Passive listening: Listening to another person's message without verbally responding; the obvious limitation is the speakers do not know whether they have been understood, only that they have been heard.

Professional development plan (*PDP*): An individualized document used to record an employee's current training needs or desires and short- and long-term career goals; a written plan for developing knowledge, skills, and competencies that support both the employer's objectives and the employee's needs and goals.

Two-way communication: A variety of methods for obtaining feedback from others, ranging from granting opportunities for presenting viewpoints to inviting those concerned to participate actively in the decision-making process.

Stakeholder: Specific people or groups who have a stake in the outcome of the project, including internal clients, management, employees, administrators, and external stakeholders, including suppliers, investors, community groups, and government organizations.

MODULE 44103-08 ◆ INTERPERSONAL SKILLS 3.25

Resources & Acknowledgments

References and Resources

Bennett, Jeffrey, "Diversity of Construction Workforce Presents Employee Benefits, Challenge," The Business Review (Albany), accessed at www.bizjournals.com/albany/stories/2004/07/26/focus7.html.

Blake, R. and J. Mouton, *The Managerial Grid: The Key to Leadership Excellence.* Houston: Gulf Pubishing Co., 1964.

Construction Managers, Occupational Outlook Handbook, U.S. Bureau of Labor Statistics, accessed at stats.bls.gov/oco/ocos005.htm.

Construction Managers—Job Description, accessed at www.careerplanner.com/Job-Descriptions/Construction-Managers.cfm.

Employee Selection and Hiring Strategies—Get the Right People on Your Bus. Available at www.therainmakergroupinc.com/services/Item.asp?ID=23. Accessed Aug. 28, 2007.

FMI Corporation, Raleigh, N.C.

Mochal, Tom, "Manage These Three Aspects of Change in Your Project," TechRepublic, June 4, 2007. Accessed at articles.techrepublic.com.com/5100-10878_11618851.html.

Nesby, Tom, "Diversity: New Realities for Construction Companies," Nesby & Associates, Seattle Daily Journal of Commerce, www.djc.com/special/construct99/10050621.htm.

Project Management Institute, www.pmi.org

Workforce Central Florida. Accessed at www.workforcecentralflorida.com.

Instructor's Notes:

3.26 PROJECT MANAGEMENT

The following are suggested activities or instructional methods to help you teach the material in this module.

General

When you call on someone to answer a question, the rest of the class relaxes or even tunes out because they expect that the question and answer will take place only between you and the participant you called on. Instead, use this technique to involve more participants in answering questions and to keep them on their toes.

1. Ask participants to define a term or explain a concept.
2. After one participant has answered, ask a participant seated nearby if the answer is right. Then ask whether a participant in the back of the room agrees.
3. Ask participants to explain why they think an answer is right or wrong.
4. Use the session to clear up incorrect ideas and encourage participants to learn from their mistakes.

Section 4.0.0 Communication

This exercise will help participants fully understand how easily information can be misinterpreted. It also demonstrates the importance of listening and asking the right questions. The following pages contain the shape worksheets used in this activity. The Instructor will need enough copies for half the class, and blank paper and pencils for the other half. Allow 25 minutes for this exercise.

1. Have participants partner up, designating one partner "A" and one partner "B."
2. Participants should sit back-to-back so that they cannot see each other.
3. Give the "A's" a shape worksheet. You may choose to distribute the same shape to the entire class or randomly hand out the four shapes throughout the class.
4. While sitting back-to-back, "A" must describe the shape on the worksheet, and "B" must try to draw the shape on a blank piece of paper.
5. Instruct "B's" to listen only to their partners' instructions. Each "B" will use the oral instructions to try to recreate his or her partner's drawing. Allow 5 minutes for drawing.
6. Once the time is up, have participants compare the original drawings to the re-created ones. How many were accurate?
7. You may choose to perform the activity again, so that each partner has an opportunity to draw. If this is the case, use a different shape for each group. This time, the person explaining the drawing can give directions, and the person drawing can obtain information about the picture by asking "yes" or "no" questions. Allow 5 minutes for drawing.
8. Emphasize that the point of the exercise is that poor communication can lead to problems when messages get misconstrued.

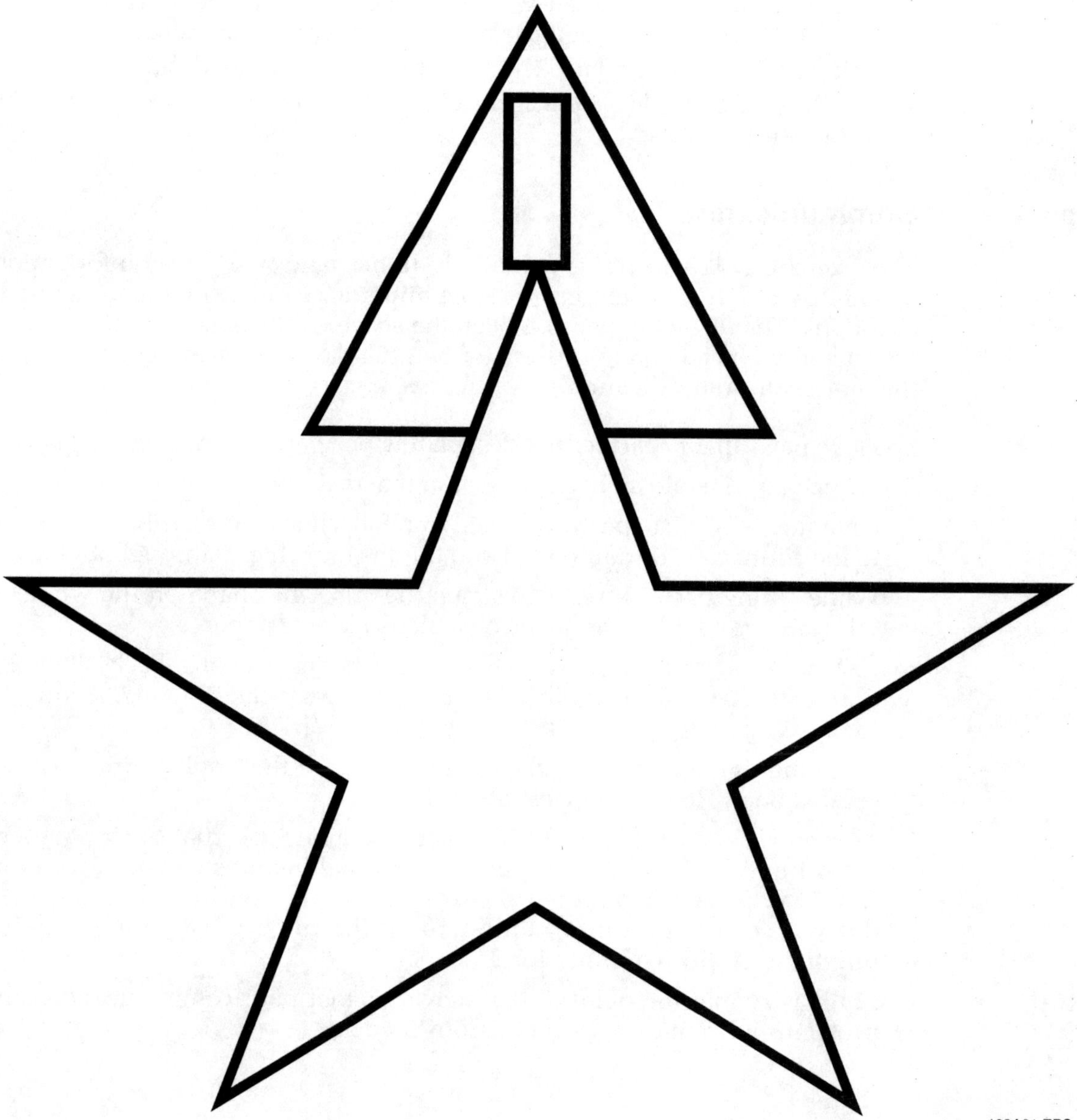

103A01.EPS

Copyright © 2008 National Center for Construction Education and Research. Permission is granted to reproduce this page provided that copies are for local use only and that each copy contains this notice.

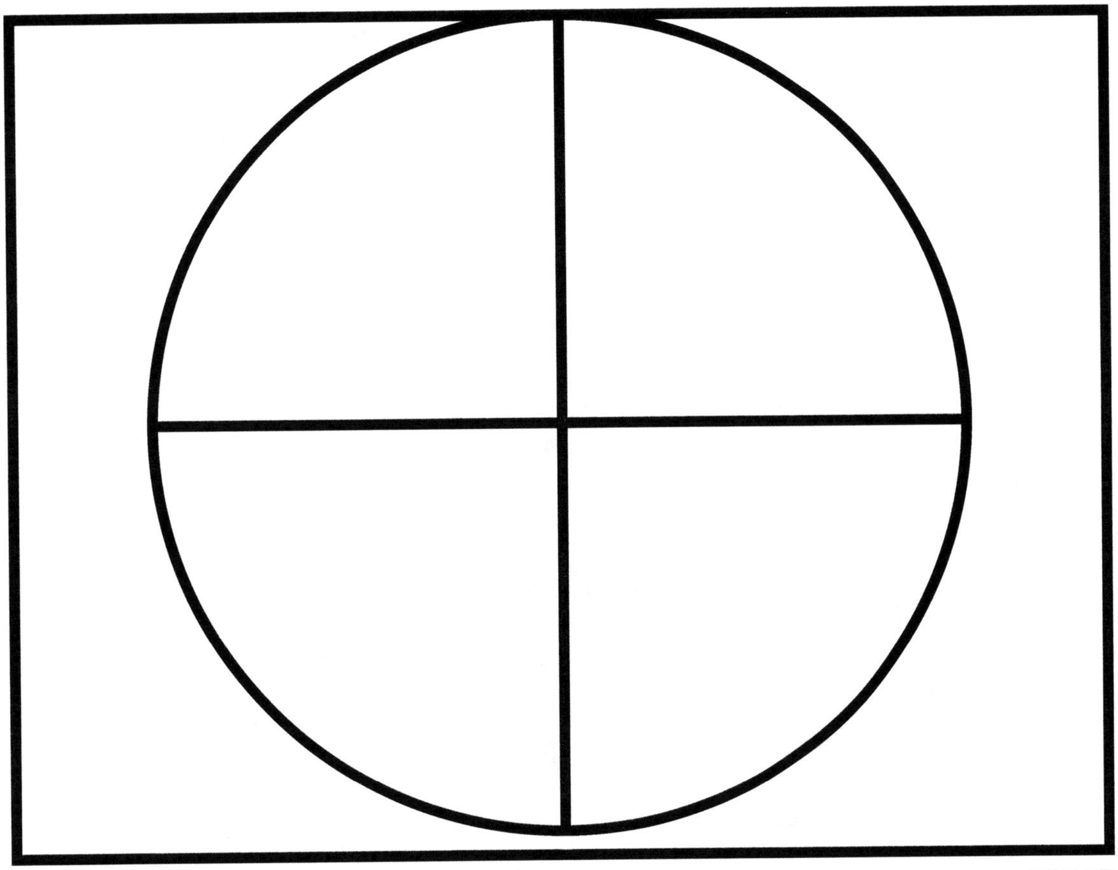

103A02.EPS

Copyright © 2008 National Center for Construction Education and Research. Permission is granted to reproduce this page provided that copies are for local use only and that each copy contains this notice.

103A03.EPS

Copyright © 2008 National Center for Construction Education and Research. Permission is granted to reproduce this page provided that copies are for local use only and that each copy contains this notice.

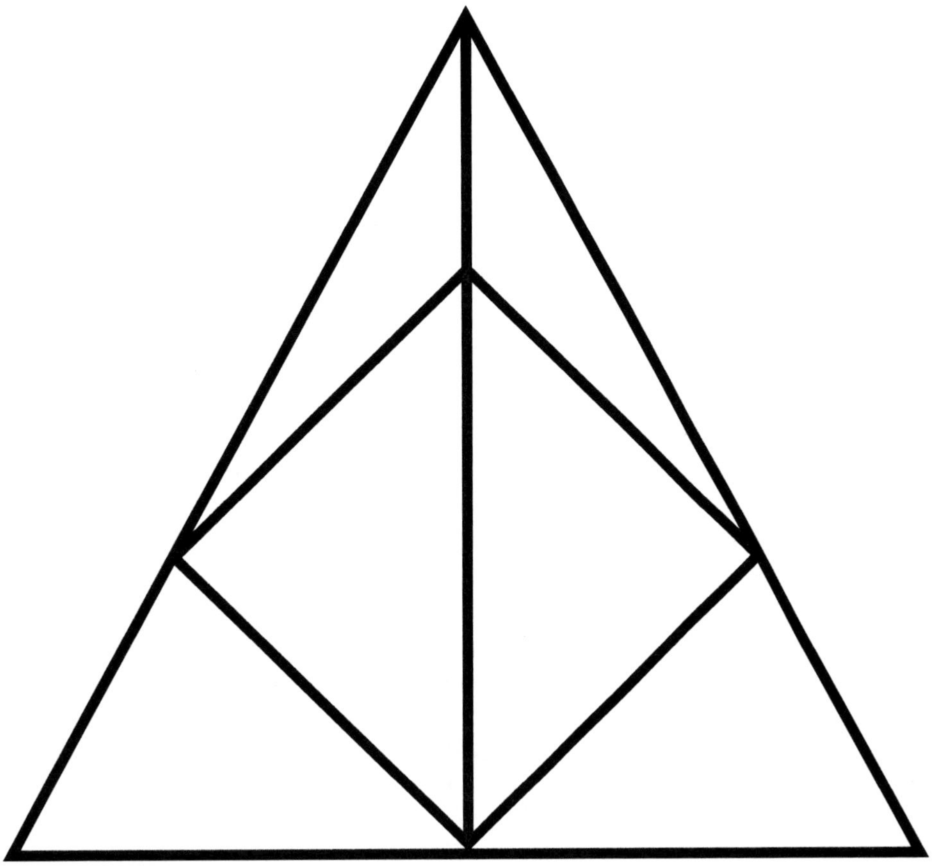

103A04.EPS

Copyright © 2008 National Center for Construction Education and Research. Permission is granted to reproduce this page provided that copies are for local use only and that each copy contains this notice.

Sections 5.4.0 *Managerial Styles*

This exercise will help participants assess their personal managerial style. Participants will need their Participant Guides and pencil and paper to takes notes for discussion. The Instructor will need an overhead projector, screen, and Transparency 5 (Figure 4, Managerial Grid). Allow 45 to 60 minutes for this exercise.

1. Display Transparency 5 (Figure 4, Managerial Grid).
2. Uncover each section with a blank sheet of paper as you present. To see how the grid can be used to assess a project manager's management style, consider this example:

   Assume that Project Manager X is a "9-1 Manager": a manager who is intensely task-oriented (a 9 on the horizontal scale) but practically blind to the concerns of people (a 1 on the vertical scale). This score would place the manager in the lower right corner of the grid, in the area labeled "Bull of the Woods."

   Allow 30 minutes for this part of the exercise.

   a. Explain "Bull of the Woods"

   Managers who use the "Bull of the Woods" style are most likely to respond to questions about their decisions or methods with the response: "Do it my way or hit the highway!" Punishment appears to be the only motivation they know. The "Bull of the Woods" approach achieves some degree of success, but the success is usually short-lived. Sabotage, pay-back and goofing off when managers are not around are the most common ways employees get revenge on "Bulls of the Woods."

   b. Explain "The Sarge"

   In the same quadrant of the Managerial Grid is "The Sarge." This type is also highly task-oriented. They generally will ask, tell, then yell if they do not observe a proper sense of urgency in getting a job done. They rank about a 3 on people and a 6 or 7 on tasks. They use punishment and reward as motivational tools. This style of management is effective with new employees and participants. Seasoned foremen or craftspeople do not usually respond positively to this style.

   c. Explain "The Abandoner"

   In the lower left quadrant of the grid is 1-1 style of management, the style that demonstrates a low concern for both tasks and people. It is known as "The Abandoner." This style of manager would rather jump ship than rock the boat. When faced with conflict, they abandon their responsibilities as leaders. They often use statements like "I don't care how you run the pipe, but the architect said…" or "I don't agree with the office either, but we have to follow what they said."

   d. Explain "The Bureaucrat"

   Within the same quadrant as "The Abandoner" is "The Bureaucrat," which ranks about 4 on tasks and 4 on people. "The Bureaucrat" follows the book. If the schedule says do something this week, you do it this week. They are equally fair and consistent with all employees and seldom play favorites. Theirs is an effective style of management because it is fair, consistent, and predictable. However, this inflexibility can lead them into the trap of following a procedure or a schedule blindly, without paying proper attention to circumstances.

   e. Explain "The Sweetheart"

   In the upper left quadrant of the grid is the 1-9 style: the "Sweetheart". This type has a high concern for relationships with people but a low concern for tasks. They want their employees to like them but seldom earn employee respect. If asked how they want a job done or why, they try to portray the task as a personal favor to them – "Hey, buddy, do this for me, will ya?" – or worse, sidestep the question with a joke. Employees and peers usually pay back "The Sweetheart" by manipulating them into doing jobs that they do not want or like to do.

f. Explain "The Processor"

In the same quadrant, however, is an effective style: "The Processor." This type ranks 3 or 4 on tasks and 6 or 7 on people. They generally see a question or a complaint for what it is – a sign of legitimate concern. If they see an apprentice not pulling his weight, they might recognize that training is needed. If they see a supervisor falling behind schedule, they would look for the root cause. Although "The Processor" is effective, this style is not always appropriate. For instance, if the foreman's problem is personal, she might not want to discuss it. Inappropriate probing by The Processor could lead to hard feelings.

g. Explain "The Juggler"

In the center of the grid is the "5-5" manager: "The Juggler." This type shows a reasonable concern for both tasks and people. It is the most common style of management in the construction industry. This style pushes for productivity but is also concerned with employee feelings. "The Juggler" generally assesses a situation in terms of what needs to be done immediately. Managers with this style frequently rely on snap decisions, which are effective only if the majority of their decisions are correct. If they make bad decisions too often, their employees perceive them as manipulators and withhold their trust. There is a danger that the flexibility of "The Juggler" will be seen as inconsistent or unfair. As a result, this type often achieves short-term goals but fails to achieve long-term goals.

h. Explain "The Team Leader"

In the upper right corner of the grid is the "9-9" manager. A "9-9" manager shows an extremely high level of concern for both people and tasks and is known as "The Team Leader." "Team Leaders" are effective when the jobs they manage require continuous communication and interaction, when there is time available to talk things over, and there is a great deal of trust on the job and within the company.

3. Read each of the bullet points below and have each Participant choose the one that best describes their own management style. Participants may refer to page 3.7 in the Participant Guide for this exercise.

   Chose the *one* statement that best describes your personal management style:

   - It's my way or the highway. (Bull of the Woods)
   - I'm highly task-oriented. I ask first, then tell, and then if I have to, yell. (Sarge)
   - I strive to get along at all costs and will not rock the boat. (Abandoner)
   - I'm fair and consistent, and follow the book to a "t." (Bureaucrat)
   - I want all of my employees to like me. (Sweetheart)
   - I see others' problems as legitimate concerns and deal with them directly, looking for root causes. (Processor)
   - I assess each situation in terms of what needs to be done immediately. I push hard for productivity, but am still concerned with people's feelings. (Juggler)
   - I have a high level of concern for both people and tasks, and understand the jobs I manage require continuous communication and interaction. (Team Leader)

4. Ask participants to assess where they fit on the grid.

MODULE 44103-08 — ANSWERS TO REVIEW QUESTIONS

Answer	Section
1. Try to obtain feedback.	(4.1.0)
2 1. Pay close attention to what people are talking about. 2. Don't be too quick to judge. Listen to all the facts before forming an opinion. 3. Where appropriate, ask questions or, in your own words, restate the facts or concepts you heard. And ask the speaker if you are correct in how you understood his message. 4. Internalize what has been said by making a mental review as well as a few notes about what was presented.	(4.3.0)
3. The following answers apply: 1. Technical knowledge and skills 2. Clear thinking skills 3. Ethical behavior 4. Management style 5. Skills in managing change 6. Skills for conducting effective meetings 7. Team-building skills	(5.0.0)
4. d.	(5.5.0)
5. 1. Goals 2. How the goals will be met 3. Rewards connected with change	(5.5.0)
6. d.	(5.6.0)
7. 1. Fear 2. Politics 3. Title	(6.1.1)
8. Coaching – Provides feedback, critique, and what to do and not do Mentoring – Performed by an experienced person who can help with company culture, politics and business processes	(7.2.2)
9. 1. Content skills 2. Functional (transferable skills) 3. Adaptive (self management skills)	(8.2.0)
10. Productivity and capabilities	(9.0.0)

MODULE 44103-08 — ANSWERS TO TRADE TERMS QUIZ

1. two-way communication
2. Ethics
3. Change
4. active listening
5. professional development plan (PDP)
6. behavioral interview techniques
7. Management style
8. Stakeholders
9. Passive listening

CONTREN® LEARNING SERIES — USER UPDATE

NCCER makes every effort to keep these textbooks up-to-date and free of technical errors. We appreciate your help in this process. If you have an idea for improving this textbook, or if you find an error, a typographical mistake, or an inaccuracy in NCCER's Contren® textbooks, please write us, using this form or a photocopy. Be sure to include the exact module number, page number, a detailed description, and the correction, if applicable. Your input will be brought to the attention of the Technical Review Committee. Thank you for your assistance.

Instructors – If you found that additional materials were necessary in order to teach this module effectively, please let us know so that we may include them in the Equipment/Materials list in the Annotated Instructor's Guide.

Write: Product Development and Revision
National Center for Construction Education and Research
3600 NW 43rd St, Bldg G, Gainesville, FL 32606

Fax: 352-334-0932

E-mail: curriculum@nccer.org

Craft __________ Module Name __________

Copyright Date __________ Module Number __________ Page Number(s) __________

Description __________

(Optional) Correction __________

(Optional) Your Name and Address __________

Module 44104-08

Issues and Resolutions

NCCER STANDARDIZED CRAFT TRAINING PROGRAM

The National Center for Construction Education and Research (NCCER) provides a standardized national program of accredited craft training. Key features of the program include instructor certification, competency-based training, and performance testing. The program provides trainees, instructors, and companies with a standard form of recognition through a National Craft Training Registry. The program is described in full in the *Guidelines for Accreditation*, published by NCCER. For more information on standardized craft training, contact the NCCER by writing us at 3600 NW 43rd St., Bldg. G, Gainesville, FL 32606; calling 352-334-0911; or emailing info@nccer.org. More information may be found at our website, www.nccer.org.

HOW TO USE THIS ANNOTATED INSTRUCTOR'S GUIDE

Each page presents two sections of information. The larger section displays each page exactly as it appears in the Trainee Module. The narrow column ties suggested trainee and instructor actions to each page and provides icons (detailed below) to call your attention to material, safety, audiovisual, or testing requirements. The bottom of each page includes space for your notes.

The **Audiovisual** icon indicates an appropriate time to show a transparency or other audiovisual aid.

The **DVD** icon indicates an appropriate time to discuss or show segments of the video.

The **Classroom** icon prompts you to define a term, stress a point, ask trainees to explain a concept, or give examples.

The **Demonstration** icon directs you to show trainees how to perform tasks.

The **Examination** icon tells you to administer the written module examination.

The **Homework** icon is placed where you may wish to assign reading for the next class, assign a project, or advise trainees to prepare for an examination.

The **Laboratory** icon is used when trainees are to practice performing tasks.

The **Materials** icon is a reminder for you to gather materials needed for classes, labs, and testing.

The **Performance Testing** icon tells you to administer a performance test or a portion thereof.

The **Safety** icon is used to emphasize safety issues. It is often keyed to *Caution* and *Warning!* statements in the Trainee Module.

The **Teaching Tip** icon indicates additional guidance is available, such as how to conduct an exercise, get the most educational value from a field trip, or encourage class participation. Teaching Tips may expand on a feature (*Think About It, Did You Know?*) or provide *Quick Quizzes* or similar exercises. You will be referred to the Teaching Tips section at the back of the module if there is additional material.

The **Combination** icon indicates that the laboratory listed corresponds with a performance task. If desired, you can note the proficiency of the trainees during the laboratory, and use it to satisfy performance testing requirements.

PREPARATION

Before teaching this module, you should review the Objectives, Performance Tasks, Materials and Equipment List, and Module Outline. Be sure to allow ample time to prepare your own training or lesson plan and gather all required materials and equipment.

MODULE OVERVIEW

This module describes key elements of successful negotiations, including a review of negotiating techniques and styles. It explains how to recognize nonverbal signals, use negotiating tools, and apply conflict resolution strategies. Participants will learn how to identfy barriers to solving problems, and how to apply problem-solving techniques such as brainstorming and identifying root cause consequences. Also included are tools, techniques, and processes for dispute resolution.

PREREQUISITES

Please refer to the Course Map in the Participant Module. Prior to training with this module, it is recommended that the participant shall have successfully completed *Project Management*, Modules 44101-08 through 44103-08.

OBJECTIVES

Upon completion of this module, the participant will be able to do the following:

1. Identify signs of incompetent problem solving and negative problem identification climates.
2. Identify four major barriers to problem solving.
3. Demonstrate these problem-solving techniques:
   - Eight-step ladder
   - Fact-finding
   - Root cause diagram
   - Brainstorming
4. Name five key elements of successful negotiation.
5. List four universal truths of negotiation.
6. Cite the four phases of negotiation.
7. Identify and explain the eight negotiating techniques and how to respond to them.
8. Recognize communication cues.
9. Describe the stages of dispute resolution.

MATERIALS AND EQUIPMENT LIST

Desktop or laptop computer or DVD player

Multimedia projector and screen

Minor Decisions: Major Impact – How to Deal with Real Issues in Project Management DVD

Transparencies

Blank acetate sheets

Overhead projector and screen

Transparency pens

Whiteboard/chalkboard

Markers/chalk

Pencils and scratch paper

Module Examinations (located in the Test Booklet)

ADDITIONAL RESOURCES

This module is intended to present thorough resources for task training. The following reference works are suggested for both instructors and motivated participants interested in further study. These are optional materials for continued education rather than for task training.

American Arbitration Association, 1633 Broadway, 10th Floor, New York, New York 10019. www.adr.org

The Construction Industry Institute, 3925 West Braker Lane (R4500), Austin, Texas 78759. www. construction-institute.org

Frankel, Mark A. "Med-Arb: A Valuable Settlement Strategy," *Special to Wisconsin Law Journal.* Available at www.wislawjournal.com/special /adr2004/med-arb.html

TEACHING TIME FOR THIS MODULE

An outline for use in developing your lesson plan is presented below. Note that each Roman numeral in the outline equates to one session of instruction. Each session has a suggested time period of 2½ hours. This includes 10 minutes at the beginning of each session for administrative tasks and one 10-minute break during the session. Approximately 15 hours are suggested to cover *Issues and Resolutions*.

Topic	Planned Time
Session I. Introduction to Problem Solving, Part One	
A. The Eight-Step Problem-Solving Ladder	________
B. Fact-Finding	________
Session II. Introduction to Problem Solving, Part Two	
A. Using a Root Cause Consequence Diagram	________
B. Brainstorming Methods	________
Session III. Introduction to Negotiating, Part One	
A. Key Elements	________
B. Phases of Negotiation	________
Session IV. Introduction to Negotiating, Part Two	
A. Negotiating Techniques	________
Session V. Recognizing Communication Cues	
A. Physical Cues	________
B. Verbal or Conversational Cues	________
Session VI. Dispute Resolution; Review and Testing	
A. Negotiating Tools	________
B. Alternative Dispute Resolution Techniques	________
C. Typical Process for Settling a Dispute	________
D. Review	________
E. Module Examination	________

1. Participants must score 70% or higher to receive recognition from NCCER.
2. Record the testing results on Craft Training Report Form 200, and submit the results to the Training Program Sponsor.

USING THE PM DVD WITH *ISSUES AND RESOLUTIONS*

Section 2.0.0 presents the opportunity to focus on the real issue(s) presented in the scenarios. Was the "Missing Tool", for example, the problem, or was it a symptom?

Suggested Class Exercise: Project managers need to be able to identify the root cause(s) of a problem and solve it as quickly as possible. Have the students construct a root cause (fishbone) diagram to illustrate the cause(s) and consequences of the production delay problem.

Have students consider in their analysis what caused the initial conflict and, ultimately, the cause of delay in production. What side issues contributed to production delays?

Using the blank root cause diagram provided in Transparency 5 of this manual, have students follow the steps for developing a root cause diagram to analyze and create the root cause diagram.

Note to Instructor: If the class is large enough, have the students form teams to analyze the root causes and determine consequences. Have the students present their diagrams to the entire class.

Project Management

44104-08

Issues and Resolutions

Assign reading of Module 44104-08.

44104-08
Issues and Resolutions

Topics to be presented in this module include:

Overview

In the course of planning for and managing a project, there will be issues and challenges that need to be resolved. An effective project manager will strive to resolve issues before they become serious problems. In this module you will learn two essential management skills: problem solving and negotiating.

Problem solving may be defined as the process of identifying obstacles on a project and designing ways to overcome or eliminate such obstacles. Problem solving is a core skill for project managers. The two key factors at the root of many problems encountered in construction and maintenance firms are incompetence in solving technical problems and a negative problem identification climate.

Successful negotiating depends on good communication whether it is in person, in writing, or on the phone. Project managers spend a sizeable amount of time negotiating at coordination meetings with the project team. The focus of the negotiating process is to resolve issues expediently.

Instructor's Notes:

Objectives

Upon completion of this module, you will be able to do the following:

1. Identify signs of incompetent problem solving and negative problem identification climates.
2. Identify four major barriers to problem solving.
3. Demonstrate these problem solving techniques:
   - Eight-step ladder
   - Fact-finding
   - Root cause diagram
   - Brainstorming
4. Name five key elements of successful negotiation.
5. List four universal truths of negotiation.
6. Cite the four phases of negotiation.
7. Identify and explain the eight negotiating techniques and how to respond to them.
8. Recognize communication cues.
9. Describe the stages of dispute resolution.

Trade Terms

Arbitration
Brainstorming
Conciliation
Fact-finding
Fishbone diagram
Litigation
Med-arb
Mediation
Mini-trial
Nonverbal communication cues
Problem solving
Root cause diagram
Verbal communication cues

Prerequisites

Before you begin this module, it is recommended that you successfully complete *Project Management,* Modules 44101-08 through 44103-08.

This course map shows all of the modules in the *Project Management* curriculum. The suggested training order begins at the bottom and proceeds up. Skill levels increase as you advance on the course map. The local Training Program Sponsor may adjust the training order.

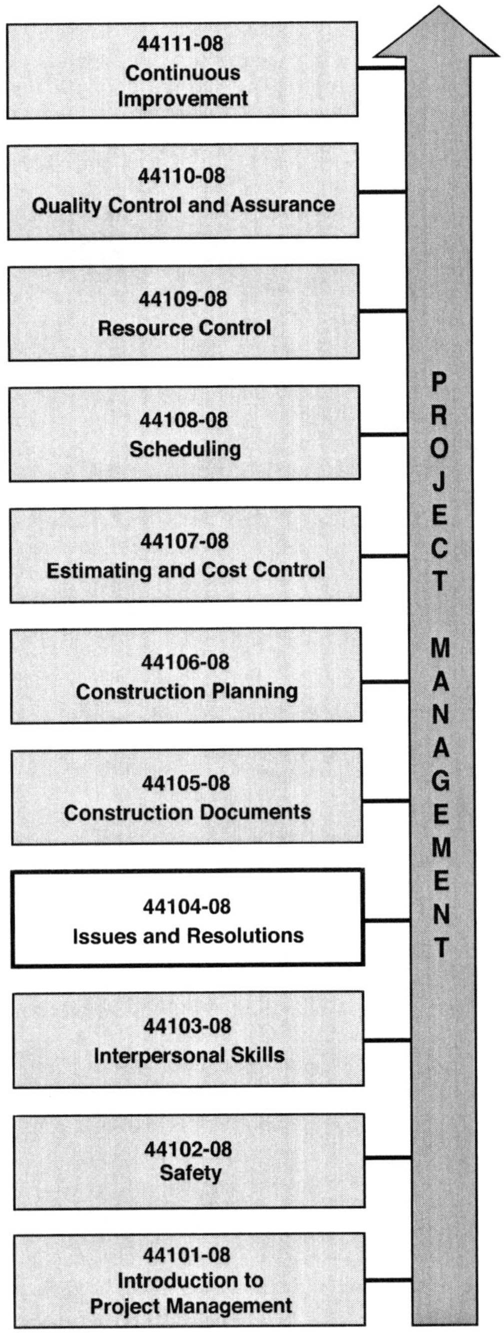

104CMAP.EPS

Ensure that you have everything required to teach the module. Check the Materials and Equipment list at the front of this module.

See the general Teaching Tip at the end of this module.

Show Transparency 1, Objectives. Review the goals of the module, and explain what will be expected of the participant.

Explain that terms shown in bold are defined in the Glossary at the back of this module.

MODULE 44104-08 ◆ ISSUES AND RESOLUTIONS 4.1

Explain the two factors that can be barriers to a smooth-running project.

Discuss symptoms of technical incompetence exhibited by employees and project managers.

Show Transparency 2, Four Barriers to Problem Solving. Discuss these barriers.

Review and discuss creative problem-solving techniques and the positive attitudes and behaviors required.

Show Transparency 3 (Figure 1). Discuss the eight steps of the problem-solving ladder.

1.0.0 ◆ INTRODUCTION

There are two key factors that are barriers to a smooth-running project: incompetent problem-solving skills, and a negative climate that prevents timely problem identification. Incompetence in solving technical problems appears in individuals, teams, and corporate culture. At the basic level, technical incompetence is demonstrated by an employee who is perceived to lack initiative, is unwilling to act, presents poorly thought-out ideas, and avoids confrontation. At the project manager level, symptoms of technical incompetence include poor planning, inadequate communication, and insufficient follow-up. At the company level, technical incompetence usually takes the form of inadequate guidance, insufficient time given to planning, and ignorance or denial of problems. Symptoms of a negative problem-solving climate are about the same at all levels: refusal to accept outside suggestions, minimal cooperation, resistance to change, and a refusal to listen.

2.0.0 ◆ PROBLEM SOLVING

The following issues may indicate incompetent problem-solving skills and/or a negative climate:

- Poor planning methods
- Hidden agendas
- Excessive management demands
- Suppression or denial of problems

Additionally, there are four barriers that may be present:

- *Psychological barriers* – including demand for a quick fix, fear of failure, unwillingness to take risks, and resistance to change
- *Physical barriers* – such as telephones and personal digital assistants (PDAs), equipment, and staff distractions
- *Cultural barriers* – inflexible attitudes, values, beliefs, and practices
- *Contractual barriers* – communication patterns defined by contract

Creative problem-solving techniques can correct technical incompetence and negative habits. Three positive attitudes or behaviors must be put in place to do so. A project manager who wants to solve problems effectively needs to believe the following:

- Anyone can be creative and solve problems through specific problem-solving techniques.
- Time and energy must be dedicated to solving problems.
- Creative solutions usually require a combination of new and old ideas.

2.1.0 The Eight-Step Problem-Solving Ladder

The problem-solving model shown in *Figure 1* can help a project manager identify problems and establish an environment where creative problem-solving techniques can flourish. This model can be used by individuals or groups for solving problems large and small.

Step 1 Admit the problem.
- Admit there is a problem to solve.
- Commit yourself and the necessary resources to solve the problem.

Step 2 Define the problem.
- Gather data.
- Check historical information.
- Observe circumstances.
- Develop a problem statement.
- Record conditions upon observation.
- Pay attention to detail. Stay aware and compare perceptions with others.

Step 3 Select a method.
- Match approach to size and difficulty.
- If a problem is confined to one person, do it yourself.
- If it involves many people, use a team to gain input and ownership.

Step 4 Generate alternatives.
- Generate as many options as possible.
- Evaluate your list.
- Combine items when appropriate.
- Prioritize the solutions you want to evaluate.

Step 5 Evaluate alternatives.
- Evaluate each alternative.
- Write objectives.
- List resources.
- Make note of the risks involved.

Step 6 Make a decision.
- Decide which alternative to choose.
- Communicate why and how.
- Designate responsibilities and roles.
- Use analytical tools.

Step 7 Implement the decision.
- Put plans to work immediately.
- Follow procedures.
- Follow up.

Step 8 Evaluate the results.
- Were objectives met?
- Maintain the gain.
- Thank people involved.
- If objectives were not met, identify why and start the process again.
- List and communicate lessons learned.

Instructor's Notes:

4.2 PROJECT MANAGEMENT

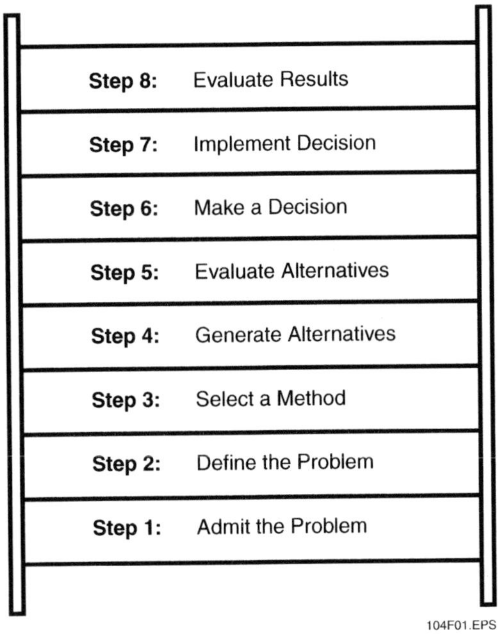

Figure 1 ◆ The eight-step problem-solving ladder.

2.2.0 Fact-Finding

Project managers are confronted with problems as they pursue their goals. Sometimes the sources of the problems come from superiors in their organizations, customers, vendors, subcontractors, general contractors, or employees. At other times, project managers create their own problems. Whatever the source, project managers are responsible for solving problems, correcting mistakes, troubleshooting urgent situations, and preventing similar problems in the future.

One of the first steps in solving a problem is **fact-finding**, which helps define the components of a problem. Sorting out objective, not subjective, events helps you isolate problems and points efforts in the right direction for possible solutions. It also allows you to see past distorted perceptions, hidden agendas, distractions, and inappropriate behavior such as:

- Minimizing or denying the problem
- Discounting ideas and abilities of self and others
- Failing to determine true circumstances
- Focusing on the symptom rather than the problem
- Perceiving a problem as a burden rather than an opportunity

Five elements of most job-site problems involve:

- People
- Circumstances
- Methods
- Materials
- Machinery

Under each of the following topics is a list of questions that you may want to ask when faced with a problem. The answers to these questions may be developed by using the worksheet in *Figure 2*.

2.2.1 People

Discover who is behind the problem to get you started on how best to deal with it.

- Who is involved? An employee, a particular craft, a subcontractor, a vendor, or someone in the office?
- How are they involved? Are they the source of the problem or did they fail to report it in a timely manner?
- Who is most affected by the problem?

2.2.2 Circumstances

When and where the problem occurs is another variable that will point you toward your best option for resolution.

- When does the problem occur? What day or days of the week? What time of the day?
- Where does it happen—at one job site or at several?
- Exactly what is happening when the problem occurs?

2.2.3 Methods

If the problem happens when a particular technique or method is being followed, these questions should be answered:

- Is the method being used applied the same way each time? If not, what varies from application to application?
- Is it the appropriate method for the circumstances?
- Are the methods used consistent with company procedures?
- Can the method be changed?

Explain that one of the first steps in problem solving is fact-finding.

Show Transparency 4 (Figure 2). Discuss the questions for each topic.

MODULE 44104-08 ◆ ISSUES AND RESOLUTIONS 4.3

People

Who is involved?
How are they involved?
Who is affected by the condition? How?

Circumstances

What is happening when the problem occurs?
How often does the problem happen?
Where does the problem occur? Is it at one site or several?

Methods

What procedures are involved?
Are the procedures followed?
Can the procedure be changed if it is found to be flawed?

Machinery/Equipment

Are the appropriate machinery, tools, or equipment being used?
Is the equipment operated properly?
Does the log show that the equipment is being maintained according to schedule?
Is the employee properly trained to do the job? Did you observe the employee?

Materials

Is the correct material in use?
Does the material meet specifications?
Is there an adequate supply of material?
Are convenient substitutions used?

Problem Statement

Following completion of fact-finding, the project manager should complete a problem statement. A written problem statement is the objective of the fact-finding process. It is a description of the problem designed to set the stage for analysis and implementation. It is brief and clear. The problem statement helps to isolate the root causes of a problem. It helps the project manager identify the people involved, the circumstances, the methods, material and machinery involved.

Enter problem statement below.

____
____
____
____

104F02.EPS

Figure 2 ◆ Fact-finding worksheet.

Instructor's Notes:

____
____
____

2.2.4 Materials

If the problem is related to materials, explore these questions:

- Do the materials in use meet specifications?
- Is there enough material?
- Do employees know how to work with it?
- When was the material use started?

2.2.5 Machinery

Machinery and the machine operators can also be sources of problems.

- What equipment is in use?
- Is it the correct equipment?
- Is it operating properly?
- Is it properly maintained?
- Is there an equipment log? Is it current?
- Does the problem lie with the equipment, the operator, or the environment?

2.3.0 Using a Root Cause Consequence Diagram

The **root cause** consequence **diagram**, sometimes called a **fishbone diagram**, illustrates relationships between some consequences and potential root causes. The consequences of the problem are entered on the right side of the diagram and the potential root causes are entered on the left. *Figure 3* shows a blank example.

2.3.1 Case Study

Figure 4 outlines the problem, or consequence, of low gas mileage and its possible root causes. Assume that the XYZ Construction Company has a fleet of 14 trucks that, according to the manufacturer's specifications, should be getting 18 miles to the gallon. However, expense receipts indicate the trucks are getting only 12.3 miles per gallon. Over the course of a year, this expense can be substantially more than planned. XYZ is considering buying 10 more trucks. The boss has asked the project manager to talk to the drivers of the trucks to find out why they are not getting the gas mileage expected. The boss has talked to a friend from another construction and maintenance firm whose company uses the same kind of vehicles. Their trucks are averaging 17.6 miles per gallon.

Using the case study as an example, the steps involved in creating the root cause consequence diagram are:

Step 1 Create a problem statement, e.g., "We are experiencing low gas mileage on our truck fleet."

Step 2 Assemble the drivers or a sample of the drivers.

Step 3 Tell the drivers why they are meeting.

Step 4 Tell them that another company is achieving 17.6 mpg and that XYZ company is getting 12.3 mpg.

Assign reading of Sections 2.3.0–2.4.0.

Show Transparency 5 (Figure 3). Discuss the purpose of a root cause consequence diagram.

See the Teaching Tip for Section 2.3.0 at the end of the module.

Show Transparency 6 (Figure 4).

Have students create a root-cause diagram of the production delay problem in the video.

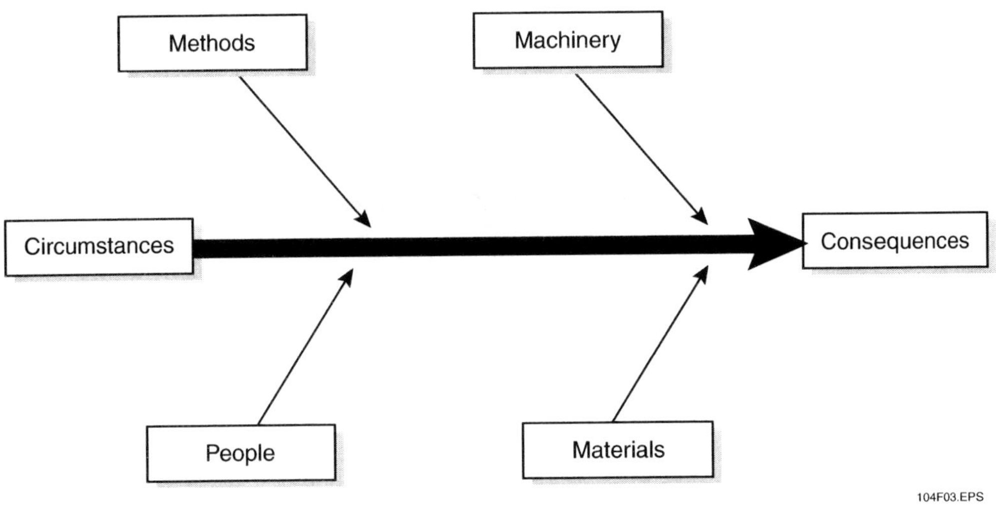

Figure 3 ◆ Root cause consequence (fishbone) diagram.

MODULE 44104-08 ◆ ISSUES AND RESOLUTIONS 4.5

Point out that when trying to solve a problem, brainstorming is a good way to collect many ideas quickly.

Discuss the code of conduct guidelines for brainstorming.

Pose a problem to the class and brainstorm solutions using the guidelines.

Assign reading of Sections 3.0.0–3.2.4.

Step 5 Enter the problem statement in the consequence box (low mileage).

Step 6 For each of the potential root causes, ask the appropriate question and enter the responses in the appropriate section:

People – What are people doing that might cause this issue?

Circumstances – When does it happen? Where does it happen?

Methods – What methods or techniques cause this to happen?

Materials – What materials cause this to happen?

Machinery – What equipment features might cause this to happen?

Step 7 Complete the diagram and discuss each of the responses by cause.

Step 8 Look for causes that appear frequently.

Step 9 Reach a team consensus.

Step 10 Gather additional data as required.

2.4.0 Brainstorming Methods

In problem solving, a team needs to focus on as many alternatives as possible. **Brainstorming** is an excellent method for generating alternatives because it helps to collect many ideas quickly. There are two primary brainstorming methods: formal and casual.

In a formal method, all team members take turns, either providing an idea or passing until the next round. This approach encourages participation while gently pressuring team members to participate.

In a casual method, team members state their ideas as they come to mind. This approach is more relaxed and works well as long as the most vocal team members do not dominate the discussion.

Both methods require a code of conduct in order to be successful. The following points may be used as guidelines.

- Everyone must agree on the question posed.
- Do not criticize other ideas.
- Do brainstorming quickly.
- Write every idea on flip chart or blackboard.
- Record ideas as they are presented, without interpretation.

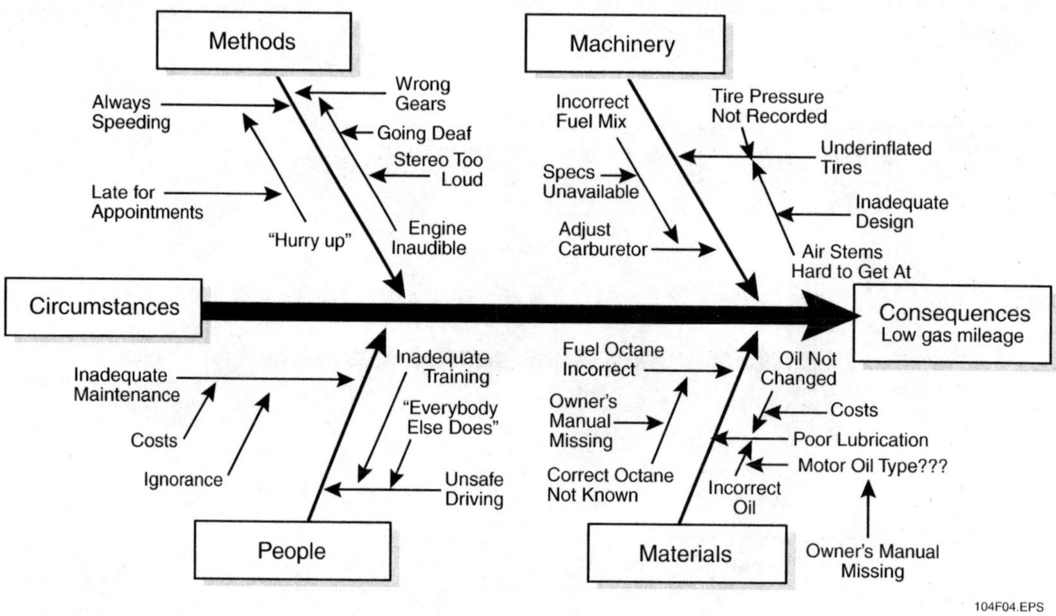

Figure 4 ◆ Completed root cause consequence (fishbone) diagram.

Instructor's Notes:

4.6 PROJECT MANAGEMENT

3.0.0 ◆ INTRODUCTION TO NEGOTIATING

Novice negotiators are frequently overwhelmed by the pressure they feel as they enter into the negotiating process. All too often they believe they are the only ones feeling the pressure. Few feelings are stronger and few less trustworthy.

Both parties feel pressure in a negotiation. The subcontractor wants the work. The general contractor wants the work to be done. Both have to answer to someone, and the pressure is on both to be successful. The ability to negotiate is fundamental to your success as a project manager, and the basic process is learnable. Practicing the methods and techniques discussed in this module will not make you a perfect negotiator, but it can make you more successful as a manager.

3.1.0 Key Elements

Four key elements to successful negotiations are sensible and often obvious.

- Each party in the negotiation must believe it has achieved its stated goals. In other words, both must feel they were successful.
- Each cannot forget the other party is human and needs to feel that their interests are valued.
- Fairness is essential to success. Both parties in the negotiation must perceive that they were treated fairly.
- A deal is a deal. Both parties need to honor their sides of the agreement.

The sum and substance of the so-called universal truths of negotiations are as follows:

- Everything is negotiable.
- People negotiate continually.
- The process is predictable.
- Information is crucial to success.
- Time constraints affect the outcome.

Among the typical topics that are negotiable are:

- Salaries
- Benefits
- Time schedules
- Sequences of work
- Fees
- Method of payment
- Payment schedules
- Change orders
- Start time and quitting time

After the bid is negotiated and awarded, negotiations truly begin. The general contractor may need to change the schedule because of owner demands or subcontractor requirements, or because another subcontractor cannot meet the schedule. Changing requirements means the need to negotiate is constant. Although the process is predictable, it is not simple. As with the mastery of any skill, the mastery of negotiation techniques will provide a reasonable level of predictability.

Gathering information before and during the process is crucial to success—that is, don't negotiate with strangers. Time is also a critical factor in the negotiation process. How soon does the deal need to be closed? When does the work need to be done? When is payment required? All of these factors contribute to the outcome of the process.

3.2.0 Phases of Negotiation

There are four phases of negotiation. Pre-planning, planning, the person-to-person meeting, and following-up to keep open lines of communication are the typical steps in the process.

3.2.1 Pre-Planning

Planning ahead for a negotiation is crucial. If you don't know what you want, when you need it, and how much you will pay or accept for your work, you're not ready to negotiate and you will lose. Define what your needs are during the pre-planning phase and know what your walk away point is. Define your goals and set your criteria for success during this phase. Who is going to be on your team? Who is going to be on their team? Once again, do not negotiate with strangers. Find out everything you can about the other parties involved.

3.2.2 Planning

The planning phase of negotiation means you need to identify the other party's needs. What do they want? Remember, it is not always money. Does the other person have the authority to make a deal? Gather more information. The more you know about the other negotiators, the easier it is to meet both sides' needs. To gather more information, ask open-ended questions, talk to other subcontractors or general contractors about how the other party operates, converse with the other party's employees, and have both parties talk to each other about daily practices. One danger of mixing your employees with their employees is that the other party may get better information than you've been able to find out about them. Predict their goals.

Emphasize that good negotiating skills are essential for a successful project manager.

Ask participants for examples of their own negotiation experiences, and discuss the outcomes.

Ask participants to list the four key elements to successful negotiations.

Show Transparency 7, Universal Truths of Negotiations. Review and discuss the list.

Ask participants to give examples of negotiable topics.

Describe the four phases of negotiation.

Explain how pre-planning allows you to prepare and set goals for the meeting.

Explain that the planning phase is for gathering information and identifying the other party's needs.

MODULE 44104-08 ◆ ISSUES AND RESOLUTIONS 4.7

Explain that the person-to-person phase allows both parties to identify areas of agreement and disagreement.

Explain that the follow-up phase involves abiding by your side of the agreement and ensuring that the other party abides by theirs.

Assign reading of Sections 3.3.0–3.3.7.

Review and discuss negotiating techniques and give examples of each.

See the Teaching Tips for Sections 3.3.0–3.3.7 at the end of the module.

3.2.3 Person-to-Person

After the pre-planning and planning phases, meet with your counterpart on the other team. Spend a little time visiting to acquire more information. But if they want to get to business immediately, go ahead. As this phase begins, identify areas of agreement such as, "Let's close these change orders," or "What can we do to help each other?" As a general rule, the sooner you can establish areas of agreement, the easier the person-to-person phase will go.

Then identify areas of disagreement. Find out what your differences are. Listen carefully; you might not be so far apart after all. Restate the other negotiator's exact words. This communicates that you listened and offers a chance to clarify matters for both of you if you did not understand the message, or if the other party wasn't as clear as necessary.

Reconcile disagreements as you proceed. Clarify all points and state all agreements. Be sure to state who is responsible for what, when, and the cost.

3.2.4 Follow-Up

When you and the other party agree, make sure you follow up on your side and that your people are doing what you agreed to do. If your team is not abiding by the agreement, get back to the other party immediately. Don't take the chance that the other party will not notice—that's unlikely to happen.

Follow up with the other party to ensure they keep to the agreement. Provide the other party with immediate feedback. Don't indict, inquire instead. The other party may not know that follow-up on their side has broken down. Give them an opportunity to correct the situation. While lawyers and arbitrators are always an option, they're an expensive one. Do your best to keep lines of communication open and expectations clear.

3.3.0 Negotiating Techniques

Few negotiating sessions sail through the phases of the negotiation process without one or both parties employing at least a few common techniques and tactics.

3.3.1 Dumb is Smart

Some participants purposefully play dumb to gain time or information. This person asks you to explain and define virtually every statement. This tactic is useful to stop the other party from changing the agenda or to keep the other party on track. It is an effective method when you are outmaneuvered or when negotiating with a superior or higher-level member of management. It may also be used to stall for time if the other party is aware that the time available to you is brief. If the technique is directed at you, a good response is to carefully explain, check the agenda, and reduce your goals. Be cautious about using this technique excessively. The other party may lose patience, harbor bad feelings, or believe that you are ill-prepared and so avoid doing business with you.

3.3.2 I Am in Control

In this technique, the practitioners act as though they have the power to authorize expenditures until the final moment. At this point, they say they don't have sign-off authority for the sum under study, or that they need a final sign-off by a higher authority. This tactic is used as a basis for further negotiation, but it doesn't always benefit the other party. Suggested responses to this technique include clarifying authority at the beginning of the person-to-person meeting or in the planning phase. If clarification is not possible, include slack in your proposal and state that you, too, must take the proposition to a higher authority. A danger in using this tactic is that it may cause bad feelings in the other party, and it is time-consuming, since it requires additional time to gain an agreement.

3.3.3 The Flinch

This participant appears startled and recoils at the other party's offer. A suggested response is to say, "Where did you learn to flinch like that?" or "Great flinch!" Be cautious before trying those rebuttals, the other party might actually be startled and may be offended at a glib response.

3.3.4 Crazy Bucks

In using this tactic, the participant acts as though the difference in dollars, time schedule, and other negotiable factors between the two sides is either minuscule or large depending upon perspective. For example, if the dollar difference were $36,500, and the project's duration was a year, the party employing this tactic might say, "We're only talking $100 a day or a little over $4 per hour difference. Why are you counting pennies?" A suggested response would be "$36,500 is over seven percent of the total package. Do you know what the average profit margin is for a subcontractor like me? Hey, I'm in the business for profit—cut me a break." This is not a technique to use lightly because it may cause ill will.

Instructor's Notes:

4.8 PROJECT MANAGEMENT

3.3.5 *I'm Outta Here*

This participant acts insulted or pretends that the other party is unreasonable, and literally walks out of the person-to-person meeting. If you're on the receiving end, a suggested response is to follow up with a telephone call if you want to do business with the other party. The danger in using it yourself is that no one will call back and you will lose an opportunity to gain business.

3.3.6 *Dr. Jekyll and Mr. Hyde*

Participants act like the good cop/bad cop team seen on so many television dramas. One person is easily offended and leaves the meeting. The partner remains and offers to intercede on behalf of the other party if certain conditions were changed. You might respond by finding out what the changes are, and if reasonable, play along. Alternatively, you could walk away and tell them to call when both parties are more reasonable. Caution: Using this technique can cause ill will.

3.3.7 *Feel, Felt, Found*

Use this counter-tactic to fend off either a highly emotional response or a strongly held belief without insulting or ridiculing the other party. For example, you could say something like: "I believe that you think that what you are proposing is fair. In fact, over the years I often believed that similar responses were fair myself. However, our experience has found..." If the other party uses this technique, you should listen carefully, offer more current data, or accept the statement as it is. There is little harm in using the technique yourself, as long as your data is fair and accurate.

4.0.0 ◆ RECOGNIZING COMMUNICATION CUES

Negotiations do not take place in a vacuum. Feedback occurs constantly; your actions, behavior and words affect the other party. Conversely, their actions, behavior, and words affect you. This interaction between the two parties is what allows for win-win solutions, even though there may have been an impasse before the in-person meeting. Recognizing how your words, tactics, and actions change meaning in the person-to-person phase is key to a successful negotiation.

The meaning of virtually any communication is measured by the response it elicits. As a negotiator, you plan your goals; however, success is measured by the meaning you convey to the other party and the feedback you receive from them. You can detect whether your message has been received and if it is acceptable to the other party via both nonverbal and verbal cues. If you sense that your initiative is not being received well, you can do or say something to recover the advantage.

In a diverse population such as that found in the United States, there are no hard and fast rules. Gender, race, ethnic background, and age are among the elements that affect our verbal and nonverbal responses. There are cues that help us to read the responses of others. Do not attempt to assign only the meanings suggested in this text. These clues are only clues; no more, no less.

4.1.0 Physical Cues

There are **verbal** and **nonverbal communication cues** that are conveyed and can be interpreted in a negotiating situation. The following nonverbal cues usually indicate irritation, impatience, discomfort or other negative feelings. Keep in mind that these cues may not convey the same meaning in a different cultural environment.

- Drumming of fingers
- Scratching head
- Tugging at ear(s)
- Hand on back of neck
- Finger under collar
- Fiddling with eyeglasses
- Object in mouth
- Eyeglasses removed and set on table
- Eyes blinking rapidly
- Head tilted forward, knuckles under chin
- Head held straight in heel of hand
- Person unbuttons jacket

Another form of nonverbal communication that affects your success in negotiation is how you organize your team to communicate, in other words, where to sit and who sits where. Following are some rules for consideration:

- When you are negotiating with two people, sit where you can watch both.
- When you have two people on your team, sit apart so you speak with two voices.
- When you have a large group opposing their small group, keep your group together for the appearance of power.
- When they have a large group opposing your small group, intermingle to diffuse their power.

Assign reading of Sections 4.0.0–4.2.0.

Point out that during negotiations, your actions, behavior, and words affect the other party, and vice versa.

Emphasize that the success of a negotiation is measured by the meaning you convey to the other party and the feedback you receive.

Ask participants to list elements that affect verbal and non-verbal responses.

Review and demonstrate nonverbal cues that may be conveyed as negative. Note that these cues may not convey the same meaning in a different cultural environment.

Discuss the rules for organizing your team for negotiations.

MODULE 44104-08 ◆ ISSUES AND RESOLUTIONS 4.9

Discuss phrases that may indicate a shift in the negotiations.

Assign reading of Sections 5.0.0–5.3.3.

Discuss dispute resolution.

Explain that if disputes are not settled quickly, conflicts can escalate and create personal and professional problems, delay projects, and add to project costs.

Show Transparency 8, Advantages of Negotiations.

Point out that negotiation is the most common form of dispute resolution.

Explain that if a settlement cannot be achieved through negotiation, an alternative may need to be considered.

Discuss alternative types of dispute resolution.

See the Teaching Tip for Sections 5.2.0–5.2.7 at the end of the module.

4.2.0 Verbal or Conversational Cues

Verbal or conversational cues usually signal almost the opposite of the words used. The phrases that follow should raise a flag in a negotiator's mind that a change is coming.

- "In my humble opinion..."
- "To be perfectly frank..."
- "To be honest with you..."
- "I'm just an old country boy..."
- "I'll try..."
- "However..."
- "I don't want to intrude..."
- "Off the top of my head..."
- "This is very embarrassing, but..."
- "As you are aware..."

Be prepared to change your proposal, defend your proposal or hear a counter proposal when you hear the preceding conversational clues.

5.0.0 ◆ DISPUTE RESOLUTION

Contractual disputes on a construction project are bound to happen. Without quick intervention these conflicts can escalate and create personal and professional problems within the project team. Aside from personal problems that may arise, the construction schedule can also be at risk. Resolving disputes without a strategy is time-consuming and unpleasant; not only is the internal project team at risk, but subcontractors, suppliers, and, in the worst case, owners as well. Disputes not resolved will add to the cost of the contract, so it's important to react quickly with an efficient resolution procedure.

In the 1980s, increased interest materialized in the development of new and innovative procedures for dispute settlements. The American Arbitration Association (AAA) has been given the credit for creating alternative dispute resolution (ADR) systems that respond to the needs of disputing parties.

5.1.0 Negotiating Tools

Negotiation is by far the most common form of dispute resolution. The objective of sensible dispute management should be to negotiate a settlement as soon as possible. Negotiation can be, and usually is, the most efficient form of dispute resolution in terms of management time, costs and preservation of relationships. It should be seen as the preferred route in most disputes. Its advantages are:

- Speed
- Cost savings
- Confidentiality
- Preservation of relationships
- Range of possible solutions
- Control of process and outcome

If you are unable to achieve a settlement through negotiation, you will need to consider what other method or methods of dispute resolution would be suitable. But remember, it will still be possible or may be necessary to continue negotiating as part of, or alongside, other forms of dispute resolution. There is a wide range of models regarding negotiating, from win-win to just say no.

5.2.0 Alternative Dispute Resolution Techniques

Whenever a disagreement arises on a construction project, alternative dispute resolution may be the method of settlement for a disagreement. ADR techniques are designed to resolve disputes quickly, effectively, and inexpensively. They may either assist the parties in their direct negotiations or resolve impasses when discussions have failed to produce a settlement.

5.2.1 Mediation

Mediation involves an attempt by the parties to resolve their dispute with the aid of a neutral third party. The mediator's role is advisory; he may offer suggestions and point out issues that the parties might have overlooked, but resolution of the dispute rests with the parties themselves. Mediation proceedings are confidential and private.

5.2.2 Arbitration

Arbitration is the submission of a dispute to one or more impartial persons for a final and binding decision. The arbitrator chosen by the parties is often a business person with expertise in the construction industry. The parties, by mutual agreement, fully control the range of issues to be resolved by arbitration, the scope of the relief to be awarded, and many of the procedural aspects of the process. Arbitration is less formal than the court system and in most instances the hearing is private. Few awards are reviewed or overturned by the courts because the parties have agreed to be bound by the decision of the arbitrator(s).

Instructor's Notes:

5.2.3 Mediation Arbitration (Med-Arb)

Med-arb begins with a written commitment from both parties to participate in the med-arb process and to bind themselves to either a mediated or an arbitrated outcome. Sometimes a neutral third party is selected to serve as both mediator and arbitrator in a dispute. Med-arb combines the voluntary techniques of persuasion and discussion, as in mediation, with an arbitrator's authority to issue a final and binding decision when necessary.

Because med-arb is less bitter than **litigation,** it also offers the potential for maintaining ongoing business relationships while providing a method for resolving disputes. Research cited in the *Wisconsin Law Journal* has shown that only a very small percentage of cases submitted to med-arb actually wind up being resolved in arbitration.

5.2.4 Mini-Trial

The **mini-trial** is a confidential, nonbinding exchange of information intended to facilitate settlement. The goal of the mini-trial is to encourage prompt, cost-effective resolution of complex litigation. The mini-trial seeks to narrow the areas of controversy, dispose of collateral issues, and encourage a fair and equitable settlement.

5.2.5 Fact-Finding

In dispute resolution, fact-finding is the investigation of a dispute by a neutral third party who issues a report on the findings, usually recommending a basis for settlement. The fact-finder's report can assist further negotiations between the parties.

5.2.6 Conciliation

Conciliation is a process whereby a neutral third party brings the disputing parties together in order for them to solve their problems. The conciliator does not take part in the resolution process or settlement discussions. The conciliator's primary role is to reduce the parties' inflammatory rhetoric and tension, open channels of communication, and facilitate continued negotiations.

5.2.7 Litigation

Litigation is used when the parties require the court system to settle their dispute. In litigation an impartial judge or jury will hear witnesses, review evidence, weigh the facts, and impose a decision. The decision, if not satisfactory to either party, may be appealed before a higher court.

5.2.8 The Difference Between Litigation and Arbitration

Contractors should avoid litigation (lawsuits) or arbitration (third-party decision-making) to settle a dispute. Disputes should be settled between the parties in a professional manner, but if a dispute must be resolved by litigation or arbitration, the contractor should be awarded the choice of the two methods.

Most contractors do not want to become involved in a lawsuit to settle their disputes. The lawsuit will be heard by a judge or jury in a formal courtroom setting. Since the judge and jury are not familiar with the construction process and terminology, a great deal of time may be spent educating the judge and jury.

The litigation process can be lengthy with continued delays, become very costly over time, and cause ill will between the parties. Therefore, arbitration is usually faster, less expensive, and conclusive. The following list explains these and some other advantages of arbitration:

- *Speed* – Prompt scheduling, expeditious procedures, and established time frames for the arbitration hearing reduces the time required to resolve the disagreement.
- *Economy* – Many of the costs associated with litigation can be eliminated or reduced in arbitration.
- *Privacy* – The arbitration hearings and awards are private and confidential. The privacy helps to preserve goodwill and positive working relationships between the parties.
- *Neutral* – The parties select the arbitrators.
- *Expertise* – The arbitrators are familiar with the construction process and terminology.
- *Flexibility* – The arbitration procedures are flexible.
- *Informality* – The arbitration hearing is conducted in an atmosphere that is less formal than a court proceeding.
- *Finality* – The arbitration awards are final, binding, and legally enforceable, subject only to limited review by the courts. The court does not second-guess the arbitrator's decision on facts and the law.

5.3.0 Typical Process for Settling a Dispute

The following sections explain typical steps to process a dispute in litigation, arbitration, and mediation.

Explain that litigation is typically expensive, lengthy, and can cause ill will between parties.

Explain that arbitration is usually faster, less expensive, and conclusive.

Discuss other advantages of arbitration.

MODULE 44104-08 ◆ ISSUES AND RESOLUTIONS 4.11

Review a typical mediation process.

Review a typical arbitration process.

Review a typical litigation process.

5.3.1 Mediation Process

Step 1 Either the parties or the judge presiding over the case may request mediation. The mediator may be court-appointed or selected by the parties through an independent mediation association.

Step 2 After a mediator has been selected, the first meeting date is arranged.

Step 3 The parties meet with the mediator, who guides the discussion and helps to clarify the issues.

Step 4 Private caucuses may be held between the mediator and each party in an attempt to resolve the dispute and bring the parties closer to an agreement.

Step 5 If the mediator cannot assist the parties in settling the dispute, the parties may submit the dispute to litigation or arbitration.

5.3.2 Arbitration Process

Step 1 A party files a demand for arbitration with an American Arbitration Association (AAA) regional office, and a tribunal administrator is assigned to the case.

Step 2 Other parties named in the demand are notified and replies are requested.

Step 3 The tribunal administrator reviews the panel qualifications and selects a number of individuals suitable for the particular case. Depending on the nature of the case, the panel may consist of either one or three arbitrators.

Step 4 The names of potential arbitrators are sent to the parties. Each party will review the arbitrators' qualifications and select those persons they find acceptable. If more than one arbitrator is selected, then the parties will rank the arbitrators in order of preference.

Step 5 Arbitrators are selected by the administrator according to the mutual desires of the parties. If the parties are unable to agree on an arbitrator, the tribunal administrator may appoint an arbitrator.

Step 6 A hearing date and location convenient to the parties and the arbitrator are arranged.

Step 7 At the hearing, testimony and documents are submitted to the arbitrator, and witnesses are questioned and cross-examined by the parties' attorneys. The arbitrator may question the witness. The hearing time will depend on the complexity of the case.

Step 8 Within 30 days after the close of the hearing, the arbitrator will issue a binding award; copies are sent to the parties by the tribunal administrator.

Step 9 The prevailing party will then present the award to the courts, where a judge will issue a judgment in favor of the prevailing party. The award by the arbitrator is final, binding, and legally enforceable, subject only to limited review by the courts.

5.3.3 Litigation

Step 1 A lawsuit regarding the dispute is filed by one of the parties in a local court jurisdiction.

Step 2 The case is assigned to a judge.

Step 3 Preliminary hearings are held in the judge's chambers.

Step 4 Motions are filed; hearing date is set.

Step 5 The case is heard by the judge and/or jury.

Step 6 A decision is rendered.

Step 7 Either party may appeal the decision.

Step 8 Appeal date is set.

Step 9 Appeal case is heard.

Step 10 A decision is rendered.

Step 11 In some states, the decision can be appealed again to the next highest court in the state.

Instructor's Notes:

4.12 PROJECT MANAGEMENT

Review Questions

1. Define problem solving.

____

2. List four barriers to effective problem solving.
   1. ______________________________
   2. ______________________________
   3. ______________________________
   4. ______________________________

3. State two key factors that are at the root of most problems in construction and maintenance firms.
   1. ______________________________
   2. ______________________________

4. List the eight steps of the problem-solving ladder.
   1. ______________________________
   2. ______________________________
   3. ______________________________
   4. ______________________________
   5. ______________________________
   6. ______________________________
   7. ______________________________
   8. ______________________________

5. List five major elements related to job-site problems.
   1. ______________________________
   2. ______________________________
   3. ______________________________
   4. ______________________________
   5. ______________________________

6. The first step in solving problems is to _____.
   a. define the problem
   b. generate alternative lists
   c. make a decision
   d. admit the problem

7. Symptoms of a negative problem-solving climate include poor planning methods, hidden agendas, excessive management demands, and _____.
   a. finger pointing
   b. malfunctioning phones
   c. excessive courtesy
   d. denial of problems

8. The problem-solving tool developed to illustrate relationships between consequence and root causes is referred to as a _____.
   a. CPM display
   b. flow chart scat diagram
   c. fishbone diagram
   d. logarithmic comparisons logic diagram

9. In a fishbone diagram, problems associated with specifications and quantities are related to _____.
   a. people
   b. methods
   c. circumstances
   d. materials

10. During brainstorming, everyone should agree to _____.
    a. take time to think carefully before speaking
    b. interpret ideas as they are presented
    c. focus on one question
    d. only write down the best ideas

11. List the four key elements to successful negotiation.
    1. ______________________________
    2. ______________________________
    3. ______________________________
    4. ______________________________

12. What are the five universal truths of negotiating?
    1. ______________________________
    2. ______________________________
    3. ______________________________
    4. ______________________________
    5. ______________________________

Have the participants complete the Review Questions, and go over the answers prior to administering the Module Examination.

Review Questions

13. List the four methods of gathering information when preparing for negotiation.
    1. ______
    2. ______
    3. ______
    4. ______

14. Negotiations really begin _____.
    a. during the scope meeting
    b. before the contract is signed
    c. after the bid is awarded
    d. during phase four

15. Describe the four phases of negotiation.
    1. ______
    2. ______
    3. ______
    4. ______

16. When should arbitration be used?

______

______

______

17. List the six advantages of negotiating to resolve a dispute.
    1. ______
    2. ______
    3. ______
    4. ______
    5. ______
    6. ______

18. Describe the purpose of mediation.

______

______

______

19. When is litigation used?

______

______

______

20. Describe the purpose of arbitration.

______

______

______

Instructor's Notes:

______

______

______

4.14 PROJECT MANAGEMENT

Summarize the major concepts presented in the module.

Summary

Project managers are responsible for completing projects using the skills and talent of their project team. They must also endeavor to resolve serious problems and issues by negotiating an acceptable outcome with the stakeholders. When negotiations break down, the resolution of the problem will often be determined by a third party through mediation, arbitration, or litigation.

While the project manager needs to monitor a dizzying array of variables and be prepared to make adjustments, having the right management style and communication skills can sometimes outweigh technical abilities. Problem solving is a core competency for a project manager. Being able to identify and resolve problems quickly will have a direct effect on a project's bottom line.

The negotiating process should resolve issues and not alienate others. The goal of negotiating should be to resolve those issues expediently to avoid affecting workers' performance and the project objectives. During the negotiation process you may be able to determine participants' positions by observing body language or other nonverbal cues. However, there are cues that help us to read the responses of others.

When the negotiating process reaches an impasse, everyone involved will move to mediation and, if necessary, to arbitration. Mediation is the first step in dispute resolution. If mediation fails, then the parties must go to arbitration, which is more formal and the result is a binding ruling.

Notes

MODULE 44104-08 ◆ ISSUES AND RESOLUTIONS 4.15

Have participants complete the Trade Terms Quiz, and go over the answers.

Administer the Module Examination. Be sure to record the results of the testing on Craft Training Report Form 200, and submit the results to the Training Program Sponsor.

Trade Terms Quiz

1. Defining the components of a problem is ____________.
2. ____________ is the thought processes and techniques involved to find a solution.
3. A ____________ is used to illustrate the relationships between a problem and its potential root causes.
4. ____________ is a group problem solving technique that involves the spontaneous contribution of ideas from all group members.
5. A legal dispute between parties argued in a court is called ____________.
6. A voluntary process in which the parties work with a neutral third party to find a mutually acceptable solution is called ____________.
7. A legal alternative to litigation whereby the parties to a dispute agree to submit their respective positions to a neutral third party for resolution is called ____________.
8. ____________ go beyond the words spoken to include intonation, stress, rate of speech, and pauses or hesitations.
9. A ____________ is an articulated form of mediation that includes each party to the dispute presenting a summary of its case to the other side, with a neutral party taking the role of the judge but not making any decisions.
10. ____________ is a dispute resolution procedure where the mediator is armed with the power to settle unresolved issues by binding arbitration in the event they are not settled through mediation.
11. An investigation of a dispute by an impartial third person who examines the issues and facts in the case is called ____________.
12. Communication that occurs as a result of appearance, posture, gesture, eye contact, facial expressions, and other nonlinguistic factors are known as ____________.
13. ____________ is an alternative dispute resolution process in which the parties to a dispute agree to use the services of a conciliator.
14. ____________ is the investigation of a dispute by an impartial third party who examines the issues in the case and may recommend settlement.

Trade Terms List

Arbitration
Brainstorming
Conciliation
Fact-finding
Fishbone diagram
Litigation
Med-arb
Mediation
Mini-trial
Nonverbal communication cues
Problem solving
Root cause diagram
Verbal communication cues

Instructor's Notes:

____

____

____

4.16 PROJECT MANAGEMENT

Trade Terms Introduced in This Module

Arbitration: A form of alternative dispute resolution; a legal alternative to litigation in which the parties to a dispute agree to submit their respective positions to a neutral third party for resolution.

Brainstorming: A group problem-solving technique that involves the spontaneous contribution of ideas from all members of the group.

Conciliation: An alternative dispute resolution process in which the parties to a dispute agree to use the services of a conciliator, who then meets with the parties separately in an attempt to resolve their differences.

Fact-finding: (1) A process designed to find information or ascertain facts, e.g., "a fact-finding committee." (2) An investigation of a dispute by an impartial third person who examines the issues and facts in the case, and may issue a report and recommended settlement.

Fishbone diagram: A graphic technique for identifying cause-and-effect relationships among factors in a given situation or problem; also called Ishikawa Diagramming.

Litigation: A legal dispute between parties argued in a court.

Med-arb: A dispute resolution procedure where the mediator is armed with the power to settle unresolved issues by binding arbitration in the event they are not settled through mediation.

Mediation: A voluntary process in which the parties involved in a dispute work with a neutral third party to find a mutually acceptable solution.

Mini-trial: An articulated form of mediation that includes each party to the dispute presenting a summary of its case to the other side, with a neutral party taking the role of the judge but not making any decisions.

Nonverbal communication cues: Communication that occurs as a result of appearance, posture, gesture, eye contact, facial expressions, and other nonlinguistic factors.

Problem solving: The thought processes and techniques involved in finding a solution to an issue.

Root cause diagram: A diagram used to illustrate the relationships between a problem and its potential root causes.

Verbal communication cues: Communication that goes beyond the words spoken to include intonation, stress, rate of speech, and pauses or hesitations.

MODULE 44104-08 ◆ ISSUES AND RESOLUTIONS 4.17

Resources & Acknowledgments

Reference

Frankel, Mark A. "Med-Arb: A Valuable Settlement Strategy," *Special to Wisconsin Law Journal.* Available at www.wislawjournal.com/special/adr2004/med-arb.html

Additional Resources

This module is intended to be a thorough resource for task training. The following reference works are suggested for further study. These are optional materials for continued education rather than for task training.

American Arbitration Association, www.adr.org, 1633 Broadway, 10th Floor, New York, New York 10019.

The Construction Industry Institute, www.construction-institute.org, 3925 West Braker Lane (R4500), Austin, Texas 78759.

Figure Credit

M. C. Dean, Inc., module opener image

Instructor's Notes:

4.18 PROJECT MANAGEMENT

The following are suggested activities or instructional methods to help you teach the material in this module.

General

When you call on someone to answer a question, the rest of the class relaxes or even tunes out because they expect that the question and answer will take place only between you and the participant you called on. Instead, use this technique to involve more participants in answering questions and to keep them on their toes.

1. Ask participants to define a term or explain a concept.
2. After one participant has answered, ask a participant seated nearby if the answer is right. Then ask whether a participant in the back of the room agrees.
3. Ask participants to explain why they think an answer is right or wrong.
4. Use the session to clear up incorrect ideas and encourage participants to learn from their mistakes.

Section 2.3.0 *Root Cause Consequence Diagram*

This exercise will help participants better understand the purpose of the root cause consequence diagram. Participants will need their Participant Guides and pencil and paper to takes notes for discussion. The Instructor will need an overhead projector, screen, dry-erase marker, and Transparencies 4, 5 and 6 (Figures 2 - 4). Allow 30 to 40 minutes for this exercise.

1. Hand out copies of the blank root cause consequence diagram or display Transparency 5 on an overhead projector.
2. Review the case study found in Section 2.3.1.
3. Follow the steps in the case study to create a completed root cause diagram. Refer to Transparency 4 (Figure 2) to answer questions regarding circumstances, people, methods, machinery, and materials to determine the root cause.
4. After completion, compare the class root cause consequence diagram to Transparency 6 (Figure 4).
5. Note any differences. What did you leave out? Did you include anything not mentioned in Figure 4?

Sections 3.3.0–3.3.7

Negotiation Techniques

This exercise will help participants become familiar with different negotiation techniques. Participants will need their Participant Guide and pencil and paper to takes notes for discussion. Allow 30 minutes for this exercise.

1. Review and discuss negotiation styles.
2. Ask participants to name the negotiation technique associated with the following statements:

   - "Well, when you really think about it, you're only spending a few dollars more per day. What's the big deal?" (Crazy Bucks)
   - "Oh my goodness! Are you serious?" (The Flinch)
   - "I can guarantee we can make that happen . . . I just have to get my supervisor to sign off on it." (I Am in Control)
   - "I believe that you think that what you are proposing is fair. In fact, over the 0years I often believed that similar responses were fair myself. However, our experience has found . . ." (Feel, Felt, Found)
   - "You are being completely unreasonable. I'm going back to my office." (I'm Outta Here)
   - "I don't quite understand what you mean. Could you please explain that in better detail?" (Dumb is Smart)
   - Mr. Smith: "This is ridiculous . . . I refuse to sit here and listen to this!"

     Ms. Jones: "I'm sorry for Mr. Smith's outburst. None of us want to leave here without some sort of resolution. Let's go back to the budget and see if we can work something out." (Dr. Jekyll and Mr. Hyde)

Sections 3.3.0–3.3.7 *Negotiation Styles*

This exercise will help participants become familiar with different negotiation styles. Participants will need their Participant Guide and pencil and paper to takes notes for discussion. Allow 30 to 45 minutes for this exercise.

Explain to participants that in addition to negotiation techniques, there are different styles that may be used during negotiations. These styles are referred to by the animals that display attributes similar to them. Review and discuss the negotiating styles below:

- *The Shark* – The shark negotiator is a competitor. This type of negotiator is power-oriented and pursues his own interests at the expense of others. The shark negotiation style may be used when the person knows he is right, when a quick decision is necessary, or when a person is dealing with an intimidator and feels the need to defend his own rights.
- *The Teddy Bear* – The teddy bear is almost the exact opposite of the shark in that the teddy bear feels the need to satisfy the other party, often setting aside his own concerns to do so. The teddy bear style of negotiation is often used when a person realizes he is wrong, the issue is not that important to him, or when the priority shifts from the issue to trying to keep the peace.
- *The Turtle* – The turtle prefers to avoid confrontation at all costs and does not pursue his own interest or the interest of the other party. This type of negotiating style is used when a person is taken by surprise, is unprepared, and needs to collect more information, when there are more important issues to address, or when dealing with an angry, short-tempered person.
- *The Owl* – The owl is the opposite of the turtle, in that the owl attempts to work with the other party to find a solution that works best to fully satisfy both parties' concerns. The owl uses the collaboration style of negotiating when there is a need to work through hard feelings and resentment, when there is a high level of trust between parties, or when other's lives are involved.
- *The Fox* – The fox tries to find a suitable, mutually agreed-upon solution that partially satisfies both parties. The fox uses compromise to find a middle ground between parties. This type of negotiating style is used to reach a temporary agreement for complicated or complex issues, when the goals for each party are important but not worth becoming assertive over, or when competition and collaboration methods do not work.

Section
5.2.0–5.2.7

Alternate Methods of Dispute Resolution

This exercise will help participants become familiar with alternative methods of dispute resolution. Participants will need their Participant Guide and pencil and paper to take notes for discussion. Allow 30 minutes for this exercise.

1. Review and discuss alternative dispute resolution techniques.
2. Ask participants to name the dispute resolution technique based on its description below:
   - The investigation of a dispute by a neutral third party who issues a report on the findings, usually recommending a basis for settlement. (Fact Finding)
   - Involves an attempt by the parties to resolve their dispute with the aid of a neutral third party, whose role is advisory. (Mediation)
   - A third neutral party brings the disputing parties together in order to reduce the parties' inflammatory rhetoric and tension, open channels of communication, and facilitate continued negotiations, but does not take part in the resolution process or settlement discussions. (Conciliation)
   - The submission of a dispute to one or more impartial persons for a final and binding decision. (Arbitration)
   - A confidential, nonbinding exchange of information intended to narrow the areas of controversy, dispose of collateral issues, and encourage a fair and equitable settlement. (Mini-Trial)
   - Begins with a written commitment from both parties to participate and to bind themselves to a mediated or arbitrated outcome. (Mediation-Arbitration)
   - Used when the parties require the court system to settle their dispute. (Litigation)

MODULE 44104-08 — ANSWERS TO REVIEW QUESTIONS

	Answer	Section
1.	To identify obstacles to progress on a project and design ways to overcome or eliminate obstacles.	(Trade Terms)
2.	1. Psychological 2. Physical 3. Cultural 4. Contractual	(2.0.0)
3.	1. Incompetence in solving technical problems 2. A negative problem identification climate	(1.0.0)
4.	1. Admit the problem 2. Define the problem 3. Select a method 4. Generate alternatives 5. Evaluate alternatives 6. Make a decision 7. Implement the decision 8. Evaluate results	(2.1.0)
5.	1. People 2. Circumstances 3. Methods 4. Materials 5. Machinery	(2.2.0)
6.	d.	(2.1.0)
7.	d.	(2.0.0)
8.	c.	(2.3.0)
9.	d.	(2.3.1)
10.	c.	(2.4.0)
11.	1. Accomplishments for both sides 2. The perception that each of the negotiators cared about the other side's needs 3. Both sides were treated fairly 4. Both parties will honor the agreement	(3.1.0)
12.	1. Everything is negotiable 2. We need to continually negotiate 3. The negotiations process is predictable 4. Gathering of information is crucial to success 5. Timeliness of the negotiation affects the outcome of the process	(3.1.0)
13.	1. Using open-ended questions 2. Asking other subcontractors or general contractors about your opponent 3. Visiting with their people 4. Mixing your people with theirs	(3.2.2)
14.	c.	(3.1.0)

	Answer	Section
15.	1. Preplanning – define your needs and goals 2. Planning – identify other party's needs 3. Person-to-person – meet and discuss 4. Follow-up – ensure that you and the other party abide by agreements	(3.2.0)
16.	Any of the following: speed, economy, privacy, neutrality, expertise, flexibility, informality, finality	(5.2.8)
17.	1. Speed 2. Cost savings 3. Confidentiality 4. Preservation of relationships 5. Range of possible solutions 6. Control of process and outcome	(5.1.0)
18.	Involves an attempt by the parties to resolve their dispute with the aid of a neutral third party.	(5.2.1)
19.	When the parties require the court system to weigh the facts and impose a decision.	(5.2.7)
20.	The submission of a dispute to one or more impartial persons for a final and binding decision.	(5.2.2)

MODULE 44104-08 — ANSWERS TO TRADE TERMS QUIZ

1. Fact-finding
2. Problem solving
3. root cause diagram
4. Brainstorming
5. litigation
6. mediation
7. arbitration
8. Verbal communication cues
9. mini-trial
10. Med-arb
11. fact-finding
12. nonverbal communication cues
13. Conciliation
14. Fact-finding

CONTREN® LEARNING SERIES — USER UPDATE

NCCER makes every effort to keep these textbooks up-to-date and free of technical errors. We appreciate your help in this process. If you have an idea for improving this textbook, or if you find an error, a typographical mistake, or an inaccuracy in NCCER's Contren® textbooks, please write us, using this form or a photocopy. Be sure to include the exact module number, page number, a detailed description, and the correction, if applicable. Your input will be brought to the attention of the Technical Review Committee. Thank you for your assistance.

Instructors – If you found that additional materials were necessary in order to teach this module effectively, please let us know so that we may include them in the Equipment/Materials list in the Annotated Instructor's Guide.

Write: Product Development and Revision
National Center for Construction Education and Research
3600 NW 43rd St, Bldg G, Gainesville, FL 32606

Fax: 352-334-0932

E-mail: curriculum@nccer.org

Craft ____________ Module Name ____________

Copyright Date ____________ Module Number ____________ Page Number(s) ____________

Description ____________

(Optional) Correction ____________

(Optional) Your Name and Address ____________

Module 44105-08

Construction Documents

NCCER STANDARDIZED CRAFT TRAINING PROGRAM

The National Center for Construction Education and Research (NCCER) provides a standardized national program of accredited craft training. Key features of the program include instructor certification, competency-based training, and performance testing. The program provides trainees, instructors, and companies with a standard form of recognition through a National Craft Training Registry. The program is described in full in the *Guidelines for Accreditation*, published by NCCER. For more information on standardized craft training, contact the NCCER by writing us at 3600 NW 43rd St., Bldg. G, Gainesville, FL 32606; calling 352-334-0911; or emailing info@nccer.org. More information may be found at our website, www.nccer.org.

HOW TO USE THIS ANNOTATED INSTRUCTOR'S GUIDE

Each page presents two sections of information. The larger section displays each page exactly as it appears in the Trainee Module. The narrow column ties suggested trainee and instructor actions to each page and provides icons (detailed below) to call your attention to material, safety, audiovisual, or testing requirements. The bottom of each page includes space for your notes.

The **Audiovisual** icon indicates an appropriate time to show a transparency or other audiovisual aid.

The **DVD** icon indicates an appropriate time to discuss or show segments of the video.

The **Classroom** icon prompts you to define a term, stress a point, ask trainees to explain a concept, or give examples.

The **Demonstration** icon directs you to show trainees how to perform tasks.

The **Examination** icon tells you to administer the written module examination.

The **Homework** icon is placed where you may wish to assign reading for the next class, assign a project, or advise trainees to prepare for an examination.

The **Laboratory** icon is used when trainees are to practice performing tasks.

The **Materials** icon is a reminder for you to gather materials needed for classes, labs, and testing.

The **Performance Testing** icon tells you to administer a performance test or a portion thereof.

The **Safety** icon is used to emphasize safety issues. It is often keyed to *Caution* and *Warning!* statements in the Trainee Module.

The **Teaching Tip** icon indicates additional guidance is available, such as how to conduct an exercise, get the most educational value from a field trip, or encourage class participation. Teaching Tips may expand on a feature (*Think About It, Did You Know?*) or provide *Quick Quizzes* or similar exercises. You will be referred to the Teaching Tips section at the back of the module if there is additional material.

The **Combination** icon indicates that the laboratory listed corresponds with a performance task. If desired, you can note the proficiency of the trainees during the laboratory, and use it to satisfy performance testing requirements.

PREPARATION

Before teaching this module, you should review the Objectives, Performance Tasks, Materials and Equipment List, and Module Outline. Be sure to allow ample time to prepare your own training or lesson plan and gather all required materials and equipment.

Construction Documents
Annotated Instructor's Guide

Module 44105-08

MODULE OVERVIEW

This module emphasizes the importance of project documentation and explains the types of documents, drawings, and specifications used on a typical construction project. It explains methods of obtaining work in the industry, types of contracts, and insurance requirements. Finally, it describes the change order process and the documents required to close out a project.

PREREQUISITES

Please refer to the Course Map in the Participant Module. Prior to training with this module, it is recommended that the participant shall have successfully completed *Project Management,* Modules 44101-08 through 44104-08.

OBJECTIVES

Upon completion of this module, the participant will be able to do the following:

1. Explain the need for documentation on a project.
2. State the various approaches for obtaining work in the construction industry.
3. Identify the parts of a typical project manual.
4. Identify the various types of drawings and format specifications.
5. Discuss the types of contracts used in the construction industry.
6. Discuss insurance requirements for a company and a project.
7. List the types of documents used on a project.
8. Describe the change order process.
9. List the documents necessary to close out a project.

MATERIALS AND EQUIPMENT LIST

Desktop or laptop computer or DVD player
Multimedia projector and screen
Minor Decisions: Major Impact – How to Deal with Real Issues in Project Management DVD
Transparencies
Blank acetate sheets
Overhead projector and screen
Transparency pens
Whiteboard/chalkboard
Markers/chalk
Pencils and scratch paper
Example of project meeting minutes
Examples of construction drawings (if available)
Examples of an Invitation to Bid document
Sample punch list (if available)
Module Examination (located in the Test Booklet)

ADDITIONAL RESOURCES

This module is intended to present thorough resources for task training. The following reference works are suggested for both instructors and motivated participants interested in further study. These are optional materials for continued education rather than for task training.

PMI Standards Committee, *A Guide to the Project Management Body of Knowledge*. PMI Publications, Newton Square, Pa. (2004).

The American Institute of Architects, 1735 New York Ave., NW, Washington, DC 20006-5292.

The Construction Specifications Institute, 99 Canal Center Plaza, Suite 300, Alexandria, VA 22314.

TEACHING TIME FOR THIS MODULE

An outline for use in developing your lesson plan is presented below. Note that each Roman numeral in the outline equates to one session of instruction. Each session has a suggested time period of 2½ hours. This includes 10 minutes at the beginning of each session for administrative tasks and one 10-minute break during the session. Approximately 10 hours are suggested to cover *Construction Documents*.

Topic	Planned Time
Session I. Obtaining Work, Project Manual, and Drawings and Specifications	
A. Obtaining Work in the Construction Industry	________
B. Project Manual	________
C. Drawings and Specifications	________
Session II. Types of Contracts and Insurance Requirements	
A. Types of Contracts	________
B. Insurance Requirements	________
Session III. Project Correspondence and Change Orders	
C. Project Correspondence	________
D. Change Orders	________
Session IV. Contractor Payments and Closeouts; Review and Testing	
A. Contractor Payments	________
B. Closeouts	________
C. Review	________
D. Module Examination	________

1. Participants must score 70% or higher to receive recognition from NCCER.
2. Record the testing results on Craft Training Report Form 200, and submit the results to the Training Program Sponsor.

USING THE PM DVD WITH *CONSTRUCTION DOCUMENTS*

There are several opportunities within the course of the video to discuss contract documents related to personnel or to the job itself. Daily reports are important records of what has taken place within a given day. Samantha uses them so her foremen can get back on track with their schedules.

Referring to Section 7.0.0, ask the class to consider what other kinds of reports might be generated as a result of the delays presented by Samantha's company. Back charges, perhaps?

Project Management

44105-08

Construction Documents

Assign reading of Module 44105-08.

44105-08
Construction Documents

Topics to be presented in this module include:

Overview

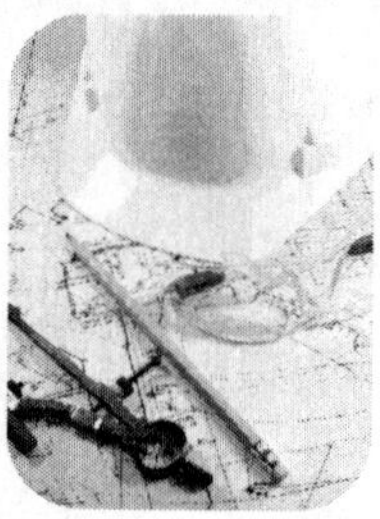

Documentation is vital for a successful business. Numerous studies indicate that poor documentation causes the most problems for industry, followed by poor administration.

Documentation is the who, what, where, how, and when of a construction project. It is a collection of facts that records and documents the actual history of the project. The specific documentation you'll need to maintain depends on the type of project and the needs of the owner, the architect/engineer, the contractors, and the project team members. Without exception, all projects should have some form of documentation.

Instructor's Notes:

Objectives

Upon completion of this module, you will be able to do the following:

1. Explain the need for documentation on a project.
2. State the various approaches for obtaining work in the construction industry.
3. Identify the parts of a typical project manual.
4. Identify the various types of drawings and format specifications.
5. Discuss the types of contracts used in the construction industry.
6. Discuss insurance requirements for a company and a project.
7. List the types of documents used on a project.
8. Describe the change order process.
9. List the documents necessary to close out a project.

Trade Terms

Addenda
Bid bond
Change order
Construction Specifications Institute (CSI)
Punch list
Scope
Specifications
Shop drawings

Prerequisites

Before you begin this module, it is recommended that you successfully complete *Project Management,* Modules 44101-08 through 44104-08.

This course map shows all of the modules in the *Project Management* curriculum. The suggested training order begins at the bottom and proceeds up. Skill levels increase as you advance on the course map. The local Training Program Sponsor may adjust the training order.

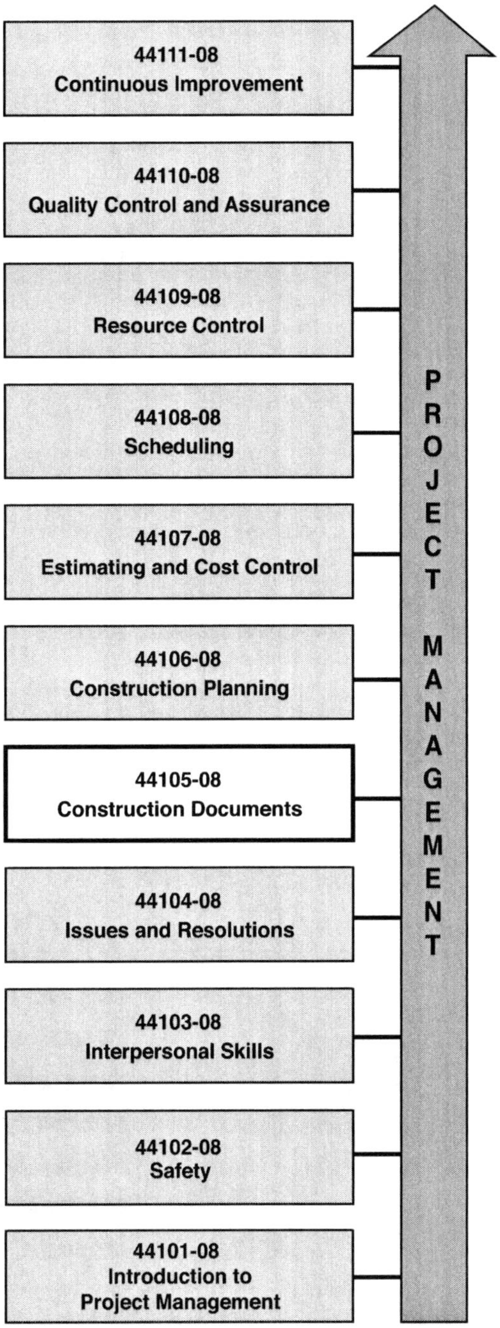

105CMAP.EPS

Ensure that you have everything required to teach the course. Check the Materials and Equipment list at the front of this module.

See the general Teaching Tip at the end of this module.

Explain that terms shown in bold are defined in the Glossary at the back of the module.

Show Transparency 1, Objectives. Review the goals of the module, and explain what will be expected of the participant.

MODULE 44105-08 ◆ CONSTRUCTION DOCUMENTS 5.1

Define documentation and explain why documentation is important.

See the Teaching Tip for Section 1.0.0 at the end of the module.

Show Transparency 2 (Figure 1).

Show Transparency 3, Five Ways of Obtaining Bids. Discuss each of the methods.

See the Teaching Tip for Sections 2.1.0–2.6.0 at the end of this module.

1.0.0 ◆ INTRODUCTION

Although documentation was once a secondary activity for the project manager, during the past decade it has grown to become one of the most important tasks the project manager will face. These days, careful documentation is vital for a successful business. A study conducted by Accenture (a global consulting firm) shows that poor documentation causes the most problems for industry, followed by poor administration. *Figure 1* illustrates the results of this survey.

Documentation is a collection of records that details the history of the project, including who was there, what was done, and where, how, and when events occurred that affected the project. The type and extent of documentation required depends on the project, the owner, the architect/engineer, the contractors, and the project team members. Without exception, all projects must have documentation, although there is some flexibility in exactly how records are kept.

Documentation records the progress of the project as a way to control the amount of time spent and to keep track of costs. Documentation strategies should be developed at the start of the project and continued throughout the life of the project.

The types of documents created will vary from a daily log to provide project managers with the day-to-day progress of the project, to photographs, which give visual details of project activities, to project correspondence, which includes the whole range of written communication among all parties involved on a construction project.

2.0.0 ◆ OBTAINING WORK IN THE CONSTRUCTION INDUSTRY

Contractors and subcontractors who want to obtain work in the construction industry must submit a bid or cost proposal to an owner or architect/engineer. There are several ways a contractor can obtain work:

- Competitive bid
- Invited bid
- Negotiation
- Design-build
- Construction management

2.1.0 Competitive Bid

The most common way of obtaining work in the construction industry is through a method known as competitive bidding. In competitive bidding, the owner will seek bids from several general contractors and will award a contract to the general contractor who has submitted the lowest responsive bid.

While preparing a bid, the general contractor may seek prices from specialized contractors (known as subcontractors), vendors, and material suppliers that have more knowledge and experience to perform the required tasks. The general contractor will review the **scope** of work and bids received, and may select the lowest responsive bid from subcontractors, vendors, and material suppliers to determine the price that will be submitted to the owner.

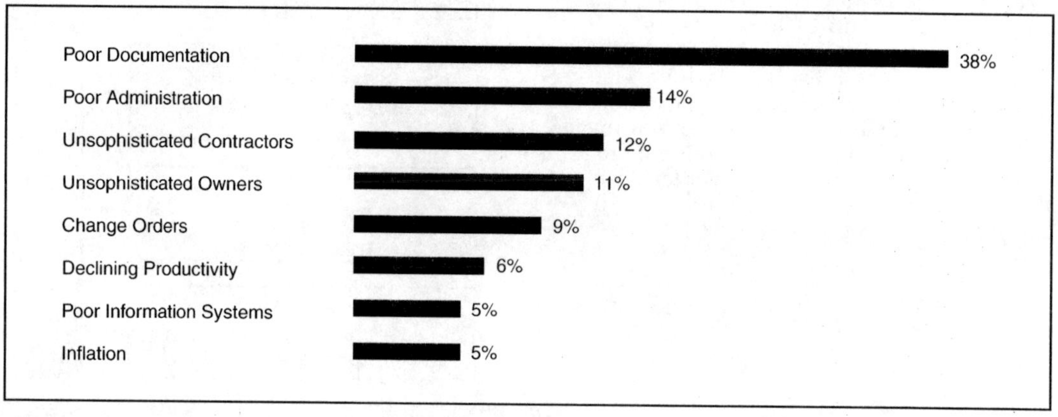

105F01.EPS

Figure 1 ◆ Construction contract hang-ups (based on a survey of 790 industry firms).

Instructor's Notes:

5.2 PROJECT MANAGEMENT

Bids are submitted to the owner and architect/engineer at a predetermined location, date, and time. On public projects and in many private projects, the bids or cost proposals are opened and read aloud. This allows all interested parties to know the results of the bids or cost proposals immediately.

2.2.0 Invited Bid

The invited bid method of obtaining work is very similar to competitive bidding, except the owner will pre-qualify the general contractors. Pre-qualified general contractors are selected based on their experience, financial strength, present and future work loads, and management team. An advantage for being selected to bid is that the contractor will be competing against other contractors with similar experience and business philosophies.

2.3.0 Negotiations

The negotiation method of obtaining work occurs when the owner selects several general contractors to prepare and submit bids or cost proposals. The owner has the opportunity to interview and discuss with each contractor its proposal, work experience, construction schedule, work sequence, and project cost before awarding a contract. The general contractor also has the same opportunity to discuss with the owner its approach and cost of the project.

2.4.0 Design-Build

The design-build method requires the owner to select a contractor that will be responsible for both the design and construction of the project. The owner and the contractor will work closely together from the very beginning of the project. Both will be involved in developing design guidelines, selecting materials, and establishing the project estimate.

2.5.0 Construction Management

Construction management is used by an owner who lacks the expertise and knowledge of construction. The owner will employ a firm that will be responsible for reviewing the contract documents and managing the construction process. The construction management firm will have experience in both the technical aspects and administration of projects.

2.6.0 Summary of Methods

Many contractors will split their methods of obtaining work between competitive bidding, invited bidding, and negotiated work. However work is obtained, each method requires the contractor to compile estimates for the project, including seeking bids from subcontractors, vendors, and material suppliers. The estimating procedure takes considerable time, company resources, and finances; therefore, contractors should examine the various approaches they use to obtain work to make the best use of their estimating and management staff.

3.0.0 ◆ PROJECT MANUAL

The term "project manual" was coined by the American Institute of Architects (AIA) in 1964 when it adopted the concept and title of *Project Manual* to replace the common title **specification** (spec) book used by the industry up to that time.

The project manual consists of written documents prepared by the architect/engineer for the owner in order to bid and construct the project. The project manual contains a variety of separate documents that a project manager must thoroughly understand. These documents include:

- Bidding documents
- Various contract forms
- Specifications
- List of drawings
- Other information required to construct the project

The project manual has been arranged into two categories: the bidding requirements for the project and the specifications. The use of bidding requirements is to assure that all bidders will have the same opportunity to review the documents, prepare a bid, and submit the bid for fair evaluation by the owner and architect/engineer.

A typical project manual will contain bidding requirement documents which include an invitation to bid, instructions to bidders, bid form, **bid bond**, performance and payment bond, form of agreement, general conditions, supplementary conditions, specifications, list of drawings, and **addenda**.

3.1.0 Invitation to Bid

The invitation to bid advises prospective bidders about a project. It contains information bidders need in order to determine if they wish to submit a bid or proposal for the project.

Show Transparency 4, Project Manual Documentation.

Point out the five broad areas of documentation that make up a project manual.

Show Transparency 5, Invitation To Bid Contents.

Point out the contents of an Invitation to Bid package and define the purpose of the Invitation to Bid document.

If available, show an example of an Invitation to Bid document.

MODULE 44105-08 ◆ CONSTRUCTION DOCUMENTS 5.3

Show Transparency 6, Instructions to Bidders.

Discuss the requirements and content of an Instruction to Bidders document and the Bid Form document.

Explain that the bid form provides information about the bidding procedure and the types of standard forms to be used.

Show Transparency 7, Bid Bond or Bid Security.

Discuss the requirements and content of a Bid Bond or Bid Security document.

Show Transparency 8, Performance and Payment Bond.

Discuss the requirements and content of a Performance and Payment Bond.

The following information is contained in the invitation to bid section:

- *Project identification* – Name of the owner, architect/engineer, or organization issuing the bid documents
- *Description of work* – Construction type, project type, and size of the project
- *Types of bids required* – Single contract or multiple contracts for various phases of the project
- *Time of completion* – May be either the number of calendar days or a calendar date
- *Time and location of bid* – The date, time, and location of the bid opening will be outlined, including whether it is a public or private opening
- *Review of documents* – Lists the names and addresses of sources where bidding documents may be examined and obtained
- *Bid bond or bid security* – States if a bid bond or bid security is required, in what amount, and the bond form
- *Bidder pre-qualifications* – Owner or architect/engineer may require the bidder to provide proof of experience on a similar construction project
- *Right to reject bids* – Statement that the owner reserves the right to waive irregularities or to reject all bids
- *Guaranty bonds* – States whether the owner will require a performance and payment bond from the winning bidder

3.2.0 Instructions to Bidders

This document contains information that the bidder must comply with in preparation of the bid. In addition to the information outlined in the invitation to bid, the instructions to bidders may contain the following:

- *Resolution of discrepancies and ambiguities* – Outlines the method by which issues will be resolved during the bidding process
- *Substitution of products* – Requirements, procedures, and time limit under which substitutions of products will be considered during the bidding period
- *Bid submittal* – Requirements for submitting a bid including type of form, number of copies, identification of bid, required signatures, list of subcontractors, and other related information
- *Governing laws and regulations* – Outlines applicable regulatory requirements, labor agreements, safety requirements, and hiring requirements

3.3.0 Bid Form

The bid form is the document used by each bidder to submit its price (or bid) for doing the work specified in the contract documents. The uniformity of the bid form makes it easier for the owner and architect/engineer to evaluate and compare bids.

3.4.0 Bid Bond or Bid Security

The bid bond or bid security is a legal document posted by the bidder as an assurance that the lowest bidder will enter into a contract with the owner if awarded the project. The bid bond or bid security is required at the time of the bid and is usually represented as a percentage of the total bid amount. The bid bond or bid security is usually furnished by the contractor's surety (bonding) company, which will also issue the performance and payment bond, if required by the contract documents.

Many owners and architects/engineers will use the bid bond or bid security requirements as a method for pre-qualifying contractors. Many times, only contractors who can show evidence of bond ability will be permitted to bid on a project. If a contractor's construction record or financial condition is such that a surety company will not provide a bond, the owner or architect/engineer may be reluctant to award a contract to a contractor that may not have the construction experience or the financial capacity to complete the project.

3.5.0 Performance and Payment Bond

Performance and payment bond is a document guaranteeing that the contractor will complete the work in accordance with the terms of the contract and make payments to their subcontractors and vendors. Also, the payment portion of the bond indemnifies the owner for liens filed against the project. The performance and payment bond is issued for the full amount of the contract.

If the contractor defaults (that is, becomes unable to complete the work or make payments) the surety (bonding) company becomes responsible for both the completion of the work and payment of all bills. Usually, the surety company will assume the financial responsibility and will retain another contractor to fulfill the construction portion of the project. Defaults by the contractor may result in construction delays and additional claims against the surety company from the owner and other parties involved on the project. Waiving the performance and payment

Instructor's Notes:

5.4 PROJECT MANAGEMENT

bond requirement may be considered when the owner has knowledge of the contractor's reputation for performance and financial stability.

3.6.0 Form of Agreement

An example of the form of agreement (or contract) between the owner and contractor will be included in the project manual. It may vary from a standard agreement developed by a professional association, the owner, or architect/engineer. It is important that the form of agreement presented in the project manual is reviewed by both the construction management team and its legal counsel.

3.7.0 General Conditions

General conditions define the standard contractual relationships, particularly the interrelationship of the owner, architect/engineer, contractor, and subcontractors. It also details how the contract will be administered and the procedures relative to the project, including rights and responsibilities of the parties. The most commonly used general conditions for private projects is the American Institute of Architects (AIA) *A201* document. For government-related and certain large projects, the architect/engineer or owner will develop general conditions specific to the requirements of a particular project.

3.8.0 Supplementary Conditions

The supplementary conditions are modifications, deletions, and additions to the standard, or prepared, general conditions. Supplementary conditions are developed to meet the project's specific needs or to supplement the requirements of a project.

3.9.0 Specifications

The specifications include information that supports and describes the construction drawings. The role and structure of specifications will be presented in detail in another section of this module.

3.10.0 Addenda

Addenda are documents used to modify the contract prior to receiving the bid. They are issued by the architect/engineer when errors or discrepancies are discovered during the review of the bid documents. When modifications to the contract documents occur after the award of the contract, the document issued by the architect or contractor is known as a **change order**.

4.0.0 ◆ DRAWINGS AND SPECIFICATIONS

Those outside of the construction industry are likely bewildered by the maze of lines, dimensions, and notations associated with construction drawings. Project managers, however, must be able to read and understand construction drawings and specifications if they are to communicate the information to others.

4.1.0 Types of Drawings

The various types of construction drawings that project managers may see are:

- *Conceptual drawings* – Prepared by the architect/engineer during the initial phase of the project, these drawings provide the owner with an architect's perspective of the project, including elevations, building layout, interior finishes, and other information necessary for the owner to evaluate the project.
- *Preliminary drawings* – After the owner has evaluated the project concept drawing, the owner will authorize the architect/engineer to develop drawings that will contain construction details, project requirements, and information necessary to submit to governmental agencies for review, to financial institutes for construction loans, and for development of an estimate of the project cost. The preliminary drawings may go through several stages of development before the final drawings have been accepted by the owner.
- *Working or construction drawings* – These drawings provide the information for the contractors to prepare a bid or proposal to be submitted to the owner.

4.2.0 Reading Drawings

The first step in studying a set of drawings is to become familiar with the general features of the drawings. Project managers should review the methods the architect/engineer used for notation, legends, sections, details, and other information.

After reviewing the general features of the drawings, project managers should review the drawings systematically. One suggested method for reviewing drawings is:

Step 1 Review the site drawings for building layout and location of roadways, utilities, site and building grades, and other site-specific information.

Show Transparency 9, Form of Agreement and General Conditions.

Point out that the Form of Agreement should be reviewed by the firm's legal counsel.

Explain that the General Conditions establish ground rules and relations between the owners and all parties involved in the project.

Show Transparency 10, Supplementary Conditions, Specifications, and Addenda.

Explain that Supplementary Conditions are additional modifications to the contract.

Explain that addenda are modifications issued by the architect/engineer when errors or discrepancies are discovered.

Show Transparency 11, Drawing Types.

Show Transparency 12, Reviewing Drawings.

MODULE 44105-08 ◆ CONSTRUCTION DOCUMENTS 5.5

See the Teaching Tip for Sections 4.0.0–4.2.0 at the end of this module.

Describe the three purposes of specifications.

Explain that specifications provide detailed information that is not included in the drawings, such as finishing materials, fittings, pumps, and so forth.

Step 2 Review the structure drawings for footing sizes, column pads, foundation walls, reinforcement placement, floor system, superstructure, and other specific structural information.

Step 3 Review architectural drawings for floor plans, individual room layouts, elevations, wall sections, finish details, and other specific architectural information.

Step 4 Review the mechanical drawings for plumbing, heating, ventilation, air conditioning, fire sprinkler layout, equipment, and other specific mechanical information.

Step 5 Review electrical drawings for lighting layout, power distribution, equipment requirements, and other specific electrical information.

Step 6 Review all notes, details, and specific instructions.

Step 7 Review all drawings for possible interference of trades and coordination problems.

Step 8 Finally, review all drawings to ensure a complete understanding of the project and the requirements of the architect/engineer.

After becoming familiar with the drawings, project managers should review the specifications.

4.3.0 Specifications

Since there is never enough space on the drawings to include all the required technical information regarding the project, the architect/engineer will develop a set of specifications that provide important technical information about the work to be accomplished and the materials to be used.

To illustrate the need for specifications, consider the job of building custom cabinets in a particular project. The drawings would detail the locations of the units, the elevation, and the arrangement of cabinet doors, drawers, and shelving. The specifications would describe the type of wood to be used, hardware requirements, finishes, and the installation instructions, all in accordance with performance guidelines established by a trade or manufacturers' association. Therefore, in order to meet the requirements of the cabinets for the project, both drawings and specifications must be used.

Specifications serve three purposes:

- Define the responsibilities of the architect/engineer, the owner, the contractor, and the subcontractor
- Supplement the working drawings with detailed technical information regarding the quality of work and the material to be used
- Support the drawings as part of the contract documents

The architect/engineer is responsible for developing and resolving any conflicts between the drawings and specifications. If a conflict occurs between the drawings and specifications, the technical specifications will usually take precedence.

Specifications usually follow a format created by the **Construction Specifications Institute (CSI)** that divides the technical data into divisions. Each division is divided into several sections for materials, products, and installation requirements. The project manager should be familiar with the CSI divisions, sections, subsections, and format. The current format is the MasterFormat™ 2004 Edition. *Figure 2* shows an excerpt from the MasterFormat™ 2004 Edition.

CSI has established the following format for subsections of specifications:

- *Scope* – Describes the scope of work to be performed
- *Material* – Describes the specific materials that are to be used
- *Installation* – Specifies how the material is to be applied
- *Guarantee* – States the quality parameters for which the contractor, subcontractor, and suppliers are responsible

Each division is organized in a consistent manner and contains a number of sections and subsections. For example, Division 03, Concrete, a part of the Facility Services Subgroup, has the following breakdown:

Division	03 00 00	Concrete
Section	03 06 00	Schedules for Concrete
Subsection	03 06 10	Schedules for Concrete Forming and Accessories
Subsection	03 06 20	Schedules for Concrete Reinforcing
Subsection	03 06 20.13	Concrete Beam Reinforcing Schedule
Subsection	03 06 20.16	Concrete Slab Reinforcing Schedule

Instructor's Notes:

5.6 PROJECT MANAGEMENT

MasterFormat™ 2004 Edition – Numbers & Titles

DIVISION 00 – PROCUREMENT AND CONTRACTING REQUIREMENTS

00 00 00 PROCUREMENT AND CONTRACTING REQUIREMENTS

INTRODUCTORY INFORMATION
00 01 01 Project Title Page
00 01 05 Certifications Page
00 01 07 Seals Page
00 01 10 Table of Contents
00 01 15 List of Drawing Sheets
00 01 20 List of Schedules

PROCUREMENT REQUIREMENTS

00 10 00 SOLICITATION
00 11 00 Advertisements and Invitations
00 11 13 Advertisement for Bids
00 11 16 Invitation to Bid
00 11 19 Request for Proposal
00 11 53 Request for Qualifications

00 20 00 INSTRUCTIONS FOR PROCUREMENT
00 21 00 Instructions
00 21 13 Instructions to Bidders
00 21 16 Instructions to Proposers
00 22 00 Supplementary Instructions
00 22 13 Supplementary Instructions to Bidders
00 22 16 Supplementary Instructions to Proposers
00 23 00 Procurement Definitions
00 24 00 Procurement Scopes
00 24 13 Scopes of Bids
00 24 13.13 Scopes of Bids (Multiple Contracts)
00 24 13.16 Scopes of Bids (Multiple-Prime Contract)
00 24 16 Scopes of Proposals
00 24 16.13 Scopes of Proposals (Multiple Contracts)
00 24 16.16 Scopes of Proposals (Multiple-Prime Contract)
00 25 00 Procurement Meetings
00 25 13 Pre-Bid Meetings
00 25 16 Pre-Proposal Meetings
00 26 00 Procurement Substitution Procedures

00 30 00 AVAILABLE INFORMATION
00 31 00 Available Project Information
00 31 13 Preliminary Schedules

DIVISION 32 – EXTERIOR IMPROVEMENTS

32 00 00 EXTERIOR IMPROVEMENTS
32 01 00 Operation and Maintenance of Exterior Improvements
32 01 11 Paving Cleaning
32 01 11.51 Rubber and Paint Removal from Paving
32 01 11.52 Rubber Removal from Paving
32 01 11.53 Paint Removal from Paving
32 01 13 Flexible Paving Surface Treatment
32 01 13.61 Slurry Seal (Latex Modified)
32 01 13.62 Asphalt Surface Treatment
32 01 16 Flexible Paving Rehabilitation
32 01 16.71 Cold Milling Asphalt Paving
32 01 16.72 Asphalt Paving Reuse
32 01 16.73 In Place Cold Reused Asphalt Paving
32 01 16.74 In Place Hot Reused Asphalt Paving
32 01 16.75 Heater Scarifying of Asphalt Paving
32 01 17 Flexible Paving Repair
32 01 17.61 Sealing Cracks in Asphalt Paving
32 01 17.62 Stress-Absorbing Membrane Interlayer
32 01 19 Rigid Paving Surface Treatment
32 01 19.61 Sealing of Joints in Rigid Paving
32 01 19.62 Patching of Rigid Paving
32 01 23 Base Course Reconditioning
32 01 26 Rigid Paving Rehabilitation
32 01 26.71 Grooving of Concrete Paving
32 01 26.72 Grinding of Concrete Paving
32 01 26.73 Milling of Concrete Paving
32 01 26.74 Concrete Overlays
32 01 26.75 Concrete Paving Reuse
32 01 29 Rigid Paving Repair
32 01 29.61 Partial Depth Patching of Rigid Paving
32 01 29.62 Concrete Paving Raising
32 01 29.63 Subsealing and Stabilization
32 01 30 Operation and Maintenance of Site Improvements
32 01 80 Operation and Maintenance of Irrigation
32 01 90 Operation and Maintenance of Planting
32 01 90.13 Fertilizing
32 01 90.16 Amending Soils
32 01 90.19 Mowing
32 01 90.23 Pruning
32 01 90.26 Watering
32 01 90.29 Topsoil Preservation
32 01 90.33 Tree and Shrub Preservation
32 05 00 Common Work Results for Exterior Improvements
32 05 13 Soils for Exterior Improvements
32 05 16 Aggregates for Exterior Improvements
32 05 19 Geosynthetics for Exterior Improvements

105F02.EPS

Figure 2 ◆ CSI MasterFormat™ 2004 Edition.

The CSI MasterFormat™ 2004 Edition has been revised to meet today's complex requirements that have evolved since the original MasterFormat™ 1995 Edition. The revised edition has increased the divisions from 16 to 49, and includes a better defined sub-classification system. The added divisions also allow for future expansion. The differences between the revised and current versions include the following:

- There are four subgroups with new topics and new sections under each subgroup.
- The first 16 divisions (00–15) cover general construction subjects, similar to the 1995 edition.
- There are separate divisions for plumbing, fire suppression, electrical, communications, integrated building systems, safety and security systems, and HVAC.
- A site infrastructure group has been added for unique civil construction (dams, roads, bridges, tunnels, utilities, and so forth) and industrial construction (power plants and factories).
- There are placeholders reserved for future information.
- The five-digit format has been replaced with a six-digit numbering system for sections to be added at a later date.

Show Transparency 13 (Figure 2).

Show Transparency 14, CSI Format Subsections.

Introduce the CSI MasterFormat™ 2004 Edition.

Review the subsections.

Explain that some companies still use the MasterFormat™ 1995 Edition, and participants may need to be familiar with that edition.

Ask participants which edition their companies use.

Assign reading of Sections 5.0.0-6.6.0.

MODULE 44105-08 ◆ CONSTRUCTION DOCUMENTS 5.7

Show Transparency 15, Contract Types.

Review each contract type described in Sections 5.1.0–5.8.0.

Provide job-related examples for each contract type.

5.0.0 ◆ TYPES OF CONTRACTS

A contract, whether between an owner and general contractor or general contractor and subcontractor, is the legal agreement, either written or oral, between two parties in which one party agrees to perform some task or work for the other party in return of some type of payment or compensation.

For any contract to be considered legal and binding, it must contain the following four elements:

- An offer and acceptance
- An agreement between two or more parties
- A benefit to each party named in the contract
- A legal agreement (illegal agreements are not enforceable)

The project manager should be aware of the legal terms and specific clauses contained in each contract. Since no two contracts are ever the same, project managers should review the following clauses of a contract:

- Name of the parties
- Date of the agreement
- Scope of the work to be performed
- Start and completion dates (the schedule)
- Insurance requirements
- Bonding requirements
- Payment terms and conditions
- Indemnifications (the protection of another party against damage or loss)
- Cancellation or termination notices
- Liquidated damages or bonus/penalty clauses
- Special conditions
- Safety requirements
- Guarantees/warranties
- Change order procedures
- Employment requirements

Contracts should be fair to both parties, and each contract clause should be written so that its intent is clear to the parties. One-sided contracts, where terms are vague and benefit only one of the parties, should be reviewed carefully to determine the risk before signing.

Contracting covers a variety of situations. It is important to use the proper type of contract for specific construction requirements. Typical contracts awarded by an owner to a contractor or by a contractor to a subcontractor are:

- Fixed-price/lump sum
- Cost plus
- Target contracts
- Letter of intent or letter contracts
- Fast-tracking contract
- Subcontracts
- Purchase orders

5.1.0 Fixed-Price/Lump-Sum Contract

The most common type of construction contract is the fixed-price, also known as a lump-sum contract. Under this type of contract, the contractor agrees to perform all the work specified at a price agreed to at the time the contract is signed. As noted in the previous section, the price is usually decided through competitive bidding, invited bidding, or negotiated work.

The central characteristic of the fixed-price contract is that the contractor agrees to do the work at a pre-determined price and assumes the risk of all unforeseen conditions (for example, increased labor cost, low productivity, and adverse weather). The owner agrees to pay the price whether the contractor makes a profit or suffers a loss. The fixed-price contract provides the owner with the cost of the project before work starts.

There are circumstances under which a contractor may be entitled to an adjustment to the fixed-price contract. The contract amount may be increased or decreased by issuing change orders triggered by changes in the design or material used or unforeseen site conditions.

Many fixed-price contracts contain a liquidated damage clause requiring the contractor to pay the owner a stipulated amount, usually per day, if the work is not completed on the date stated in the contract. Other contracts will have a bonus provision, which allows the contractor to receive stated bonus payments for completion of the work before the specified contract completion date.

5.2.0 Cost-Plus Contract

A cost-plus contract is one in which the project cost is decided by the actual costs as they are incurred, plus an agreed fee to cover the contractor's overhead and profits. These contracts eliminate or at least minimize many problems associated with fixed-price contracts. For example, inclement weather, escalation of wages or material prices, or changes in work are usually handled without any controversy or claim procedures, since the cost of performing the work will be measured against the actual cost of the entire project.

Many owners favor a cost-plus contract since it creates incentives for the contractor when costs are reduced through efficiency or obtaining the best price from subcontractors or suppliers. A cost-plus contract requires documentation of all costs. In addition, the cost of reimbursable and

Instructor's Notes:

5.8 PROJECT MANAGEMENT

non-reimbursable costs must be clearly defined in the contract.

Examples of typical reimbursable and non-reimbursable costs include:

- *Reimbursable cost-plus contract items* – Items such as cost of materials, cost of labor, and salaries related to the job. Other items include taxes for labor and materials, subcontractor payments, equipment costs, permits and fees, bonds, insurance, and safety costs.
- *Non-reimbursable cost-plus contract items* – These include salaries of personnel working at the contractor's principal or branch office, overhead and general expenses, capital expenses, or other costs such as disposal of materials, damage to property and other issues related to negligence.

The following are various types of cost-plus contracts that can be used to determine contractor overheads and fees (profit):

- *Cost-plus a fixed fee* – This contract reimburses contractors for their actual construction costs and their compensation for overhead and fee (profit) that is a fixed amount determined at the time the contract is signed. The fee is frequently determined by a percentage based on the original estimated cost of the project.
- *Cost-plus a percentage of cost fee* – Contractors are reimbursed for their actual construction costs and receive a fee based on an agreed percentage applied against the actual costs. The final fee will be determined when the actual cost has been specified.
- *Cost-plus an incentive fee* – Contractors are reimbursed for their actual construction costs plus compensation based on a special formula for sharing actual costs over or under the estimated costs.
- *Cost-plus with guaranteed maximum fee* – Contractors are reimbursed for their actual construction costs including overhead and profit up to but not exceeding the agreed contract amount. The contract may allow for sharing of any cost savings if the project comes in under the guaranteed maximum.

5.3.0 Target Contract

A variation of the fixed-price and cost-plus type contracts is the target contract. Although there are various types of target contracts, they all have some common elements including the following:

- An agreement between the parties on estimated target costs
- Target profit or fee
- Ceiling price
- Other combinations to establish a fair cost target.

Upon completion of the project, the final cost is decided and the contractor's fee will be adjusted if the target costs were met.

5.4.0 Unit-Price or Unit-Cost Contract

The unit-price contract, also known as the unit-cost contract, fixes the price per unit at the time the contract is signed and that price remains the same throughout the life of the contract. Several examples of unit-cost items are electrical (add ceiling outlet, add wall switch, wire and connect exhaust fans), mechanical (add water heater, add 4-inch floor drain), and finishes (add wall tile, add texture to stucco finish).

Sometimes a problem arises with a unit-price contract when there is a substantial overrun or underrun in quantities. Contractors develop their unit price based on the estimated quantities listed in the bid documents, which should be reasonably close to the final measured quantities. If the quantities vary substantially, then the contractor may be entitled to a price adjustment.

5.5.0 Letter of Intent or Letter Contract

A letter of intent or letter contract is a preliminary contract issued by the owner under which the contractor may start work prior to the execution of a formal contract. Letters of intent or letter contracts outline the liability of both parties including maximum costs, fees, scope of work, payment, and a cancellation provision if a formal contract cannot be executed.

5.6.0 Fast-Tracking Contract

A fast-tracking contract, also known as a design and construction contract, allows construction to proceed in stages or phases by issuing contracts when the design of each phase is sufficiently completed.

5.7.0 Subcontract

A subcontract is a contract between a general contractor, or prime contractor, and another contractor, usually a specialty contractor. The subcontractor agrees to do the labor and material portion of the work that the general contractor has agreed to perform under the same agreement with the owner.

A subcontract can be fixed or cost price, and is subject to the same laws that govern normal contracts. A subcontract ordinarily does not create any

Discuss participants' experiences with contracts, including the advantages and disadvantages of different types.

See the Teaching Tip for Sections 5.0.0–5.8.0 at the end of this module.

MODULE 44105-08 ◆ CONSTRUCTION DOCUMENTS 5.9

Show Transparency 16, Basic Types of Insurance.

Explain that a company requires considerable insurance coverage.

State that a deductible is usually stipulated in each policy and that accidents or losses may cause premiums to increase.

Stress that the project manager is responsible for managing risk through careful planning and safety awareness in order to minimize accidents and losses.

Review each type of insurance required by contract documents, and the types maintained by a typical construction company.

See the Teaching Tip for Section 6.0.0 at the end of this module.

Assign reading of Sections 7.0.0-8.5.0.

contractual relationship between the subcontractor and the owner. This relationship is important in deciding rights under state and construction lien laws and with respect to payment and performance bonds.

5.8.0 Purchase Order

A purchase order is used to cover the purchase of materials from a supplier and contains provisions appropriated to such transactions that usually are governed by the law of sales, which may differ in various respects from the law of contracts.

6.0.0 ◆ INSURANCE REQUIREMENTS

An extensive amount of insurance of various types is necessary for a company to operate. Policy acquisition and maintenance is needed to protect the employees, owners, public, and the company from losses.

Insurance does not protect the company from all losses, since a deductible is often stipulated in each policy. When an accident or loss occurs, the company's premiums may increase. The project manager should understand the company's insurance requirements, be aware of the company's risk exposure, and help reduce premiums through careful planning and safety awareness.

Insurance is an extremely complicated business and the project manager must be able to recognize any potential problems related to insurance coverage. Generally the home office will handle all the administration necessary for filing a claim but relies on the project manager or field supervisor for notification and written reports. The project manager must realize one of his primary responsibilities is to minimize accidents and losses.

The basic types of insurance a company should carry include: workers' compensation, general liability, vehicle, builder's risk or installation floater, equipment floater, and property insurance.

6.1.0 Workers' Compensation Insurance

Workers' compensation insurance pays employees who are injured on company time. It is usually required by a state statute, and the company is required to maintain coverage in each state in which it works or risk financial penalty. The field supervisor must report all injuries to the project manager. Failure to report injuries may result in the injured employee losing his benefits and the company being penalized.

6.2.0 General Liability Insurance

General liability insurance provides coverage for claims that the company is obligated to pay as a result of bodily injury or property damages. General liability insurance may include the following coverage:

- Premises owned by the company
- Project construction activities
- Liability incurred as a result of injury or damage caused by a subcontractor
- Completed work
- Liability assumed under written contracts
- Libel, slander, false detention, false entry and similar personal injury claims

General liability insurance policies may not cover all possible types of losses. The project manager should be aware of the items covered by a general liability policy, as well as the standard exclusions. If an incident occurs that may involve a standard exclusion, the project manager should notify the home office immediately to either verify the coverage or advise management of a potential problem. Some standard exclusions include the following:

- Property in care, custody, or control of the company
- Losses as a result of design
- Losses from the use of watercraft, aircraft, or autos
- Losses occurring from underground work
- Losses caused by explosions
- Losses caused by collapse

The premium for general liability insurance may be very high. The premium cost is incorporated into the company overhead or project costs. As losses occur, premiums rise, increasing the company's cost of doing business and therefore making the company less competitive.

6.3.0 Vehicle Insurance

Vehicle insurance covers losses the company incurs from ownership and usage of automobiles or trucks and usually includes physical damage to company vehicles due to collision, fire, theft, and vandalism. Many companies also carry insurance on leased or employee-owned vehicles used for company business.

If an accident involving a company-insured vehicle occurs, the project manager should notify the home office. The employee involved in the accident must complete an accident report with

Instructor's Notes:

5.10 PROJECT MANAGEMENT

information on the cause of the accident, names of witnesses, names of parties involved, photographs, and sheriff or police reports. All pertinent documents should be sent to the home office as soon as possible, so the insurance company can be informed.

6.4.0 Builder's Risk/Installation Floater Insurance

Builder's risk/installation floater insurance provides protection for damage to the work being performed and for losses of materials and labor due to fire, theft, vandalism, or other causes. Unless contractually provided by the owner, all builder's risk insurance is maintained by the contractor in an amount equal to the cost of the project. An installation floater policy is usually written in the amount of the subcontractor's contract or the general contract. The policy is in effect when the contractor gains possession of the materials, during temporary storage on or off site, while in transit, and during installation. Coverage ceases upon acceptance of the work performed.

Builder's risk/installation floater insurance generally provides protection against most losses of materials, labor, fixtures, and other items due to the following causes:

- Fire
- Theft
- Malicious mischief
- Vandalism
- Wind storm
- Flood and earthquake

When a loss occurs on the project, the project manager must notify the home office immediately, so the insurance company can be notified. A report that contains a list of the material lost or damaged, photographs, names of any witnesses, and a copy of the sheriff or police report should also be prepared by the project manager.

6.5.0 Equipment Floater Insurance

Equipment floater insurance provides coverage against loss or damage of company-owned or leased equipment and tools that are not covered under the builder's risk insurance policy. The policy covers only the equipment that is listed on the insurance policy schedule of equipment. It is up to the project manager and field supervisors to advise the home office of any additions of equipment so the policy can be adjusted to reflect the changes. The coverage includes damage or loss to equipment due to fire, theft, or accidents.

6.6.0 Property Insurance

Property insurance provides coverage of all company-owned and -leased property that is not already covered under another insurance policy, including the contents of the main office building and job-site offices.

Show Transparency 17 (Figure 3).

7.0.0 ◆ PROJECT CORRESPONDENCE

Discuss how a typical project generates a large amount of correspondence.

Discuss how problems can occur when people are unaware of their responsibilities.

Most projects generate a considerable amount of written correspondence, and project managers occupy the center of the correspondence flow. They determine who shall communicate to whom, proper relationship channels, and secondary parties who must be informed. The project manager is responsible for channeling information to the proper parties, so theycan quickly perform any required action or response. *Figure 3* illustrates the typical roles and authority of key personnel on a project.

Correspondence includes all drafted, written, and printed material relating to the project. To handle this vast amount of material, project managers should develop a filing system and maintain it throughout the life of the project. *Figure 4* shows a suggested filing system format; however, users will generally define their own numbering system. A successful filing system will meet two basic requirements:

Show Transparency 18 (Figure 4).

Show Transparency 19, Project Manager Correspondence.

- Provide easy recovery of any document at any time during the life of the project
- Provide a method of follow-up to verify that a given correspondence has achieved its intent

See the Teaching Tip for Sections 7.0.0–7.1.0 at the end of this module.

Correspondence that is usually prepared by the project manager includes the following:

- Formal letters
- Memorandums
- Speed messages
- Emails
- Directives

When preparing any correspondence, the project manager should ensure that it addresses the subject and reflects the company business philosophy. One poorly executed letter, a hasty email, or a poorly thought-out comment can destroy the goodwill a contractor has been building for months, or even years. When composing a letter or any other correspondence, make every effort to consider the recipient's reaction to the message. Write in a clear, concise manner, giving all essential information that will accomplish the purpose in the fewest possible words, and always remember to be courteous.

MODULE 44105-08 ◆ CONSTRUCTION DOCUMENTS 5.11

Show Transparency 20 (Figure 5). Stress the importance of accurate daily reports.

Show Transparency 21 (Figure 6).

Show Transparency 22 (Figure 7).

See the Teaching Tip for Section 7.2.0 at the end of this module.

Discuss the importance of daily reports, and that they should be routine, not punishment, as in Anthony's case.

Item	Project Manager	Superintendent	Other
Schedule			
A. Long Term	x	x	
B. Short Term		x	
Decision on Field Methods		x	
Responsible For:			
A. Subcontracts	x		
B. Submittals			X (Engineer)
C. Change Orders	x	x	x
Field Documents			
A. Quantity Reports		x	x
B. Equipment Reports		x	x
Owner/Client Relationship	x		
A/E Relationship	x		
Equipment Repairs & Maintenance		x	x
Purchase of Tools		x	x
Storage of Tools		x	
Equipment Rental	x	x	
Storage of Materials			x
Personnel Decisions	x	x	
Budget Tracking and Analysis	x	x	
Cost Control	x	x	
Review Subcontractors' Status	x	x	
Communication Procedures	x	x	

105F03.EPS

Figure 3 ◆ Project chart of authority.

7.1.0 Daily Reports

Daily reports or logs catalogue activities that occur on the project on a given day. It is a permanent written record of the project recorded by the field supervisor.

The format of a daily report will vary with the type of information each contractor requires from the field supervisor, but the basic intent of all reports is to document, for the project manager, what has taken place within the project on a given day.

The project manager should insist that the field supervisor complete the daily report by the end of each working day. The field supervisor who conscientiously writes a daily report will avoid struggling to recall (with inevitable omissions and errors) activities that occurred days or weeks earlier. The project manager is responsible for regularly reviewing the daily reports. *Figure 5* illustrates a typical daily report and information that may be required.

7.2.0 Meetings

There are two types of formal meetings that a project manager may conduct or attend. One is the supervisor's meeting, usually held weekly, and the other is the owner or architect/engineer meeting, usually held monthly.

Each type of meeting should be well planned. The agenda will advise the attendees of the structure and objective of the meeting. *Figure 6* illustrates a typical meeting agenda.

During the meeting, minutes should be taken by at least one person, with this duty rotating to a different staff member each time. When minutes are officially recorded and distributed, they must be reviewed immediately. The accuracy of the minutes is extremely important because they become part of the project documentation. Errors in the minutes should be corrected to ensure they represent an accurate record of the meeting. *Figure 7* illustrates typical project minutes.

7.3.0 Project Photographs

Photographs are one of the best visual forms of documentation. Photographs should be taken for the following reasons:

- Show daily progress
- Verify existing conditions
- Illustrate quality of work
- Record materials and equipment stored on site
- Record construction techniques
- Communicate problems to the architect/engineer or contractor's home office

Since photographs document the project, the following information should be recorded on the reverse side of each photograph:

- Job name and job number
- Date the photograph was taken
- Name of the person taking the photograph
- Subject or purpose of the photograph
- Any other comments describing the photograph

Each job site should have a still camera, digital camera, or video camera, and it should be used on a regular basis. The subjects and number of photographs that the project managers will require field personnel to take vary from project

Instructor's Notes:

5.12 PROJECT MANAGEMENT

Pre-Bid Files:
1.A. Project Solicitation
1.B. Original Specifications
1.B.1. Addenda
1.B.2. Current Amended Specifications
1.C. Original Drawings
1.C.1. Revision Drawings
1.D. Site Visit Report
1.D.1. Step-bid Site Photographs
1.D.2. Owner-furnishes Site Reports
1.D.3. Contractor-procured Site Reports
1.H. Transmittals
1.E. Owner's Finances
1.E.1. Credit Report
1.E.2. References
1.F. Subcontractor Quotations
1.G. Unit Price Calculations
1.I. Bid Sheets
1.J. Submitted Document

Pre-Construction Files:
2.A. Meeting Minutes
2.B. Modifications/Proposals
2.C. Notice of Award
2.D. Permits
2.E. Signed Contracts
2.E.1. Contractor/Subcontractor
2.E.1. Purchase Orders
2.F. Bonds (Performance and Payments)
2.F.1. Contractor/Owner
2.F.2. Subcontractor/Contractor
2.G. Insurance Certificates
2.H. Schedules
2.H.1. Contractor
2.H.2. Subcontractor
2.H.3. Suppliers
2.H.4. Schedule Update

Construction Files:
3.A. Notice to Proceed
3.B. Daily Superintendent Report
3.C. Disruption/Interference Reports
3.D. Change Order Files
3.E. Claim Files
3.F. Shop Drawing Log
3.G. Submittal Log
3.H. Transmittals
3.I. Revised Drawings
3.J. Progress Photographs
3.K. Schedule
3.K.1. Original Schedule
3.K.2. Revisions/Updates
3.K.3. Correspondence
3.L. Job Meeting Minutes
3.M. Inspection Reports
3.N. Payment Requisitions
3.P. Correspondence
3.P.1. Owner
3.P.2. Construction Manager
3.P.3. Architect
3.P.4. Subcontractors
3.P.5. Suppliers
3.Q. Intraoffice Memoranda
3.R. Accident Report
3.S. Insurance Claims

Close-Out Files
4.T.1 Lien Waivers
4.T.2. Turn over to Customer
4.T.3. Punch List
4.T.4. Architect's Approval
4.T.5. Warranties and Guarantees
4.T.6. Final Releases
4.T.7. Final Requisition

105F04.EPS

Figure 4 ◆ Correspondence filing system.

to project. Since the advent of more user-friendly video cameras, many contractors now use these devices to record and document projects. They can provide more information than a photograph because they can record actions and verbal comments.

The project manager must maintain a master log listing each photograph or video by sequence. *Figure 8* illustrates a typical job-site photograph.

7.4.0 Schedule

The project schedule is an estimate of the time and the sequence of activities necessary to get a project completed on time. After the schedule has been approved and issued, project managers should review and update the schedule on a regular basis. All activities that affect the schedule, whether they are activities controlled by the con-

Refer to Figure 8 in the Participant Guide.

Explain the value of using photographs to document job-site activities and situations.

Point out that every job site should have a still camera, digital camera, or video camera.

Provide examples where job-site photographs helped resolve a situation.

Discuss the importance of maintaining a master log of photos and videos.

Explain the importance of a project schedule.

Emphasize that the project manager should review the schedule daily and update as needed.

Inform participants that developing, maintaining, and resolving schedule conflicts will be covered in the Scheduling module.

MODULE 44105-08 ◆ CONSTRUCTION DOCUMENTS 5.13

Explain that supervisors use time cards to verify the accuracy of hours worked.

Point out that time cards are used by the project manager to estimate and measure productivity.

Explain the purpose of shop drawings and describe common types.

Show Transparency 23 (Figure 9).

See the Teaching Tip for Section 7.6.0 at the end of this module.

DAILY PROJECT REPORT

PROJECT ____________ **DATE** March 3

REPORT BY John Doe **REPORT NO.** 043

WEATHER AVG TEMP: AM:50 PM:80 **Sunny, Cloudy, Rainy:** sunny

WORKFORCE:

Plumbers	6 incl. 1 Foreman	Sheet Metal	4	Cooling Tower	
Fitters	7 incl. 1 Foreman	Insulation	1	Refrigeration	
Laborers		Temperature Controls	1	Other	
Operators		Painter			
Others					

JOB PROGRESS REPORT:
3 Plumbers working basement pump room
2 Plumbers roughing in 3rrd floor
3 Pipefitters in main equipment room installing chilled and condenser water lines
2 Fitters working on HVAC risers – 4th floor
1 Fitter fabricating coil connections
4 chilled and condenser water pumps delivered today

SPECIAL DECISIONS OR INSTRUCTIONS:
General Contractor requested that we pull off 3rd floor so drywall contractor can stock floor with sheetrock. (This will affect sheet metal sub and will cause us to shift a crew of plumbers to another work area.)

ITEMS NEEDING EXPEDITING:
Air handling units were scheduled to ship February 15, but I've heard no word. Please check with supplier and expedite. Need badly or will have to cut back on pipefitter crew and will be covered up by sheetrock.

105F05.EPS

Figure 5 ◆ Sample daily project report.

tractor or not, should be documented on the schedule. The role of project managers in developing, maintaining, and resolving scheduling problems is discussed in the *Scheduling* module.

7.5.0 Time Cards

Time cards are important documents because they record labor and material costs as they are expended. Time cards can verify additional costs resulting from change orders or directives from the architect/engineer or owner. Time cards should record hours worked, areas or activities worked, work accomplished, and material used. Since time cards track the cost of work, project managers should verify that the field supervisors are completing the time cards properly.

7.6.0 Shop Drawings

Contract drawings and specifications provide sufficient information to estimate a project but may lack the level of detail needed to produce all the various components of a project. The information necessary to produce or provide components is contained in **shop drawings**. Shop drawings encompass drawings, schedules, lists, performance data, manufacturers' literature, samples, vendors' drawings, and submittals.

The architect/engineer normally requires that shop drawings be submitted before material or equipment is delivered to the project, and that the material or equipment meets the requirements contained in the contract documents. Consequently, shop drawings are provided for almost

Instructor's Notes:

5.14 PROJECT MANAGEMENT

XYZ Company, Inc.
Project: Mercy Hospital

Meeting Agenda
June 12

1.	Review last meeting minutes.	9:00 a.m.
2.	Review status of action items.	9:15 a.m.
3.	Discuss job safety requirements.	9:30 a.m.
4.	Update schedule, comparing actual to planned.	10:00 a.m.
5.	Develop a plan to improve schedule, recover lost time, improve productivity.	10:30 a.m.
6.	Review field observation on quality and work standard.	10:45 a.m.
7.	Review outstanding change orders.	11:00 a.m.
8.	Review submittal schedule status.	11:15 a.m.
9.	Develop goals and assign action items.	11:30 a.m.
10.	Discuss potential problems, delays, conflicts that will affect the progress of the project.	11:45 a.m.
11.	Other business.	12:15 a.m.

105F06.EPS

Figure 6 ◆ Sample meeting agenda.

every component of a project. The general contractor's shop drawings may range from reinforcing steel and millwork to finished hardware. The mechanical contractor's shop drawings may include ductwork, piping, and equipment data. The electrical contractor's shop drawings may range from electric panels and power distribution to a single light fixture.

Although the procedure for processing shop drawings may vary from contractor to contractor, there are basic steps that should be followed:

- Log in shop drawings when received.
- Verify quantities of required copies and samples.
- Review drawings and submittals for accuracy.
- Submit shop drawings to the architect/engineer.
- Review drawings when returned from the architect/engineer for notes and comments.

If shop drawings are returned approved or approved as noted, the contractor returns them to the supplier or subcontractor and distributes the information to the supervisor and other members of the project team. If the shop drawings are rejected by the architect/engineer, the contractor returns the information to the supplier or subcontractor for correction and resubmission. Rejected drawings should not be distributed to the field to avoid using inaccurate information in construction.

All shop drawings received from a supplier or subcontractor must be carefully checked against the contract drawings and specifications before being forwarded to the architect/engineer for review and approval. The architect/engineer is responsible for checking the manufacturer, style, and quality of the material or equipment to make sure that they meet project requirements, then approving or rejecting the shop drawings and returning the information to the contractor.

The contractor should note any comments or changes noted by the architect/engineer. Remember, the architect/engineer can give the only qualified approval; the contractor is ultimately responsible for verifying quantities and dimensions.

Shop drawings are important because of the information they provide. The responsibility for verifying that information should not be taken lightly. Usually the project manager or project engineer is responsible for checking shop drawings, but the superintendent should be involved when dimensions must be verified.

Over the years, the responsibility for checking shop drawings has become a major area of disagreement. Generally, contract documents stipulate that the approval of shop drawings by the architect/engineer does not relieve the contractor of responsibility for furnishing materials and equipment in accordance with the contract documents. Nor does such approval relieve the contractor from responsibility for comments not noted on the shop drawings. As a result, maintaining records of transmittal, receipt, and approval of shop drawings is critical.

Since it is the project manager's responsibility to keep track of all transactions related to shop drawings, an organized system should be developed. The shop drawing log in *Figure 9* is an example of such a system.

7.7.0 Back Charges

During the course of a project, the project manager may be requested to perform work that is outside the contractor's normal **scope** of work such as furnishing labor, material, or equipment to assist another trade, or to repair work damaged by another contractor. These situations may lead to back charges, companies billing each other for extra costs incurred. The project manager should recognize that back charges are a form of change order and, as such, increase or decrease the company's costs. Before a project manager can approve a back charge, the applicable documents should be checked to verify that the back charge is justifiable.

Explain that sometimes work may be requested that is outside the normal scope of work and provide examples.

Explain that back charges are similar to a change order.

Stress the importance of reviewing all information to justify any back charges.

Discuss how all of the production delays in Samantha's company could have initiated back charges and how back charges are a form of change order.

MODULE 44105-08 ◆ CONSTRUCTION DOCUMENTS 5.15

Explain that change orders are a major issue for the project manager because they affect the contractor and subcontractors.

Emphasize that the project manager needs to know the authorization process so that work will not be delayed.

Stress the importance of ensuring that the change order is within the original intent of the work.

Provide a job-related example of a change order.

XYZ Company Minutes

May 27

Subject: Subcontractors' Meeting

Time: 9:00 A.M.

Attendees:

John Reid	Reid's Mechanical Maintenance
Jane Smith	J and R's Formwork Company
Larry Jones	Zap! Electric Company
Richard Graham	Bare Walls Drywall & Painting
Jake Robinson	Grounded Flooring, Inc.
Dan Stone	Builder's Best Company, Inc.
Walt Meyers	Walter and Sons Glass Installation

1. Safety: All subcontractors were reminded that this project is a "HARD HAT" JOB. Hard hats must be worn at all times.
2. Schedule: The total project is nine days ahead of schedule. All subcontractors must maintain their present manpower and productivity in order to continue ahead of schedule.
3. General Notes:
   a. Next window delivery is scheduled for June 2.
   b. Metal studs installation on 5th floor will be completed by June 5.
   c. Domestic water rough-in on 5th floor will be completed on June 9.
   d. Electric rough-in on 5th floor will start June 9.
   e. Flooring for top floors will start next week.
4. Change Order: The owner is revising the first floor layout. All work on the first floor is on "HOLD" until new drawings are released by the architect.
5. Clean-up: Each subcontractor must clean up its own trash. A dumpster is provided at the south end of the property for all contractors to use.

Meeting Adjourned at 9:32 A.M.

The meeting minutes are transcribed and will become part of the project record. If you disagree with the minutes, contact this office within seven days with your corrections or comments.

Minutes recorded and transcribed by: ______________________

105F07.EPS

Figure 7 ◈ Sample project meeting minutes.

The following list gives examples of back charges that can occur on construction projects:

- A contractor furnishes scaffolding to another contractor
- A contractor provides saw cutting and patching for another contractor
- A contractor furnishes a pump to another contractor to remove water from a trench
- A contractor provides labor and equipment to remove another contractor's trash
- A supplier furnishes materials to another contractor
- A contractor reworks material furnished by the supplier

8.0.0 ◈ CHANGE ORDERS

Few items in a project manager's daily schedule cause more work and anxiety than change orders. Change orders reflect alterations in the planning, design, or execution of a job. As authorized by the owner or architect/engineer, change orders increase or decrease the cost of the project, resulting in an increase in administrative time and cost. Dealing with change orders is unavoidable because construction projects are bound to encounter changes to the original scope of work.

The bid documents normally include contract clauses that permit the owner and architect/engineer the authority to request that the contractor carry out changes to the work. However, these

Instructor's Notes:

5.16 PROJECT MANAGEMENT

105F08.EPS

Figure 8 ◆ Sample project job-site photograph.

changes must be within the intent of the contract; a contractor cannot be required to supply craftsmen, equipment, and material not originally contemplated. When the owner requests a change, the contractor must determine if it is within the original intent of the work. If the change is outside the original scope, then the contractor should submit a request for the additional work.

8.1.0 Reasons for Creating a Change Order

During the course of the project, items may be identified by the architect/engineer or contractor that may become the subjects of change orders. Change orders may be a result of many factors to include the following:

- Document errors or omissions
- Scope of work changes requested by owner
- Adjustments to the project schedule
- Building code or regulation changes
- Changes in materials specified for the project

8.2.0 The Change Order Process

The change order process should begin with the pre-construction phase of a project. By reviewing the change order process early in the project, the manager can save time during construction and avoid disputes. The change order procedure should address how the contractor, owner, and architect/engineer will review the change order for scope of work and cost. Necessary change orders should be approved as quickly as possible since contractors are not authorized to proceed with work until the change order has been approved. *Figure 10* illustrates a typical change order.

Change order procedures should include the following information:

- The scope of work
- Justification for the change orders
- The amount requested by the contractor
- The start and completion dates of the change
- The effect the change has on the entire project

8.3.0 Change Order Review Meetings

Regular change order meetings should be scheduled to review and identify the changes to the work. Part of the meeting should involve the contractor reporting on the progress of pending, existing, and future change orders. This provides the owner with an early notification of possible additional costs and an assessment of the final construction cost.

8.4.0 Pricing of Change Orders

The cost of a change will vary depending on the phase of construction. During the beginning of a project, when only a few trades are involved, the relative cost may be lower than later in the con-

Show Transparency 24, Factors Related to Change Orders.

Discuss reasons for change orders and provide examples (change in scope, site conditions, code revisions, material substitutions, etc.).

Emphasize that changes usually affect the schedule and productivity, and contribute to overhead costs.

Show Transparency 25 (Figure 10).

Discuss and provide examples of the change order process, which is part of the construction contract.

Emphasize that any change requests should be in writing, not verbal.

Explain the need for change order review meetings.

	Job Name	Warehouse Bldg.		Job No.	86006						Add'l Dwgs						
Spec Station	Contractor	No	Drawing Number or Item	Sent Dt.	Ret'd Dt.	Arct.	Sub	Files	Job	Other	Remarks	Need	File	Job	Mach'l	Elec.	Mis. Iron
2660	Utilities, Inc.	1	Water Distribution Pipe	1/4/xx	1/18/xx		xx		x		Approved						
3100	Salt	2	Construction JT Loc.	1/7/xx	1/22/xx			xx			Approved						
3200	AB Steel	3	Footing & Column Steel	1/7/xx	1/20/xx		xx		x		Approved						
3300	Concrete, Inc.	4	Concrete Mix	1/12/xx	1/20/xx		x				Not Approved						
3345	L&B Chem.	5	Curing Compound	1/13/xx	2/2/xx		xx		x		Approved						
3300	Concrete, Inc.	6	Concrete Mix-Resub.	2/1/xx	2/15/xx		xx		x		Approved						
3200	AB Steel	7	2nd Floor Steel	2/6/xx	2/16/xx		xx		x		Approved						
1210	Up/Down Elv.	8	Hydraulic	2/14/xx	3/1/xx		xx		x		Approved						
15250	Ted's Insul.	9	Pipe Insulation	2/20/xx	3/1/xx		x				Not Approved						
9340	Townsend	10	Paver Tile	2/26/xx	3/14/xx		xx		xx		Approved						

105F09.EPS

Figure 9 ◆ Sample shop drawing log.

MODULE 44105-08 ◆ CONSTRUCTION DOCUMENTS 5.17

Show Transparency 26 (Figure 11).

Review the general conditions checklist for change orders.

Ask participants to provide examples of costs for some of the items in the checklist and total the dollar amounts to illustrate the impact of change orders.

Emphasize that even simple changes can be lengthy and costly.

Show Transparency 27, Change Order Steps.

Review and discuss the seven steps for reviewing a change order.

Stress the importance of using a consistent approach when reviewing change orders.

Address the bullet points in the text and provide job related examples where applicable.

Review and explain items that are included in a construction claim.

Assign reading of Sections 9.0.0-10.10.0.

CHANGE ORDER

PROJECT: (name, address) Hospital USA, Anywhere, USA 12345

CHANGE ORDER NUMBER: 003
INITIATION DATE: Oct. 5, 20xx

TO: XYZ Construction, PO Box 1620, Perth, NJ 04321

ARCHITECT'S PROJECT NO: ____
CONTRACT FOR: Renovation Work

Contract Date: April 1, 20xx

You are directed to make the following changes to this contract:

Authorized extra work approved by owner per Change Order no. 24.

1.	Install acoustical ceiling in utility room	$503.80
2.	Install cabinet in Room 104	$3751.00
3.	Install access panel Room 104	$497.00
4.	Install lock & dead bolt in pair doors at Admin office	$215.00
5.	Additional partition work at stair #3	$183.00
6.	Revise medical records wall per C.O.R. 112	$2148.00
	Total:	$7297.00

Not valid until signed by both the Owner and Architect. Signature of the contractor indicates agreement herewith, including any adjustments in the Contract Sum or Contract Time.

The original (Contract Sum) (Guaranteed Maximum Cost) was	$214,800.00
Net change by previously authorized Change Orders	$17,888.20
The (Contract Sum) (Guaranteed Maximum Cost) prior to this Change order was	$232,688.20
The (Contract Sum) (Guaranteed Maximum Cost) will be (increased) (decreased) (unchanged) by this Change Order	$7297.80
The new (Contract Sum) (Guaranteed Maximum Cost) including the Change Order will be	$239,986.00
The Contract Time will be (increased) (decreased) (unchanged) by	( ) Days

The Date of Substantial completion as of the date of this Change Order therefore is __________

Design Firm, Inc.	Ajax Construction, Inc.	Hospital USA
Architect	Contractor	Owner
100 Design Avenue Orlando, FL 32801	PO Box 1620 Perth, NJ 04331	High Street Anywhere, USA 12345
Address	Address	Address
By ________	By ________	By ________
Date ________	Date ________	Date ________

105F10.EPS

Figure 10 ◆ Change order.

5.18 PROJECT MANAGEMENT

Instructor's Notes:

struction process, when many more trades are employed on the project.

Contract language normally states that the contractor shall not proceed with the change without written approval from the owner or architect/engineer. Some changes are discovered just prior to required implementation; therefore, timing often becomes a problem. The pricing of the change can be a lengthy process because quotations are required from suppliers and/or subcontractors.

In addition to the costs directly related to the changes, it is important to ensure that the price quote includes all the costs related to any delays, cost impacts from loss of productivity, additional temporary facilities, or additional supervision. If these costs are not included in the cost of the change order, the contractor may have to submit a claim at the completion of the project to recover the costs. A general conditions checklist can be used to identify items that may be included when preparing a change order (*Figure 11*).

The following steps should be used while reviewing a change order.

Step 1 Define the scope of work.

Step 2 Determine the quantity of materials.

Step 3 Determine acceptable unit cost for materials, labor, equipment, and any necessary subcontractors.

Step 4 Conduct an analysis of impact cost for the field/home office, idle time, sequence, or other costs.

Step 5 Determine the amount of allowable overhead.

Step 6 Determine the amount of allowable markup/fee.

Step 7 Determine bonding requirements.

8.5.0 Change Orders and Claims

Change orders often become the subject of claims and disputes. Therefore, information regarding change orders must be documented carefully. The project manager should treat each change order as though it might eventually become part of a future dispute or claim. The manager must keep written records because they may become the foundation for pursuing or defending a claim.

The project may be subject to an audit. If a dispute arises, the documents will be subject to review. In either case, well-organized, thorough, accurate records and documentation are essential. Records and documents include the following:

- Minutes of meetings
- Discussion notes
- Change order procedures
- Weather conditions
- Workforce, material, and equipment status
- Schedule updates and impacts
- Revised drawings, specifications, and sketches
- Photographs

Change order oversight is a requirement of project management. By implementing sound change order principles, the project manager can complete the change order more effectively within his or her budget, schedule, and scope of work. *Figure 12* illustrates the items that may be included in a construction claim.

9.0.0 ◆ CONTRACTOR PAYMENTS

Along with determining the right contract to meet project requirements, the project manager is usually responsible for preparing and submitting payment requests to the contractor or owner on a monthly basis. The payment documentation includes:

- Request for payment
- Waivers and affidavits
- Certificate for payment
- Final payment request

9.1.0 Request for Payment

The contractor submits a request for payment based on completed work, material, and equipment purchased up to a specific date. A date of payment is set to allow for review, approval, certification, and payment processing.

Contractors should prepare a schedule of value for their work and the work of subcontractors to facilitate progress payments before the first request for payment. The schedule of value should reflect the value of the work and should be submitted to the owner or architect/engineer for review and approval. The request for payment format can vary from a simple letter to detailed forms.

The owner and architect/engineer should review the request for payment with the contractor, including verification of the percentage of completion for each item listed in the request. In many cases, the percentage of completion is a subjective judgment and may be negotiated to a percentage acceptable to the parties involved. If the amount requested corresponds to the actual percentage of completion, and the contractor's

Show Transparency 28, Payment Documentation.

Explain the request for payment process.

MODULE 44105-08 ◆ CONSTRUCTION DOCUMENTS 5.19

GENERAL CONDITIONS CHECKLIST

Project ______________________ No.__________

Change Estimate No. _____________ Owner Bulletin No.________

	Material	*Labor*	*Total*
1. Supervision			
A. Project manager			
B. Superintendent			
C. Project and office engineer			
D. Field engineer			
E. Additional foremen			
F. Accountant/timekeeping/material check			
G. Home office supervision			
H. ________________			
2. Temporary Facilities			
A. Field office			
B. Material trailers/sheds			
C. Temporary toilets			
D. Temporary roads			
E. Safety protection/equipment			
F. ________________			
3. Field Support			
A. Office/first-aid supplies			
B. Blueprinting/copying/photos			
C. Telephone			
D. Fire/theft alarm			
E. Insurance			
F. Home office expense			
4. Temporary Utilities			
A. Heat			
B. Light and power			
C. Water			
D. Elevators/lifting/moving			
E. Tests/inspections			
F. ________________			
5. Construction Equipment			
A. Small tools			
B. Trash removal/light trucking			
6. Special conditions			
A. Winter conditions			
B. Snow removal			
C. Cutting and patching			
D. Final cleanup			
E. ________________			

Total Change Order General Conditions $__________

105F11.EPS

Figure 11 ◆ General conditions checklist.

Instructor's Notes:

performance has been satisfactory, the request for payment will be approved and forwarded to the person or lending institution (if one is involved) responsible for construction finances.

Show Transparency 29 (Figure 12).

9.2.0 Waivers and Affidavits

Discuss the process of waivers and affidavits. Explain that these are legal documents requiring notarization.

Explain the use of liens and how they can be used to secure payment.

Discuss certificates for payment and the final payment request.

Explain project retainage.

Lien laws are designed to protect workers, materials suppliers, and, under certain conditions, general contractors and subcontractors from nonpayment by the owner or general contractor. The underlying legal principal is that of unjust enrichment. In this case, the owner or general contractor would unfairly benefit at the expense of the party that performed the work without compensation.

With each request for payment, the contractor may have to submit to the owner a waiver for the total amount received to date along with an affidavit that all suppliers and subcontractors have been paid to date. Contractors may have to submit waivers from their suppliers and subcontractors verifying the amounts have been received.

When work is completed, all final waivers must accompany the contractor's request for full and final payment. All waivers and affidavits should be signed, sealed, and notarized. They should also be examined for accuracy and compliance with the terms of the contract before the payment request is approved.

In many states contractors can file a mechanic's or construction lien, which is a legal process to secure payment for work performed and materials furnished in the improvement of land. This statutory right attaches to the land itself in much the same way as a mortgage. The purpose of a lien statute is to permit a claim on the premises where the value or condition of real property has been increased or improved and where suitable payment has not been made by the owner. Lien laws are strictly construed by the courts, and full compliance with all provisions of the local statute is mandatory.

9.3.0 Certificate for Payment

As part of a request for payment, the architect/engineer or owner completes a certificate for payment, authorizing the lending institution to make payment. The information on the certificate should match the information on the request for payment, the waivers, and affidavits.

9.4.0 Final Payment Request

Construction contracts usually allow the owner to withhold a percentage of each request for payment amount (known as retainage) to protect it against unsettled claims and uncompleted work. Retainage will usually range between 5 and 10 percent. The contractor will eventually submit a final request for payment for the total balance due, including retention, along with final waivers and affidavits from all its subcontractors and suppliers.

A. Increased cost due to:
- Labor rates escalation
- Acceleration
- Manpower
- Field overhead
- Office overhead
- Inefficiency
- Overtime

B. Cost due to the parties failure to:
- Resolve conflicts between Regulatory Agencies
- Coordination
- Meet milestone dates
- Obtain permits
- Commence on schedule
- Perform work
- Provide access
- Pay promptly

C. Cost related to changes in:
- Original scope of work
- Changed conditions
- Quality or details
- Dimension
- Nature of work
- Work schedule
- Starting or ending dates
- Contract documents

D. Cost related to third party:
- Interest rate increase
- Finance charges
- Errors & omissions
- Breach of contract
- Code of conflicts
- Bidding errors

105F12.EPS

Figure 12 ◆ Types of construction claims.

MODULE 44105-08 ◆ CONSTRUCTION DOCUMENTS 5.21

Describe project close-out.

Explain "substantial completion."

Show Transparency 30 (Figure 13). Discuss project close-out procedures.

Show Transparency 31, Project Inspection. Discuss project inspection steps.

Show Transparency 32, Punch List. Discuss the purpose of the punch list.

Stress that the punch list is a construction document, not just a checklist.

Provide a sample punch list, if available.

Discuss the process leading to owner acceptance.

Discuss certificate of occupancy.

Emphasize that health and safety requirements must be met before occupancy.

10.0.0 ◆ CLOSE-OUT

Project close-out begins at substantial completion, the point in time where the building is ready for occupancy. The purpose of project close-out is to ensure that the owner is receiving a finished product that conforms to the contract documents. The project manager should provide the documentation that provides this assurance.

10.1.0 Substantial Completion

According to the American Institute of Architects, as shown in *AIA Document A201*, substantial completion is the stage in the progress of the project when the project or a designated portion thereof is sufficiently completed in accordance with the contract documents so that the owner can occupy or use the structure for its intended purpose.

10.2.0 Close-Out Procedures

The project manager should formally organize project close-out no later than 30 days prior to substantial completion. This involves sending all subcontractors and material suppliers a copy of the project's close-out requirements. An example of a contractor's project close-out requirements is illustrated in *Figure 13*.

As all mechanical, electrical, and other systems are completed, the contractor may be required to test their operation. If a representative of the manufacturer conducts the test for proprietary systems, the project manager and the owner's representative should be present to observe. Test records should be submitted to the proper authorities, with copies going to the project manager, the architect/engineer, and the owner's representative.

Tests are time-consuming and should be scheduled in advance so that all involved parties can attend. The owner's representative should have every opportunity to ask questions about the process. Testing is not complete until it demonstrates that the system functions properly. Testing procedures are often videotaped to document the results.

10.3.0 Project Inspection

The project manager should arrange for all parties, particularly the owner, the architect/engineer, and prime contractors, to meet and conduct an inspection tour of the project well before the date of substantial completion. Written notification of the inspection tour date should be sent to all involved parties well in advance. At the conclusion of the inspection tour, the project manager should have each party sign the certificate to verify the acceptance of the work. In preparation for the inspection tour, project managers should conduct inspections about two weeks before substantial completion.

10.4.0 Punch List

As the work is completed, it should be inspected and a list of deficiencies (known as a **punch list**) is prepared. Generally, the architect/engineer conducts the inspection, but it is not unusual for the project manager and the owner to undertake their own inspections. A punch list is given to the contractors for completion prior to a stated date. A re-inspection is conducted to ensure the work has been completed. There are no established guidelines for punch lists, but they should be clear and specific.

10.5.0 Owner Acceptance

As soon as the deficiencies on the final punch list have been corrected, the project manager notifies the owner that the project is ready for acceptance. At this time, applications for required permits or certificates of occupancy should be processed. The architect/engineer prepares the certificate of substantial completion for the owner and attaches a list of any remaining deficiencies, making the substantial completion conditional on the correction of those deficiencies.

10.6.0 Certificate of Occupancy

Even if the project is accepted by the owner and the architect/engineer, the project must be certified by the local building department and must meet all health and safety requirements. Once those requirements are met, the owner may occupy the building.

10.7.0 Record Drawings

Record drawings, formerly called as-built drawings, are developed to indicate changes to the project as it was being built. When changes are made, they must be recorded on the drawings.

Normally, the architect/engineer will furnish a set of drawings (either prints or sepias) for recording any changes. These drawings must be clearly identified, kept at the project, and kept separate from other project drawings. The general contractor is usually responsible for maintaining a complete set of record drawings and verifying that all subcontractors are updating their record drawings.

Instructor's Notes:

5.22 PROJECT MANAGEMENT

RE: (Project)

SUBJECT: Project Close-out Procedures

In order to satisfy the project close-out requirements and expedite the owner's release of retainage, the following items (marked X) must be submitted to this contractor.

1. ______ Warranty and Guarantee (form enclosed)
2. ______ Specific manufacturer's guarantee
3. ______ Operating and maintenance manuals
4. ______ Maintenance agreement (per specifications requirements)
5. ______ Letter certifying instruction of owner's operating/maintenance
6. ______ Inspection and field testing reports
7. ______ Record or As-built drawings (per specifications requirements)
8. ______ Certification letter stating that your firm has complete your punch list items
9. ______ Consent of surety to final contract amount and final payment
10. ______ Final Release of Lien from your firm and all other material suppliers, vendors, or sub-subcontractors who have provided Notices to Owner
11. ______ Final Application for Payment
12. ______ Furnish spares per specs
13. ______ Other:______________________________

A timely completion of the above marked items will require submission to this office within 30 days from the date of this letter. If for any reason any of the items marked above cannot be submitted within the requested time frame, please contact this writer immediately.

Sincerely,

John Doe
Project Manager

105F13.EPS

Figure 13 ◆ Sample contractor close-out procedures.

At project completion, the record drawings are submitted to the architect/engineer as part of the close-out documents. Often the architect/engineer or the owner will withhold final payment until all record drawings have been submitted. Therefore, the project manager must ensure that record drawings are maintained and submitted in a timely manner.

10.8.0 Operation and Maintenance Manuals

Operation and maintenance manuals must be furnished to the owner upon completion of the project. Manuals should contain all information necessary for operating, maintaining, repairing, dismantling, and assembling all equipment provided under the project contract.

Discuss the purpose of record drawings and that they must be kept separate from other documents.

Stress that the contractor is responsible for maintaining a complete set of record drawings and that the subcontractors keep their records updated.

Note that the drawings are submitted to the architect / engineer as part of the close-out documents.

Emphasize that final payment may be contingent upon the submittal of the record drawings.

Explain that all operation and maintenance manuals must be the most recent versions.

MODULE 44105-08 ◆ CONSTRUCTION DOCUMENTS 5.23

Explain that all warranties and guarantees are delivered to the owner on project completion.

Explain the importance of consent of surety and release of liens papers.

Have participants complete the Review Questions, and go over the answers prior to administering the Module Examination.

10.9.0 Warranties and Guarantees

Warranties and guarantees are furnished by contractors and suppliers to guarantee workmanship and material integrity for a specified length of time. As described in the project specifications, most guarantees and warranties must be in writing and on authorized forms.

Warranties and guarantees usually go into effect at substantial completion. The project manager should advise the owner of the maintenance and security responsibilities, the cost of utilities, and services related to the equipment.

10.10.0 Consent of Surety and Final Release of Liens

Two of the most important close-out documents are the consent of surety and release of liens papers. The consent of surety releases the bonding company from further liability for the project, while the final release of liens releases the owner from all claims by the general contractors or subcontractors.

Review Questions

1. The most common way to obtain work in the construction industry is through _____.
   a. referral
   b. purchase order
   c. competitive bidding
   d. teaming arrangements

2. When the owner pre-qualifies a contractor, this is known as a(n) _____ bid.
   a. procured
   b. sole source
   c. invited
   d. negotiated

3. When an owner selects several contractors to interview for the work, this is a(n) _____ bid.
   a. procured
   b. sole source
   c. invited
   d. negotiated

4. The documents that contain the information that the bidder must comply with in bid preparation is known as a(n) _____.
   a. invitation to bid
   b. instruction to bidders
   c. general conditions document
   d. specifications sheet

5. The legal document posted by the bidder as an assurance that the apparent low bidder will enter into a contract with the owner if awarded the project is known as _____.
   a. material bond
   b. bid bond
   c. payment bond
   d. Flemish bond

6. _____ supplement the working drawings with detailed technical information not contained on the drawings.
   a. Purchase orders
   b. Change orders
   c. Specifications
   d. Daily reports

7. The revised CSI MasterFormat™ 2004 has expanded from 16 Divisions to _____.
   a. 20
   b. 29
   c. 34
   d. 49

8. The most common type of construction contract is _____.
   a. lump sum/fixed price
   b. cost-plus
   c. target
   d. unit price

Instructor's Notes:

Review Questions

9. Items such as labor, materials, salaries, equipment, fees, and bonds are reimbursable in a _____ contract.
   a. lump sum
   b. unit-price
   c. back charges
   d. cost-plus

10. The contractor must maintain _____ insurance to pay employees who are injured on company time.
    a. risk
    b. general liability
    c. workers' compensation
    d. vehicle

11. The type of insurance that provides coverage for claims that the company is obligated to pay as a result of bodily injury or property damages is known as _____.
    a. worker's compensation
    b. general liability
    c. builder's risk
    d. floater

12. Which item is *not* covered by general liability insurance?
    a. completed work
    b. injury or damages caused by subcontractors
    c. loss caused by explosion
    d. project construction activity

13. _____ are important documents because they identify labor and material costs.
    a. Purchase orders
    b. Job logs
    c. Change orders
    d. Time cards

14. The change order management process begins during the _____ phase.
    a. pre-bid
    b. pre-construction
    c. scope
    d. coordination

15. The purpose of record drawings is to _____.
    a. provide the owner with the original drawings
    b. provide documentation of changes during construction
    c. provide the architect with updated drawings
    d. provide stakeholders with copies

MODULE 44105-08 ◆ CONSTRUCTION DOCUMENTS 5.25

Summarize the major concepts presented in the module.

Summary

In order to be useful, good documentation must meet the following criteria:

Accuracy – The facts should be recorded the way the facts actually occurred. If the facts are misstated in one place, it may be assumed that facts have been misstated throughout the documentation.

Objectivity – Objectivity means that the facts should be stated in an unbiased way. Sometimes it is difficult for the project manager to be unbiased because they may be directly involved in the situation being documented. Facts should be stated the way they occurred.

Completeness – If the company provides the project manager or the field supervisors with specific forms for documentation, they should be filled out completely. Incomplete forms demonstrate inaccuracy or lack of knowledge of the situation.

Uniformity – Information should be recorded in a precise, defined and routine manner.

Credibility – Documents should be completed, recorded timely and without bias. The project manager should be certain that all documentation is accurate and carefully reported

Timeliness – All occurrences should be documented immediately and thoroughly.

For a successful project close-out, the project manager must plan and begin close-out activities well ahead of the date of substantial completion. He or she should develop control documents that will allow for the tracking of close-out documents and inspections. By following proper close-out procedures, the project manager reduces overhead expenses, expedites the processing of final payments, and earns the reputation of being able to finalize a project.

Notes

Instructor's Notes:

Trade Terms Quiz

1. The written requirements for materials, equipment, construction systems, standards and workmanship are known as ______________.
2. The work done to deliver a product with specified features and functions is known as the ______________ of the project.
3. An itemized list documenting incomplete or unsatisfactory items after the contractor has notified the owner that the tenant space is substantially complete is known as the ________.
4. ______________ are the documents used to modify the contract prior to receiving the bid.
5. A ___________ is a guarantee that the contractor will enter into a contract, if it is awarded.
6. The ______________ is an organization that maintains and advances the standardization of construction language as it pertains to building specifications.
7. Detailed drawings showing how building elements will be fabricated are called _________.
8. A _____________ is a document that modifies the original contract due to changes incurred after contract award.

Trade Terms List

Addenda
Bid bond
Change order
Construction Specifications Institute (CSI)
Punch list
Scope
Specifications
Shop drawings

Have participants complete the Trade Terms Quiz, and go over the answers.

Administer the Module Examination. Record the results on Craft Training Report Form 200, and submit the results to the Training Program Sponsor.

MODULE 44105-08 ◆ CONSTRUCTION DOCUMENTS 5.27

Trade Terms Introduced in This Module

Addenda: Documents used to modify the contract prior to receiving the bid.

Bid bond: A guarantee that the contractor will enter into a contract, if it is awarded.

Change order: A document that modifies the original contract documents due to changes that occur after the award of the contract.

Construction Specifications Institute (CSI): An organization that maintains and advances the standardization of construction language as pertains to building specifications.

Punch list: An itemized list documenting incomplete or unsatisfactory items after the contractor has notified the owner that the tenant space is substantially complete.

Scope: The work that must be done to deliver a product with the specified features and functions.

Specifications: A part of the construction documents contained in the project manual consisting of written requirements for materials, equipment, construction systems, standards and workmanship.

Shop drawings: Detailed drawings showing how building elements will be fabricated

Instructor's Notes:

5.28 PROJECT MANAGEMENT

Resources & Acknowledgments

References

PMI Standards Committee, *A Guide to the Project Management Body of Knowledge*. PMI Publications, Newton Square, Pa. (2004).

The American Institute of Architects, 1735 New York Ave., NW, Washington, DC 20006-5292.

The Construction Specifications Institute, 99 Canal Center Plaza, Suite 300, Alexandria VA 22314

MODULE 44105-08 ◆ CONSTRUCTION DOCUMENTS 5.29

MODULE 44105-08 — TEACHING TIPS

The following are suggested activities or instructional methods to help you teach the material in this module.

General

When you call on someone to answer a question, the rest of the class relaxes or even tunes out because they expect that the question and answer will take place only between you and the participant you called on. Instead, use this technique to involve more participants in answering questions and to keep them on their toes.

1. Ask participants to define a term or explain a concept.
2. After one participant has answered, ask a participant seated nearby if the answer is right. Then ask whether a participant in the back of the room agrees.
3. Ask participants to explain why they think an answer is right or wrong.
4. Use the session to clear up incorrect ideas and encourage participants to learn from their mistakes.

Section 1.0.0 *Sources of Problems in the Construction Industry*

Refer to Figure 1 in the Participant Guide or show Transparency 2 to illustrate the results of the Accenture survey.

This exercise reinforces the major causes of problems in the construction industry. Ask participants if they have experienced any of the issues listed, and discuss their comments.

- *Poor Documentation* – Insufficient records, including documentation related to drawings and specifications, contracts, change orders, addenda, bid documents, etc.
- *Poor Administration* – May be a result of poor training or lack of experience in both office and field personnel.
- *Unsophisticated Contractors* – Refers to contractors who may have limited knowledge of business practices related to cash flow, insurance, accurate estimating, cost control, etc.
- *Unsophisticated Owners* – Owners who may have limited knowledge and experience in the construction industry.
- *Change Orders* – Indicates difficulties in estimating and collecting for work completed when changes have or have not been authorized.
- *Declining Productivity* – Attributed to declining production and decreasing labor experience in the construction industry.
- *Poor Information Systems* – Directly related to contractors who are not using adequate productivity tools.
- *Inflation* – Refers to contractors that do not adjust rates in relation to increases in labor, materials, and supplies.

Sections 2.1.0–2.6.0

Bidding Methods

This exercise provides a review of various bidding methods.

- *Competitive Bidding* – Briefly describe the competitive bidding process. Explain the importance of understanding the scope of work. Point out the important elements that should be in every scope of work.
- *Invited Bid* – Discuss the advantages and disadvantages of pre-qualifying for an Invited Bid.
- *Negotiations* – Discuss the advantages and disadvantages of the Negotiations method. Ask participants why this method may be advantageous to a contractor. Discuss their responses.
- *Design-Build* – Use a Design-Build project as an example to illustrate how and why the owner selected a particular contractor. Explain the services that the Design-Build firm provides. Ask the participants what type of firm would have the capabilities to participate in this method of obtaining a bid. Discuss their responses.
- *Construction Management* – Provide an example of a project that has been managed by a firm that specializes in construction management. Illustrate how such a firm has established standards, resources, supply chain, and a quality workforce along with project management experience.

Summary

Explain that many companies splitting their work among the five methods. Remind participants of the components of bidding and the need to determine which approach makes the best use of the company's estimating and management staff. Remind participants that bidding is time consuming, and that time is money. Emphasize that bid documents have very specific language and terminology. Describe key factors to take into account when considering a bid.

Ask participants which bidding methods are used by their companies.

From your own experience or actual case histories, explain that there are times when a company will "no bid" a project. Explain why this may be beneficial, in some cases.

Sections 4.0.0–4.2.0 *Reading Drawings*

This exercise provides an overview of construction drawings. To the extent possible, provide examples of the various drawings.

Show Transparency 11, Drawing Types.

Explain that there are three types of drawings: conceptual, preliminary, and working or construction drawings. Explain the purpose of each type.

Show Transparency 12, Reviewing Drawings.

Explain that a set of contract drawings has several sections. Project managers must be familiar with the notations and general features of the set of drawings. They must understand all of the drawing types and have a systematic approach to reviewing them.

Refer to Section 4.2.0. Discuss Steps 1 through 8, and point out what the project manager should review for each type of drawing discussed. Use the following points to supplement the discussion:

- *Step 1* – Explain that site drawings show civil engineering features such as drainage, utilities, and other items. Site drawings include contour drawings, and show both new and existing contours. These drawings provide important information about elevations and dimensions for use in larger architectural drawings.
- *Step 3* – Architectural drawings relate specifically to the building's location, size, form, and materials to be used. Architectural drawings are always preceded with an "A", followed by a numerical system. These drawings include information about:
  - symbols, abbreviations, locations and notes
  - demolition plans, floor and ceiling plans, and roof plans
  - building sections, wall sections, and so forth

  Discuss other information included in architectural drawings.
- *Step 4* – Explain that mechanical drawings are used by various subcontractors. Discuss the importance of reviewing Schedules concerning piping, fixtures, etc. Point out that information not found on the drawings will be found in the project manual.
- *Step 5* – Electrical drawings have unique symbols that the project manager needs to know. Inform participants that electrical drawings should not be removed from the set of drawings because there are other systems integrated with electrical drawings (alarms, communications, lighting and so forth).

Sections 5.0.0–5.8.0 *Types of Contracts*

The purpose of this activity is to have the participants determine the best type of contract to meet an owner's requirements.

1. Using a project from your area of experience, present an example situation.
2. List the requirements on a whiteboard or flip chart (use bullet points and discuss).
3. Show Transparency 14, Contract Types, for reference.
4. Inform participants that they can refer to their Participant Guides.
5. Have them select the most appropriate contract type for the situation and present their recommendation.
6. Evaluate and discuss their answers. Provide feedback.

Estimated time, 20 minutes (time may vary depending on class size).

Option: Divide participants into teams and have each team present its findings.

Section 6.0.0 *Insurance Coverage*

This activity will enable the participants to evaluate insurance coverage as it relates to a job-site accident. They will take the role of the project manager and determine the actions that need to taken.

1. Select an actual job site accident (or a case study) from your area of experience.
2. Present the situation and list the particulars on the white board or flip chart. For example:
   a. A crane cable snapped, and the roof top fell and hit part of the structure.
   b. Structural damage occurred.
   c. Six workers were injured.
3. Ask participants – "As a project manager, what action should be taken first?"
   a. Call home office.
   b. Contact field supervisor.
   c. Determine injuries, etc.
4. Having assessed the situation, have participants write a report of the occurrence.
5. Discuss how different insurance coverage will be required.
6. Emphasize the need to have knowledge of insurance options and remind participants that job site safety is the best insurance.

Sections 7.0.0–7.1.0

Project Correspondence

Show Transparency 18 (Figure 4).

Review the need for documentation control, and restate the two basic reasons for controlling documents.

- To provide easy recovery of documents during project life cycle
- To provide procedures for follow-up and verification

Cite examples of problems caused by a poor document management system. Inform participants that today's project managers rely on software tools and online methods to manage documentation.

Show Transparency 19, Project Manager Correspondence.

The project manager is responsible for a large amount of correspondence. Reinforce the need for:

- Timely response
- Complying with company standards
- Being clear, concise, and accurate
- Maintaining a professional attitude

From your experience, cite an example where poor correspondence had an adverse effect on a project.

Show Transparency 20 (Figure 5).

Stress the importance of accurate daily reports. Provide an example of how a project manager can solve problems by reviewing reports. Explain that the daily report is a job-site report prepared by the site supervisor or engineer, but which may have input from all levels.

Point out types of information that should be included in a daily report. Emphasize that accurate reports contain valuable information and may be used to justify change orders, or be used for insurance and litigation.

Show Transparency 21 (Figure 6).

Discuss the rationale and expected outcomes of the two types of formal meetings that project managers attend. Explain that many meetings occur during the construction process, including those that are scheduled and others that may occur to resolve immediate problems.

Explain the importance of reviewing the meeting agenda and point out that agenda items may have a direct impact on project management issues. Emphasize that the project manager must be prepared to explain and resolve issues.

Show Transparency 22 (Figure 7).

Review an example of project meeting minutes. Using a job-related example, discuss how an error in the meeting minutes can have a direct impact on a project.

Section 7.2.0 *Meeting Agendas*

This activity will enable participants to prepare a meeting agenda to resolve a conflict between electrical and mechanical trades.

1. Select a typical coordination issue from an existing project.
2. Have the participants create an agenda addressing the issue for an upcoming meeting.
3. Have the participants present their agendas and discuss their rationale in class.
4. Ask for class reaction to the agendas presented.
5. Provide a critique of their presentations.

Estimated time, 20 minutes (time may vary depending on class size).

Section 7.6.0 *Shop Drawings*

Show Transparency 23 (Figure 9).

Explain the purpose of shop drawings and describe the types that are common to a project. Discuss why shop drawings are submitted before materials and equipment are delivered to the project. Explain that every company has different procedures for processing shop drawings.

Describe the basic steps for processing shop drawings. Explain the review and approval process using a job-related example.

Emphasize that the project manager or project engineer is responsible for checking shop drawings. Discuss the importance of maintaining records of transmittal and receipt of all shop drawings. Review the shop drawing log.

Ask participants how their company processes shop drawings.

MODULE 44105-08 — ANSWERS TO REVIEW QUESTIONS

Answer	Section Reference
1. c.	2.1.0
2. c.	2.2.0
3. d.	2.3.0
4. b.	3.2.0
5. b.	3.4.0
6. c.	4.3.0
7. d.	4.3.0
8. a.	5.1.0
9. d.	5.2.0
10. c.	6.1.0
11. b.	6.2.0
12. c.	6.2.0
13. d.	7.5.0
14. b.	8.2.0
15. b.	10.7.0

MODULE 44105-08 — ANSWERS TO TRADE TERMS QUIZ

1. specifications
2. scope
3. punch list
4. Addenda
5. bid bond
6. Construction Specifications Institute (CSI)
7. shop drawings
8. change order

CONTREN® LEARNING SERIES — USER UPDATE

NCCER makes every effort to keep these textbooks up-to-date and free of technical errors. We appreciate your help in this process. If you have an idea for improving this textbook, or if you find an error, a typographical mistake, or an inaccuracy in NCCER's Contren® textbooks, please write us, using this form or a photocopy. Be sure to include the exact module number, page number, a detailed description, and the correction, if applicable. Your input will be brought to the attention of the Technical Review Committee. Thank you for your assistance.

Instructors – If you found that additional materials were necessary in order to teach this module effectively, please let us know so that we may include them in the Equipment/Materials list in the Annotated Instructor's Guide.

Write: Product Development and Revision
National Center for Construction Education and Research
3600 NW 43rd St, Bldg G, Gainesville, FL 32606

Fax: 352-334-0932

E-mail: curriculum@nccer.org

Craft ____________ Module Name ____________

Copyright Date ____________ Module Number ____________ Page Number(s) ____________

Description

____

____

____

____

(Optional) Correction

____

____

____

(Optional) Your Name and Address

____

____

____

Module 44106-08

Construction Planning

NCCER STANDARDIZED CRAFT TRAINING PROGRAM

The National Center for Construction Education and Research (NCCER) provides a standardized national program of accredited craft training. Key features of the program include instructor certification, competency-based training, and performance testing. The program provides trainees, instructors, and companies with a standard form of recognition through a National Craft Training Registry. The program is described in full in the *Guidelines for Accreditation*, published by NCCER. For more information on standardized craft training, contact the NCCER by writing us at 3600 NW 43rd St., Bldg. G, Gainesville, FL 32606; calling 352-334-0911; or emailing info@nccer.org. More information may be found at our website, www.nccer.org.

HOW TO USE THIS ANNOTATED INSTRUCTOR'S GUIDE

Each page presents two sections of information. The larger section displays each page exactly as it appears in the Trainee Module. The narrow column ties suggested trainee and instructor actions to each page and provides icons (detailed below) to call your attention to material, safety, audiovisual, or testing requirements. The bottom of each page includes space for your notes.

The **Audiovisual** icon indicates an appropriate time to show a transparency or other audiovisual aid.

The **DVD** icon indicates an appropriate time to discuss or show segments of the video.

The **Classroom** icon prompts you to define a term, stress a point, ask trainees to explain a concept, or give examples.

The **Demonstration** icon directs you to show trainees how to perform tasks.

The **Examination** icon tells you to administer the written module examination.

The **Homework** icon is placed where you may wish to assign reading for the next class, assign a project, or advise trainees to prepare for an examination.

The **Laboratory** icon is used when trainees are to practice performing tasks.

The **Materials** icon is a reminder for you to gather materials needed for classes, labs, and testing.

The **Performance Testing** icon tells you to administer a performance test or a portion thereof.

The **Safety** icon is used to emphasize safety issues. It is often keyed to *Caution* and *Warning!* statements in the Trainee Module.

The **Teaching Tip** icon indicates additional guidance is available, such as how to conduct an exercise, get the most educational value from a field trip, or encourage class participation. Teaching Tips may expand on a feature (*Think About It, Did You Know?*) or provide *Quick Quizzes* or similar exercises. You will be referred to the Teaching Tips section at the back of the module if there is additional material.

The **Combination** icon indicates that the laboratory listed corresponds with a performance task. If desired, you can note the proficiency of the trainees during the laboratory, and use it to satisfy performance testing requirements.

PREPARATION

Before teaching this module, you should review the Objectives, Performance Tasks, Materials and Equipment List, and Module Outline. Be sure to allow ample time to prepare your own training or lesson plan and gather all required materials and equipment.

Construction Planning

Annotated Instructor's Guide

Module 44106-08

MODULE OVERVIEW

This module explains the importance of formal job planning and creating a performance-based work environment. It discusses the Work Breakdown Structure (WBS) as the foundation that identifies deliverables, tasks, and time. It defines the roles and responsibilities of an effective team, explains how to allocate resources, and introduces the basics of quality control.

PREREQUISITES

Please refer to the Course Map in the Participant Module. Prior to training with this module, it is recommended that the participant shall have successfully completed *Project Management,* Modules 44101-08 through 44105-08.

OBJECTIVES

Upon completion of this module, the participant will be able to do the following:

1. Explain the importance of planning a job.
2. Create a performance-based work environment.
3. Explain the importance of scope and the work breakdown structure.
4. State the differences among the pre-construction, construction, and review phases of planning.
5. Describe how the planning process is carried out.
6. Define the roles and responsibilities of an effective team and how to allocate resources.
7. Define commodities, engineered equipment, construction equipment, and construction supplies.
8. Describe how to implement a plan.

MATERIALS AND EQUIPMENT LIST

Desktop or laptop computer or DVD player

Multimedia projector and screen

Minor Decisions: Major Impact – How to Deal with Real Issues in Project Management DVD

Transparencies

Blank acetate sheets

Overhead projector and screen

Transparency pens

Whiteboard/chalkboard

Markers/chalk

Pencils and scratch paper

Sample scope of work statement

Module Examination (located in the Test Booklet)

ADDITIONAL RESOURCES

This module is intended to present thorough resources for task training. The following reference work is suggested for both instructors and motivated participants interested in further study. This is optional material for continued education rather than for task training.

Robledo, Rebecca, "Unforeseen circumstances: your contract must acknowledge unforeseen construction problems, and lay out who's responsible for them or—you'll end up paying the price." *Pool & Spa News*, April 11, 2003. Accessed on Oct 3, 2007, at http://findarticles.com/p/articles/mi_m0NTB/is_8_42/ai_99909406 (03 October 2007).

TEACHING TIME FOR THIS MODULE

An outline for use in developing your lesson plan is presented below. Note that each Roman numeral in the outline equates to one session of instruction. Each session has a suggested time period of 2½ hours. This includes 10 minutes at the beginning of each session for administrative tasks and one 10-minute break during the session. Approximately 10 hours are suggested to cover *Construction Planning*.

Topic	Planned Time
Session I. Introduction and Pre-Construction Planning	
A. Introduction	________
B. Pre-Construction Planning	________
Session II. Implementing the Plan	
A. Work Analysis	________
B. Overview	________
C. Work Analysis Format	________
Session III. Planning Resources, Part One	
A. Planning Materials	________
B. The Quantity Survey	________
C. Purchasing	________
D. Pre-Purchase Considerations	________
E. Issuing the Written Purchase Order	________
F. Expediting	________
G. Receiving Materials	________
Session IV. Planning Resources, Part Two; Review and Testing	
A. Inventory	________
B. Planning Equipment	________
C. Planning Tools	________
D. Planning Labor	________
E. Coordinating the Work For Contractors	________
F. Implementation of the Plan	________
G. Review	________
H. Module Examination	________

1. Participants must score 70% or higher to receive recognition from NCCER.
2. Record the testing results on Craft Training Report Form 200, and submit the results to the Training Program Sponsor.

USING THE PM DVD WITH *CONSTRUCTION PLANNING*

Resource Planning is a key topic in the Project Management DVD.

Referring students to Section 5.9.0, Planning for Tools, ask them to consider methods they might employ to ensure tools are properly controlled and don't go missing again.

***Suggested response:** The workers (e.g., plumbers, drywall workers) were using contractor-supplied tools on Samantha's job site. However, she could have provided a small inventory of tools to prepare for instances where tools become damaged or lost. She could also have insisted that each contractor follow agreed-upon tool control procedures in order to minimize risk on the job.*

Section 5.1.0, Planning for Labor, presents an opportunity to discuss why Samantha's team leaders react negatively to her offer to add more crew members.

***Suggested response:** The team leaders feel that additional workers would not solve the problem at this stage and are particularly concerned that untrained workers would cause more delays. Samantha would need to identify the number of skilled and unskilled crew members, and determine the amount of time it will take to train unskilled workers. She also would need to determine whether there is enough space to accommodate extra workers so that all workers are able to work safely and productively. Extra workers may need extra resources.*

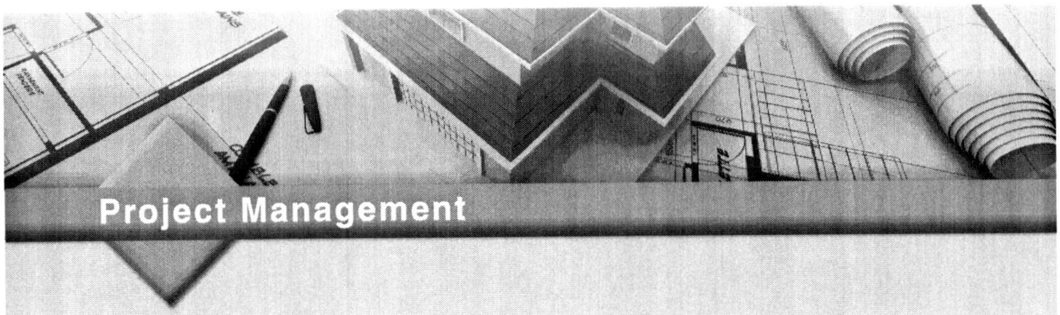

44106-08

Construction Planning

Assign reading of Module 44106-08.

44106-08
Construction Planning

Topics to be presented in this module include:

Overview

Most construction activities are actually quite simple, yet complexity results from having to perform those activities in a specific sequence and within a scheduled time frame. As project manager, it's your job to ensure that the project makes a profit for the company. This means making the most efficient use possible of all the equipment, materials, and workers and coordinating them with the resources on other projects.

Instructor's Notes:

Objectives

When you have completed this module, you will be able to do the following:

1. Explain the importance of planning a job.
2. Create a performance-based work environment.
3. Explain the importance of scope and the work breakdown structure.
4. State the differences among the pre-construction, construction, and review phases of planning.
5. Describe how the planning process is carried out.
6. Define the roles and responsibilities of an effective team and how to allocate resources.
7. Define commodities, engineered equipment, construction equipment, and construction supplies.
8. Describe how to implement a plan.

Trade Terms

Activity	Planning
Bill of material	Purchase order
Expediting	Quantity survey
Free on board (FOB)	Work analysis

Prerequisites

Before you begin this module, it is recommended that you successfully complete *Project Management*, Modules 44101-08 through 44105-08.

This course map shows all of the modules in the *Project Management* curriculum. The suggested training order begins at the bottom and proceeds up. Skill levels increase as you advance on the course map. The local Training Program Sponsor may adjust the training order.

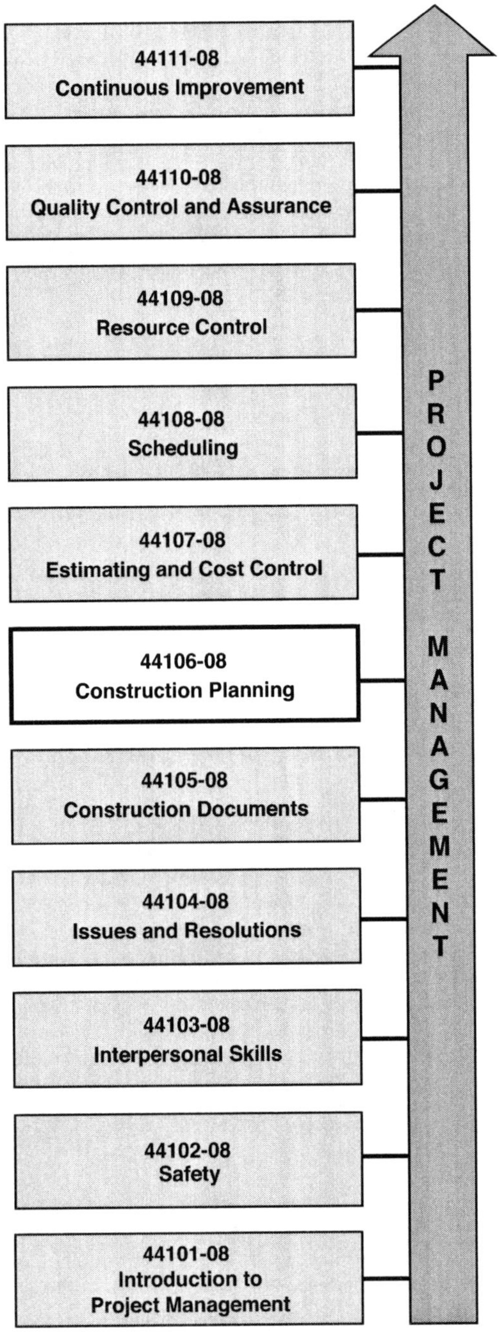

106CMAP.EPS

Ensure that you have everything required to teach the course. Check the Materials and Equipment list at the front of this module.

See the general Teaching Tip at the end of this module.

Explain that terms shown in bold are defined in the Glossary at the back of the module.

Show Transparency 1, Objectives. Review the goals of the module, and explain what will be expected of the participant.

MODULE 44106-08 ◆ CONSTRUCTION PLANNING 6.1

Explain that planning a construction project can be complex.

Define the meaning of the terms activity and task.

Introduce the topic of work analysis.

Ask participants how they define planning, and discuss their responses.

Define planning as presented in the text.

Provide an example of poor planning that had a negative effect on a project.

Discuss a job-related example where you used your planning skills.

Discuss planning as a team effort and explain which personnel are usually part of a typical planning team.

Ask participants for their experiences as part of a planning team. Discuss their responses.

Show Transparencies 2, 3, and 4, and discuss formal planning.

1.0.0 ◆ INTRODUCTION

It is common in the construction industry to use the words "job" and "project" interchangeably. However, there is a distinction between these two terms. A project includes all the work necessary to build a facility; for example, the construction of a 14-story office building is a project composed of a series of jobs such as excavation, steel erection, and plumbing.

The words **activity** and task also mean different things. Every job consists of two or more activities; a plumbing job, for example, might include activities such as installing a one-inch copper pipeline and setting sinks. The activity of setting the sink includes the tasks of installing a trap and tightening nuts.

A construction project is composed of many different jobs and each job is composed of many different activities. **Planning** a typical construction project is a complex task. A study of the construction of a 10-story office building showed that it involved more than 85,000 different materials and activities.

This module examines the basics of planning, how to plan for various job conditions, and how to perform a **work analysis** that indicates how well a project plan is working.

2.0.0 ◆ DEFINITION OF PLANNING

Planning can be defined as the act of selecting a course of action through analysis and evaluation of all possible alternatives. More specifically, planning determines the amount of time a particular job or project will take as well as the resources available and cost limitations.

Effective planning anticipates problems and organizes the construction operation to overcome them. It focuses on the most significant cost factors in the construction process (labor and equipment) and minimizes the unproductive time of both. It forecasts material, labor, and equipment needs and provides a control device for measuring progress during the project and evaluating it after completion.

2.1.0 The Planning Team

Planning should involve all supervisory and appropriate non-supervisory personnel who will be responsible for carrying out the plan. This may include project managers, estimators, schedulers, supervisors, and even key craftworkers. Failing to include everyone involved and capable of contributing to the plan is similar to assigning responsibility without authority, and the result is often exasperation, delays, and cost overruns. Studies have shown that people involved in planning change develop a kinship toward the change process, resulting in a greater sense of team effort. Cultivating this sense of teamwork throughout the company results in the entire staff, both office and field personnel, working toward common goals, thus minimizing problems.

2.2.0 Formal Planning

A formal plan is a system that helps not only to complete a project but also to create a more performance-based work environment. Formal planning determines:

- What has to be done
- When activities must be accomplished
- How long it will take to perform them
- How they are going to be done
- Who will perform them
- How results will be measured
- How performance gaps will be identified and corrected

Every project, regardless of its size and complexity, consists of the same basic elements that are addressed in the formal plan. These elements are as follows:

- Goals and objectives that must be met within a given time
- A series of jobs composed of activities, which must be performed to reach the objective
- Resource requirements, specifically the labor, material, equipment, time, and space needed to complete each job and each activity
- A sequence of activities that identifies which ones must be completed before others can begin

To meet planning responsibilities, the project manager must do the following:

- Understand the planning process and be able to follow it in a logical manner, while documenting all efforts.
- Focus on the sequence of jobs within the project rather than on the activities within those jobs, and also be able to analyze job plans when necessary.
- Coordinate the plan for each project with one or more people, especially when the various projects share common resources such as labor.

Instructor's Notes:

6.2 PROJECT MANAGEMENT

2.2.1 Work Breakdown Structure (WBS)

Before you, as project manager, can accurately provide estimates and implement planning, you need to get a more complete scope of what will be involved. The **work breakdown structure (WBS)** is used to divide the scope of work into manageable units. The WBS is a valuable tool for detailed planning, scheduling, and cost control.

A more defined scope statement evolves soon after the contract award. The scope includes information obtained from project documents that describe the work and resources needed to complete the project. At this point, you will likely involve different personnel assigned to the project. Supervisors will provide input about resources, methods of construction, production rates, and so forth.

Using the scope statement as a baseline helps you to define the project's scope in terms of deliverables and breaks down those deliverables into components. The phased approach is usually the preferred choice, enabling these phases to be broken down into tasks. Tasks are then assigned to personnel or subcontractors, and deliverables are determined.

The WBS organizes the project into detailed tasks and summary tasks, and is similar to a traditional company organizational chart. However, today's project managers use software to generate a WBS and the graphical approach to making a WBS is rarely used. The preferred approach is sometimes referred to as an indented list format, which is similar to a content outline. The WBS needs to identify every project task using two steps: identify the major deliverables and create tasks for every deliverable. The objective of the WBS is to break down the project into its main categories, such as pre-construction, installation, finishing work, and so forth. After the major phases are in place, they are broken down into finer details or activities as shown in *Figure 1*.

Having a detailed WBS will help you to ensure that all resources remain within the prepared estimate. A detailed and well-prepared WBS will allow you to select scheduled activities and assign that portion to the subcontractor responsible for completing the task.

The WBS can also plot inspections, order and recover materials, and other construction activities. The project manager making the schedule has leeway regarding the level of detail in a WBS. This becomes a judgment call; in the field, specific detail is preferred.

The amount of detail provided by the project manager or planner in the WBS and the amount of detail required on site could lead to problems.

WBS Outline

1.0 Project Kickoff
2.0 Preconstruction Phase
  2.1 Obtain permits
  2.2 Obtain bonds
  2.3 Obtain insurance
3.0 Engineering Phase
  3.1 Shop drawings
  3.2 Review shop drawings
  3.3 Fabrication
4.0 Job Mobilization Phase
  4.1 Set up office complex
  4.2 Install temporary utilities
  4.3 Set up site security
  4.4 Communications hookup
    4.4.1 Telephone
    4.4.2 Telecommunications
5.0 Site Preparation Phase
  5.1 Survey
  5.2 Install site drainage system
6.0 Foundation Phase
  6.1 Excavate footings
  6.2 Form and reinforce footings
  6.3 Pour footings
7.0 Structural Phase
  7.1 Steel erection
  7.2 Concrete placement

106F01.EPS

Figure 1 ◆ WBS outline.

For example, the planner or project manager may be very experienced and able to operate without much detail in the WBS. However, the on-site supervisor or other workers who use the WBS may need more detail. Be sure to structure the WBS for the level of detail the person using it needs. Additional information about the WBS can be found in the *PMBOK Guide*.

3.0.0 ◆ PLANNING PROCESS

The project manager is required to complete the work safely, in a quality manner, by a given date, and within the established budget. You reach this goal by completing a series of jobs, each comprised of a series of activities. The activities, then, are the small steps leading to the goal, and you must identify and sequence these activities and assign them the resources they require.

Review the process of formal planning and the basic elements of a formal plan.

Go over the project manager's responsibilities in relation to planning.

Show Transparency 5 (Figure 1). Explain how the WBS is used to divide the work into manageable units.

Discuss how project managers develop a scope statement to accurately identify information about jobs, activities, and tasks.

Review the process for creating the WBS and explain how it is structured.

See the Teaching Tip for Section 2.2.1 at the end of the module.

Show Transparency 6, The Planning Process, and discuss.

MODULE 44106-08 ◆ CONSTRUCTION PLANNING 6.3

Identify factors that can affect planning.

Emphasize that good planning helps to determine how time and resources can best be allocated.

Point out the various constraints that can affect a project, and that constraints must be considered when planning.

Explain that quality is part of the planning process and that it is the project manager's responsibility to implement quality standards.

Remind participants that safety measures are also the responsibility of the project manager.

In breaking a project down into jobs and activities, you must first define how much detail is required. A general rule is that each activity should be identifiable and definable. An identifiable activity is one that can be assigned specific types and amounts of resources. A definable activity is one that can have a specific time assigned to it.

An efficient plan is a simple one. Complexity should enter planning only on a job that has never been done before or one that requires a detailed breakdown to achieve efficiency. In the case of construction of a one-story school building, the entire project can be broken down into the following jobs, though not necessarily in the order shown:

- Lay out site
- Perform rough earthwork
- Install site utilities
- Install foundations
- Erect structural steel frame
- Erect external walls
- Erect internal walls
- Install insulation
- Install roof
- Install plumbing
- Install electrical
- Install HVAC
- Install finishes
- Install special-condition items
- Install external pavements
- Landscape site

A more detailed list might look something like this:

- Obtain permits
- Mobilize to site
- Lay out site
- Strip/stockpile topsoil
- Rough-excavate for foundation
- Finish excavate
- Place form work for foundation
- Place reinforcing steel for foundation
- Place concrete for foundation

The second list contains considerably more detail than the first, and it takes a project manager more time to develop such a list. However, this time is offset by the fact that more detail provides a better control standard against which to measure progress. When dividing a project into jobs, a good rule is to use the amount of detail shown in the project estimate as a guide.

When developing a project or job plan, you must recognize the constraints that affect the process. Some very specific constraints might be built into the plan itself, while others can only be anticipated.

Some constraints are technological; for instance, the activity of placing concrete in a wall is dictated by when the formwork is completed. Other constraints relate to the availability of job resources such as specific types and quantities of door closures and coordination of work among all contractors. A third type of constraint includes factors which are external to the project, such as weather conditions and governmental regulations. By considering these constraints during planning, the project manager will be more effective. The more details the project manager can identify, the better prepared he will be when starting pre-construction.

3.1.0 Importance of Quality

Along with planning tasks, you also need to develop and put into place a project management quality process. Maintaining quality both in the product and management deliverables creates pride in work and denotes a performance-based work environment. All work must be done correctly while meeting drawings and specifications. The project manager is responsible for knowing the company's quality standards and implementing them at all levels. Another important component of quality is safety. The project manager needs to ensure that all safety measures and regulations will be enforced by supervisors at the job site.

The long-range benefits of project management quality include satisfied customers, motivated employees, and profit for the company. The *Quality Control and Assurance* module provides more detailed information on this topic.

3.2.0 Pre-Construction Planning

Pre-construction planning involves two distinct phases. The first occurs when the bid, or negotiated price, for the project is being developed. This is sometimes called pre-bid planning—it is the time when the estimator (along with others, such as the project manager) develops a preliminary idea of how the work will be done. This picture is developed through experience and training and includes what methods, personnel, tools, and equipment will be used and what level of productivity can be expected from them.

Depending on the type and size of the project, the estimator may or may not make a thorough

Instructor's Notes:

6.4 PROJECT MANAGEMENT

study of factors specific to the project. If such a study seems advisable, he usually requires assistance in completing it, and the project manager may be responsible for providing him with information. In some companies, the project manager may also be the estimator. *Figure 2* lists a summary of activities that would typically be included in pre-bid planning.

The second phase of pre-construction planning occurs after the contractor gets the project, but before construction begins. During this stage, the actual work methods and resources needed to perform the work are selected. However, because the total cost of performing the work has already been determined by this time, the work methods and resources identified here must fit within the earlier estimate if the cost of performing each task, and thus the total cost to build the project, are not to be exceeded.

The project manager may want to develop a pre-construction planning checklist such as the one shown in *Figure 3*. Of course, the items on the list will vary somewhat from project to project, and the project manager should revise the list accordingly.

3.3.0 Construction Planning

Planning is an important part of any construction project. The project manager needs to identify the means to define project scope, and the appropriate methods for completing the project. At project start-up, an effective manager applies the ability to reduce each job into simpler parts and organize a plan for handling each of them. Since the project manager is directly responsible for planning during construction, he must care-

Show Transparencies 7 and 8 (Figures 2 and 3) and review the items on the checklists.

Discuss the two phases of pre-construction planning.

Pre-Bid Planning Checklist

- ______ Site visit
- ______ Review new specifications and drawings
- ______ Site access
- ______ Availability of utilities
- ______ Disposal of excavated materials
- ______ Labor availability
- ______ Local weather conditions
- ______ Availability of subcontractors
- ______ Staff to be used on job
- ______ Bonding capacity
- ______ Company experience
- ______ Equipment availability
- ______ Possible competition
- ______ Preliminary planning
- ______ Overhead and profit
- ______ Activity analysis
- ______ Selection of planning team
- ______ Optimization of the construction methods as they apply to the particular job
- ______ Financing
- ______ Purchasing critical activities
- ______ Work the contractor will carry out
- ______ Work to be subcontracted

106F02.EPS

Figure 2 ◆ Pre-bid planning checklist.

Pre-Construction Planning Checklist

- ______ Superintendent—who/when
- ______ Distribute copies of drawings—subs and suppliers
- ______ Office space—define needs and obtain
- ______ Phone—order as early as possible
- ______ Permit—city, county, other jurisdictions
- ______ Temporary power—usually electrical subcontractor
- ______ Toilets—rent and deliver
- ______ Security—look at whole picture
- ______ Water—usually plumbing subcontractor
- ______ Site planning—where to set trailers, where to park, etc.
- ______ Signs—what will owner allow, obtain signs
- ______ First aid—make provisions
- ______ Tools and equipment—superintendent plans this
- ______ Trash—arrange for garbage pick-up
- ______ Office equipment—look into needs, if any
- ______ Surveying instruments—look into needs, if any
- ______ Subs/suppliers list—from estimating
- ______ Get initial survey completed

106F03.EPS

Figure 3 ◆ Pre-construction planning checklist.

MODULE 44106-08 ◆ CONSTRUCTION PLANNING 6.5

Show Transparency 10, Construction Planning Factors, and Transparency 11, Major Goals of Planning.

Discuss the factors that a project manager must consider when planning.

Discuss the major goals of construction planning and provide examples.

Emphasize the importance of cost control.

Show Transparency 11, Planning Review Factors, and Transparency 12 (Figure 4).

Discuss planning review factors that should be considered.

Discuss the importance of regular follow-up on projects and progress reviews.

Point out who should be included in post-project reviews and stress the importance of examining lessons learned.

Review Figure 4 and discuss how a checklist can help guide the review process.

Assign reading of Sections 4.0.0-4.2.0.

fully analyze each of the factors which affect the job, primarily:

- Timing of all phases of work
- Weather
- Types of material to be installed and their availability
- Equipment and tools required and their availability
- Personnel requirements and availability
- Subcontractor availability
- Relationships with the other contractors and their representatives on the project

On a simple job, these items can be handled almost automatically. But larger or more complex jobs force the project manager to give these factors more formal consideration and study. Regardless of the size or complexity of a job, the project manager must keep in mind the following major goals of planning:

- Determine the best method for doing the job.
- Identify what each work crew is to do.
- Make sure the required materials are at the work site when needed.
- Identify what tools will be needed and ensure that they are at the jobsite when needed.
- Make sure heavy construction equipment is available when required.
- Identify the other contractors that may have an effect on the completion of the work, and hold meetings to discuss possible conflicts to prevent any future problems.
- Include safety considerations in the plan.

When planning, you should review the results of any critique performed on a similar project previously completed. Furthermore, you must develop alternate plans. On every construction project, it is inevitable that unforeseen events will upset original expectations. The best way to plan for unforeseen events is to make sure that safeguards are written into the contract. As any builder can attest, you never actually know the scope of a project until it's completed.

Many things can happen along the way. In the time it takes to hear that telltale "ting" when the backhoe hits rock, the project price can go up thousands of dollars. Discovering problem soil during the excavation can increase the cost by tens of thousands of dollars if the structure requires pylons or caissons.

When these things happen and clients find out the extraordinary measures you must take to build their pool, they always pose that tense first question: "Who's going to pay for it?" That's when a properly written contract becomes your best friend.

Having alternate plans readily enables you to deal with such situations without incurring lengthy delays and/or costly overruns. There is often not enough time during planning to develop alternate approaches for each activity; therefore, to maximize efficiency, you should develop alternate plans only for those activities or jobs in which the company's experience is limited or the availability of adequate resources is not certain. By planning as early in the project as possible, you will have more alternatives available.

By using an effective project reporting system such as that described in the *Estimating and Cost Control* module, you can stay abreast of progress and make changes as needed, before the project gets out of control. This helps you avoid falling behind schedule and exceeding the project cost estimates.

3.4.0 Planning Review

Once a plan is implemented, follow up on a regular basis to ensure that it is being done. If you have 28 weeks to complete a project, you cannot wait until week 15 to determine whether it will be completed on time and within the estimate. Planning reviews produce information that might be useful to you in the future, including:

- Identification of problems before they occur
- Documentation of problems so that they may be avoided in the future
- Evaluation of the effectiveness of the plan

Any progress review should include documenting changes made in work methods or job resources that weren't used the way the original plan specified due to problems or unforeseen circumstances. Even if the original plan is working flawlessly, you should document which work methods and resources are being used. Write the documentation in a project diary or similar format so the information can be used on similar projects in the future. Keeping thorough records of problems that arise and successful and unsuccessful solutions provides a valuable resource that is well worth the time it takes to prepare. In addition, the documented information may serve as evidence for future job claims.

3.4.1 Lessons Learned

The planning review continues after the project is completed and includes the project manager sitting down with others involved with the work

Instructor's Notes:

and reviewing the project in detail. This is often referred to as a post-project review. Those involved in the review process besides the project manager are key field supervisory personnel, others from inside and outside the company such as the estimator, other contractor representatives, the architect, and the client. This debriefing session should be held somewhere away from the interruptions of the daily work routine. It should focus on both the project's successes and failures, and analyze both the successful and unsuccessful solutions used to resolve problems that arose. The results of the discussion, including recommendations for similar projects in the future, should be documented for future use. A checklist like the one shown in *Figure 4* can help your review process.

4.0.0 ◆ PERFORMING A WORK ANALYSIS

When all of the planning is completed, the project manager and the team must perform a detailed work analysis to implement the plan. This section presents examples of how to conduct a work analysis; keep in mind that every project may not need to include the information presented.

Post-Project Review Checklist

_____ Sequencing of activities
_____ Innovations in construction methods
_____ Productivity
_____ Quality
_____ Cost
_____ Vendor performance
_____ Performance of other contractors
_____ Lead times
_____ Type of schedule
_____ Effectiveness of schedule
_____ Labor, production, and control
_____ Tool control
_____ Material use and control
_____ Equipment use and control
_____ Overhead items
_____ Safety program
_____ Documentation
_____ Changes
_____ Absenteeism and turnover
_____ Lessons learned

106F04.EPS

Figure 4 ◆ Post-project review checklist.

Work analysis is a study of each job, aimed at effectively and efficiently integrating all tasks and activities. It is most often used when situations such as inclement weather or late material deliveries threaten to delay work. The project manager, possibly in conjunction with other managerial or non-managerial personnel on the job, must then identify the remaining activities and, if necessary, alter resources and work methods to complete the job on time and without additional expense. A work analysis can also be conducted during the pre-construction phase to determine the most efficient and effective job plan.

The work analysis objectives reduce time and costs required to accomplish a task by identifying more productive ways to do a job. To understand work analysis, consider the following two methods of changing a light bulb shown in *Table 1*.

The second method is more efficient and could have resulted from an analysis of the work in question. There are several ways of doing a job, and the way a person does so depends on experience, authority, responsibility, and creativity.

On the job, sometimes you are taught how to do a particular task. At other times, you perform a task as you always have. But if the task is new, or a particular situation adds an unusual twist to a job, you must plan how to perform effectively, and this plan involves work analysis.

4.1.0 Steps of a Work Analysis

Work analysis begins with a brief written description of the job along with a list of the various identifiable and definable activities involved and the needed resources. Then the activities and resources are arranged in such a way that they

Table 1 Sample Work Analysis for Changing a Light Bulb

Method One	Method Two
Obtain ladder	Obtain ladder
Set up ladder	Obtain new light bulb
Climb ladder	Set up ladder
Remove burned out light bulb	Climb ladder
Climb down ladder	Remove burned out bulb
Discard old light bulb	Install new light bulb
Procure new light bulb	Climb down ladder
Climb up ladder	Take down ladder
Install new light bulb	Put ladder away
Climb down ladder	Discard old light bulb
Take down ladder	
Put ladder away	

106T01.EPS

Show Transparency 13 (Table 1) and discuss how performing a work analysis will reduce time and cost.

Show Transparency 14 (Table 2). Review the questions that can be used to analyze the details of a job.

Explain when a work analysis should be conducted, and provide job-related examples.

Demonstrate how a work analysis is performed.

Explain and provide tips on creating a more efficient work analysis.

See the Teaching Tip for Section 4.1.0 at the end of the module.

MODULE 44106-08 ◆ CONSTRUCTION PLANNING 6.7

Show Transparencies 15 and 16 (Figures 5 and 6).

Review how project managers may restructure a task.

Explain the differences between the two work analysis formats.

Discuss the importance of measuring actual and estimated production for activities.

Explain that work analysis charts provide important job data, both during and after construction.

meet the desired goal or objective in the manner that takes the least amount of time and effort, and costs the least amount of money.

Details of a job can be analyzed by asking the following questions and writing down the answers:

- What is the job?
- What is the purpose of the job?
- What are the activities within the job?
- What is the purpose of each activity?
- Could the job be done another way?
- Is every listed activity necessary?
- Who does the work?
- Who could do it better?
- Where is the work done?
- Could it be done somewhere else for less?
- When is the work to be done?
- Would it be better to do it some other time?
- How is the work done?
- Could more efficient methods be used?

These questions can be applied to virtually all elements of any job, physical layout, tools, equipment, material, labor, and coordination. Answering them gives the project manager a better understanding of the job goals, job activities, variables involved, and possible alternatives. This information then allows the project manager to restructure by:

- Eliminating unnecessary detail
- Rearranging activities for optimum sequence
- Providing better tools, material, and equipment
- Simplifying, where possible, to make things easier to accomplish
- Keeping safety in mind
- Using different people, different skills, and/or a different number of people
- Consulting with others involved in the job, such as supervisors or other crew members
- Eliminating unproductive time spent searching for materials, tools, equipment, and so forth

A work analysis that takes these steps into account will more than pay for the time spent preparing it. Of course, the amount of time spent performing a work analysis should be in proportion to the size and complexity of the job. However, even small jobs benefit from work analysis and the project manager should keep in mind that even a minimal work analysis is better than none at all.

A work analysis of a typical job is provided as an example in *Table 2*.

4.2.0 Work Analysis Format

A work analysis chart, which can take many forms, documents the results of a work analysis. Some, like the one in *Figure 5*, are quite simple. Others are more detailed as shown in *Figure 6*.

Both formats have certain features in common, including the following items:

- A chronological list of activities included in the job. The list should allow for delays and unproductive activities such as waiting time, set-up, take-down, clean-up, and similar activities.
- Identification of the labor and equipment involved in performing each activity.
- The estimate or actual time to perform each activity.
- Hourly labor rates and equipment costs.
- Total cost of performing each activity.

Although only labor and equipment costs are shown in the illustrations, material quantities and costs can also be included on a work analysis chart to allow the project manager to calculate the total direct costs of performing the work. This makes the work analysis chart a valuable resource for pre-construction planning.

During construction, the work analysis chart can be used to compare actual production with estimated production for each activity on the project. This is a particular advantage on jobs in which the contractor has little or no experience. If the job does not meet the production rates estimated by the work analysis, immediate action can be taken to correct the situation and get the job back on schedule without incurring cost overruns. A special work analysis can be prepared to deal with a particular problem that arises; such an analysis allows the project manager to identify and evaluate alternative courses of action.

After construction, the work analysis chart can be used as a guide for reviewing the job. Comparing the chart prepared during pre-construction planning with one illustrating the actual course of the job serves as the basis for identifying and evaluating problems that occurred in the field and solutions that were tried. A work analysis chart is an excellent control device throughout the project and should be considered one of the project manager's most valuable management tools.

Instructor's Notes:

6.8 PROJECT MANAGEMENT

Table 2 Tasks in Constructing Forms

Construct forms for a poured-in-place reinforced concrete beam on the third floor of a building.	
1. What is the job to be done?	Construct job-built form work for a reinforced concrete beam.
2. What is the purpose of the job?	Provide a structurally stable form to retain the wet concrete and reinforcing steel until the concrete cures.
3. What are the activities within the job?	a. Lay out beam location. b. Erect support scaffolding, braces, and so forth. c. Obtain materials for the form work. d. Cut material to size for the beam bottom and place in position. e. Cut material to size for the beam sides and erect in place. f. Install braces for the beam sides. g. Make ready forms to receive reinforcing steel and concrete.
4. What is the purpose of each activity?	a. To ensure the beam is in correct location per drawings. b. To support form work and workers. c. To build the forms. d. To construct the beam bottom per the drawings. e. To construct the beam sides per the drawings. f. Provide support for the beam sides so they do not move outward when the concrete is placed in the form. g. To perform any other activities such as oiling the form sides and cleaning up the area so there will be no delays in placing the reinforcing steel and concrete.
5. Could the job be done in another way?	The beam bottom and sides could be constructed as a unit and placed in final position as same.
6. Is every listed activity necessary?	Yes, since the tasks depend on each other to reach the objective.
7. Who does the work?	Trades people skilled in the specific tasks.
8. Who could do it better?	It may be possible to use different individuals who the supervisor feels are more productive. Also, individuals trained in more than one craft might be used to perform a variety of tasks on the job.
9. Where is the work done?	On the third floor in the area where the final beam will be located.
10. Could it be done somewhere else for less?	The entire beam form could be built on the ground level and installed as a unit. It would then be braced and made ready to receive the reinforcing steel and concrete.
11. When is the work done?	The tasks are to be performed in accordance with the job schedule.
12. Would it be better to do it some other time?	Not unless the schedule is revised.
13. How is the work done?	The tasks are to be performed using current acceptable methods, implemented by skilled craftspeople under the direction of their supervisor.
14. Could more efficient methods be used?	It appears that the methods to be used to perform the tasks are the most efficient. However, the procedure will be studied to determine if a more productive way exists.

106T02.EPS

Assign reading of Sections 5.0.0-5.6.0.

MODULE 44106-08 ◆ CONSTRUCTION PLANNING 6.9

Show Transparency 17, Resource Planning Considerations.

Provide an overview of resources that the project manager needs to consider for a project.

Explain that the project manager must have knowledge about pricing, materials, trends and so forth.

Discuss the importance of vendor and supply chain relationships.

Describe the quantity survey method.

Provide examples of a bill of materials and explain how it is used to issue purchase orders.

Discuss the importance of knowing company purchasing procedures.

Date: ______________________________

Job: ______________________________

		Labor				Equipment					
Date	Activity	Class	Hours	Rate	Total	Class	Hours	Rate	Total	Total L&E	Comments

106F05.EPS

Figure 5 ◆ Work analysis chart.

5.0.0 ◆ RESOURCE PLANNING

Once the project manager has broken down the project into jobs and activities, the next step is to identify the resources needed to perform them.

5.1.0 Materials Planning

In planning material requirements, the project manager must consider prices, price trends, optimum times to buy, quantities, dealers or manufacturers, transportation and delivery, amounts to be kept on hand, inspections, tests, and insurance.

5.1.1 The Quantity Survey

The basis for material planning is the **quantity survey**, the estimate of the material quantity. The survey is used to develop a **bill of materials** and begin the ordering process. Some companies use the estimate as the bill of materials and write **purchase orders** directly from it. This is acceptable as long as the format of the survey is self-explanatory regarding the types, qualities, and quantities of material. There are many different formats available for the bill of material.

5.2.0 Purchasing

The project manager will frequently have to purchase materials, tools, or commodities such as storage space or equipment rental, so it is important for the manager to have a working knowledge and understanding of purchasing and its related responsibilities.

A proper understanding of procedures, responsibilities, and an attitude of fairness are critical. The project manager's understanding will not only minimize the risk of exposure, but also reduce the likelihood of litigation if problems arise.

Instructor's Notes:

____

____

____

6.10 PROJECT MANAGEMENT

Reg. No. 2854
Subject: Concrete Pumping—Beam Repair
Date: 10/19/2008 Cont. No. 500
Work Cycle: Place conc. by pipe line
Lst. Quantity: 180 Est. Cost: 2/26/CY
Plant Equipment: Contractor Provides
Gen. Supv. ______________ Supervisor: ______________

Location: Loading Dock
Charted by: RWL
Unit: Cu. Yd.
Actual Cost: ______________
Present: ______________
Proposed: For Estimate ______________

Hourly Rates			20.00	16.00	10.00	8.00	8.00	8.00	8.00	8.00	75.00	50.00				
Distance	Time		Supervisor	Pump Operator	Hopper Worker	Laborer	Laborer	Laborer	Laborer	Vibrator Worker	Vibrator Worker	Pump & Fittings	Truck & Driver	Costs	Number	Actual Activities
200	60		1		1		1	1	1	1	1			83.00	1	Make ready—tools.
150	15		2	3	2	2	2	2	2	2	2			19.50	2	Move to job site.
2 mi	45		3									3	3	108.75	3	Pump mach. to job.
	90		4	4	4	4	4	4	4	4	4	4		342.50	4	Set up mach. time.
	30		5	1	1	4	4	4	4	4	9	9		20.00	5	Inspection.
	30		6	6	6	4	4	4	4	4	4	9		23.00	6	Mach. adjusting.
	20		7	7	7	7	7	7	7	7	7	7		56.33	7	Delay—ready-mix.
115	360		8	8	8	8	8	8	8	8	8	8		1,032.00	8	Pump concrete to forms.
	80		9	9	9	9	9	9	9	9	9	9		304.33	9	50 min. hr. normal delay.
	45			10	10	10	8	8	10	10	8	10		119.00	10	Dismantle pump.
150	15			10	10	11	11	11	11	11	11	10		12.00	11	Tools to shed.
2 mi	30		5	12	12	13	13	13	13	13	13	12	12	75.50	12	Return pump.
	10													8.00	13	Final cleanup.
	5					14	14	14	14	14	14			4.00	14	Crew—check out.
															15	
															16	
Vibrators and Tools in Plant Setup.															17	
															18	
															Totals	Summary
Min.			760	715	775	780	780	780	780	780	780	745	75			
Hrs.			12.67	11.92	12.92	13.00	13.00	13.00	13.00	13.00	13.00	12.42	1.25	129.18		Quantity: 180 c.y. Unit Cost: 12.17
Costs			253.40	190.72	129.70	104.00	104.00	104.00	104.00	104.00	104.00	931.50	62.50	2,191.32		

106F06.EPS

Figure 6 ◆ Sample work analysis chart.

MODULE 44106-08 ◆ CONSTRUCTION PLANNING 6.11

Review the three areas of material acquisition requiring purchase orders.

Emphasize that the purchase order is considered a contract and that a purchase order should be priced.

Discuss why quotes should be obtained from several vendors, and explain the criteria for selecting a vendor.

Point out the impact of the cost of miscellaneous materials on the job's profitability.

Emphasize the importance of professional business ethics when dealing with vendors.

Provide examples of professional and unprofessional actions from either party that may impair future business relations.

Show Transparency 18, Information Required for Quotations. Briefly review each item.

Review and discuss the information that should appear in a quotation.

5.2.1 Purchase Orders and Contracts

A purchase order is issued because a specific product is required for a specific price in a specific time frame, with certain conditions such as delivery or guarantee. A subcontract differs from a purchase order only by the degree of labor involvement removed from the original point of manufacture. If it is necessary for the supplier to perform labor on-site, it is appropriate to issue a subcontract (see the *Issues and Resolutions* module). This may provide a more effective way of making the vendor fully perform and comply with the original specifications, as well as proceeding with payments to ensure complete and proper performance.

There are three categories of material acquisition that require the use of purchase orders:

- Over-the-counter or common-stock materials – These are often smaller quantities, or materials that are needed immediately.
- Major purchase materials that are usually quoted by vendors – This category consists of either large quantities of common material or select packages of specified material.
- Specialized material involving elements of service or support, such as process controls or generators.

Regardless of the category of material acquisition, there are certain elements in common with all purchase orders. The purchase order is a confirmation of understanding between two parties for supplying goods in exchange for an amount of money. This purchase order is, in fact, a binding contract, enforceable on both parties.

Although states apply the Uniform Commercial Code with varying interpretations, it is generally accepted that the price is the price in effect at time of delivery unless stated otherwise. What this means is that unless a price is stipulated, the vendor may charge whatever it wants, including a price that is in fact higher than a published list price. A responsible vendor will not abuse an unpriced purchase order, knowing that it may result in the loss of future business. Unless your company is in the habit of issuing blank checks, purchase orders should be priced.

There are many reasons for choosing one vendor over another, including the following factors:

- Availability of the desired product
- Price
- Time constraints
- Quality of service
- Dependability
- Pre-existing relationship

Just as the project manager's responsibility is to make a profit for the company, the vendor's representative is equally responsible for making profitable sales for the vendor. For either party to abuse the other for the sake of short-term profit may in fact have the opposite long-term effect.

What impact does the cost of miscellaneous material have on a project, or more importantly, on the anticipated profit on a specific project? Everything from concrete to labor to light bulbs has a great bearing on cost. Consideration must not only be given to specific item costs, but also the cost effect on the project due to lost production while waiting on material, or the liquidated damage assessment due to delayed completion.

5.3.0 Pre-Purchase Considerations

All purchase orders are issued following the determination of required goods. This determination usually follows a common chain of events, beginning with a specified bill of material as noted earlier in this section.

Once a specified bill of material has been developed, a quote should be sought from one or more vendors. If at all possible, this quote should be in writing; if it is not possible to get a written quote, and a considerable amount of time will pass between the quote and the issuance of a purchase order, confirmation of the quote should be sent. This form of communication may actually demonstrate the contractor's intent to purchase.

All quotes, whether received in writing or over the telephone, should contain at minimum the following information:

- The vendor's proper and complete name
- The name of the individual issuing the quotation
- The date and time the quotation is received
- Any exclusions or variances from the bill of material for which a quotation was requested
- Specificity (per plans and specifications, approved equal, non-approved equal, or nonequal alternative)
- Price (whether lot pricing based on specified quantities, or unit pricing)
- Delivery (time and freight charges)
- Tax (included or excluded)
- Terms and conditions of payment that are not always specified at time of quotation

Any quotation that is received over the telephone should be written on a telephone quotation form or a piece of 8½ × 11 paper. Take care to avoid writing quotes on smaller pieces of paper that may be incompletely filled out or misplaced.

Instructor's Notes:

6.12 PROJECT MANAGEMENT

There is a negotiation stage between receiving the quotation and issuing the purchase order, and it is the most critical one in the purchasing process. It is at this point that the body of the purchase order is agreed to and accepted by the parties. It must be remembered that until the purchase order is issued and accepted, all responsibilities and liabilities are negotiable. If a contractor attempts to issue a purchase order with terms and conditions not previously agreed upon by the vendor, he not only risks rejection, but at the very least will dispel the cooperative attitude necessary to guarantee the lowest quotes and the best performance from the vendor.

Once the specific material is defined, and price and terms are agreed upon, a purchase order is issued. It then becomes a contract, commonly thought of as a specific performance contract. To avoid unnecessary costs and frustrations, take care to make the purchase contract as thorough and complete as possible.

5.4.0 Issuing the Written Purchase Order

Although every company has its own purchase order forms, all purchase orders have a spot for (and should include) the date, vendor's name, job name and number, requisition number, date required, terms, shipping destination, FOB point, method of shipping, and written confirmation.

5.4.1 Date

Two dates are important: the date the purchase order was drawn up, and the date the order is accepted by the vendor. The date the purchase order was drawn up cannot be omitted if there will ever be a question concerning timeliness of issuing the purchase order following receipt of the quotation, or in the case of multiple quotes from the same vendor, identifying the sequence of quotes. The acceptance date is what verifies the timely performance of the vendor.

5.4.2 Vendor's Name

The vendor's full and proper name, address, and phone number (toll-free, if available), as well as the name of the individual vendor's employee should appear on the order. Legal consequences may arise from an incomplete or improper name. Assume, for instance, that a purchase order is made out to "G.E., Attn: Joe Smith," for purchase of a panel board. It is possible that G.E. Supply Company could claim they were accepting the order as agents for General Electric Company (the named vendor which they represent.) However, General Electric Company could claim that there was no direct contractual relationship.

5.4.3 Job Name and Job Number

The job name and number are important for in-house accounting, and they also provide an additional reference for the vendor.

5.4.4 Requisition Number

The requisition number permits cross reference between field control and purchasing, and becomes an **expediting** tool.

5.4.5 Date Required

The required date should be mandatory, and should be inserted as a specific date rather than vague descriptions such as RUSH or ASAP. The purchaser has little hope of recovering for damages due to late delivery by a vendor unless there is a specific date that has been accepted by that vendor.

5.4.6 Terms

Payment terms "Net 30," "2% 10, Net 30," or "50% Upon Order, Balance Net 30" each have a dramatically different effect on cash flow and profit. Assuming a volume of $1 million annual purchase of material, with material purchase of 2% 10, prompt payment will translate to additional savings of $20,000. If, on the other hand, material is ordered 50% on order, balance net on the same $1 million, additional financial costs of $12,000 to $24,000 will result. For many construction firms, this may equal the total year-end profit or loss. Close attention to payment terms is extremely critical for maintaining proper profitability and cash flow.

5.4.7 Shipping Destination

In all cases, the ship-to address must start with the name of the company, in care of a specific individual, then the project name, and project address (or shop address, if applicable). If this is not specified, anyone can sign for the material. The signer does not have to be an employee of the company. Then the vendor can substantiate delivery as addressed, leaving the company liable for payment as well as for finding the material. The shipping destination should be detailed with a proper address.

Emphasize that final pricing and terms and conditions are negotiated at this stage.

Show Transparency 19, Information Required on a Purchase Order. Briefly review each item.

Explain the consequences of lost time and tracking problems that can occur if purchase orders are missing critical information.

MODULE 44106-08 ◆ CONSTRUCTION PLANNING 6.13

Show Transparency 20 (Table 3).

Discuss the four key elements of successful expediting and provide a job-related example for each one.

Show Transparency 21 (Figure 7).

If this is not possible (as is often the case with new construction), it should be detailed with adequate directions such as: "On Highway 12, $6\frac{1}{3}$ miles east of intersection with Highway 25, at the site of XYZ College Construction Project, Mainstream, New York." Proper directions will minimize lost and misdirected shipments. Any special tagging requirements (as applicable to the entire purchase order) may be noted here. This will not only aid in the receipt and identification of the material, but also in controlling the purchase order.

5.4.8 FOB Point

As with the ship to address, the FOB point also has serious ramifications. In shipping, **FOB** stands for **free on board**, and identifies the location at which title for the merchandise passes from the seller to the buyer. Ideally, this point should be the destination, where the contractor needs the material. This transfers all risk of loss and freight damage to the vendor or shipper, as ownership and payment liability do not occur until the material is received at its destination and verified as proper. In truth, most vendors will not accept purchase orders with this designation unless it is over-the-counter material that they are delivering locally with their own truck and employees.

The most common FOB that is quoted or accepted by vendors is free on board—point of shipment—full freight allowed (FOB-POS-FFA). Occasionally, vendors will agree to special conditions that transfer some of the risk of freight claims onto either themselves or the shipper. This must be negotiated at the time the purchase order is issued. If a purchase order is written as FOB-factory, a note should be made as to who pays the freight charges, and exactly how the merchandise is to be shipped.

An alternative to FOB-factory is to ask the vendor to bring the merchandise into the vendor's warehouse and then dispatch it to the job site. Most vendors will accept this method of handling; however, it generally adds a price premium of from 1 to 3 percent. If it is a high-damage shipment such as custom glass, light fixtures, or medical equipment, this may be a cost-effective option.

5.4.9 Method of Shipping

This is often a negotiable item. However, while the purchase order specifies FOB-destination (and the vendor has accepted this) with an acknowledged delivery date, the vendor has control of shipping means and sources. This does not mean that this item should be left blank. It is common to insert Best Way, or Shipper's Discretion. In some cases, it is advisable to specify the means of shipping, such as: FedEx priority, UPS next day air, express mail, or cargo air freight. In this example, all four methods of shipping are "air," but not all four have the same traceability (see the expediting topic in this module).

On a critical project, the ability to track merchandise from production to delivery is extremely important. Another thing to think about when choosing shipping method is customer consideration. If your customer is the United Parcel Service, it would be a tactical blunder to allow a vendor to ship a small parcel by Federal Express. Similarly, if the customer is XYZ Trucking, motor freight deliveries might be specified as to be "Interlined with XYZ."

5.4.10 Written Confirmation

Telephone, email, and the Internet are indispensable tools in today's construction world. **Purchase orders** are issued using all of these options and most vendors will either email or fax confirmation. Also, many companies have internal processes and software that track orders and other project correspondence.

5.5.0 Expediting

The entire process of confirming that purchased materials and equipment reach the project when they are expected, in their entirety and ready for use, is known as expediting. There are four key elements that will lead to successful expediting:

- Timeliness of order
- Planning
- Follow-up
- Documentation

5.5.1 Timeliness of Order

The most critical element of timing occurs when expediting starts at the time the order is originally placed. For materials to arrive at the proper time, they must be ordered with sufficient time for the vendor (and all secondary and tertiary suppliers) to process and ensure delivery.

5.5.2 Planning

The sole purpose of planning is to anticipate what will happen, and to remove the harmful effects of things that may happen contrary to what is desired. A critical function of planning,

Instructor's Notes:

6.14 PROJECT MANAGEMENT

therefore, is to make allowances for things that will go wrong. If a function of timeliness is to allow sufficient time for the full supply chain, proper planning may require allowing additional lead time for breakdowns within this chain.

Also involved in planning is handling. Will there be sufficient personnel and equipment on site at the time of delivery to unload the merchandise? Will there be adequate (including secure and environmentally proper) storage space?

What if merchandise doesn't arrive on time? This possibility is frequently overlooked, but can be avoided with proper planning. Critical merchandise should be handled differently. More important than price consideration should be traceability. What if you were told that merchandise had been shipped, but had not arrived yet, or it took two weeks for a shipment to be transported 250 miles? Different forms of shipment have different response times for locating lost shipments (see *Table 3*).

As a precaution, it's important to check all sources regarding freight carriers. Times and prices may vary and change.

Another item to be considered as part of planning is the reporting of material receipt, specifically who is authorized to receive the material, and what form of documentation is required. This is necessary for payment verification, discrepancies, and handling of shortages or damaged goods. Although company policies vary, it is recommended that if a shipment arrives without corresponding paperwork a packing slip should be made immediately upon receipt.

5.5.3 Follow-Up

Although seemingly an elementary item, more orders break down due to lack of follow-up than all other factors. Follow-up begins after the order is placed, with the first task being verification of vendor's acceptance, and the second being the pursuit of shipping confirmation.

Purchasing is an expression of trust in the performance of the vendor. Follow-up ensures the performance by not allowing that specific order to get lost in the shuffle. In most cases, it is accomplished with phone calls and other forms of project correspondence. The basic principle of proper follow-up is to do whatever is necessary to assure the proper and timely delivery of merchandise.

Different methods of shipping allow easier access to information concerning shipment whereabouts. Remember that tracing shipments requires knowing exactly how, when, from where, and by whom the material was shipped, as well as the freight bill number.

5.5.4 Documentation

It is a common misconception that if everything goes well, documentation will not be needed; much to the contrary, this belief cannot be farther from the truth. The purchase order is not only a means for documenting your material needs, but also a document that authorizes the accounting department to pay a specific amount once performance is achieved.

Describe the types of documentation and procedures required to verify orders.

Assign reading of Sections 5.6.0-5.12.0.

Table 3 Freight Carriers and Response Times for Lost Shipments

Freight Carrier	Response Time
Commercial Carrier Air Freight	up to a week with tracking number
Commercial Carrier Motor Freight	up to a week with tracking number
Federal Express	48 hours with tracking number
U.S. Postal Service – Express Mail	24 hours with tracking number
U.P.S. (Ground)	1 to 2 weeks with tracking number
1st Class Mail	Cannot be traced

106T03.EPS

MODULE 44106-08 ◆ CONSTRUCTION PLANNING 6.15

Explain that the project manager must assign someone to receive materials and ensure that they know the procedures for doing so.

Not all orders will be processed as a single function, and so there will be back orders that must be tracked. Also, not all vendors will communicate the quoted prices to their accounting departments, meaning that there will be billing discrepancies that must be controlled.

The first step to proper documentation is establishing an adequate filing system that is simple, easy to use, and provides all necessary information with minimal effort. The system must additionally document notations of receipt. Since most companies require packing slips to be turned in to the home office, it becomes necessary to have a means to note when material was received, and often by whom, and how it was handled.

The system must also log all telephone calls, faxes, emails, and responses. In most cases, this will require paperwork in addition to the purchase order. There should be separation within the files, permitting active files to be kept readily accessible and separate from closed files.

Closed files, which contain completed purchase orders, must be maintained. Warranties as well as pricing excess material (whether for return or for stocking) will require review of the original purchase order. In addition, purchase files are invaluable aids in analyzing completed projects.

Another critical facet of documentation for the project manager is the cross-referencing of unusual material as well as the vendors. A properly established documentation system will allow the project manager to establish a cross-reference file for future projects, with a resulting increase in efficiency and profitability. *Figure 7* is an example document format for controlling the expediting function.

5.6.0 Receiving Materials

The project manager must assign someone the responsibility of receiving materials. Material is delivered either to the project site, where it is installed immediately or stored for future installation, or to a storage area such as a warehouse. In either case, the person responsible for receiving it (which may involve unloading it as well) must be instructed on how to receive it, including how to inspect each delivery to be sure that the type and quantity of materials as listed on the delivery ticket are actually delivered. The project manager must also be aware that suppliers often send incomplete or partial orders. This makes it critical to compare the information on the delivery ticket to the information on the original purchase order. If a discrepancy is found, the appropriate parties should be notified and corrective actions taken.

If a complete inspection cannot be made, it should be noted on the delivery ticket that the material is being accepted pending future inspection. Also note any visible problems, such as damaged containers, which might indicate dam-

Expediting Sheet

Activity ____________ Network No. ____________

CP? ____________ Dwg./No. ____________

Location ____________ Project ____________

P.O. ____________ Vendor ____________

Address ____________

Delivery Promised ____________

Via Freight ____________ Routing ____________ LCL ____________

Date Shipper Advised on Routing ____________ Trucker ____________

106F07.EPS

Figure 7 ◆ Sample expediting sheet.

Instructor's Notes:

6.16 PROJECT MANAGEMENT

aged contents. Most states have laws that dictate how long a company has to inspect shipments and file claims if the contents are damaged or incomplete. The project manager and the person he assigns to receiving materials must know these laws and tailor the receiving procedure to accommodate them.

At times, receiving materials includes testing them, or at least taking samples for future testing, as in the case of ready-mix concrete. Here again, responsibility for testing or sampling must be assigned to a specific individual or individuals who should be thoroughly trained in performing the required procedures. The project manager should develop a list of those materials that must be tested upon receipt; this will help him assign adequate resources to the testing program and follow-up to be sure required tests are completed.

The project manager should also know about demurrage, a surcharge placed on items being used. Oxygen tanks are often surcharged because they are being held by the contractor while he is using the gas. The surcharge continues to mount until the empty tanks are returned. For this reason, surcharged items should be inventoried on delivery and monitored to make sure they are returned to the vendor immediately after final use.

Large orders of materials that are shipped by rail and then placed on a siding to be unloaded are usually surcharged if the materials are not unloaded from the rail car in a specific amount of time. A daily material received report, such as the one shown in *Figure 8*, can help the project manager effectively control and monitor the receipt of materials.

5.7.0 Material Control

The company should have formal inventory procedures for controlling all stored material and should be certain that everyone involved with material storage understands and follows them. Once materials are on the job site or at a storage area, control becomes essential to ensure their security and proper use. Material control is covered in the *Estimating and Cost Control* module.

5.8.0 Planning for Equipment

Planning equipment consists of the following:

- Identifying what major construction equipment is needed
- Determining the work the equipment is to do
- Establishing the date and time when each piece of equipment is needed
- Identifying what is available from the company and what must be rented or leased
- Coordinating equipment use so that the various pieces work together efficiently
- Determining how the equipment will get to the job site
- Planning mobilization (set-up) and demobilization (take-down) of the equipment

To plan equipment properly, the project manager must be familiar with the types and sizes of materials to be installed and any excavation or other earthwork to be done. He must also verify that the equipment is in good working condition and will be available when needed. The location of the project must be considered in the planning.

It is usually most cost-effective to use equipment that the company owns. However, if the company does not own the needed equipment, it must be purchased or leased. The decision of whether to purchase or lease usually lies with upper management and is based on company-wide needs and experience.

Field supervisors should be instructed to immediately contact the project manager if equipment breaks down during the job, which allows the manager to obtain alternate equipment and minimize down time. If a particular piece of equipment requires regular maintenance during its time on the job site, the maintenance should be scheduled like any other project task so that it fits smoothly into the project schedule instead of causing delays.

Coordination is another important consideration. If several pieces are to work in conjunction (such as a loader and dump truck or bulldozer), these activities should be carefully planned to ensure that the area is large enough to accommodate them. The project manager should also have an alternate plan ready so that if one piece breaks down the others are not left inactive. Finally, if rental charges are being paid for equipment not being used, it is best to return the equipment or use it on another job.

5.9.0 Planning for Tools

Most job activities require small hand and power tools, and it is the project manager's responsibility to identify which tools are needed. This may be a very easy task or one that requires considerable thought. Studies show that the lack of well-maintained tools is a major cause of job slowdown. Consequently, the project manager should be sure that tools are on the job site when needed.

The project manager needs to know his company's policy and procedures for providing tools.

Describe the testing and inspection process, and review acceptable and unacceptable procedures.

Explain demurrage and emphasize the need to inventory and monitor items that must be returned to the vendor after use.

Discuss different shipment methods. Provide examples of how certain materials should be shipped.

Show Transparency 22 (Figure 8). Review a typical materials received report.

Explain the importance of inventory and storage control.

Emphasize the role of field supervisors in planning for equipment.

Explain the importance of coordinating equipment.

Ask participants to provide an example of a situation where equipment was not coordinated well.

Describe a typical company's procedures for providing tools.

MODULE 44106-08 ◆ CONSTRUCTION PLANNING 6.17

Discuss with students how Samantha could have had a better tool management system.

Daily Materials Received Report

Date: ____/____/____

Office ________________________ W.O. No. ____________________

Instructions: Do not accept any local deliveries without delivery slips
____________________ material clerk

Quantity	Description	Received From	Storage Location

106F08.EPS

Figure 8 ◆ Daily materials received report.

Instructor's Notes:

____

____

____

If the policy is to have the workers provide their own tools, then the manager should make it clear to the field supervisor that he is responsible for making sure that the workers bring them to the job and they are in safe and good working condition.

If the company provides all or a portion of the tools to the workers, then the job requires more planning. The project manager is responsible for ensuring that all the needed tools are available and on the job site at the proper time. This may involve acquiring the tools from other sources if they are not available in-house. The greater the quantity or types of tools needed, the greater the lead time the project manager needs to make adequate arrangements.

Bar coding is a means to keep track of assets (tools) and manage a tool inventory. Most companies have a bar coding system in place. There are many software packages and readers that are available. The following are some of the advantages of using bar codes:

- Track valuable tool assets by location or workers
- Maintain a history of changes or upgrades to tool assets
- Provide for handheld computer collection of current asset data
- Audit assets in an assigned location
- Allow tool assets to be reserved for future use
- User level security for the use of certain tools
- Automatic order generation
- Advanced cycle count features

Many times, tools on the job site are inoperable. To avoid this situation, the project manager must make sure that every tool delivered to the job site is in safe working order and that inoperable and unsafe tools are repaired or replaced. To avoid delays caused by damaged tools, a standby set of tools should be kept on hand.

The project manager should also make sure that the tool control procedure does not waste time. Many companies' tool control programs require each craftworker to check out the tools from a central location on the job site. This provides excellent control of tool inventory, but it also takes time away from getting the job done. A more effective plan is to have the tools available at the workstation and have workers check them out as needed. This plan gives effective control with less wasted time. In order to be effective, the field supervisor must take the time in advance to identify the tools required for a task and get them to the workstation.

If employees are required to provide their own tools, provisions must be made to replace or repair them if they break down on the job. Having a small inventory of tools available for such emergencies is an innovative idea. The field supervisor must make sure that workers get their own tools repaired as soon as possible and that company spares are returned.

Maintaining tool inventory and maintenance programs is also the responsibility of the project manager. He relies on feedback from the field supervisor if these programs are not working effectively. The project manager should also make it clear to the field supervisor that he is responsible for ensuring that workers use tools properly and safely. If this requires on-the-job training, it should be included in the job plan.

The following are major responsibilities in planning for tools and their uses:

- Establish a complete list of tools required for the job
- Establish who provides which tools, in accordance with company policy
- Establish standards for tool purchases
- Establish controlled and organized tool storage areas so that tools are readily available to the workers with minimum delay
- Establish and maintain a tool inventory
- Provide instruction for safe use of tools
- Stamp and identify company tools
- Maintain a spare tool and tool parts inventory
- Provide sufficient job supervision to ensure that tools are properly used

5.10.0 Planning for Labor

Every task on a project requires some labor, and the project manager's responsibilities regarding planning of labor are as follows:

- Determine the number of skilled and unskilled workers needed on the project.
- Whenever possible, use the company's permanent employees, former employees, and local labor, and import labor from another locality only as a last resort.
- Watch labor trends and wages. In general, when work is plentiful, the labor supply is low, wages are high, and labor efficiency is low. Labor wages depend not only on the amount of work and the labor supply available but also upon several other items such as government regulations, general prosperity of the country or locality, and work available in other professions.
- Identify each type of work involved in the project and the time required to do it.

Describe the project manager's responsibility regarding tool control.

Ask participants to discuss their company's procedures for acquiring and tracking tools.

Review the advantages of using bar coding versus other methods.

Emphasize that any tool control program should be efficient but not cumbersome.

Stress the importance of proper maintenance for tools to ensure they are operationally safe.

Show Transparency 23, Labor Planning Factors, and discuss each point.

Provide examples of your experience with labor planning.

Emphasize the importance of acquiring skilled and experienced personnel.

Discuss the problems associated with turnover, absenteeism, illness, and injuries.

Discuss ways that the project manager can plan for these situations.

Stress that the project manager must delegate responsibility and authority to field supervisors to ensure that all work is done correctly.

MODULE 44106-08 ◆ CONSTRUCTION PLANNING 6.19

Explain why task coordination is critical to every project.

Emphasize that coordination should be done in such a way that working relationships are maintained.

See the Teaching Tip for Section 5.11.0 at the end of the module.

Emphasize that field supervisors need to understand and implement the project plan.

Explain the need for adequate resources. Provide examples.

Describe methods to make sure that communication and the work flow proceed smoothly.

Discuss with students why Samantha's offer of "more people" to address schedule deficits may not be the best solution.

- Establish a schedule for each type of work to be performed.
- Be sure everyone knows what to do and where to go.
- Make sure each work crew has its work properly completed in accordance with the project schedule.

To meet labor requirements, the project manager needs to know if there are qualified employees available within the company or if they must be hired from outside. The project manager must also consider how much time must be spent on training and include that time in the job plan. Finally, the project manager must plan labor requirements far enough in advance so that the required personnel can be on site when the job is scheduled to begin.

Unexpected labor problems such as absenteeism, turnover, and illness often occur, and the project manager must be ready to deal with them. This involves anticipating such problems and having contingency plans.

The project manager should also analyze the space available for workers to do their jobs. Sometimes too many people are assigned to too small a space, resulting in unsafe conditions and low productivity. Some refer to such a study as a labor density analysis. Responsibility for making sure that each worker knows what kind of work to do and where it is going to be done must be delegated by the field supervisor.

5.11.0 Coordinating the Work For Contractors

No task stands alone, and so the work performed by one contractor must always be coordinated with that of other contractors on the job site. The most effective way for the project manager to coordinate the work is to review the work plans of all the contractors on the project each week and then discuss potential or anticipated problems with them. It is also in everyone's best interest for all project managers and supervisors on the project to establish a good working relationship with one another.

5.12.0 Implementation of the Plan

After the project manager has developed and analyzed a plan, it must be implemented. The first step in this process is to communicate to the field supervisors and other project personnel all the information they need to follow the plan, and to do so far enough in advance so that they have ample time to prepare. The next step is to make sure all the required resources are available so work can proceed as scheduled. It cannot be overemphasized that the key to effective planning lies in taking the time to be sure those resources are available when needed. This involves knowing who to contact to obtain tools, equipment, materials, and manpower, and what the various company policies are in this regard.

The next step in implementation is to make sure that the work flows smoothly. The project manager should spend as much time as necessary with the field supervisors and, when problems arise, should be available to assist the supervisors in solving them. Project managers can also coordinate the company's efforts with those of other contractors on the job. Communication and cooperation, combined with careful advanced planning, is the clearest road to completing a job on schedule and within the estimated cost.

Instructor's Notes:

6.20 PROJECT MANAGEMENT

Review Questions

1. Effective ________ exposes problems that might occur during the course of a project.
2. Planning focuses on the two most significant cost factors in the project: ________ and ________.
3. Only the project manager should make plans.
   a. True
   b. False
4. Effective planning can be used as a ________ device for measuring job progress.
5. Planning can be defined as an attempt to determine the amount of ________ a project will take.
6. Planning done before construction begins is referred to as ________ planning.
7. Debriefing is done at what stage of planning?
   ____
8. List five factors that can affect a job.
   1. ________
   2. ________
   3. ________
   4. ________
   5. ________
9. Why should the project manager consider alternative plans?
   ____
   ____
   ____
10. What three types of information are identified in planning review?
    1. ________
    2. ________
    3. ________
11. List the four elements that comprise every formal job plan.
    1. ________
    2. ________
    3. ________
    4. ________
12. The goal of every project manager is to complete a job within the ________ given and the ________ established by the estimate.
13. A project manager should be able to break down a project into ________ and those components into ________.
14. In determining the level of detail to use in determining project jobs and activities, the project manager should use the ________ as a guide.
15. Define an identifiable activity.
    ____
    ____
16. What four resources need to be planned?
    1. ________
    2. ________
    3. ________
    4. ________
17. The basis for ________ planning is the quantity survey.
18. Define the phrase "purchase order."
    ____
    ____
19. What is expediting?
    ____
    ____

Have participants complete the Review Questions, and go over the answers prior to administering the Module Examination.

MODULE 44106-08 ◆ CONSTRUCTION PLANNING 6.21

Review Questions

20. When planning for materials, what are some items the project manager should consider?
    1. ______________________
    2. ______________________
    3. ______________________
    4. ______________________
    5. ______________________

21. When planning for equipment, a project manager needs to know the ________ that are needed and ________ they should be on the job site.

22. Planning for tools entails determining:
    1. ______________________
    2. ______________________
    3. ______________________
    4. ______________________

23. One step in planning labor is to identify the ________ needed to perform each job.

24. The most effective way to coordinate work among the various contractors on the job site is to ________ and ________ regularly.

25. Receiving materials may require testing or ________ for future testing.

26. Define work analysis.

    ____

    ____

27. The objectives of making a work analysis are:
    1. ______________________
    2. ______________________
    3. ______________________
    4. ______________________

28. A work analysis begins by breaking a job into its ________.

29. The overall goal of a work analysis is to reduce the ________ and/or time of doing the work.

30. What is the purpose of the questions used in a work analysis?

    ____

    ____

Instructor's Notes:

____

____

____

Summary

Summarize the major concepts presented in this module.

Effective planning aims at establishing work methods that will get the work done in a safe and quality manner, in a minimum amount of time, at optimum cost savings, and with available resources. The project manager's responsibilities for planning include:

- Determining what has to be done; this involves carefully studying the contract documents and determining the work involved in each project activity in the construction plan.
- Identifying how each activity is to be performed; this involves the selection of construction methods and related support and management procedures.
- Determining required resources and their availability.
- Specifying the order in which things are to be done; this involves an understanding of the technological processes involved in the construction methods.
- Making decisions as to when things are to be done and establishing the rate at which project work is to proceed.
- Making sure that the proper resources are available at the right place at the right time.
- Ensuring that field supervisors are properly briefed as to what has to be done, when, and how.

The goal of planning should be idealistic, yet realistic. There is no rule that indicates exactly how much time and money should be spent to plan a job; this comes from experience and a complete understanding of job management.

Notes

MODULE 44106-08 ◆ CONSTRUCTION PLANNING 6.23

Have participants complete the Trade Terms Quiz, and go over the answers.

Administer the Module Examination. Record the results on Craft Training Report Form 200, and submit the results to the Training Program Sponsor.

Trade Terms Quiz

1. The process of creating a project or part of a project by the employment of construction resources is called ____________.
2. If something is ____________ , a quoted price includes the cost of loading the goods into transport at the specified place.
3. ____________ is the process of anticipating future occurrences and problems.
4. ____________ is a listing of all of the components and other materials used in the assembly of a project or system.
5. The estimate of material quantity needed for a project is the ____________.
6. The process of system by which suppliers are encouraged to meet the due date for delivery of outstanding purchase orders is know as ____________.
7. A ____________ includes item, quantity, price, discounts, vendor information, and ship-to information.
8. A ____________ is a study of each job, aimed at effectively and efficiently integrating all tasks and activities.

Trade Term List

Activity
Bill of material
Expediting
Free on board (FOB)
Planning
Purchase order
Quantity survey
Work analysis

Instructor's Notes:

6.24 PROJECT MANAGEMENT

Trade Terms Introduced in This Module

Activity: The process of creating a project, or part of a project, by the employment of construction resources.

Bill of material: A listing of all of the components and other materials used in the assembly of a project or system.

Expediting: The follow-up of purchase orders which are overdue, or are required by a prescribed deadline. The process or system by which suppliers are encouraged to meet the due date for delivery of outstanding purchase orders or to effect immediate delivery of overdue orders.

Free on board (***FOB***): A pricing term indicating that the quoted price includes the cost of loading the goods into transport at the specified place.

Planning: The process of anticipating future occurrences and problems, exploring their probable impact, and detailing policies, goals, objectives, and strategies to solve the problems.

Purchase order: Information sent to a vendor to request a product or service; typically includes item, quantity, price, discounts, vendor information, and ship-to information.

Quantity survey: The estimate of material quantity needed for a project.

Work analysis: A study of each job, aimed at effectively and efficiently integrating all tasks and activities.

MODULE 44106-08 ◆ CONSTRUCTION PLANNING 6.25

Resources & Acknowledgments

Reference

Robledo, Rebecca, "Unforeseen circumstances: your contract must acknowledge unforeseen construction problems, and lay out who's responsible for them or—you'll end up paying the price." *Pool & Spa News*, April 11, 2003. Accessed on Oct 3, 2007 at http://findarticles.com/p/articles/mi_m0NTB/is_8_42/ai_99909406 (03 October 2007).

Figure Credit

M. C. Dean, Inc., module opener

Instructor's Notes:

6.26 PROJECT MANAGEMENT

MODULE 44106-08 — TEACHING TIPS

The following are suggested activities or instructional methods to help you teach the material in this module.

General

When you call on someone to answer a question, the rest of the class relaxes or even tunes out because they expect that the question and answer will take place only between you and the participant you called on. Instead, use this technique to involve more participants in answering questions and to keep them on their toes.

1. Ask participants to define a term or explain a concept.
2. After one participant has answered, ask a participant seated nearby if the answer is right. Then ask whether a participant in the back of the room agrees.
3. Ask participants to explain why they think an answer is right or wrong.
4. Use the session to clear up incorrect ideas and encourage participants to learn from their mistakes.

Section 2.2.1 *WBS Outline*

The purpose of this activity is to allow the participants to create a WBS outline.

1. Distribute a portion of a sample Scope of Work statement.
2. Ask the participants to prepare a WBS outline based on the Scope of Work statement.
3. Have the participants present their results.
4. Critique the results and ask the class for their comments.

Estimated time: 20 minutes.

Section 4.1.0 *Work Analysis*

The purpose of this activity is to provide practice in developing a work analysis.

1. Present a typical construction-related job to the class.
2. Refer to Table 1 in the Participant Guide, or show Transparency 13. This Table may be used as an example for this exercise. They may also use Section 4.1.0 in the Participant Guide as a reference.
3. Instruct the Participants to use two columns (Method One and Method Two) on a sheet of paper.
4. In Column One, have them list in detail the activities and resources required for the sample job.
5. Then have them think about steps that can be eliminated, or how they might re-arrange the resources and activities to reduce time and effort, and save money.
6. Have them list the rearranged tasks in Column Two.
7. Have the participants present their results.
8. Critique the results with the class. Ask if some of the work analyses presented could be reduced further.

Estimated time: 30 minutes.

Section 5.11.0 *Coordinating with Other Contractors*

This Teaching Tip will provide practice in coordinating with other contractors. Introduce a mock weekly coordination meeting and present an example of an onsite job-related conflict for discussion. Ensure that the situation is not complicated. The objective is to get participants to communicate and compromise on their positions.

1. Assign different contractor roles to the participants.
2. Take the role of the general contractor and chair the meeting.
3. Present the sample conflict.
4. Have the participants discuss problems that may arise from the conflict.
5. Ask them for possible resolutions.
6. Have them list actions they would take to minimize the conflict.
8. Critique the results.

Estimated time: 30 minutes.

MODULE 44106-08 — ANSWERS TO REVIEW QUESTIONS

	Answer	Section
1.	Planning	(2.0.0)
2.	Labor, equipment	(2.0.0)
3.	False	(2.1.0)
4.	Control	(2.0.0)
5.	Time	(2.0.0)
6.	Pre-construction	(3.2.0)
7.	Planning review	(3.4.1)
8.	Five of the following: 1. Timing phases of work 2. Weather 3. Types of materials and availability 4. Equipment and tools availability 5. Personnel requirements and availability 6. Subcontractor availability 7. Relationships	(3.3.0)
9.	To be able to deal with situation and not incur delays or cost overruns.	(3.3.0)
10.	1. Identification of problems before they occur 2. Documentation of problems so they may be avoided in the future 3. Evaluation of plan effectiveness	(3.4.0)
11.	1. Goals and objectives to be met 2. Series of jobs to be done 3. Resource requirements 4. Sequence of activities for each job	(2.2.0)
12.	Time, cost	(3.0.0)
13.	Jobs, activities	(3.0.0)
14.	Estimate	(3.0.0)
15.	One that can be assigned specific types and amounts of resources.	(3.0.0)
16.	1. Materials 2. Equipment 3. Tools 4. Labor	(5.0.0)
17.	Material	(5.1.1)
18.	Document used to order material; a specific performance contract	(5.2.1)
19.	Following up to ensure that material arrives in good condition at the place specified, at the proper time	(5.5.0)

	Answer	Section
20.	Possible answers include: 1. Prices and price trends 2. Time to buy 3. Quantities 4. Vendors (dealers and manufacturers) 5. Transport and delivery 6. Amount to be kept on hand 7. Inspection and tests 8. Insurance	(5.1.0)
21.	Types, when	(5.8.0)
22.	Possible answers include: 1. What tools are needed 2. Who provides what tools? 3. Maintaining a tools inventory and parts inventory 4. Maintaining the tools in good condition 5. Standard for purchases 6. Storage areas 7. Instruction in tool use 8. Identification of company tools 9. Tool supervision	(5.9.0)
23.	Type of work/craft/skills	(5.10.0)
24.	Review work plans and meet with contractors	(5.11.0)
25.	a. True	(5.10.1)
26.	Study of each job to effectively and efficiently integrate all activities	(4.0.0)
27.	1. Reduce time 2. Reduce cost 3. Identify more effective ways of performing the job 4. Improve productivity	(4.0.0)
28.	Activities	(4.1.0)
29.	Cost	(4.0.0)
30.	To identify potential improvements in the way the job is done	(4.1.0)

MODULE 44106-08 — ANSWERS TO TRADE TERMS QUIZ

1. activity
2. free on board (FOB)
3. Planning
4. Bill of material
5. survey
6. expediting
7. purchase order
8. work analysis

CONTREN® LEARNING SERIES — USER UPDATE

NCCER makes every effort to keep these textbooks up-to-date and free of technical errors. We appreciate your help in this process. If you have an idea for improving this textbook, or if you find an error, a typographical mistake, or an inaccuracy in NCCER's Contren® textbooks, please write us, using this form or a photocopy. Be sure to include the exact module number, page number, a detailed description, and the correction, if applicable. Your input will be brought to the attention of the Technical Review Committee. Thank you for your assistance.

Instructors – If you found that additional materials were necessary in order to teach this module effectively, please let us know so that we may include them in the Equipment/Materials list in the Annotated Instructor's Guide.

Write: Product Development and Revision
National Center for Construction Education and Research
3600 NW 43rd St, Bldg G, Gainesville, FL 32606

Fax: 352-334-0932

E-mail: curriculum@nccer.org

Craft ____________ Module Name ____________

Copyright Date ____________ Module Number ____________ Page Number(s) ____________

Description

____

____

____

____

(Optional) Correction

____

____

____

(Optional) Your Name and Address

____

____

____

Module 44107-08

Estimating and Cost Control

NCCER STANDARDIZED CRAFT TRAINING PROGRAM

The National Center for Construction Education and Research (NCCER) provides a standardized national program of accredited craft training. Key features of the program include instructor certification, competency-based training, and performance testing. The program provides trainees, instructors, and companies with a standard form of recognition through a National Craft Training Registry. The program is described in full in the *Guidelines for Accreditation*, published by NCCER. For more information on standardized craft training, contact the NCCER by writing us at 3600 NW 43rd St., Bldg. G, Gainesville, FL 32606; calling 352-334-0911; or emailing info@nccer.org. More information may be found at our website, www.nccer.org.

HOW TO USE THIS ANNOTATED INSTRUCTOR'S GUIDE

Each page presents two sections of information. The larger section displays each page exactly as it appears in the Trainee Module. The narrow column ties suggested trainee and instructor actions to each page and provides icons (detailed below) to call your attention to material, safety, audiovisual, or testing requirements. The bottom of each page includes space for your notes.

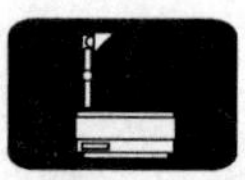
The **Audiovisual** icon indicates an appropriate time to show a transparency or other audiovisual aid.

The **DVD** icon indicates an appropriate time to discuss or show segments of the video.

The **Classroom** icon prompts you to define a term, stress a point, ask trainees to explain a concept, or give examples.

The **Demonstration** icon directs you to show trainees how to perform tasks.

The **Examination** icon tells you to administer the written module examination.

The **Homework** icon is placed where you may wish to assign reading for the next class, assign a project, or advise trainees to prepare for an examination.

The **Laboratory** icon is used when trainees are to practice performing tasks.

The **Materials** icon is a reminder for you to gather materials needed for classes, labs, and testing.

The **Performance Testing** icon tells you to administer a performance test or a portion thereof.

The **Safety** icon is used to emphasize safety issues. It is often keyed to *Caution* and *Warning!* statements in the Trainee Module.

The **Teaching Tip** icon indicates additional guidance is available, such as how to conduct an exercise, get the most educational value from a field trip, or encourage class participation. Teaching Tips may expand on a feature (*Think About It, Did You Know?*) or provide *Quick Quizzes* or similar exercises. You will be referred to the Teaching Tips section at the back of the module if there is additional material.

The **Combination** icon indicates that the laboratory listed corresponds with a performance task. If desired, you can note the proficiency of the trainees during the laboratory, and use it to satisfy performance testing requirements.

PREPARATION

Before teaching this module, you should review the Objectives, Performance Tasks, Materials and Equipment List, and Module Outline. Be sure to allow ample time to prepare your own training or lesson plan and gather all required materials and equipment.

MODULE OVERVIEW

This module emphasizes the importance of accurate estimating. It summarizes the estimating process and the steps for developing an estimate. It defines the purpose of a cost control methodology, explains how to perform simple cost analysis, and covers the project manager's role in controlling cost and tracking rework cost.

PREREQUISITES

Please refer to the Course Map in the Participant Module. Prior to training with this module, it is recommended that the participant shall have successfully completed *Project Management,* Modules 44101-08 through 44106-08.

OBJECTIVES

Upon completion of this module, the participant will be able to do the following:

1. Define cost control and identify the purpose of a cost control system.
2. Define budgeted (estimated) cost, actual cost, and projected cost.
3. Define the importance of accurate estimates
4. Explain the project manager's role in controlling cost.
5. Describe what a reporting system is and how it functions in a cost control system.
6. Explain the process of making a cost analysis.
7. Perform a simple cost analysis.
8. Describe how to track and document the causes and costs of rework.

MATERIALS AND EQUIPMENT LIST

Desktop or laptop computer or DVD player

Multimedia projector and screen

Minor Decisions: Major Impact – How to Deal with Real Issues in Project Management DVD

Transparencies

Blank acetate sheets

Overhead projector and screen

Transparency pens

Whiteboard/chalkboard

Markers/chalk

Pencils and scratch paper

Module Examination (located in the Test Booklet)

ADDITIONAL RESOURCES

This module is intended to present thorough resources for task training. The following reference works are suggested for both instructors and motivated participants interested in further study. These are optional materials for continued education rather than for task training.

Excerpts from About.com 20th Century History, Jennifer Rosenberg. Retrieved from http://history1900s.about.com/od/1930s/a/empirestatebldg_3.htm.

PMI Standards Committee, *A Guide to the Project Management Body of Knowledge.* PMI Publications, Newton Square, Pa. (2004).

Jonathan Goldman, *The Empire State Building Book* (New York: St. Martin's Press, 1980) 30.

John Tauranac, *The Empire State Building: The Making of a Landmark* (New York: Scribner, 1995) 156.

TEACHING TIME FOR THIS MODULE

An outline for use in developing your lesson plan is presented below. Note that each Roman numeral in the outline equates to one session of instruction. Each session has a suggested time period of 2½ hours. This includes 10 minutes at the beginning of each session for administrative tasks and one 10-minute break during the session. Approximately 15 hours are suggested to cover *Estimating and Cost Control.*

Topic	Planned Time
Session I. Introduction and Estimating	
A. Case Studies in Costs	________
B. Importance of Accurate Estimating	________
C. Estimating Process	________
D. Organizing and Developing the Estimate	________
E. Steps Involved in Developing the Estimate	________
Session II. An Overview of Job Costs	
A. Estimated Cost	________
B. Actual Costs	________
C. Projected Cost	________
Session III. Cost Control, Part One	
A. Purposes of Cost Control	________
B. Cost-Control System	________
C. Job Cost-Coding	________
Session IV. Cost Control, Part Two	
A. The Reporting System	________
B. Developing a Reporting System	________
C. Field Reporting System	________
D. Office Reporting System	________
Session V. Cost Analysis	
A. Analysis Process	________
B. Cost of Changes	________
Session VI. Impact of Improper Reporting; Historical Records; Review and Testing	
A. Impact of Improper Reporting	________
B. Historical Records	________
C. Review	________
D. Module Examination	________
1. Participants must score 70% or higher to receive recognition from NCCER.	
2. Record the testing results on Craft Training Report Form 200, and submit the results to the Training Program Sponsor.	

USING THE PM DVD WITH *ESTIMATING AND COST CONTROL*

Section 3.0.0, An Overview of Job Costs, presents an opportunity to discuss the challenges project managers sometimes face trying to get the job done both on time and within budget.

If more labor resources were added, Samantha's attempt to keep costs under control might be thwarted. Think of some steps she could take or changes she might make that would help offset these added labor costs. What might the team do to help Samantha keep costs in check?

Suggested response: Review the original plan and budget to determine if the remaining tasks can be re-organized and/or simplified to increase productivity (including minimizing downtime) and reduce additional labor requirements. Get the job site organized and cleaned up to reduce lost time looking for tools and to eliminate lost time due to prospective injuries.

Section 5.0.0, Cost Analysis, presents an opportunity to discuss how Samantha might plan her next project when factoring in the lessons from this one.

Suggested response: Samantha should review the actual costs of this job and compare it to the contract costs (estimated or budgetary) to determine which elements contributed to any substantial variances. She should consider whether her next bid should include additional labor and equipment resources to reduce the potential for delays that cause cost and schedule overruns. This is also an opportune time to examine workflow and the team's productivity and assess where gains could be achieved to reduce costs.

Project Management

44107-08

Estimating and Cost Control

Assign reading of Module 44107-08.

44107-08
Estimating and Cost Control

Topics to be presented in this module include:

Overview

Your goal as project manager is to construct projects to the standards in the drawings and specifications, and to do it within the estimated time and cost to meet contract requirements. You must understand budgeted, actual, and projected costs and perform cost analysis. Using these skills effectively allows you to be aware of the cost of doing the work, the cost impact of rework, and how to exercise controls to keep costs in line.

Instructor's Notes:

Objectives

When you have completed this module, you will be able to do the following:

1. Define cost control and identify the purpose of a cost control system.
2. Define budgeted (estimated) cost, actual cost, and projected cost.
3. Define the importance of accurate estimates.
4. Explain the project manager's role in controlling cost.
5. Describe what a reporting system is and how it functions in a cost control system.
6. Explain the process of making a cost analysis.
7. Perform a simple cost analysis.
8. Describe how to track and document the causes and costs of rework.

Trade Terms

Actual cost	Cost analysis
Bid	Projected cost
Budget	Reporting system
Budgeted cost	Spreading

Prerequisites

Before you begin this module, it is recommended that you successfully complete *Project Management,* Modules 44101-08 through 44106-08.

This course map shows all of the modules in the *Project Management* curriculum. The suggested training order begins at the bottom and proceeds up. Skill levels increase as you advance on the course map. The local Training Program Sponsor may adjust the training order.

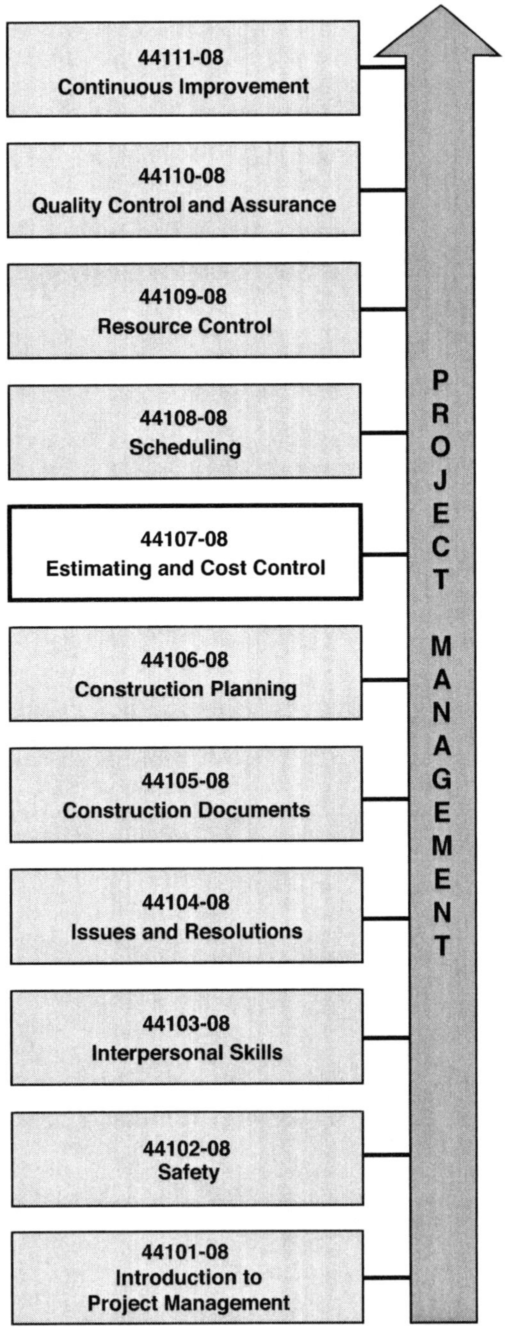

107CMAP.EPS

Ensure that you have everything required to teach the course. Check the Materials and Equipment list at the front of this module.

See the general Teaching Tip at the end of this module.

Explain that terms shown in bold are defined in the Glossary at the back of the module.

Show Transparency 1, Objectives. Review the goals of the module, and explain what will be expected of the participant.

Briefly discuss the importance of estimating and controlling costs.

Show Transparency 2, Cost Overrun Factors.

Discuss participants' reactions to the Montreal case study.

Discuss the major factors contributing to the cost overruns in this case study.

Compare and contrast the Montreal and Empire State Building case studies.

Ask the participants whether planning techniques used in 1929 are applicable today. Ask for examples.

Stress that a project manager must understand the scope of work and carefully review plans and specifications.

See the Teaching Tip for Section 1.1.0 at the end of the module.

1.0.0 ◆ INTRODUCTION

This module presents information on estimating, project cost awareness, and cost control. In addition, it offers the project manager the tools necessary for analyzing job costs. As project manager, you should always keep in mind that you are responsible for ensuring your company makes a profit.

1.1.0 A Case Study – Cost Overruns

As an example of construction costs out of control, consider one of the most famous (or infamous) construction projects of all time, the construction of the Montreal Olympics Complex for the 1976 Summer Olympics. Planning for the construction of the Olympics Complex began in 1970 and consisted of numerous structures including the main stadium. The cost of the stadium was estimated to be $40 million, just a portion of the estimated total cost of $120 million for the entire complex. In the end, the stadium cost over $836 million and the cost for the entire complex exceeded $1.5 billion.

What happened to create this financial disaster? The following factors contributed to the overrun:

- *Design* – The ultramodern designs required new and complex construction techniques to be employed by the contractors. For the main stadium, the structural members were designed as complex ribs assembled in the field by gluing the parts together with epoxy and post-tensioning a number of precast units. Each rib was a different size and had to be perfectly aligned. Cranes had to be used to hold the ribs in place and hoist workers and materials to the overhead positions. Another design problem was that the original drawings were metric measurements and, since the contractors were unaccustomed to the metric system, the drawings had to be converted.
- *Labor* – An estimated 80 days were lost to strikes and the equivalent of another 20 days from slowdowns caused by the union labor force. There were no agreements between the unions and management to restrict strikes or other union activity during the construction. Eventually, police had to control access into the site to prevent sabotage.
- *New construction technique* – The use of the epoxy-glued, post-tensioned structural members was a completely new technique to the contractors. The learning process proved to be slow and expensive.
- *Resource shortages* – The local sources of labor, materials, and equipment were exhausted by the heavy concentration of construction in the Montreal area. The cost of importing these resources to Montreal carried a high premium.
- *Weather* – Wintertime construction was hampered by the cold temperatures. At its peak, the cost of protective heating measures was around $400,000 per day.
- *Scheduling* – The critical path network schedule was poorly prepared and unrealistic. Many activities were required to be performed simultaneously that were physically impossible to do in one area. The schedule was eventually abandoned and the project operated under a daily crash schedule.
- *Fixed Deadline* – With the fixed deadline, the planning of the project should have been started about two years earlier. The city of Montreal was so late in preparing the contracts that competitive **bids** could not be solicited.
- *Crowded Work Space* – With the fixed deadline, crews were doubled and double shifts and overtime were required; however, the project could not accommodate the numbers of workers and equipment. At one time, 80 cranes were used inside the main stadium.
- *Inflation* – Inflation rates during the construction went into double-digits.

1.2.0 A Case Study – When Things Go Right

In contrast to the Montreal Olympics Complex case, this case study of the Empire State Building demonstrates what happens when a project is well-planned. The construction of the Empire State Building is a masterpiece in planning, scheduling, cost control, and a coordinated work force.

The Empire State Building is one of the most iconic buildings of the 20th century. It was built as a challenge to the French to match the 984-foot Eiffel Tower. It was developed by former General Motors president, John Jacob Raskob, and partners. In 1929, they purchased property on Fifth Avenue in New York City. Then, Raskob and his architects proceeded to develop a unique plan for the new skyscraper.

- *Design* – Raskob challenged his architect, William Lamb, to make the Empire State building the tallest it could be "without falling down." Lamb devised a simple plan based upon a pyramidal design with a compact cen-

Instructor's Notes:

ter core that contains the vertical circulation, mail chutes, toilets, shafts and corridors. Office space would be built around the core in a 28-foot deep space. Along the way, the plan to build the tallest skyscraper was challenged by the 1,046-foot tall, 77-story Chrysler Building, so Lamb increased the number of stories from 80 to 85. To be sure it would surpass Walter Chrysler's building, Raskob asked Lamb to add a dirigible mooring mast to make the building 1,250-feet tall.

- *Resource Planning and Scheduling* – Planning the workforce and materials resources would be a crucial factor in completing the building on schedule and budget. The builders hired dependable contractors who would be able to follow through with quality work that met the schedule. Due to the size of the job, most of the supplies had to be ordered to spec and built at the plant to minimize work at the construction site. To meet critical timing, building processes overlapped.
- *Construction Techniques* – In contrast to the Montreal Olympic Complex, the Empire State Building was a more straightforward design. While there were no major obstacles, raising the steel skeleton had to be carefully planned and managed. There were safety risks involved. The vertical frame contained 210 steel columns, ranging from the entire height of the building to six to eight stories in length. Large cranes were used to pass steel girders up to the higher floors, and the beams were then riveted into place.
- *Efficiency and Innovation* – The builders implemented efficient procedures and innovations to save time, money, and manpower. A railway was built at the construction site to move materials quickly and with less effort. Bricks were dropped down a chute to a hopper in the basement, rather than being dumped in the street. They were then hoisted up to the appropriate floor. Seven banks of elevators were staggered to service the floors. The Otis Elevator Company designed passenger and service elevators that traveled 1,200 feet per minute—faster than the existing building code allowed. The builders took a chance and installed the faster, more expensive elevators (running them at the allowable speed) under the premise that the building code would change. A month after the Empire State Building was opened, the building code was changed to 1,200 feet per minute, and the elevators in the Empire State Building were sped up.
- *Reduced Costs and Rapid Completion* – The Empire State Building was constructed in just one year and 45 days. Because the Great Depression significantly lowered labor costs, the cost of the building was $10 million less than the $50 million expected cost.

2.0.0 ◆ ESTIMATING

Estimating is a very important part of construction. Estimates determine the probable costs of construction as well as provide the owner with **budget** figures. Estimates are also made in the field to determine costs related to changes and resources required to implement the changes. Estimates need to be accurate because they can affect profit and cost control. This section will focus on the importance of accurate and ethical estimates, types of estimates, and what makes a quality estimate.

2.1.0 Importance of Accurate Estimating

Every construction project is different, and each estimate is a unique entity. It is a complete and particular summary of the best available cost information. Each estimate is separate and different from all previous estimates and must be treated as such. Realistically, creating a completely accurate estimate is nearly impossible. To compensate for this, many companies will use levels of estimating such as:

- Project initiation, where logistics and organization are in flux.
- Project planning, where estimates are more accurate as more details are identified.
- Project execution, where estimates are often reviewed and revised.

The purposes of estimating are to:

- *Get future work* – by predicting what the cost of construction will be at the time the companies bid or negotiate to acquire work (a typical contractor gets one out of every 6 to 10 jobs that it bids).
- *Control present work* – the estimate is used as the project budget and actual progress and costs are compared against it.
- *Control the scope of work* – often the scope changes as work progresses and new estimates are required in order to present the owner with the cost of making changes.

Show Transparency 3, Purpose of Estimating.

Explain the importance of accurate estimates.

Describe the three levels of estimating commonly used to make estimates as accurate as possible.

Discuss the four purposes of estimating.

MODULE 44107-08 ◆ ESTIMATING AND COST CONTROL 7.3

Provide examples of the different variables that can affect estimates.

- *Establish and control risk* – the estimate reflects the amount of risk the contractor has agreed to take on by assuming responsibility for the project.

Project costs are subject to variables including weather, transportation, soil conditions, labor strikes, material availability, subcontractor availability, and the quality of project management. The estimate must be prepared in advance to anticipate the nature and effect of these variables.

One of the main causes of contractor failure is the lack of business ability and training to estimate costs accurately. The importance of accurate estimating cannot be overemphasized. Estimates that are too high will not get the work needed for the company to remain in business, while those that are too low will get the work but will not allow the company to make a reasonable profit or may even cause the company to lose money. There is an old saying among estimators that "the only thing worse than not getting a job is getting a job that you can't make money on."

See the Teaching Tip for Section 2.2.0 at the end of the module.

Show Transparency 4 (Figure 1).

2.2.0 Estimating Process

Explain why estimating can be a complex process.

Provide examples of fixed price costs on an estimate.

Provide examples of variable costs.

Estimates are your best attempt to predict time and cost. Be aware of the types of tasks involved and always refer back to the work breakdown structure (WBS) to double check tasks (see the *Construction Planning* module). The project manager also needs to be able to estimate both fixed and variable costs. Fixed costs are those that usually don't change immediately, such as supplies, submittals from vendors, or fixed bids. Variable costs are associated with wages, overhead, and any other conditions that create change. Refer to the company's historical data of **actual costs** and consult with other project managers for input.

Discuss the differences between preliminary and detailed pre-construction estimates.

The types of estimates that arise in a typical project include pre-construction estimates and field estimates. Both are extremely important; pre-construction estimating is normally associated with costs, while field estimating may be concerned with costs, quantities, or production.

2.2.1 Pre-Construction Estimates

Pre-construction estimating is the process of determining probable construction costs for a given project. Many items contribute to the cost of a project, and each must be identified and analyzed. Since the estimate is prepared before construction begins, an accurate estimate relies on a thorough study of the construction documents. Pre-construction estimating may be classified as either preliminary or detailed.

A preliminary estimate, sometimes referred to as a conceptual estimate, is used to determine a project budget or an expected project cost, normally before the construction drawings and specifications are complete. It may be based on little more than the scope of the project or the preliminary project plans or sketches. Preliminary estimates are developed by using project information such as the production capacity of an industrial facility, the square footage of office space in a commercial building, the volume of a warehouse, or the number of bedrooms and bathrooms in a residence.

Cost records of similar, previous projects are then applied and modification factors incorporated to develop the estimate. Modification factors are used to account for items such as inflation, geographical location, and project size differences.

For example, a proposed hospital is planned to house 500 beds. In a similar job performed five years ago, the cost was $10,000 per bed for a 300-bed hospital. The preliminary estimate is $10,000 per bed × 500 beds × 1.15 modification factor = $5,750,000.

An experienced estimator can develop certain types of preliminary estimates within a matter of minutes. However, the accuracy of this estimate may vary as much as 20 percent.

The most common method of developing a preliminary estimate is based on square footage since it's easy to obtain costs per square foot from past job records or from commercially published square foot cost manuals.

Detailed estimates are prepared from the completed working drawings and specifications. Using these documents, the estimator must visualize all the different phases of construction and determine their cost.

A detailed estimate takes much longer to prepare than a preliminary estimate. It includes quantities and costs of everything required to complete the work—materials, labor, equipment, insurance, bonds, overhead, and profit. During the course of construction, the project manager can use this information to compare the estimated and actual costs of doing the work.

Explain why field estimates are required.

2.2.4 Field Estimates

Field estimating is performed after the construction process begins. While the project manager may or may not be involved with the pre-construction estimate (depending on the organization of the construction company), the project manager will always be involved with field estimating. Field estimates are invaluable to field

Instructor's Notes:

personnel for planning and scheduling the work to be performed. They are required for a variety of reasons, most importantly for determining:

- Costs of changes in the work
- Resource requirements
- Information for payment requests

Changes are a part of any construction project and result from differing conditions or alterations to the original plan. Each change requires the execution of a change order, and each change order must be carefully and fully documented. This requires starting with an estimate of the cost of the change, so the project budget, schedule, and any effect on the scope of work can be modified accordingly.

The pre-construction estimate may overlook the use of resources. If so, the project manager must estimate his labor, materials, and equipment needs to ensure that these resources are available when needed.

As construction progresses, the contractor must periodically (usually monthly) request payment from the owner in order to be paid for completed work. This involves estimating the stage of completion of all work items in progress.

2.3.0 Organizing and Developing the Estimate

The estimate must be organized in an orderly, systematic fashion, so it is neat, clear, and easy to follow. If something unforeseen occurs and the original estimator cannot complete what was started, another estimator should be able to take over and easily pick up where the first estimator left off. As such, the best question for judging the organization of an estimate is, "Could someone else come in and find exactly what they are looking for?" If the answer is yes, the estimate is well-organized.

2.3.1 Steps Involved in Developing the Estimate

Figure 1 illustrates the typical estimating process. The steps involved in developing an estimate include these central activities:

Step 1 Select and review contract documents to determine whether to bid.

Step 2 Determine what work will be self-performed and what will be subcontracted.

Step 3 Visit the job site.

Step 4 Determine the material quantities for the work.

Step 5 Determine the labor and equipment requirements and expected productivity for the project.

Step 6 Price the materials, equipment, and labor.

Step 7 Determine all job overhead cost items based on the general conditions of the job.

Step 8 Obtain and analyze subcontract bids and vendor quotations.

Step 9 Summarize and review the final costs of all items.

Step 10 Recap the estimate and apply an adequate mark-up factor to cover all general overhead items and an appropriate profit margin.

Step 11 Apply the bid strategy and determine the final bid.

More guidelines on estimating can be found in the *PMBOK Guide*. This resource advises the project manager to make a distinction between estimating and pricing. Estimating is an assessment, but pricing is a business decision. This guide provides inputs, tools, techniques, and outputs to effective estimates.

3.0.0 ◆ AN OVERVIEW OF JOB COSTS

It is the objective of every contractor to build projects for a profit. During difficult economic times, when the number of available jobs decreases and competition for work increases, contractors are often forced to submit bids that allow them little or no profit. But this is the exception rather than the rule. The rule is that profit is the standard by which the success of a project, and its project manager, is measured.

To make a job profitable, the project manager has to be continually aware of the costs of doing the work and exercise control over those costs. The three major types of costs involved are estimated cost, actual cost, and **projected cost**.

3.1.0 Estimated Cost

The estimated cost of completing the project establishes the budget for the job and serves as a tool for measuring progress and profitability. In most companies, the project manager is involved to some degree in the development of the estimate, or may even be the estimator. If neither is the case, project managers should have a copy of the estimate for each project under their control

See the Teaching Tip for Section 2.3.0 at the end of this module.

Explain that an estimate needs to be organized in a systematic manner.

Assign reading of Sections 3.0.0-3.3.0

Show Transparency 5, Types of Costs.

Explain that profit is the standard that defines a successful project and project manager.

Emphasize that profit is attained by understanding the cost of doing the work.

Discuss the three types of costs and review the calculations for each type.

Review the role of the project manager in each cost category.

Explain how Samantha might offset costs of additional labor.

MODULE 44107-08 ◆ ESTIMATING AND COST CONTROL 7.5

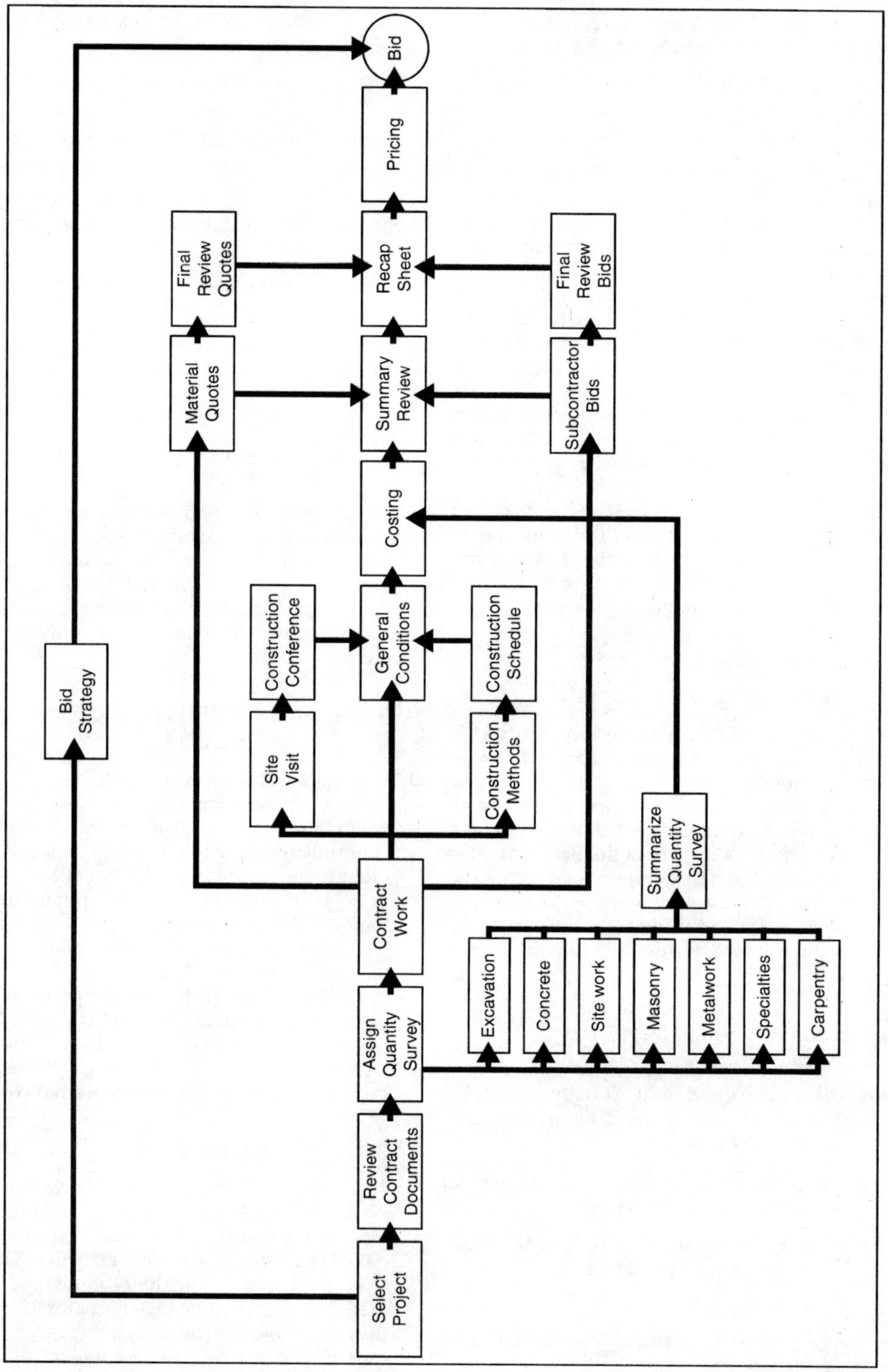

Figure 1 ◆ Typical estimating process.

Instructor's Notes:

7.6 PROJECT MANAGEMENT

and thoroughly understand its makeup and how to use it as a measuring tool.

The estimate is the standard against which cost-effectiveness is measured. Most estimates state either a total direct cost of performing a specific job, such as placing a given number of bricks, or a unit price based on the total quantity of material to be installed. For placing brick, for example, the estimate might state that 15,000 bricks will be installed for a total cost of $21,000, or $1.40 per brick. This statement becomes the project manager's goal. It sets the limit of the amount of time and materials that can be spent in completing that particular job task. If during construction the cost of doing work exceeds the estimate, the project manager must find out why costs are higher and bring costs back in line by pinpointing and correcting the problem.

3.2.0 Actual Costs

Actual costs are those incurred in getting the job built. The ideal situation is to finish the job with actual costs equal to or below estimated costs, but there are many factors that can lead to cost overruns (when actual costs exceed estimated costs). The project manager can understand and control overruns by regularly tracking what the job is costing the company to build while it is under way. Waiting until the job is finished to analyze actual costs versus estimated ones may help with bidding the next job, but cannot correct work already completed.

Consider again the job of placing bricks. The estimate called for the installation of 15,000 bricks at a total cost of $21,000, or $1.40 per brick. As work proceeds, the project manager discovers by tracking costs and work progress that half of the bricks, 7,500, have been placed for a total cost of $12,000, or $1.60 per brick. Recognizing a potential overrun, the project manager takes steps to bring the actual cost in line with the estimate. These steps might include improving worker productivity or altering the work method.

If, on the other hand, the actual cost of placing the bricks is discovered to be less than the estimated cost, the project manager should be able to determine why. Is it because worker productivity is exceptionally high? Is the work being done quickly and sloppily? Is the original estimate in error? The project manager needs to know why the actual costs don't match the estimate. It may be necessary to take corrective action if quality is compromised. Alternatively, if high-quality work is being completed quicker than expected, the project manager will want to alert the estimator to adjust calculations when bidding future jobs.

Continually tracking costs requires an effective job **reporting system** that provides regular comparisons of actual and estimated costs. Establishing and using an effective cost-control system is discussed later.

3.3.0 Projected Cost

Projected cost is the project manager's estimate of the total cost of the project, based on analysis of actual costs and job progress. It is a measuring tool for forecasting the final outcome of the project.

Consider the bricklaying job. If the project manager knows that it cost $12,000 or $1.60 per brick to place the first 7,500 bricks, the projected cost of completing the job can be calculated by multiplying the actual unit cost ($1.60) by the total number of bricks (15,000). In this case, the projected cost of the job is $24,000 which is $3,000 more than the budget.

Obviously, the project manager can do nothing about the money already spent, but can set a new goal that will bring the job costs back in line. Knowing that $9,000 of the original budget for placing the brick remains and that 7,500 bricks must still be installed, the balance of the bricks must be placed for a unit cost of $1.20 per brick. This gives the project manager, the field supervisor, and the crew a target to shoot for. And if the project team reaches that goal, the entire job will meet the original estimate, despite the overrun incurred during the first half of the work.

4.0.0 ◆ COST CONTROL

Cost control is the process of keeping expenditures within an authorized estimate while providing the desired quality of work within a specified time. An alternate definition is to identify deviations from the estimate and take appropriate steps to correct them.

4.1.0 Purposes of Cost Control

Cost control has two important purposes. They are to assist in keeping project costs within the established budget and to produce historical data for future estimates.

By establishing a system to retrieve and analyze actual cost information as work progresses, the project manager has the opportunity to correct deviations from the **budgeted costs** before the work item is completed. It is important that the project cost control system is a company information system that allows for the identification of trouble where it is occurring. Knowing that costs are being exceeded is of no use if the source cannot be identified.

Emphasize that the estimate is the standard against which cost-effectiveness is measured.

Show Transparency 6, Activity: Estimating Costs.

See the Teaching Tip for Sections 3.1.0 through 3.3.0 at the end of this module.

Assign reading of Sections 4.0.0-4.3.0.

Explain that cost control is identifying and correcting deviations from an estimate.

Discuss the purposes of cost control and why it is important.

MODULE 44107-08 ◆ ESTIMATING AND COST CONTROL 7.7

Point out that the objective of a cost control system is to keep projects on budget.

Discuss the other advantages and features of a cost-control system.

Review the factors that should be considered for an effective job cost control system.

Discuss the project manager's role in cost control.

Show Transparency 7 (Figure 2).

Discuss the use of job codes to track costs.

Explain the relationship between account numbers and job codes.

Provide an example of cost codes used by your company, if available.

Assign reading of Sections 4.4.0-4.5.3.

The pricing of an estimate, particularly labor and equipment, is based on past experience. A company's historical data is its track record and the best source for producing future estimates. Using a cost-control system can provide the estimators with current productivity information that can, in turn, be used to produce better estimates.

4.2.0 Cost-Control System

The objectives of a cost-control system are to track and compare estimated costs against actual costs to keep projects on budget. Other advantages and features of a cost-control system include the following:

- Creating historical cost data for future use in estimating
- Comparing actual costs with estimated costs
- Motivating the project team to achieve lower costs in the field
- Placing responsibility and credit where they belong
- Alerting management to the need for corrective action during the job
- Simplifying work and identifying methods for reducing costs
- Enabling the estimator to check outside contractor bids

To be effective and comprehensive, a job cost control system must take into account the following factors:

- Job location
- Procurement of labor and materials
- Lead time
- Equipment utilization
- Management personnel
- Financial requirements
- Purchasing
- Expediting
- Accounting and recordkeeping
- Cost control procedure during job
- Estimating and cost accounting
- Cost control records for material, labor, and equipment

Job cost control begins with developing accurate, concise, and clear designs, drawings, specifications, and labor instructions, but for the project manager the basis for the cost control program is the estimate. The work categories in the estimate should match those used in the field, so that the two can be accurately compared.

The labor cost units in the estimate should be the same as those used by the company in preparing payroll. The same holds for inventory and control systems for material and equipment. The estimate and the field should work on the same basis, so direct comparisons can be made.

4.3.0 Job Cost-Coding

The basis of any cost-control program is a cost-coding system. The system is based on the codes the estimator assigns to each work item as he prepares the estimate. A code can be either a number or letter designation, but the code for each work item must be unique.

For example, the code 1900 might designate installing incandescent lamps, 20 indicates office areas, and 0.5 designates that the electrician does the work. Cost code 1920.5, then, is the work item in which the electrician installs incandescent lamps in a building's office areas. Where numbers and letters are used, 100 might indicate paint, 25 the cooling tower, and A that the job is to be done by a painting crew. So 125A is used for the work item in which a painting crew paints the cooling tower. An example of cost codes is shown in *Figure 2*.

Each job cost code represents an account against which labor hours, material, equipment, and other cost items are charged. For the system to work properly, every hour and every penny spent on doing the work must be assigned to the proper account. If the cost-coding system is not taken seriously and not used appropriately by all project personnel, the information it generates will be useless and the project manager will lose control of project costs. Consequently, all company personnel involved in cost control must be trained in the proper use of the codes.

In designing a useful job cost accounting system, the contractor must be certain that it is:

- Logical and understandable.
- Simple to use.
- Complete enough to forestall charging items to a miscellaneous account; in fact, there should be no miscellaneous cost code.

4.4.0 The Reporting System

A reporting system is the collection of reports that provides information on the hours and costs incurred in the construction of a job. In a small construction company, the system may consist of a few simple reports. In a larger company, it may include many reports, each requiring comprehen-

Instructor's Notes:

7.8 PROJECT MANAGEMENT

Account Number	Designation
0100	Fired Heaters and Boilers
0200	Stacks
0400	Reactors and Internals
0500	Towers and Internals
0600	Heat-Exchange Equipment
0700	Cooling Towers
0800	Vessels, Tanks, Drums and Internals
0900	Pumps and Drivers
1000	Blowers and Compressors
1100	Elevators, Conveyors, Materials Handling Equipment
1200	Miscellaneous Mechanical Equipment
1300	Piping
1400	Sewers
1500	Instrumentation
1600	Electrical
1700	Concrete
1800	Structural Steel
1900	Fireproofing
2000	Buildings
2100	Site Development
2200	Insulation
2300	Painting and Protective Coatings
2400	Field Testing
2500	Tankage
2600	Chemicals and Catalyst
2700	Piling
2800	Filters, Centrifuges, Separation Equipment
2900	Agitators and Mixers
3000	Scrubbers and Entertainment Separators
3100	Machine Tools and Machine Shop Equipment
3200	Heating, Ventilation, Air Conditioning, Dust Control (process only)
3300	Fire Protection
3400	Package Units
3500	Miscellaneous Furniture
3700	Miscellaneous Direct Charges
3900	Construction Supplies and Petty Tools
4000	Field Extra World
5000	Insurance and Taxes
6000	Field Supervision and Field Office
6500	Construction Equipment and Tools
7000	Engineering Department
7300	Estimating and Project Cost Control
7400	Purchasing Department
7500	Construction Department Planning and Scheduling
7900	General Services and Expenses
8000	Main Office Expense
9000	Branch Office Expense

107F02.EPS

Figure 2 ◆ Example of cost codes.

MODULE 44107-08 ◆ ESTIMATING AND COST CONTROL 7.9

Show Transparency 8, Requirements for Cost Control Reports.

Stress that reports need to be timely, simple, detailed, and related to the estimate.

Discuss the importance of building a cost database that can be used for bidding future work.

Explain why accurate and complete records are essential for substantiating claims in a legal dispute.

Show Transparencies 9 – 12 (Figures 3 – 6). Discuss the information contained in these documents.

Explain the project manager's role in developing and maintaining a reporting system.

Ask participants to describe their company's reporting system.

Discuss the essential data to be gathered for a field reporting system.

Review the information found in an office reporting system.

sive information and requiring considerable time to complete. Although the size of the reporting system and the nature of the reports vary from company to company, the prime objectives of any good reporting system are the same: to provide information for cost and production control and to provide records of production and costs

Reports detail the current status of specific jobs or pieces of work. The sum of the reports on a project provides the current status of the project as a whole, making it possible to measure actual costs and performance against those in the estimate. Reports document change orders and changes in job conditions, to ensure payment for the work done. They can also be used to compare the costs and effectiveness of different procedures for accomplishing a particular task.

For reports to be useful in cost control, they must meet certain requirements for speed, clarity, and detail. Reports must be:

- *Timely* – They must be available when needed for planning and decision making, especially on short-term projects.
- *Simple to prepare and read* – Reports should be easy to complete quickly, so that time spent in reporting doesn't outweigh the benefits of the reports.
- *Reasonably detailed* – They should be complete enough to provide useful information on labor, materials, and equipment.
- *Related to the project estimate* – Reports should be set up according to line items and other descriptors used in the estimate, so the report and the estimate can be readily compared.

Reports become part of the project's records. Project records, in turn, are used primarily for bidding on future work and substantiating claims.

Building a cost database provides historical data that can be used for bidding future work. Information on current and completed work is invaluable for bidding new work. Accurate records, provided by accurate reports, provide some of the best information a company can use in assembling a bid.

If claims or legal suits arise, being able to substantiate claims is essential. Reports that provide accurate and complete records are essential whenever legal problems arise.

4.5.0 Developing a Reporting System

The project manager's responsibilities in developing and maintaining a reporting system depend on the company's size, type, and internal policies. In some cases, the project manager may only have to interpret data that is generated from an existing system designed and operated by the company. In other situations, the project manager may have to develop and implement a personal system.

A complete project reporting system can be divided into two parts: reports completed in the field and reports developed in the home or regional office using the field-generated data.

4.5.1 Field Reporting System

The purpose of the field reporting system is to acquire information about the actual work performed. The essential data to be gathered includes labor time, equipment time, and material amounts.

- Labor time is spent performing each coded work item, which is usually obtained using labor time cards. A sample daily labor time card is shown in *Figure 3.*
- Equipment time is spent performing each coded work item. This information can be recorded on a daily equipment log such as the sample shown in *Figure 4.*
- Amount of material installed in a designated amount of time is information that should be obtained through measurement, not guessing. The project manager should verify the reported quantities, especially if the information appears to be out of line with the estimate. *Figure 5* illustrates a sample form for recording the amount of material installed during the course of a week.

Whatever field reporting forms are used, they should be complete enough to provide the data needed to perform the following calculations for each coded item.

- Total actual cost to date
- Unit cost to date
- Productivity to date

4.5.2 Office Reporting System

The office reporting system is composed of a series of control reports that show labor costs, equipment costs, material costs, and a project summary. Labor, equipment, and material cost reports are usually prepared weekly and contain the following information for each coded activity:

- Amount of material installed during the week and to date.
- Unit production for the week and to date (for example, how many worker-hours it took to install one unit of material or how much was installed per day).

Instructor's Notes:

	DAILY LABOR TIME CARD	Page: 1 of 1
Job # 9515	Project: XYZ Project	Day: Tuesday
Prepared by: GRC	Weather: Warm and Windy	Date: 29-Aug 20xx

EMPLOYEE ID	RATE	COST CODE 3411	3414	3417	3418	3456	3496			TOTAL WK-HRS	GROSS AMOUNT
Anderson, T.L. 01-4142	RT $25.00	4.0		2.0		2.0				8.0	$200.00
	OT										
Cook, F.R. 01-1892	RT $23.00						8.0			8.0	$184.00
	OT										
Davis, W. 01-3757	RT $14.25				4.0		4.0			8.0	$114.00
	OT										
Harrington, G.H. 01-1194	RT $14.25		4.0	2.0						6.0	$85.50
	OT										
Mack, B. 02-1158	RT $15.00		4.0	2.0						6.0	$90.00
	OT										
Manning, T. 02-1687	RT $15.00				4.0	4.0				8.0	$120.00
	OT										
Pepper, S. 01-8469	RT $15.00		4.0	4.0						8.0	$120.00
	OT										
Taylor, M.T. 01-2495	RT $15.00				4.0	4.0				8.0	$120.00
	OT										
REGULAR HOURS		4.0	12.0	10.0	12.0	10.0	12.0			60.0	
OVERTIME HOURS											
TOTAL WORKER-HOURS		4.0	12.0	10.0	12.0	10.0	12.0			60.0	
TOTAL LABOR COST		$100.00	$177.00	$168.50	$177.00	$170.00	$241.00				$1,033.50

107F03.EPS

Figure 3 ◆ Daily labor time card.

	DAILY EQUIPMENT LOG	Page: 1 of 1
Job # 9515	Project: XYZ Project	Day: Tuesday
Prepared by: GRC	Weather: Warm and Windy	Date: 29-Aug 20xx

EQUIP. ID	DESCRIPTION	RATE/HOUR	COST CODE		TOTAL HOURS			TOTAL COST
			1016	1055	Working	Repair	Idle	
DT-1	Dump Truck	$60.00	8.0		7.0	1.0		$480.00
DT-2	Dump Truck	$60.00	8.0		8.0			$480.00
DT-3	Dump Truck	$60.00	8.0		8.0			$480.00
S-3	2-1/2 cy Shovel	$48.00		8.0	8.0			$384.00
WP-10	Tractor	$40.00	4.0	4.0	8.0			$320.00
AC-1	Air Compressor	$15.00		8.0	7.0		1.0	$120.00
TOTAL EQUIPMENT-HOURS			28.0	20.0	46.0	1.0	1.0	
TOTAL EQUIPMENT COST			$1,600.00	$664.00				$2,264.00

107F04.EPS

Figure 4 ◆ Daily equipment log.

MODULE 44107-08 ◆ ESTIMATING AND COST CONTROL 7.11

Discuss the project manager's role in reporting.

Assign reading of Sections 5.0.0-5.2.0.

Describe the process of cost analysis.

Emphasize that collecting data at the right time is critical for an accurate analysis.

Explain how Samantha could apply the "lessons learned" in the video to her next job estimate.

- Total cost for the week and to date.
- Unit cost for the week and to date.

Figure 6 shows a sample cost control report.

4.5.3 The Project Manager's Responsibilities

The project manager's responsibility for reporting is to ensure that field reports are properly and quickly prepared. The project manager needs to check with supervisors and use interpersonal skills to make sure people perform their duties as contracted. Additionally, the project manager needs to analyze data on the control reports, make appropriate decisions to correct cost deviations, and implement changes that will keep the project on track.

5.0.0 ◆ COST ANALYSIS

Cost analysis is the process of computing actual cost and projected costs of performing an activity like a job task or a project to determine if that activity is being performed within budget. If a project's job costs are exceeding the budget, the project manager must form a plan to correct the problem.

5.1.0 Analysis Process

A cost analysis is performed at specified intervals during the progress of the job, unless the job is very small, in which case it can be performed after the job is completed. Selecting the right time during the job to collect data for the analysis is crucial to an accurate analysis.

No job progresses at a steady, average pace. Most jobs start out relatively slowly, pick up speed as labor and equipment productivity increases, and slow again near the end due to less productive activities like clean-up and demobilization. Plotting the time spent on a job against the resources used would yield a rough S-shaped curve, as shown in *Figure 7*. The curve is based on the project schedule and estimate and is plotted before construction starts.

WEEKLY QUANTITY REPORT

Job # 9515 | Project: XYZ Project | Page: 1 of 1
Prepared by: GRC | Week Ending: 2-Sep | Date: 2-Sep 20xx

COST CODE	WORK DESCRIPTION	UNIT	TOTAL LAST REPORT	Mon. 28-Aug	Tues. 29-Aug	Wed. 30-Aug	Thur. 31-Aug	Fri. 1-Sep	Sat./Sun. 2-Sep	TOTAL THIS WEEK	TOTAL TO DATE
3102	Beam Concrete	cy	0					60		60	60.00
3124	Strip Footing Concrete	cy	479	35	41	44	40	36		196	675.00
3132	Grade Beam Concrete	cy	208	38	42	44	42	42		208	416.00
3146	Slab Concrete	cy	595			65				65	660.00
3203	Slab Trowel Finish	sf	50,595				2,865			2865	53,460.00
3253	Slab Curing	sf	50,595				2,865			2865	53,460.00

107F05.EPS

Figure 5 ◆ Weekly quantity report.

LABOR COST CONTROL REPORT

Job # 9515 | Project: XYZ Project | Page: 1 of 1
Prepared by: GRC | Week Ending: 2-Sep | Date: 2-Sep 20xx

COST CODE NO.	WORK DESCRIPTION	UNIT	ESTIMATED				THIS WEEK DATE				TO DATE				SAVINGS/LOSS	
			Total Qty.	Total Wk-hr	Labor Total $	Labor $/unit	Total Qty.	Total Wk-hr	Labor Total $	Labor $/unit	Total Qty.	Total Wk-hr	Labor Total $	Labor $/unit	To Date	Projected
3102	Beam Concrete	cy	508	542	$11,176	$22.00	60	70	$1,634	$27.23	60	70	$1,634	$27.23	($314)	($2,657)
3124	Strip Footing Concrete	cy	1040	416	$8,320	$8.00	193	48	$965	$5.00	675	256	$3,483	$5.16	$1,917	$2,954
3132	Grade Beam Concrete	cy	920	294	$5,888	$6.40	208	65	$1,308	$6.29	416	132	$2,617	$6.29	$46	$101
3146	Slab Concrete	cy	2772	807	$16,216	$5.85	65	21	$423	$6.50	660	185	$4,250	$6.44	($389)	($1,635)
3203	Slab Trowel Finish	sf	12800	192	$4,352	$0.34	2865	33	$745	$0.26	5346	90	$1,390	$0.26	$428	$1,024
3253	Slab Curing	sf	12800	38	$384	$0.03	2865	6	$57	$0.02	5346	11	$107	$0.02	$53	$128

107F06.EPS

Figure 6 ◆ Labor cost control report.

Instructor's Notes:

7.12 PROJECT MANAGEMENT

During construction, the same sort of curve can be plotted, using data from actual conditions. Comparing the shape and location of the curve using job data relative to the curve based on the estimate provides information that the project manager can use to control the pace of the project. The curve is constructed by plotting the total amount of a specific resource, such as dollars, hours worked, and units of material used at a specific time in the project.

The accumulated amount of resources used or estimated to be used from the beginning of the project is plotted on the vertical axis, and the time into the project for which the quantity of resource was determined is plotted on the horizontal axis. For the curve shown in *Figure 7*, at the beginning of the job no time and no resources had been used. After three time units (either days or weeks) into the job, 1.9 resource units were used (or are estimated to be used), depending when the curve is developed. Both the time unit scale and resource unit scale represents accumulated use of the units from the beginning of the project.

See, for example, the graphic showing progress on a formwork task (*Figure 8*). Assume that the resource in question is carpenter worker-hours estimated for completing the formwork. The eight-month job is estimated to require 10,000 worker-hours to erect 40,000 sfca (square feet of contact area) of formwork.

After 4.3 months, 4,700 worker-hours have been spent (based on labor cost control reports). This is plotted as "Actual Resource Usage." Since this figure represents 47 percent (4700/10,000 × 10%) of the activity, the project manager expects that 47 percent of the formwork is in place at this point. A field inspection shows, however, that only 12,500 sfca (31 percent) is in place. This is plotted as "Amount of Forms Erected."

The vertical distance between actual usage and forms erected represents the gain or (in this case) the loss in productivity. As *Figure 8* shows, the discrepancy between the worker-hours used and the percentage of the activity completed represents a loss of 1,600 worker-hours of productivity (4,700 minus 3,100).

The horizontal distance between actual forms erected and scheduled usage shows where the formwork stands in comparison with the schedule. *Figure 8* indicates that, at 4.3 months, 68 percent of the formwork should have been completed in order to make the original eight-month schedule. The 12,500 sfca completed to date should have been finished at the 2.4-month mark. Therefore, the formwork is 1.9 months (4.3 minus 2.4) behind schedule.

When making a cost analysis, the project manager must take into consideration the point at which the data is gathered. If it is early in the job, the actual unit cost can be expected to be higher than that anticipated by the job estimated. As the job progresses, the actual unit cost drops, perhaps

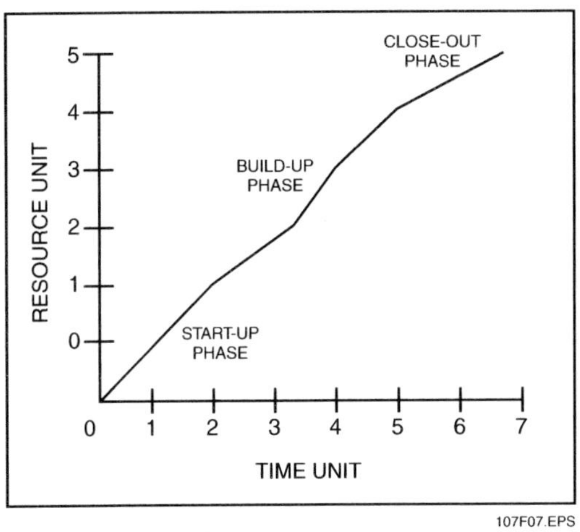

Figure 7 ◆ S-curve: time versus resource use.

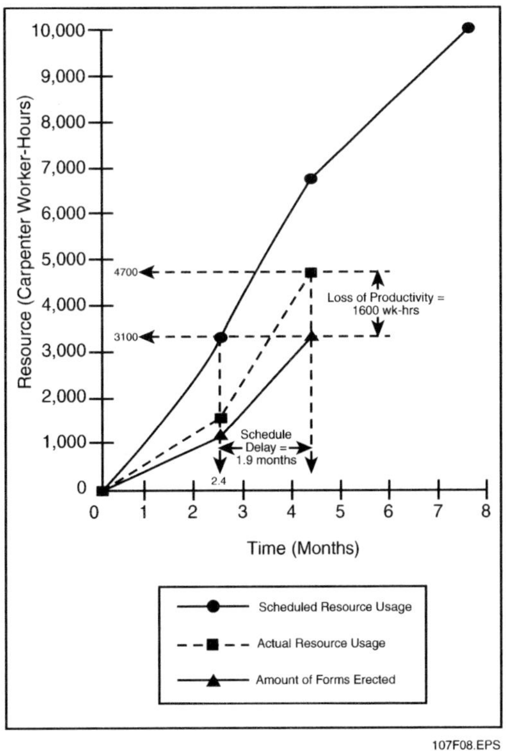

Figure 8 ◆ Time resource relationship for formwork.

Show Transparency 13 (Figure 7). Explain the graphic plotting of resource use in time units.

Show Transparency 14 (Figure 8). Demonstrate how plotting resources used over time gives an indication of the level of productivity.

MODULE 44107-08 ◆ ESTIMATING AND COST CONTROL 7.13

Explain how the project manager utilizes collected data to make decisions and adjustments.

Review the examples and discuss factors that can contribute to cost overruns.

even to the point of dipping below estimated cost. Then, toward the end of the job, actual unit costs again increase. These fluctuations are acceptable as long as the final goal is met, which is to have the job's total actual unit cost equal to the total estimated unit cost.

What about cases where the total actual cost of a job falls below the total estimated costs? If it occurs occasionally, there is usually no problem. However, if actual costs consistently fall below estimated costs, the project manager should be concerned. The manager and other team members should examine whether the figures being used to develop the estimates are accurate. If the cost difference occurs during similar jobs, perhaps the estimated production rates are wrong. If the discrepancy occurs on only one job, the difference might be due to an error in estimating production rate or a mistake made in taking-off quantities.

Consider this example: Five weeks into an estimated 12-week project, the reporting system shows that a specific job is being constructed for $5.60 per unit. The estimate indicates the job should be installed for $6.25 per unit. The project manager examines these figures and makes a number of preliminary decisions:

- The job is well underway, so the actual unit cost should come close to the estimate.
- The job is below the estimate at this point, but the situation needs to be monitored.
- There might be an error in the estimate. The actual total job cost needs to be monitored in the event it comes out substantially below the estimated cost.

Seven weeks later, the job is finished. Reviewing the final figures from the reporting system, the project manager notes the following:

- Estimated total cost: $36,000
- Final actual cost: $31,600

The difference between the two is significant enough that the project manager should review how the estimate was derived and answer the following questions:

- Was the quantity take-off about the same as what was installed?
- Were the estimated labor and equipment productivity figures valid for the job?
- Was the estimated cost of materials the same as was billed?
- Was the hourly cost of equipment estimated correctly?

The project manager may also want to examine whether the field reports were accurate and that the control reports were calculated correctly.

Why is doing work consistently under estimate such a big problem? Because it indicates that the contractor's bids are not as competitive as they could be.

More commonly, the project manager finds that actual costs exceed estimated costs. When this occurs, the manager must immediately find out why and take corrective action. When the actual unit cost exceeds the estimated unit cost, any of the following could be the cause:

- Material costs are more than estimated
- Labor wage rates are higher than estimated
- Equipment costs are higher than estimated
- Labor or equipment productivity factors are lower than estimated

The project manager usually has very little control over material and equipment costs or wages. However, if cost overruns occur, the project manager should look for ways to lower the direct costs of material, labor, and equipment. When the actual unit cost exceeds the estimated unit cost, in most cases the problem lies with the labor or equipment production.

In the following example, a project manager reviews the summary report during the job of placing electrical conduit:

Units placed to date:	800 lf (linear feet)
Unit cost to date:	$35.00 per unit
Units placed this week:	130 lf
Unit cost this week:	$36.00 per unit

Reviewing the estimate (or the estimate summary sheet), the project manager finds:

Total units to be placed:	1800 lf
Unit cost for activity:	$30.50 per unit

The actual cost of the work ($35.00 per unit) far exceeds the estimate ($30.50 per unit). Furthermore, the past week's unit cost ($36.00) is even higher than the unit price to date, which includes this week's quantity of 130 lf. If this trend is allowed to continue, the job will come in far over budget.

What are the problems in this situation? There could be an issue with the estimate. However, the project manager finds that the original estimate is accurate, so attention must be turned to the field, where there are innumerable variables that can cause high unit costs, such as the following factors:

- Inclement weather
- Late material deliveries
- Unmotivated workers
- Incorrect work methods
- Breakdown of equipment

Instructor's Notes:

7.14 PROJECT MANAGEMENT

It is the project manager's responsibility to uncover the cause of the overruns and take appropriate action to correct the situation. Steps the manager can take might include:

- Changing work methods
- Improving labor or equipment productivity
- Adding or changing resources
- Working overtime

In changing work methods or improving labor productivity, the project manager must be careful not to increase the total cost of the job. Adding or changing resources and working overtime obviously add to the cost of the coded work items; however, they may be necessary when the activity is a critical one that threatens to cascade into delays affecting the entire project.

5.2.0 Cost of Changes

Most contractors would rather have a project proceed without change orders, because each change order can slow down a job, disrupt the schedule, and cause other problems.

Changes to the project result in changes to the project cost. Thus, it is important to keep track of expenses incurred in doing any change work, including both direct and indirect costs. For example, if a footing must be torn out and replaced, it is important to track not only the costs of materials, labor, and equipment, but also the costs due to a change in the schedule, the loss of productivity, and more factors that affect the project budget.

The project manager must realize that even small, relatively inexpensive changes add up to a large impact on the project's bottom line. To illustrate this point, consider how much additional work a company must do in order to retain a specific profit margin in the face of changes:

Total project direct costs:	\$ 5,000,000
Including material & equipment:	\$ 3,000,000
Labor & other costs:	\$ 2,000,000

In the course of the project, the changes amounted to five percent of the labor cost and three percent of the material and equipment costs. This comes to a total of:

0.05 × \$2,000,000 =	\$100,000
0.03 × \$3,000,000 =	\$ 90,000
Total	\$190,000

If the contractor does not collect for the changes, profit is lost. How much work must be sold and performed to make up the loss? (Assume a five percent target profit.)

Lost profit: \$190,000 × 0.05 = \$9,500

The contractor has to sell another \$190,000 to recapture the lost profit of \$9,500. He wants to make a 5 percent profit (\$9,500) on the new work. The total profit to be captured is \$9,500 × 2 = \$19,000. He must then figure out of what amount \$19,000 is 5 percent. To do so, \$19,000 is divided by 0.05, which equals \$380,000.

6.0.0 ◆ IMPACT OF IMPROPER REPORTING

Faulty reporting from the field due to honest mistakes can happen and is understandable, if not excusable. Faulty reporting due to carelessness, indifference, or intent is inexcusable, and the project manager cannot allow it.

One type of faulty reporting, or **spreading**, should always be avoided. Spreading is the practice of reporting labor hours from activities which are over the estimate to labor hours for activities that are under budget. Spreading occurs when a supervisor has one or more job activities running over estimate and one or more that are well within the estimate. In order to report all activities within estimate, he decides to spread the losses into the activities that are within estimate when he reports on the project.

Spreading has several disastrous effects. The most critical is that it hides potential flaws in the estimating process. Without being aware of it, management continues to use the faulty information in estimating future jobs, and thereby unwittingly places other project managers and supervisors in the position of having to meet impossible production standards. The results are frustration, cost overruns, and unprofitable jobs that endanger job security and create more spreading.

For example, the estimate for erecting wall forms indicates 25,560 sfca and a production rate of 16 crew-hours per 100 sfca are used. After the job begins, the supervisor discovers that the actual production rate is 19 crew-hours per 100 sfca. Realizing that the estimate cannot be met, the supervisor decides to spread some of the wall form costs taking two of the 19 crew-hours per 100 sfca and places them in a different code category, thereby reporting the current production rate as 17 crew-hours per 100 sfca.

What are the results? In evaluating the performance of the first-line supervisor, the project manager sees that the wall forms are roughly within the estimate. However, when the estimator uses the 17 crew-hours per 100 sfca rate when estimating the next formwork job, the supervisor on that job is faced with an impossible production goal.

Discuss how the project manager might adjust for cost overruns without increasing the total cost of the job.

Explain the importance of keeping track of expenses resulting from changes to the project.

Review the examples to illustrate that even small changes can have a large impact on costs.

Show Transparency 15, Perform a Cost Analysis.

See the Teaching Tip for Section 5.2.0 at the end of the module.

Assign reading of Sections 6.0.0-7.0.0.

Stress that faulty reporting cannot be allowed.

Define spreading and discuss why it is a problem.

Ask participants to discuss their experiences with spreading and faulty reporting.

MODULE 44107-08 ◆ ESTIMATING AND COST CONTROL 7.15

Show Transparencies 16 and 17 (Figures 9 and 10).

Discuss the use of historical records as a reference for estimating and cost control.

Point out how labor costs and productivity records can be used for future estimating.

To continue the example, a subsequent job involves 104,300 sfca of formwork. It does not take the supervisor long to discover that the work takes 19 crew-hours per 100 sfca rather than 17. With an error of 2 crew-hours per 100 sfca built into the job, the total overrun will amount to 2,086 crew-hours (2 × 104,300/100). If the average hourly rate is $10 and the crew contains three people, the dollar amount of the overrun will total 3 × $10 × 2,086 = $62,580! On a lump-sum job, that amount comes out of project's profit.

To minimize spreading, a company may have to rely on people other than the field supervisor for its field reporting. In such a system, the supervisor reports labor hours for each worker in the respective activities, then another person measures the amount of work completed during those hours. No matter who provides the information as part of the field reporting system, it must be accurate and complete to be of value.

7.0.0 ◆ HISTORICAL RECORDS

A company's historical records are the estimator's most useful reference. No matter how closely the material quantities are measured, bad productivity information contained in the historical records will produce inaccurate estimates. Inaccurate estimates will, in turn, produce jobs that are completed with little or no profit. *Figure 9* shows how the historical records are tied to estimating and cost control.

As an estimate is prepared, the historical records are consulted for pricing and productivity. After the estimate is completed, the estimated cost is converted to budget costs and time to be used in the cost-control process. As the work proceeds, project performance is measured against the budgeted costs and time, as previously explained in this module. After the project is completed, the company's historical records should be reviewed and updated as necessary to reflect current prices and productivity.

The more cycles that are completed, the better the information is for estimating. The better the estimating is, the better the cost-control process will be, since the budgeted amounts should be more accurate. This assumes that the information used to update the historical records is accurate. Spreading during the cost-control process only corrupts the historical records and any future estimates. This is why it is so important that the project manager ensures that the information reported reflects the actual performance in the field.

Once several projects have been completed, results for individual work activities can be evaluated to see if and how much the historical records should be changed. *Figure 10* shows an example of historical records for column footing forms. While this figure shows both unit cost and worker-hours per unit information, the records can be set up to show just the information that is compatible with the company's needs and procedures.

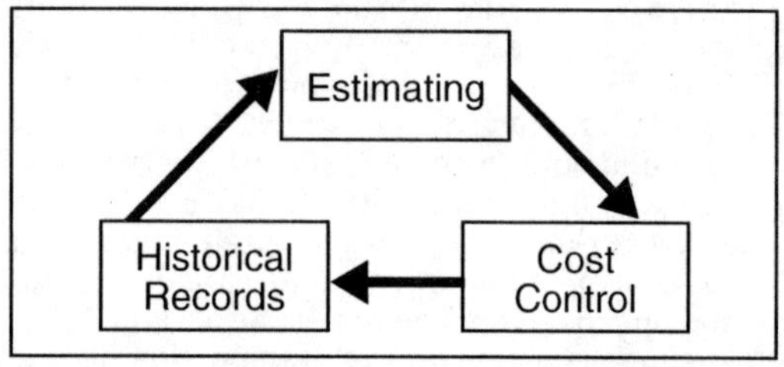

107F09.EPS

Figure 9 ◆ Estimating cost control cycle.

LABOR COST AND PRODUCTIVITY HISTORY

Item Description: Column Footing Forms **Cost Code #:** 3005

JOB #	JOB	SUPT.	JOB DATE	QUANTITY	UNIT	COST TOTAL	COST UNIT	WORKER-HOURS TOTAL	WORKER-HOURS UNIT
9406	Belleview	Joe Smith	3/10/06	2,500	cy	$2,000	$0.80	400	0.16
9408	St. Edwards	Bill Oliver	6/15/06	5,000	cy	$5,000	$1.00	1,000	0.20
9411	State Hospital	Ed Jones	9/26/06	4,000	cy	$3,800	$0.95	760	0.19
9602	City Building	Joe Smith	3/15/07	3,000	cy	$3,100	$1.03	580	0.19
9505	Johnson Building	Ed Jones	7/5/07	10,000	cy	$11,000	$1.10	1,800	0.18
9509	Federal Office	Bill Oliver	10/10/07	5,000	cy	$5,600	$1.12	870	0.17
						AVERAGES	$1.00		$0.18

107F10.EPS

Figure 10 ◆ Labor cost and productivity history.

Instructor's Notes:

7.16 PROJECT MANAGEMENT

Review Questions

1. Explain how estimated, actual, and project costs are related.

2. List five features of a cost control system.
   1. ______
   2. ______
   3. ______
   4. ______
   5. ______

3. List five factors which affect job cost control.
   1. ______
   2. ______
   3. ______
   4. ______
   5. ______

4. Define job cost coding and explain its use in construction.

5. What two parts make up a complete project reporting system?
   1. ______
   2. ______

6. List the three types of information needed from the field.
   1. ______
   2. ______
   3. ______

7. The office reporting system shows labor, equipment, and material costs and a project ______.

8. The results of the reporting system are used by the project manager to ______ the job costs.

9. Define and explain the impact of spreading on the company.

10. Explain the importance of the company's historical records.

Have participants complete the Review Questions, and go over the answers prior to administering the Module Examination.

MODULE 44107-08 ◆ ESTIMATING AND COST CONTROL 7.17

Summarize the major concepts presented in this module.

Summary

This module presented the various types of costs associated with all construction projects. It provided the techniques and tools necessary for estimating, analyzing, and controlling costs to complete a job within budget and schedule. The importance of accurate estimating and the composition of a quality estimate were discussed. The importance of maintaining accurate historical records and the project manager's responsibility for ensuring that the information gathered during the cost-control process reflects the actual work performed were also covered.

Notes

Instructor's Notes:

7.18 PROJECT MANAGEMENT

Trade Terms Quiz

1. The ____________ is the projected cost of the project, based on estimates.
2. The final cost of the project is the ____________.
3. ____________ is generally done to avoid reporting expenses or hours that bring that segment over budget.
4. The ____________ is the anticipated cost of a project based on estimates.
5. Breaking down the costs of some operation and reporting on each factor separately is called ____________.
6. A ____________ is a system for collecting data and is capable of showing details and comparisons.

Trade Term List

Actual cost
Bid
Budget
Budgeted cost
Cost analysis
Projected cost
Reporting system
Spreading

Have participants complete the Trade Terms Quiz, and go over the answers.

Administer the Module Examination. Record the results on Craft Training Report Form 200, and submit the results to the Training Program Sponsor.

MODULE 44107-08 ◆ ESTIMATING AND COST CONTROL 7.19

Trade Terms Introduced in This Module

Actual cost: Final cost of the project.

Bid: An offer to perform the work described in contract documents at a specified cost.

Budget: The dollar and time amount allocated by the owner for a project.

Budgeted cost: Projected cost of the project, based on estimates.

Cost analysis: Breaking down the costs of some operation and reporting on each factor separately.

Projected cost: Anticipated cost of a project based on estimates.

Reporting system: a system for collecting data and capable of showing details and comparisons.

Spreading: The faulty practice of reporting expenses or labor hours in a different budget area from that originally assigned; generally done to avoid reporting expenses or hours that bring that segment over budget.

Instructor's Notes:

Resources & Acknowledgments

References

Excerpts from About.com 20th Century History, Jennifer Rosenberg. Retrieved from http://history1900s.about.com/od/1930s/a/empirestatebldg_3.htm.

PMI Standards Committee, *A Guide to the Project Management Body of Knowledge*. PMI Publications, Newton Square, Pa. (2004).

Jonathan Goldman, *The Empire State Building Book* (New York: St. Martin's Press, 1980) 30.

John Tauranac, *The Empire State Building: The Making of a Landmark* (New York: Scribner, 1995) 156.

MODULE 44107-08 ◆ ESTIMATING AND COST CONTROL 7.21

MODULE 44107-08 — TEACHING TIPS

The following are suggested activities or instructional methods to help you teach the material in this module.

General

When you call on someone to answer a question, the rest of the class relaxes or even tunes out because they expect that the question and answer will take place only between you and the participant you called on. Instead, use this technique to involve more participants in answering questions and to keep them on their toes.

1. Ask participants to define a term or explain a concept.
2. After one participant has answered, ask a participant seated nearby if the answer is right. Then ask whether a participant in the back of the room agrees.
3. Ask participants to explain why they think an answer is right or wrong.
4. Use the session to clear up incorrect ideas and encourage participants to learn from their mistakes.

Section 1.1.0 *Case Study: The Montreal Olympics Complex*

The objective of this activity is to review the major cost overruns in the Montreal Olympics Complex and determine alternate solutions for each cost overrun factor.

1. Divide the class into two teams.
2. Team 1 will review the following factors:
   - Design
   - Labor
   - Construction techniques
   - Resources
   - Weather
3. Team 2 will review the following factors:
   - Design
   - Scheduling
   - Crowded workspace
   - Fixed deadline
   - Inflation
4. Instruct each team to review the details of these factors in their Participant Guides.
5. Instruct each team to develop recommendations for correcting each factor. Let them know there is no correct answer.
6. Have each team present its solutions.
7. Discuss and critique their answers.
8. Point out areas where they, as project managers, could have taken steps to resolve problems.

Estimated time: 30-45 minutes (may vary depending on class size).

As an option, it may be necessary to form more than two teams.

Section 2.2.0 *Creating a Preliminary Estimate*

The purpose of this activity is to demonstrate how to create a preliminary estimate for the construction of a structure of your choice (a warehouse, for example).

1. Present a project from your area of experience that contains past job records.
2. List the specification requirements on a whiteboard or flip chart; use bullet points and discuss each one.
3. Work through a preliminary estimate and explain the process.
4. Discuss the results.
5. Ask for questions and provide feedback.

Estimated time: 15 minutes (time may vary depending on class size).

Option: Distribute information for a similar project and ask the participants to develop a preliminary estimate. Review and discuss their results.

Section 2.3.0 *Developing an Estimate*

1. Review the steps in Section 2.3.0 for organizing and developing an estimate.
2. Use a job-related example to describe what occurs in each step.
3. Ask participants for questions. Provide feedback.

Sections 3.1.0–3.3.0 *Cost Estimating*

This activity will enable the participants to participate in a cost estimating exercise to establish the estimated and actual unit costs and the projected costs for the job.

1. Show Transparency 6, which contains the following information.
   - Estimated quantity (units) . . . . .5,000 lf (linear feet)
   - Total estimated cost $125,000
   - Actual quantity installed to date . . . . . . .3,000 lf
   - Total cost incurred to date $65,000
2. Have the participants calculate the results and discuss their anwers.
3. Work through the calculation with participants to illustrate how the answer was obtained.

Time to complete is at the discretion of the instructor.

Answer:

Estimated Unit Cost = 125,000 divided by 5,000 lf = $25 per lf

Actual Unit Cost = $65,000 divided by 3,000 lf = $21.67 per lf

Projected Costs = 5000 lf multiplied by $21.67 per lf = $108,350

Section 5.2.0 *Cost Analysis Exercises*

Exercises 1 and 2 will enable the participants to use job-related data to perform a simple cost analysis and determine any steps that need to be taken to control costs. Show Transparency 15 to provide the data for Exercises 1 and 2.

Exercise 1

Perform a cost analysis using the following information:

	Material Put in Place	Total Cost
Estimated:	8,500 Units	$11,600
Actual To Date:	4,000 Units	$4,800

Answer to Exercise 1:

- Estimated Unit Cost = $11,600 ÷ 8500 = $1.36 per unit
- Actual to Date Unit Cost = $4,800 ÷ 4000 = $1.20 per unit
- Actual Cost is below Estimated Cost. No immediate concern needed. Continue to watch figures.

Exercise 2

Perform a cost analysis for a 20-week job to install 350 plumbing fixtures. It is estimated to cost $32,600. To date, 250 fixtures have been placed for a total cost of $23,750.

Answer to Exercise 2:

- Estimated Unit Cost = $32,600 ÷ 350 = $93.13 per fixture
- Actual Unit Cost to Date = $23,750 ÷ 250 = $95.00 per fixture
- Actual Cost is above Estimated Cost. The project manager needs to determine reason(s) and take corrective steps to get unit costs in line with the estimate.

Exercise 3

This activity will enable the participants to use job-related data to perform a cost analysis for an estimate, a progress analysis, and a final evaluation.

Provide participants with copies of the handout on the following page. Explain that this is a three-part problem and that they will perform a cost analysis for each part. Have the participants perform their analysis. After they have completed their analysis, work through the calculations to illustrate how the results were obtained.

Answer to Part 1: The Estimate

- Estimated Material Cost = 10,000 x $0.80 = $8000
- Estimated Total Worker-hours = 0.060 x 10,000 = 600
- Average Crew Rate = [(2 × $15.00) + (1 × $11.50)] ÷ 3 = $13.83
- Estimated Labor Cost = 600 × $13.83 = $8298
- Estimated Total Unit Cost = ($8000 + $8298) ÷ 10,000 = $1.63 per block

Answer to Part 2: Progress Analysis

- To-date material cost = 5000 × $0.80 = $4000
- To-date total worker-hours = 3 × 8 × 10 = 240
- To-date labor cost = 240 × $13.83 = $3319.20
- To-date total unit cost = ($4,000 + $3,319.20) ÷ 5,000 = $1.46 per block
- Cost to Date is below Estimated. No immediate concern needed. Continue to watch figures.

Answer to Part 3: Final Evaluation

- Actual Material Cost = 10,000 × $0.80 = $8000
- Actual Total Worker-hours = 3 × 8 × (10,000 ÷ 350) = 685.71
- Actual Labor Cost To Date = 685.71 × $13.83 = $9483.37
- Actual Total Unit Cost to Date = ($8000 + $9483.37) ÷ 10,000 = $1.75 per block
- Actual Cost is above Estimated and Cost to Date. Project manager needs to determine reason(s) and take corrective steps so future work will not be affected.

Section 5.2.0 *Cost Analysis Exercise – Exercise 3*

Part 1: The Estimate

Find the total estimated cost given the following:

- 10,000 12-inch concrete blocks
- Estimated material cost = $0.80 per block
- Crew of two masons at $15.00 per hour and one laborer at $11.50 per hour
- Labor production = 0.060 worker-hours/block
- No equipment, scaffolding, or other items needed

Part 2: Progress Analysis

Find the savings/loss to-date and projected given the following:

- 10 working days have passed
- All block have been delivered
- 5,000 blocks have been placed

Part 3: Final Evaluation

Find the final actual cost given the following:

- Material invoice = $8,000
- Actual crew productivity was 350 blocks per day

Copyright © 2008 National Center for Construction Education and Research. Permission is granted to reproduce this page provided that copie are for local use only and that each copy contains this notice.

MODULE 44107-08 — ANSWERS TO REVIEW QUESTIONS

	Answer	Section
1.	The estimated cost is the budget and actual costs are those incurred in getting the job built. Ideally, actual costs end up as equal to or below the estimated costs. Projected costs are the project manager's estimate of the cost of the project. Projected costs are determined by analyzing actual costs and job progress to forecast the final outcome of the project.	(3.1.0-3.3.0)
2.	Any five of the following: 1. Create historical data. 2. Compare actual costs with estimated costs. 3. Motivate the team to achieve lower costs. 4. Place responsibility and credit where it belongs. 5. Alert management of problems during the job. 6. Simplify work and identify methods for reducing costs 7. Enable estimator to check contractor bids.	(4.2.0)
3.	Any five of the following: 1. Job location 2. Procurement of labor and materials 3. Lead time 4. Equipment utilization 5. Management personnel 6. Financial requirements 7. Purchasing 8. Expediting 9. Accounting and record keeping 10. Cost control procedure during job 11. Estimating and cost accounting 12. Cost control records for material, labor and equipment	(4.2.0)
4.	Assigning a separate code to each activity in order to track time and costs for each activity, i.e., control system foundation	(4.3.0)
5.	1. Field reports showing actual work performed 2. Office reports with labor, equipment, and material costs	(4.5.0)
6.	1. Labor time spent 2. Equipment time spent 3. Amount of material installed in a designated amount of time	(4.5.1)
7.	summary	(4.5.2)
8.	control	(4.4.0)
9.	The practice of reporting labor hours from activities which are over the estimate to labor hours for activities that are under budget. Spreading will result in cost data that will negatively affect future jobs.	(6.0.0)
10.	They are the most useful reference to an estimator on all new jobs. The records are used on future projects to determine the best estimate to prevent inaccuracy that would result in an unprofitable job.	(7.0.0)

MODULE 44107-08 — ANSWERS TO TRADE TERMS QUIZ

1. budgeted cost
2. actual cost
3. Spreading
4. projected cost
5. cost analysis
6. reporting system

NCCER makes every effort to keep these textbooks up-to-date and free of technical errors. We appreciate your help in this process. If you have an idea for improving this textbook, or if you find an error, a typographical mistake, or an inaccuracy in NCCER's Contren® textbooks, please write us, using this form or a photocopy. Be sure to include the exact module number, page number, a detailed description, and the correction, if applicable. Your input will be brought to the attention of the Technical Review Committee. Thank you for your assistance.

Instructors – If you found that additional materials were necessary in order to teach this module effectively, please let us know so that we may include them in the Equipment/Materials list in the Annotated Instructor's Guide.

Write: Product Development and Revision
National Center for Construction Education and Research
3600 NW 43rd St, Bldg G, Gainesville, FL 32606

Fax: 352-334-0932

E-mail: curriculum@nccer.org

Craft ______ Module Name ______

Copyright Date ______ Module Number ______ Page Number(s) ______

Description ______

(Optional) Correction ______

(Optional) Your Name and Address ______

Module 44108-08

Scheduling

NCCER STANDARDIZED CRAFT TRAINING PROGRAM

The National Center for Construction Education and Research (NCCER) provides a standardized national program of accredited craft training. Key features of the program include instructor certification, competency-based training, and performance testing. The program provides trainees, instructors, and companies with a standard form of recognition through a National Craft Training Registry. The program is described in full in the *Guidelines for Accreditation*, published by NCCER. For more information on standardized craft training, contact the NCCER by writing us at 3600 NW 43rd St., Bldg. G, Gainesville, FL 32606; calling 352-334-0911; or emailing info@nccer.org. More information may be found at our website, www.nccer.org.

HOW TO USE THIS ANNOTATED INSTRUCTOR'S GUIDE

Each page presents two sections of information. The larger section displays each page exactly as it appears in the Trainee Module. The narrow column ties suggested trainee and instructor actions to each page and provides icons (detailed below) to call your attention to material, safety, audiovisual, or testing requirements. The bottom of each page includes space for your notes.

The **Audiovisual** icon indicates an appropriate time to show a transparency or other audiovisual aid.

The **DVD** icon indicates an appropriate time to discuss or show segments of the video.

The **Classroom** icon prompts you to define a term, stress a point, ask trainees to explain a concept, or give examples.

The **Demonstration** icon directs you to show trainees how to perform tasks.

The **Examination** icon tells you to administer the written module examination.

The **Homework** icon is placed where you may wish to assign reading for the next class, assign a project, or advise trainees to prepare for an examination.

The **Laboratory** icon is used when trainees are to practice performing tasks.

The **Materials** icon is a reminder for you to gather materials needed for classes, labs, and testing.

The **Performance Testing** icon tells you to administer a performance test or a portion thereof.

The **Safety** icon is used to emphasize safety issues. It is often keyed to *Caution* and *Warning!* statements in the Trainee Module.

The **Teaching Tip** icon indicates additional guidance is available, such as how to conduct an exercise, get the most educational value from a field trip, or encourage class participation. Teaching Tips may expand on a feature (*Think About It, Did You Know?*) or provide *Quick Quizzes* or similar exercises. You will be referred to the Teaching Tips section at the back of the module if there is additional material.

The **Combination** icon indicates that the laboratory listed corresponds with a performance task. If desired, you can note the proficiency of the trainees during the laboratory, and use it to satisfy performance testing requirements.

PREPARATION

Before teaching this module, you should review the Objectives, Performance Tasks, Materials and Equipment List, and Module Outline. Be sure to allow ample time to prepare your own training or lesson plan and gather all required materials and equipment.

Scheduling
Annotated Instructor's Guide

Module 44108-08

MODULE OVERVIEW

This module provides instruction in the basics of scheduling, from simple to-do lists through bar charts, network diagrams, and methods of managing resources. It discusses the importance of formal schedules, job planning, and establishing priorities and presents alternative scheduling methods.

PREREQUISITES

Please refer to the Course Map in the Participant Module. Prior to training with this module, it is recommended that the participant shall have successfully completed *Project Management,* Modules 44101-08 through 44107-08.

OBJECTIVES

Upon completion of this module, the participant will be able to do the following:

1. Establish personal task priorities and delegate tasks.
2. Describe the purposes and benefits of using formal project schedules and why it is important to maintain schedules.
3. Identify basic project scheduling terms and inputs.
4. Develop a bar chart schedule.
5. Develop and interpret a network diagram.
6. Identify alternative scheduling methods.
7. Develop and calculate CPM schedules to include early start, early finish, late start, late finish, and total float.
8. Analyze an existing CPM schedule to optimize the project schedule.
9. Update and maintain a project schedule, including establishing baselines and targets.
10. Determine the effects of a change to the schedule.

MATERIALS AND EQUIPMENT LIST

Desktop or laptop computer or DVD player
Multimedia projector and screen
Minor Decisions: Major Impact – How to Deal with Real Issues in Project Management DVD
Transparencies
Blank acetate sheets
Overhead projector and screen
Transparency pens
Whiteboard/chalkboard
Markers/chalk
Pencils and scratch paper
Module Examinations (located in the Test Booklet)

ADDITIONAL RESOURCES

This module is intended to present thorough resources for task training. The following reference works are suggested for both instructors and motivated participants interested in further study. These are optional materials for continued education rather than for task training.

Blair, Dr. Gerard, *Starting to Manage: The Essential Skills*, published by The Institute of Electrical Engineers (IEEE)

Howell, Greg and Glenn Ballard, Lean Construction Institute, Louisville, Colorado, www.lean construction.org

International Group for Lean Construction, www. iglc.net

VIP Quality Software, Odessa, Ukraine, www.vip-qualitysoft.com/timemanagement

Womack, James P. and Daniel T. Jones, *Lean Thinking: Banish Waste and Create Wealth in Your Corporation.* New York: Simon & Schuster, 1996

TEACHING TIME FOR THIS MODULE

An outline for use in developing your lesson plan is presented below. Note that each Roman numeral in the outline equates to one session of instruction. Each session has a suggested time period of 2½ hours. This includes 10 minutes at the beginning of each session for administrative tasks and one 10-minute break during the session. Approximately 15 hours are suggested to cover *Scheduling*.

Topic	Planned Time
Session I. Time Management and Formal Schedules	
A. Introduction	________
B. Time Management	________
C. Formal Schedules	________
Session II. Project Scheduling, Part One	
A. Planning	________
B. Developing a Bar Chart	________
Session III. Project Scheduling, Part Two	
A. Developing a Critical Path Method Schedule	________
Session IV. Project Scheduling, Part Three	
A. Short Interval Scheduling	________
B. Lean Construction	________
Session V. Float Types and Lags	
A. Float Types	________
B. Lags	________
Session VI. Resource Management; Review and Testing	
A. Using the Schedule to Balance Resources	________
B. Determining the Effects of a Change Order	________
C. Enforcing a Schedule	________
D. Using the Schedule to Document Problems and Progress	________
E. Review	________
F. Module Examination	________
1. Participants must score 70% or higher to receive recognition from NCCER.	
2. Record the testing results on Craft Training Report Form 200, and submit the results to the Training Program Sponsor.	

USING THE PM DVD WITH *SCHEDULING*

When the job began to encounter delays, Samantha implemented weekly productivity quotas. Various team members reacted with frustration, resignation, and skepticism. Samantha's decision to supplement their crews with more workers was seen as ineffective since they had to be trained. In addition, some new workers were standing around with nothing to do because of schedule conflicts.

An alternative solution to implementing production quotas would have been to realign workflow and scheduling of contractors and crew tasks. Samantha eventually had to take this approach. In the end, it all came down to scheduling.

Suggested Class Exercise: Using the sample Short Interval Schedule included on the DVD, have the students prepare a short interval production schedule (SIPS) that might resolve Samantha's immediate scheduling crisis. Remind the students they would need to take into account the time it would take to orient and train new workers if they took Samantha's offer to add more labor.

Note: If the class is large enough, have the students form teams to discuss their approach and draft the schedule. Have the students present their SIPS scenarios to the entire class.

Referring to Section 4.5.0, Lean Construction, ask the class to consider how adopting lean construction principles may have improved the job site problems cited in the video (tool control, sloppy housekeeping, and safety hazards).

Suggested response: Weekly team meetings to review progress, identify problems, take action on the root causes of tasks not completed, and revise workflow where necessary can minimize or eliminate problems and slippages. The just-in-time approach can facilitate having resources (materials, labor, tools) available when specifically needed, with the goal to eliminate time wasted waiting for a job to begin or searching for needed resources. The team approach is more likely to generate cooperation, understanding of each party's issues and concerns, and mutual respect.

Referring to the section on CPM (4.3.0), ask the class to suggest methods Samantha might apply to scheduling in the future.

Suggested response: Anticipate problems that might affect schedule so that critical activities are not impacted by poor planning. In this case, there may have been inadequate resources (e.g., Anthony needing to borrow Keith's saw). It also appears that there were a number of prior problems that caused delays and Samantha needs to plan for unexpected events to keep her project on schedule. Samantha should allow time in her schedule for the team to keep the job site in order for safety and productivity considerations. She should also set aside time for crew meetings and debriefings in her schedule.

Project Management

44108-08

Scheduling

Assign reading of Module 44108-08.

44108-08
Scheduling

Topics to be presented in this module include:

Overview

Effective schedules are based on good planning. Once the project is broken down into tasks necessary to complete the work, developing a schedule will then define how the project tasks will be completed. The project manager usually develops the project schedule with input from others, including estimators, foremen, superintendents, and managers. This approach will result in a realistic schedule and a successfully completed project.

Instructor's Notes:

Objectives

Upon the completion of this module, you will be able to do the following:

1. Establish personal task priorities and delegate tasks.
2. Describe the purposes and benefits of using formal project schedules and why it is important to maintain schedules.
3. Identify basic project scheduling terms and inputs.
4. Develop a bar chart schedule.
5. Develop and interpret a network diagram.
6. Identify alternative scheduling methods.
7. Develop and calculate CPM schedules to include early start, early finish, late start, late finish, and total float.
8. Analyze an existing CPM schedule to optimize the project schedule.
9. Update and maintain a project schedule, including establishing baselines and targets.
10. Determine the effects of a change to the schedule.

Trade Terms

Critical path method (CPM)
Lags
Lean construction
Precedence diagramming method (PDM)

Prerequisites

Before you begin this module, it is recommended that you successfully complete *Project Management*, Modules 44101-08 through 44107-08.

This course map shows all of the modules in the *Project Management* curriculum. The suggested training order begins at the bottom and proceeds up. Skill levels increase as you advance on the course map. The local Training Program Sponsor may adjust the training order.

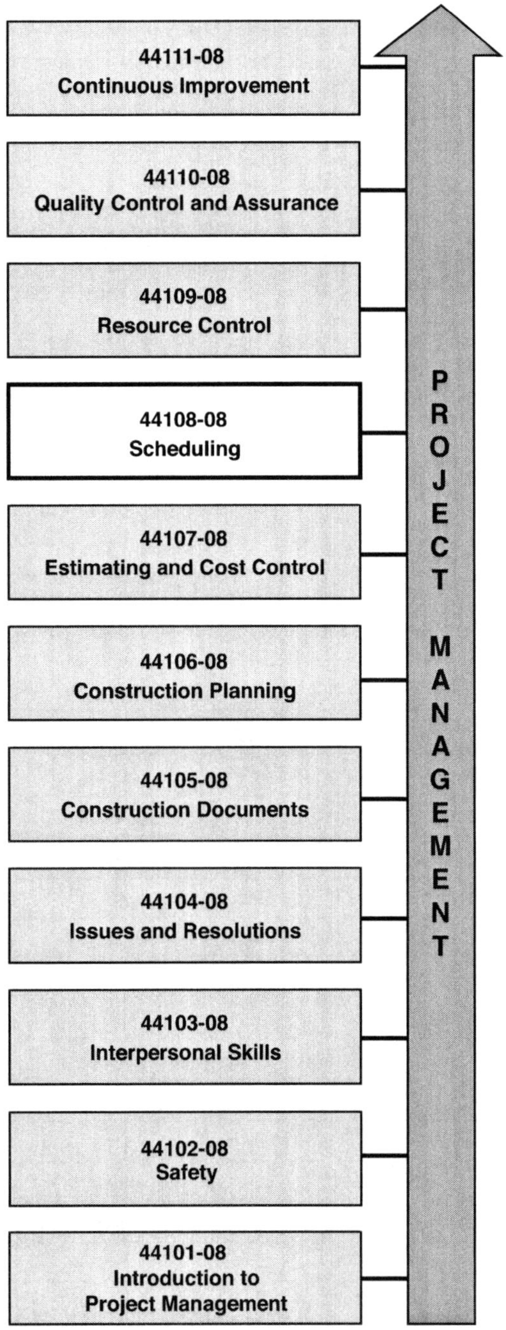

108CMAP.EPS

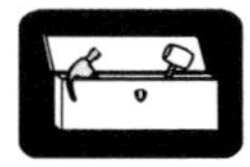

Ensure that you have everything required to teach the course. Check the Materials and Equipment list at the front of this module.

See the general Teaching Tip at the end of this module.

Explain that terms shown in bold are defined in the Glossary at the back of the module.

Show Transparency 1, Objectives. Review the goals of the module, and explain what will be expected of the participant.

MODULE 44108-08 ◆ SCHEDULING 8.1

Discuss the need for accuracy and realistic estimates when preparing schedules.

Show Transparency 2 (Figure 1). Have participants discuss reasons why "time is money."

Point out that the costs in Figure 1 reflect accumulated interest charges only and do not show other costs caused by schedule slippage.

1.0.0 ◆ INTRODUCTION

Understanding and performing scheduling is one of the essential skills a project manager must develop and practice. Scheduling requires accuracy and depends largely on the project team's ability to create realistic estimates for both hours required and the funding needed to cover time and materials.

To get an idea of how time and money can affect a project, see *Figure 1*. The amounts shown reflect just the interest costs of money owed to the bank, other loan agencies, or interest that is unavailable to make other investments because the money is tied up in other project costs.

The amounts in *Figure 1* are based on work units.

- One workday is based on a standard five-day workweek or $\frac{1}{5}$ of the interest accrued during the week.
- One work hour is $\frac{1}{40}$ of the interest accrued during the week.
- One minute is $\frac{1}{60}$ of the hourly rate.

The rates in *Figure 1* reflect the costs of accumulated interest as the result of slips in the schedule, driving home the inescapable fact that extended project durations increase the indirect costs of the project. Extra costs result from extended staff assignments, retaining project facilities, protracted project support (for example, ice, water, sanitation), extended rental of project equipment, and other costs.

TIME IS MONEY

Effect of Time Delays on Construction
Based on 9% Annual Interest Rate

Amount	Minute	Hour	Day	Week	Month	Year
$25,000,000.00	$18.03	$1,081.73	$8,653.85	$43,269.23	$187,500.00	$2,250,000.00
$20,000,000.00	$14.42	$865.38	$6,923.08	$34,615.38	$150,000.00	$1,800,000.00
$15,000,000.00	$10.82	$649.04	$5,192.31	$25,961.54	$112,500.00	$1,350,000.00
$10,000,000.00	$7.21	$432.69	$3,461.54	$17,307.69	$75,000.00	$900,000.00
$8,000,000.00	$5.77	$346.15	$2,769.23	$13,846.15	$60,000.00	$720,000.00
$5,000,000.00	$3.61	$216.35	$1,730.77	$8,653.85	$37,500.00	$450,000.00
$4,000,000.00	$2.88	$173.08	$1,384.62	$6,923.08	$30,000.00	$360,000.00
$2,000,000.00	$1.44	$86.54	$692.31	$3,461.54	$15,000.00	$180,000.00
$1,000,000.00	$0.72	$43.27	$346.15	$1,730.77	$7,500.00	$90,000.00
$800,000.00	$0.58	$34.62	$276.92	$1,384.62	$6,000.00	$72,000.00
$600,000.00	$0.43	$25.96	$207.69	$1,038.46	$4,500.00	$54,000.00
$400,000.00	$0.29	$17.31	$138.46	$692.31	$3,000.00	$36,000.00
$200,000.00	$0.14	$8.65	$69.23	$346.15	$1,500.00	$18,000.00
$100,000.00	$0.07	$4.33	$34.62	$173.08	$750.00	$9,000.00
$90,000.00	$0.06	$3.89	$31.15	$155.77	$675.00	$8,100.00
$75,000.00	$0.05	$3.25	$25.96	$129.81	$562.50	$6,750.00

108F01.EPS

Figure 1 ◆ Time is money.

Instructor's Notes:

8.2 PROJECT MANAGEMENT

2.0.0 ◆ TIME MANAGEMENT

Managing personal time is critical to your ability to produce results and deliver your projects on schedule. In today's fast-paced environment this is often a difficult task, but one you can achieve with discipline and practice. Time is arguably your most valuable resource. Time management is about controlling your time through processes that allow you to measure, plan, monitor, and improve your efficiency and productivity.

The payoff is that mastering time management will alleviate stress, improve your work/life balance, and help you keep your commitments as a project manager and team leader. Techniques and tools for managing time range from the simple, such as diaries and to-do lists, to automated or computer-based approaches. Many references, tools, and techniques are available to help you formally learn the skills of time management.

Practicing time management requires that you set priorities, distinguish between important activities and those lower on the scale in importance, and that you eliminate unproductive, or wasted time. Productivity experts have found that the 80/20 rule applies to time management and that the top 20% of activities are the ones that bring the most value. Most people spend their time on the 80% of activities that yield much less return. They do this because the top 20% are more complex or they busy themselves with the easier tasks. The smart manager will apply the 80/20 rule to make decisions about where to spend time to achieve the best value for the company and thus attain a higher level of productivity and value to the company and the team.

A basic step to gaining more time for important activities is to reduce unnecessary or time-wasting activities. It is not possible to eliminate all unproductive time—everyone needs a break now and then. Some unproductive time is outside your control; however, you can improve those areas under your control, such as personal issues and work habits. You can also try to influence and improve processes and remove external barriers that affect you and your team's productivity.

Another fundamental step is to develop a system for personal time management. You may choose to keep a log or personal diary that you can use to prioritize activities or use a software tool to help manage your activities. A log is a good tool to measure and evaluate time spent on activities because you can then use the metrics to improve your use of time. These tools can help you make an objective evaluation about how much time each activity is worth.

Planning is fundamental to personal time management. The top performers spend their time first on high-priority activities, particularly those that have strategic, long-term consequences. To put this into practice and take control of your time, sort into categories and then rank the tasks you need to complete. Start with the primary activities you must perform. Determine which activities are the most important (priorities), the most necessary (responsibilities), required or administrative (requirements), personal activities, and others that can be eliminated or delegated. This systematic ordering enables you to influence events and is the foundation of managerial skills such as effective delegation and project planning.

Some of the actions you can take to become more efficient and productive with your time at work include the following:

- Manage personal and social time, including phone calls, water cooler time, and personal business.
- Eliminate procrastination. Don't use distractions to put off unpleasant or overwhelming tasks.
- Group similar tasks to limit start-up delays.
- Manage your environment. Adhere to time allocated for meetings, paperwork, telephone, and visitors. Keep your workplace organized and efficient.
- Delegate. Don't do the work of others.
- Manage your manager and his or her effect on your time. Make sure specifications are clear and agreed upon.
- Manage your external appointments. Determine which ones can be eliminated, handled by a conference call, or delegated.
- Manage and deal with impossible deadlines. Get the deadline extended, get more resources, or redefine the deliverable so tasks are achievable.
- Balance important long-term objectives with short-term requirements.
- Set goals and priorities. It may be more interesting to work on one task, but it may interfere with your critical schedule commitments.
- Anticipate and identify potential problems. Develop an action plan to prevent their occurrence and solutions that will be required if they are unavoidable.
- Practice good decision-making. Weigh alternatives and make the choice according to the outcome needed to accomplish the goals and keep your commitments, then put the decision into action.
- Apply the 80/20 rule to determine the most important activities to perform first.

The remainder of this module will provide information to help you develop realistic project schedules and to control your projects.

Explain why a project manager must have good personal time management skills.

Explain the 80/20 rule of time management.

Show Transparency 3, Personal Time Management. Review the action steps for better personal time management.

Ask participants to give examples of personal choices they have made in order to be more effective in managing their time.

MODULE 44108-08 ◆ SCHEDULING 8.3

Show Transparency 4, Why Use Formal Schedules. Explain the benefits of formal scheduling.

3.0.0 ◆ FORMAL SCHEDULES

To bring a project in on time and within budget requires a project schedule. A well-developed schedule communicates necessary information to all the parties involved in the project. A formal project schedule has the following advantages:

Assign reading of Sections 4.0.0-4.2.0.

- Reduces construction time
- Reduces costs of labor, overhead, and interest on loans and capital
- Provides a continuous flow of work
- Increases productivity
- Provides employees and subcontractors specific milestones to work toward and indications of progress toward those goals
- Forces detailed thinking
- Improves communication
- Improves the company's image and shows professionalism
- Provides your company with a competitive edge
- Provides control and management of material, labor, money, equipment, and subcontractors
- Provides owners with relevant project data, which is increasingly important as owners are becoming more sophisticated and are requiring tighter controls

Show Transparency 5, Scheduling Considerations.

Show Transparency 6, Related Parts of an Effective Schedule.

A good scheduling system does not try to drive labor to work harder or faster, but rather to work in a more organized fashion for greater productivity and, therefore, profits. A successful scheduling system encourages detailed thinking and then gives managers the ability to communicate that thinking and planning to all those involved in the process to complete quality projects on time and within budget.

Thomas Monson underscored the importance of maintaining a schedule when he said, "When performance is measured, the performance improves. When performance is measured and reported back, the rate of improvement accelerates."

Explain that schedules are built with team input.

Explain that the first step of planning is to establish the project goal.

4.0.0 ◆ PROJECT SCHEDULING

Effective schedules are developed by the project manager and supervisors who have to use them. Large companies are likely to have a company scheduler, and computerized scheduling tools and systems are widely used in the industry. In either case, it is essential that the schedule be developed as a team effort with input from the project manager, foremen, superintendents, estimators, and other key managers who use the schedule to monitor and control the project.

When you start developing a schedule, be sure to focus on:

- Contract compliance
- Detailed scope of the project
- Job-site safety
- Cost- and resource-control measures
- Completing the project as specified
- Managing subcontractors
- Customer satisfaction expectations
- Quality issues

The size of the project will dictate the number of factors you need to track. Knowing your priorities will make the scheduling process easier. When looking at some of the core tasks of project scheduling, you'll see that a variety of approaches can be used to create a schedule. There are a number of methods and tools available to help you with scheduling.

An effective project schedule is made up of a number of related parts:

- Planning
- Work breakdown structure (WBS) – (see the *Construction Planning* module)
- Bar charts
- Network systems – **critical path method (CPM)**
- Short interval scheduling
- Periodic updates

At this point in the scheduling process, the WBS has helped to determine the scope and the best possible estimates. With this information, you can now create the project schedule by systematically gathering information. The schedule development process includes selecting a scheduling method and a scheduling tool. Project-specific data is incorporated into the scheduling tool to develop a schedule for the project and generate scheduling outputs, or reports. This process results in a model for project execution that reacts predictably to progress and changes.

The project execution team's active participation is vital in developing the project schedule model. The project team's involvement fosters buy-in and support for the schedule model.

4.1.0 Planning

The planning effort is intensive because the project manager needs to document as much information as possible about the project before any work actually begins. The first step is to establish the project goal; that is, the statement that helps to define the construction project (what is the scope of the project). Objectives are then developed to

8.4 PROJECT MANAGEMENT

Instructor's Notes:

describe the necessary steps to accomplish the goal. Project objectives define what the desired results are and help in developing the scope statement or letter. The project scope information is essentially what is going to happen, how it will get done, what will be excluded, and what will be delivered. Because projects vary in size, the planning effort will vary as well. Many companies have processes in place to help the project manager develop a planning document. The *Construction Planning* module provides the information necessary to carry out the planning process.

4.2.0 Developing a Bar Chart

The bar chart is a timeline showing the day, week, or month when certain project activities should be accomplished. The bar chart is a helpful tool for viewing a project at a macro level. These types of charts are also used to formulate schedule logic. Most project management software programs are able to generate a chart based on the tasks that are entered. *Figure 2* shows a bar chart used as a check sheet, and *Figure 3* shows a bar chart used to measure progress.

Here are some advantages and disadvantages of using a bar chart:

- *Advantages*
  - Simple and easy to understand
  - Visual clarity
  - Widely accepted
- *Limitations*
  - Doesn't show interdependency
  - Doesn't show activities critical to timely completion
  - Ineffective in forecasting effects of changes; must rethink the schedule
  - Doesn't show accurate float times unless generated from a network schedule

Due to its simplicity, the bar chart is widely used in construction. You can see exactly when the activity begins and ends, and exactly when craftworkers, subcontractors, and materials must be available.

Constructing a bar chart is simple. The left-hand column, or y axis, is a list and description of each activity in chronological order. The horizontal line at the top of the chart represents the beginning and ending times of the particular set of activities. It is divided into increments relevant to the time frame, with each square representing a specific unit of time such as hours, days, weeks, or months. The duration of the specific activity is then represented by a horizontal bar drawn from the block representing the start of the activity to the block representing the end of the activity. You can see in *Figure 2* that Settle FDN Soil starts at day 12 and ends at day 19, a duration of eight days.

The bar chart can be updated to show progress as seen in *Figure 3*. A horizontal bar representing the as-built progress is drawn below the as-planned bar, shown in *Figure 3* as a dark gray bar. The light gray bar shows the future planned activities. This chart gives a visual snapshot of the planned versus actual status of the project.

4.3.0 Developing a Critical Path Method Schedule

With the increased use of computers, network diagrams or CPM networks are now widely used in the construction industry. The following information-gathering guidelines shows the process of developing a scheduling model using the critical path method (CPM).

- Review estimate and plans; spend time to thoroughly familiarize yourself with the project
- Get an idea of the magnitude of the project
- Decide on methods of construction
- Look for restrictions (special conditions and features)
- Select subcontractors (What's their scope? What do they furnish?)
- Review special delivery dates (logistics)
- Engage key management personnel (e.g., project manager, superintendent, safety manager, quality manager)
- Discuss the overall job approach
- Outline most important areas
- Anticipate problem areas; check with the estimator
- Decide on degree of control – tight or general terms

There are a number of advantages to using CPM to document progress and problems instead of a bar chart:

- Shows dependency
- Identifies critical activities: 80/20 rule – spend 80 percent of time on 20 percent of the work; CPM tells which activities to spend time on and others will then follow
- Allows management to set priorities
- Shows effects of change
- Adaptable to any project, either a simple or complex one
- Easy to follow from a visual standpoint

Show Transparency 7, Advantages and Disadvantages of a Bar Chart.

Show Transparencies 8 and 9 (Figures 2 and 3).

Discuss bar charts and their advantages and disadvantages.

Demonstrate how to construct a bar chart and how to update progress on a bar chart.

Assign reading of Sections 4.3.0-4.3.4.

Show Transparency 10, Guidelines for Developing a CPM Schedule.

Show Transparency 11, Advantages of CPM, and Transparency 12, Disadvantages of CPM.

Explain the information-gathering guidelines for developing a CPM schedule.

Ask participants if any of their employers use CPM scheduling.

Ask students how Samantha might have used CPM to better account for project interruptions.

MODULE 44108-08 ◆ SCHEDULING 8.5

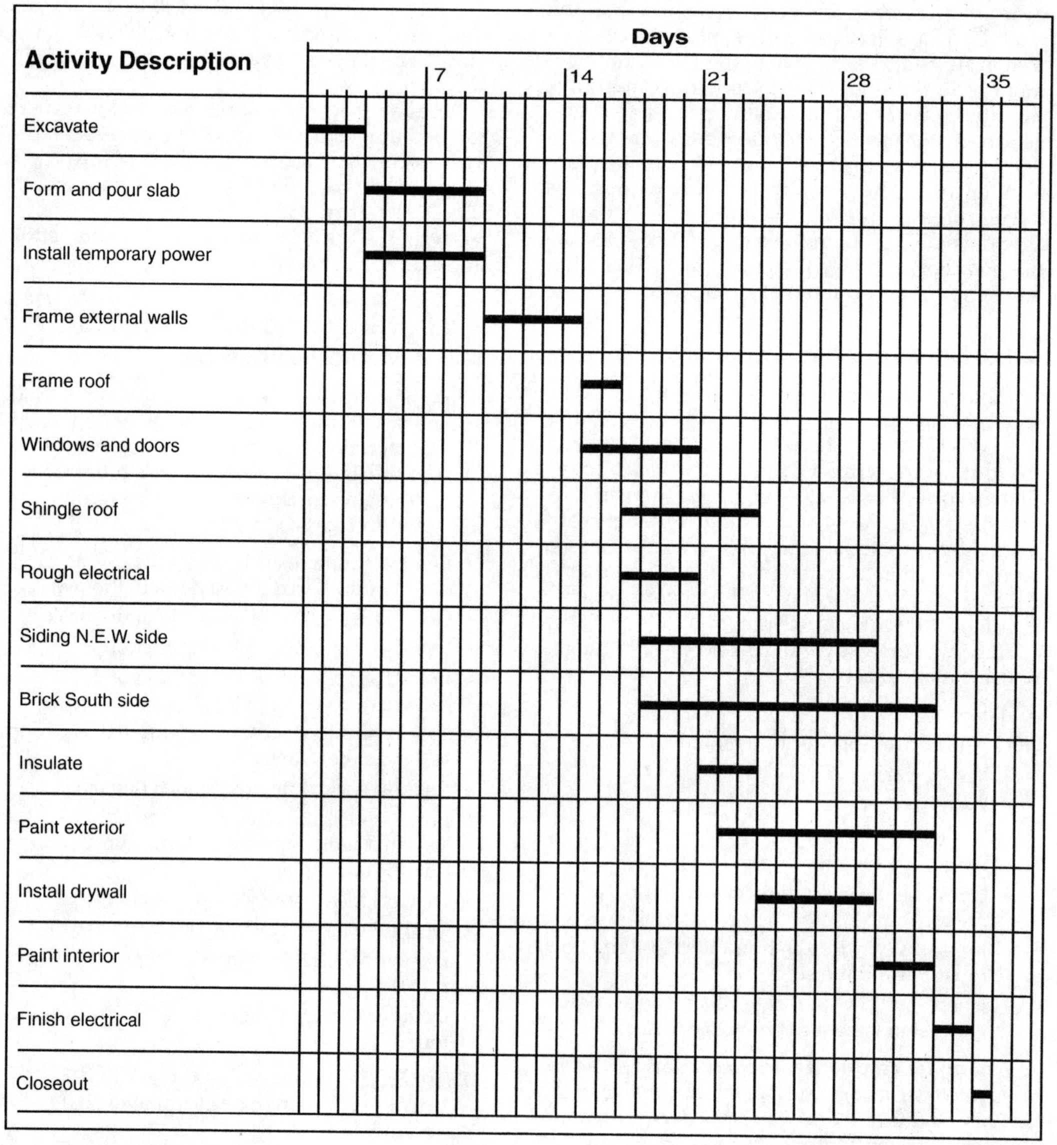

108F02.EPS

Figure 2 ◆ Bar chart check sheet.

- Potentially the most effective scheduling system
- Ideal for computer applications
- Analyze different methods or sequences of construction
- Useful for court cases; proves responsible party for delays
- Serves as a basis for systems control

There are also disadvantages to using CPM for tracking and documenting a project.

- Can become complex since it is the most sophisticated system
- Usually requires training to become an effective user
- Requires a great deal of commitment
- Is often neglected due to its complexity

Instructor's Notes:

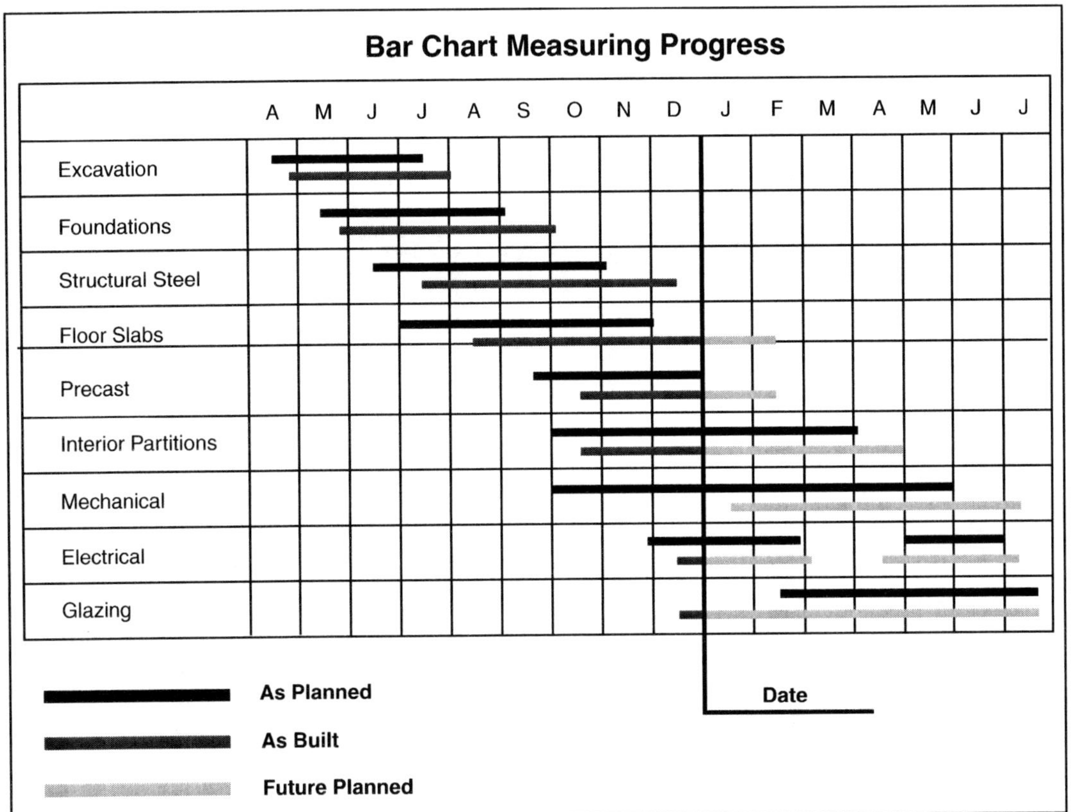

108F03.EPS

Figure 3 ◆ Bar chart measuring progress.

The wise project manager will select the best method that best supports project priorities.

4.3.1 *Activity Analysis: Divide the Project into Activities*

The activities in the work breakdown structure are arranged in a logical sequence from beginning to end in the form of a flowchart. Divide the project into activities that are easy to distinguish and plan for as project elements.

- Write a description of the system. For example:

Action Verb	**Object**	**Location**
Place, remove	foundation	Col.1 line A
Build	bearing walls	1st floor

- Keep contractual divisions separate.
- Keep trades separate.
- Consider separating crews if necessary.
- Keep field work separate from shop or office work.
- Determine level of detail needed—from precise and complex to a high-level view.
- Items commonly left off the CPM that may be helpful:
  - final bids from subcontractors, fabricators, OEMs, etc.
  - shop drawings review
  - lead time on delivery, especially items shipped from factories
  - curing time
- Use the work breakdown approach.

4.3.2 *Construct the Network Diagram*

Traditionally, two basic types of networking were practiced: Activity on arrow (AOA), sometimes called I-J format, and activity on node (AON), or precedence networks. Today's software packages generally support the **precedence diagram method** (**PDM**). PDM is defined by the Project Management Institute (PMI) as a schedule net-

Show Transparency 13, Activity Analysis. Review the fundamentals of activity analysis.

MODULE 44108-08 ◆ SCHEDULING 8.7

Define precedence and discuss the precedence diagramming method (PDM).

Show Transparency 14 (Figure 4).

Show Transparency 15 (Figure 5). Explain how the activities are linked by logical, sequential relationships.

Explain the AOA Network definitions.

Show Transparency 16 (Figure 6) and discuss how to construct the AOA activity node.

Show Transparency 17 (Figure 7) and explain an AOA dummy.

work diagramming technique in which schedule activities are represented by boxes (or nodes). Both Microsoft Project and Primavera are PDM-based scheduling systems.

Schedule activities are graphically linked by one or more logical relationships to show the sequence in which the activities are to be performed. The fundamental concept is that an activity cannot be started until the prior activity is completed (i.e., precedence). For example, a crew cannot clean brick until it has finished laying the brick. *Figure 4* illustrates a PDM representation for one mile of highway construction.

Schedule relationships that are used to construct network diagrams include:

FS—Finish to Start
FF—Finish to Finish
SS—Start to Start
SF—Start to Finish

Figure 5 defines and illustrates these relationships. They are discussed in more detail later in this module.

4.3.3 AOA Network Scheduling

While PDM is more frequently used, AOA and AON are good ways to illustrate the principles of developing network diagrams. The AOA, or arrow diagram, is the oldest of the two methods and the one which will be discussed first.

AOA Network definitions are as follows:

- *Project* – The sum of all the activities.
- *Activity* – Sub-unit of the project. Typically has a time value of greater than one day and an associated cost. The activity is represented by the arrow.
- *Node* – A circle that represents a point in time, either the beginning or ending of an activity.
  - The start node (I node) is the beginning of the activity and the end node (J node) is the end of the activity.
  - The length of the arrow is not necessarily equal to the duration of the activity; however, it could be time-scaled.

  Figure 6 shows an example of the activity node.
- *Dummy* – Zero time and zero cost. It is not an activity in and of itself, but rather it shows a relationship between activities. A dummy is generally drawn with a dotted line so that it is not confused with an actual activity, which is drawn with a solid line.

  Figure 7 shows an example of a dummy.

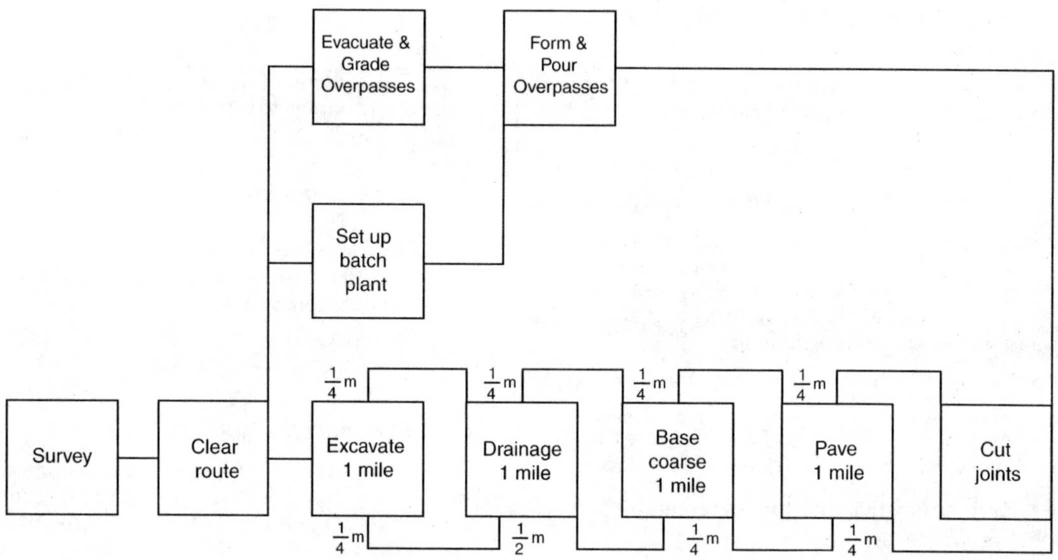

Figure 4 ◆ PDM network.

Instructor's Notes:

8.8 PROJECT MANAGEMENT

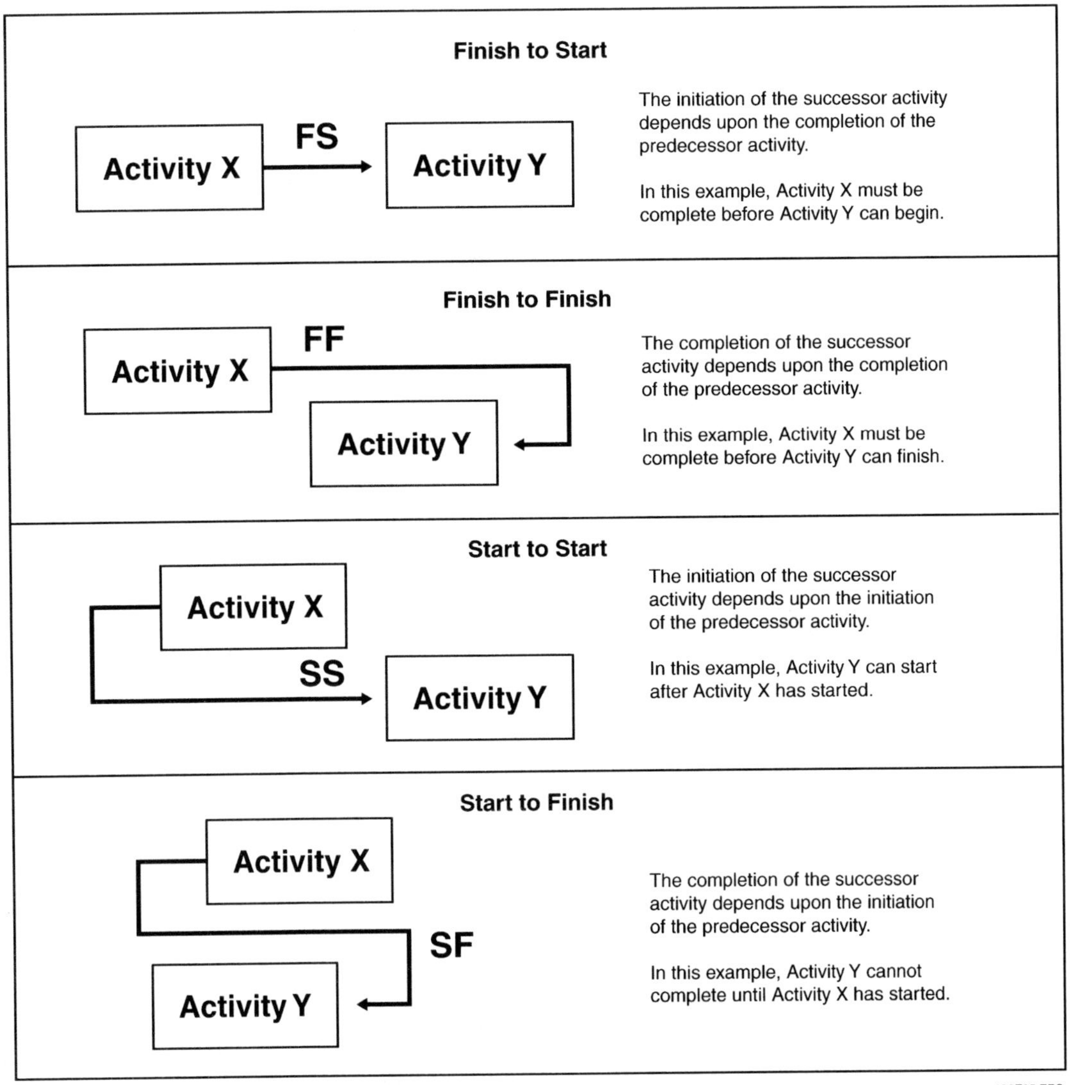

108F05.EPS

Figure 5 ◆ PDM network schedule relationships.

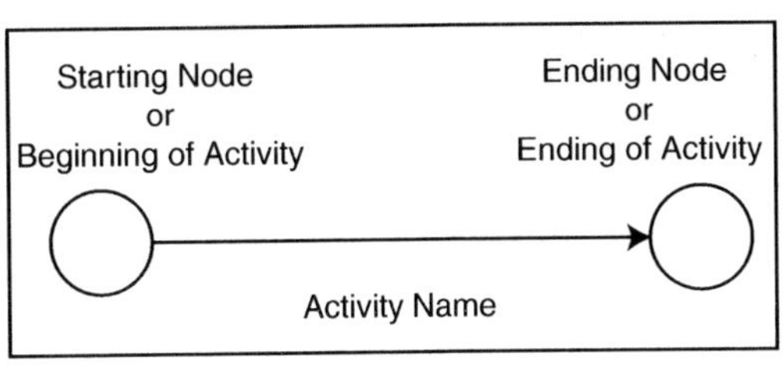

108F06.EPS

Figure 6 ◆ AOA activity node.

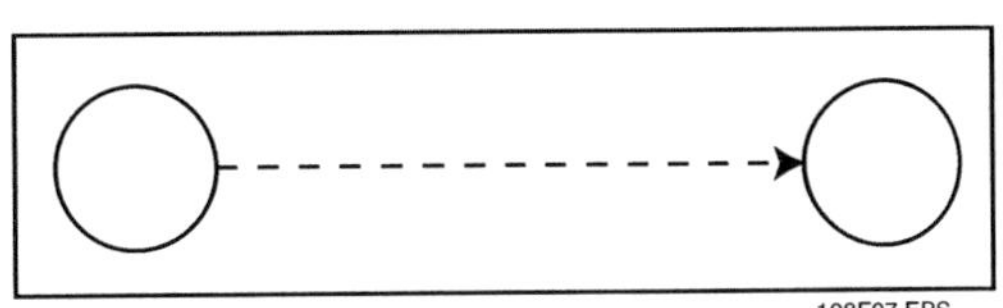

108F07.EPS

Figure 7 ◆ AOA dummy.

MODULE 44108-08 ◆ SCHEDULING 8.9

Show Transparency 18, Rules of Construction for AOA Diagrams.

Show Transparency 19 (Figure 8) and discuss the order of construction in a precedence diagram.

Show Transparency 20, Calculating Activity Duration. Use a job-related experience to explain how to determine the time required for each activity.

Define false logic relationships.

Review the questions that can be helpful tools to avoid false logic relationships.

Point out that the completion of an accurate logic diagram concludes the planning stage.

Explain that historical records provide the best source of determining durations.

Show Transparencies 21–24 (Figures 9–12). Demonstrate how to construct the AOA logic diagram. Ask the participants for input as you walk through the steps.

Show Transparency 25 (Figure 13) to illustrate a completed warehouse CPM schedule.

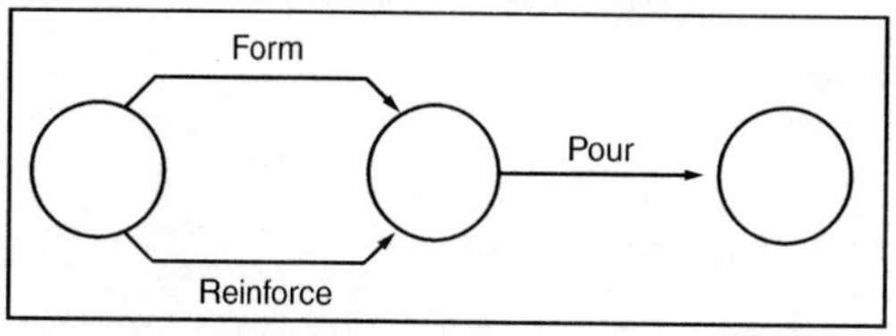

108F08.EPS

Figure 8 ◆ Predecessors, successors, and concurrent relationships.

The following are rules of construction for an AOA diagram:

- The network or project itself may have only one beginning node and one ending node.
- The logic states that all activities coming into a node must be complete before any activities going out can begin. For example, the diagram in *Figure 8* shows that both Form and Reinforce must be completed before Pour can begin.
- The logic must be exact regarding predecessors, successors, and concurrent relationships. Form and Reinforce are concurrent activities that are predecessors to Pour. Pour is a successor to Form and also a successor to Reinforce.
- There must be at least one continuous path from the beginning of the project to the end of the project without any dangling nodes.

A final note concerning the logic diagram: it is important that the logic network reflects the best planning possible for a project that is to be managed with minimum problems. Later problems are avoided by a good plan that is reflected in the logic diagram. With AOA especially, it is easy to have false logic in the network. To help avoid false logic relationships, ask the following questions of each activity in the project:

- Does this activity really have to be completed before each of the following activities can begin? Why are you doing the work?
- What other activities should be completed before this activity can begin? What does it take to start?

Once the logic diagram is completed and checked for accuracy, the planning stage is completed and it is time to start the scheduling phase: determining durations and performing the calculations for the start and finish dates of each activity.

The best method to determine durations is from your own company records. Use the estimate to get the quantities and then divide those quantities by your productivity rates. You may want to multiply that answer by a modifier to correct for logistics (material/equipment staging, mobilization/demobilization on task, etc.), weather problems, heights, or other abnormal work conditions. If you don't have your own company productivity rates, there are commercially available estimating cost guides. However, they don't take the place of historical data. Also consider sources of information such as equipment or material suppliers and subcontractors. Remember to consider workforce availability, equipment selections, special owner requirements, shop drawing approval requirements, availability of materials, shipping times, and so forth (many of which can be schedule activities in their own right).

The following CPM AOA examples (*Figures 9-12*) illustrate how to do the calculations. The first step is to add the durations to the logic diagram and number the nodes. The only rule to remember when numbering the nodes is that the beginning node of each activity, including dummies, must always be smaller than the ending node for that same activity. The purpose of numbering the nodes is for the computer to keep track of the activities. If you were doing the schedule entirely by hand, numbering of the nodes would be optional.

Figure 13 illustrates a completed warehouse CPM AOA schedule.

4.3.4 AON Network Scheduling

Activity on node (AON), or precedence scheduling, is very similar to AOA (arrow diagramming). The major difference is that now the node represents the activity rather than the arrow. The arrows are to show precedence or relationships only. There can be as many arrows leaving or entering an activity as necessary; therefore, there is no need for dummies and split nodes. A simple example of an AON project is shown in *Figure 14*.

There may still be only one beginning node and one ending node to the total project. That is the reason for the "begin" activity which may have a zero duration. This AON logic diagram states that Form and Reinforce are predecessors to Pour. Pour is a successor to Form and also a successor to Reinforce. Form and Reinforce are concurrent activities. *Figures 15* through *18* contain another example of an AON logic diagram where you learn how to perform the calculations and identify the critical activities.

Instructor's Notes:

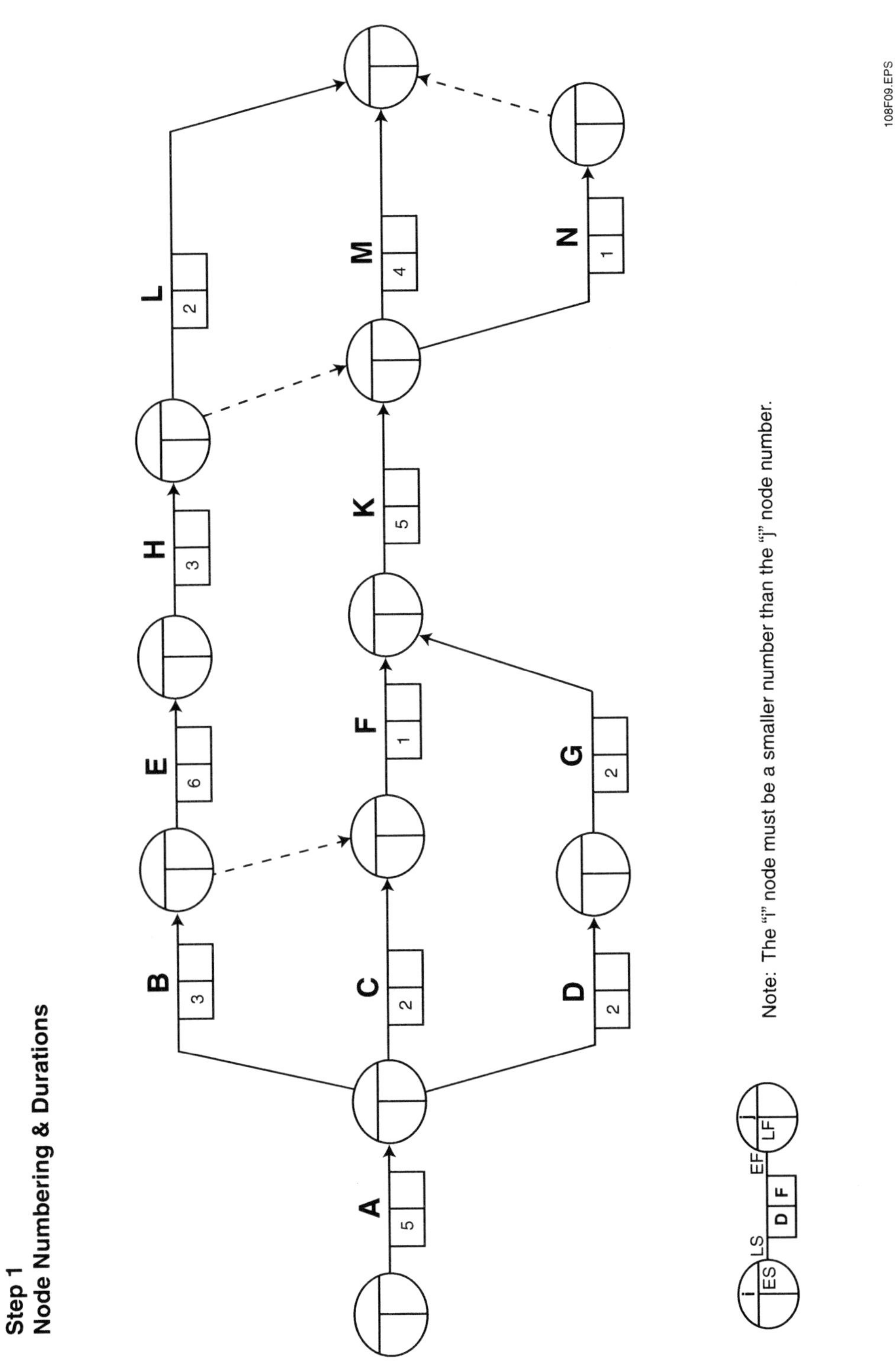

Figure 9 ◆ CPM-AOA example: node numbering and durations.

See the Teaching Tip for Sections 4.3.0–4.3.3 at the end of the module.

Show Transparency 26 (Figure 14). Explain the difference between AOA and AON diagramming.

Show Transparencies 27-30 (Figures 15–18). Briefly review how to construct an AON network diagram.

MODULE 44108-08 ◆ SCHEDULING 8.11

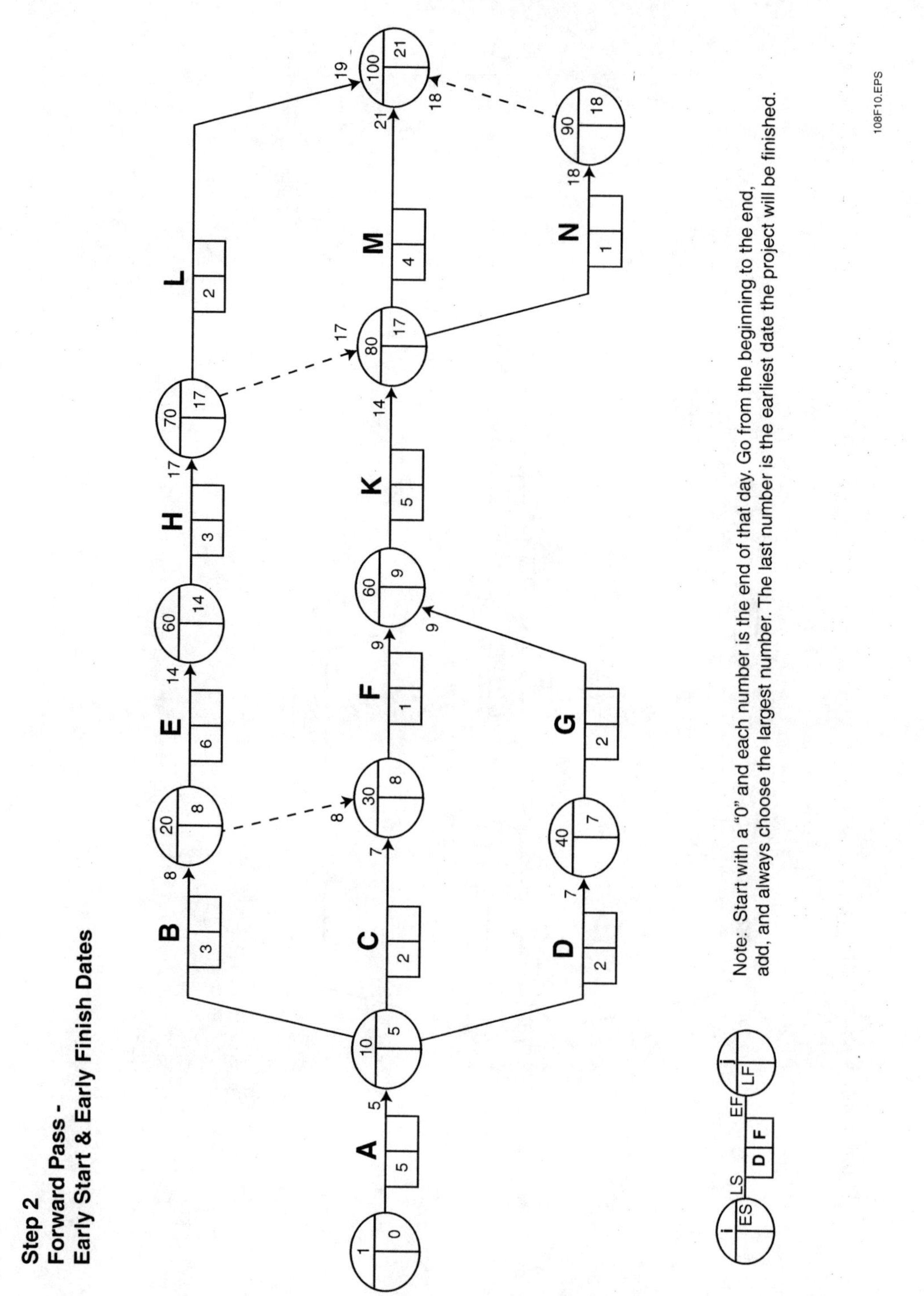

Figure 10 ◆ CPM-AOA example: forward pass – early start and early finish dates.

Instructor's Notes:

8.12 PROJECT MANAGEMENT

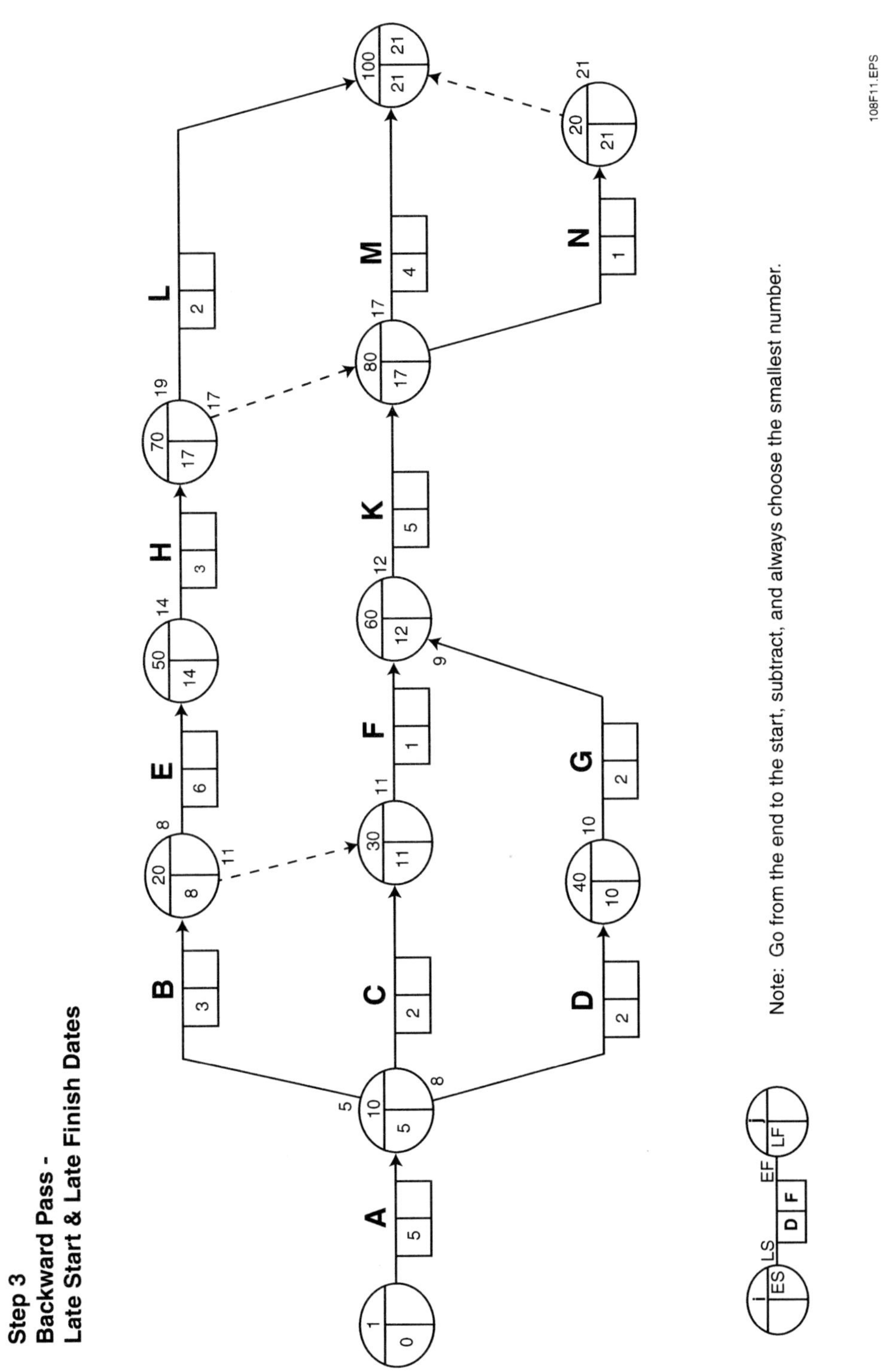

Figure 11 ◆ CPM-AOA example: backward pass – late start and late finish dates.

MODULE 44108-08 ◆ SCHEDULING 8.13

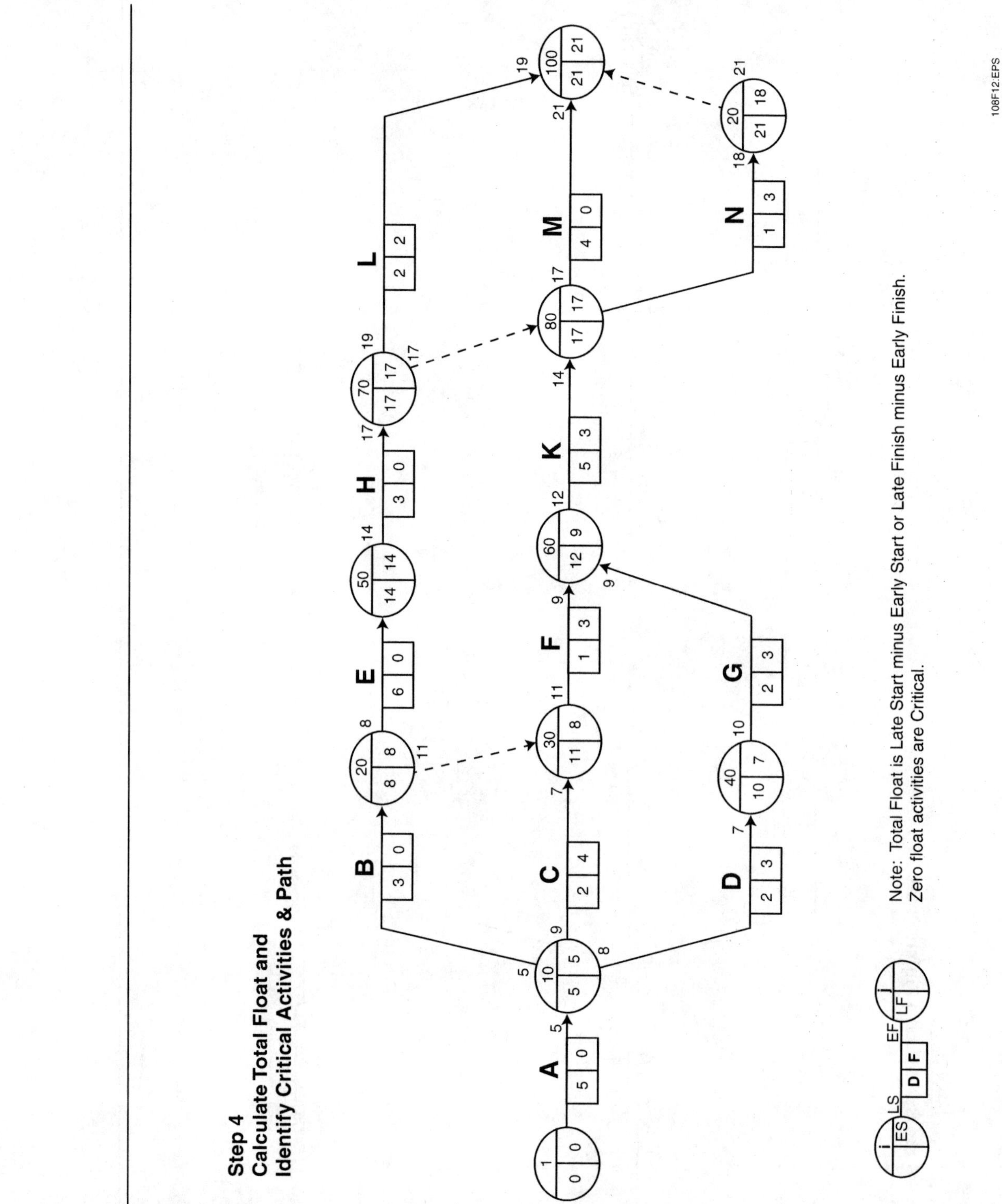

Figure 12 ◆ CPM-AOA example: calculate total float and identify critical float activities and path.

Instructor's Notes:

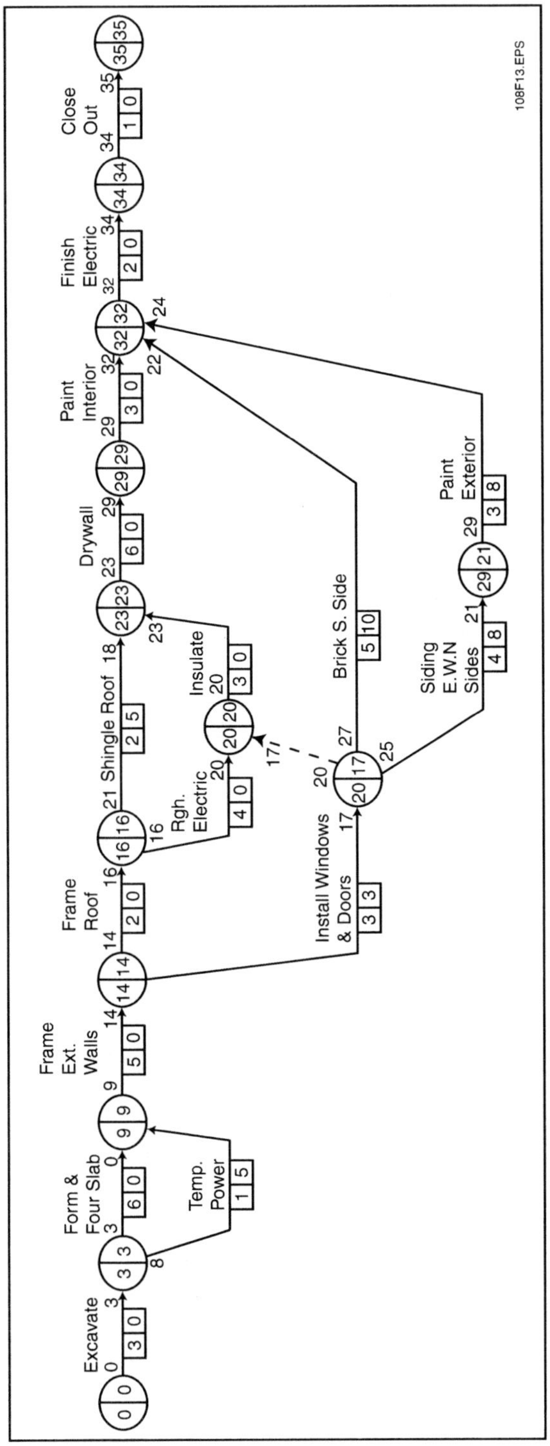

Figure 13 ◆ Warehouse CPM, AOA schedule.

MODULE 44108-08 ◆ SCHEDULING 8.15

Explain that a SIPS is a two- to four-week look-ahead generally prepared by the foreman or supervisor.

Discuss the value of short-term interval scheduling.

Introduce the topic of lean construction.

Explain that lean scheduling is becoming more common.

Discuss the process and advantages of lean scheduling.

Ask the participants if they have experience with lean construction and scheduling.

Discuss how Samantha could have used SIPS to resolve her scheduling crisis.

Discuss how lean construction principles could have improved the jobsite problems.

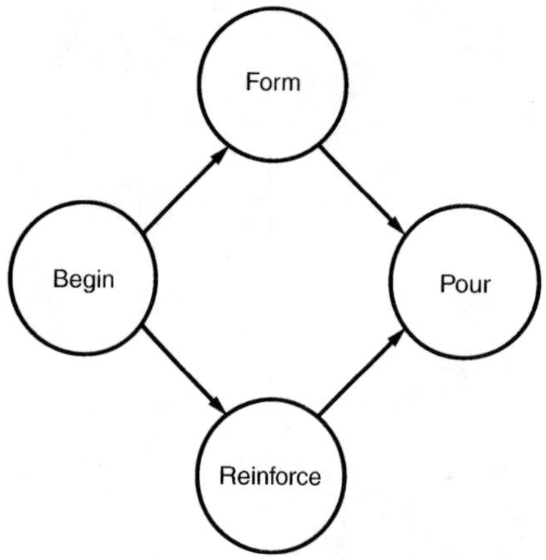

108F14.EPS

Figure 14 ◆ AON network scheduling.

4.4.0 Short-Interval Scheduling

A short-interval schedule, sometimes referred to as a short-interval production schedule (SIPS), short-term schedule, or short-term construction activity plan (CAP), is a two- to four-week look-ahead schedule. It is often called the foreman's or supervisor's schedule and is one of the most important control techniques available. The SIPS enables a contractor to react to changes with agility.

Short-interval schedules may look ahead for a maximum of three months to allow time for coordinating material acquisitions and deliveries and to meet overall project schedules. The shorter two- to four-week schedule is used to plan the use of resources and provide yardsticks to management for monitoring production. According to a study by the National Electrical Contractors Association (NECA), the ideal labor allocation and job-site inventory is three days.

The SIPS is a detailed schedule where time, manpower, material, and equipment requirements are carefully considered and other potential problems are carefully analyzed. Short-interval schedules always consider the relationship between production and cost. The schedule is derived from the estimate and the WBS to create a day-to-day schedule of events. For example, assume that a 12-inch masonry block wall has been estimated at 4,250 sq. ft. Historical records show that a daily production rate of 280 sq. ft. per crew. The short-interval schedule should include using 15 crew-days for this activity (4,250 s.f. ÷ 280 s.f. per day = 15.2 crew-days). If 15 crew-days exceed the schedule, it may be necessary to add another crew, if available.

The short-interval schedule makes it easier to compare actual production with estimated production. If actual production begins to slip behind the estimate, the problem can be identified and corrective action can be taken before it gets out of hand. With a short-interval schedule, it is easy to identify root causes for deviations and factors that affect the pace of work. Short-interval schedules are a valuable aid to continuous improvement.

4.5.0 Lean Construction

Practices in the construction industry are evolving much as they did in manufacturing in the last two decades. The construction industry is adapting models that have been proven and successfully implemented in manufacturing. One example is **lean construction**, which includes an alternative method of scheduling and is gaining acceptance in the industry.

Lean construction adapts the principles developed and implemented in lean manufacturing to focus on a new way of thinking about construction projects and project management. It requires the entire team to work to the project goals and to monitor and realign the project in order to minimize waste and produce a reliable flow of work. The project is viewed as an entire production system in contrast to the traditional activity-based or contract-based approach. Objectives are defined in customer terms, decision making is decentralized, and activities, labor, and materials are coordinated as a just-in-time model. The entire construction process is mapped as a workflow, looking first at the requirements of the completed project, then working backwards to identify each preceding step. The emphasis is always on the value being brought to the customer.

Throughout the implementation process, the project teams review progress together (for example, holding weekly team meetings), ascertaining if all tasks were completed as scheduled, and assessing exceptions and taking action on root causes for those not completed. This approach enables the team to make real time adjustments to the schedule and workflow and leads to continuous improvement to prevent future problems.

Gregory Howell, co-founder and executive director of the Lean Construction Institute (LCI, www.leanconstruction.org), says, "In a typical construction project, all the research shows workers are standing idle 20 to 30 percent of the time." He goes on to state, "At the same time, huge piles of materials often sit at construction sites untouched for weeks because they cannot be

Instructor's Notes:

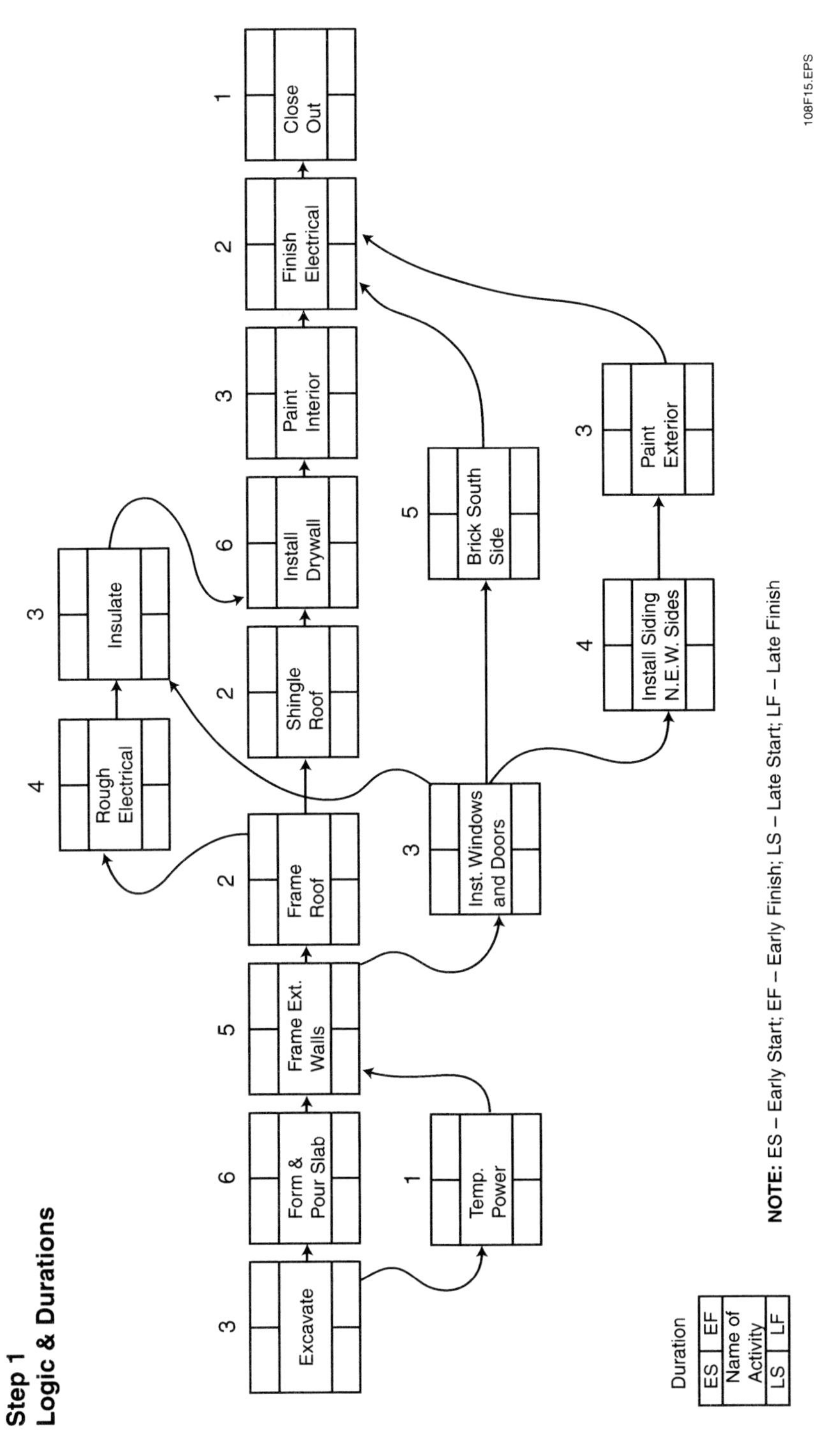

Figure 15 ◆ Warehouse CPM schedule, logic, and durations.

MODULE 44108-08 ◆ SCHEDULING 8.17

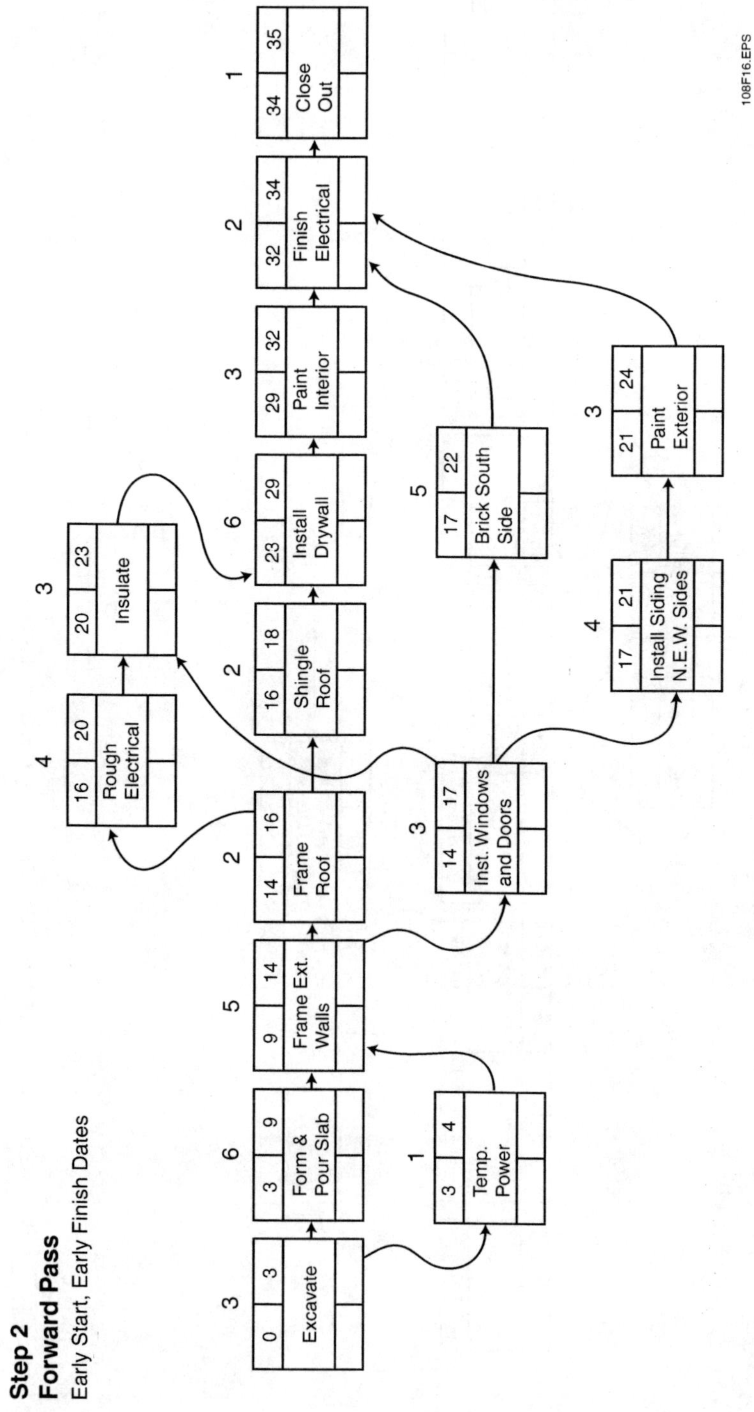

Figure 16 ◆ Warehouse CPM schedule, forward pass (ES-EF).

Instructor's Notes:

8.18 PROJECT MANAGEMENT

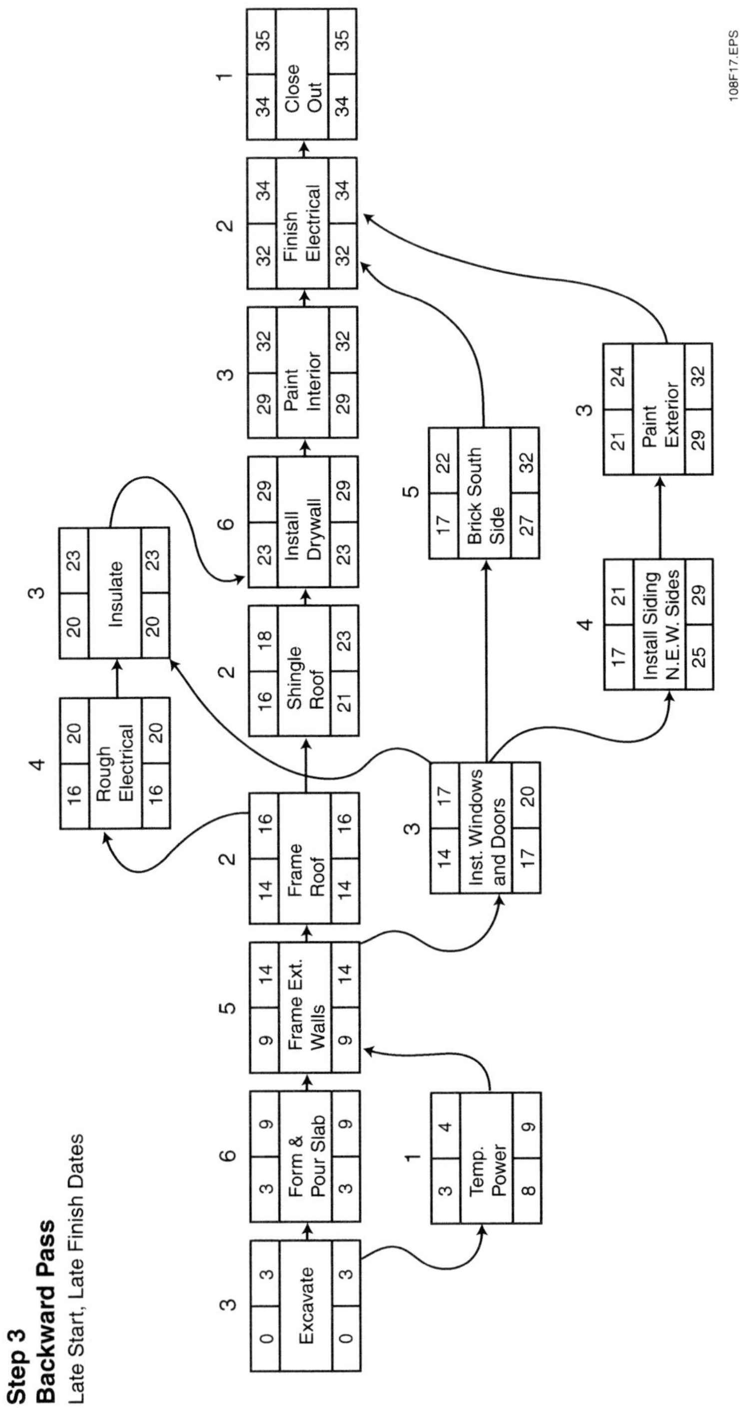

Figure 17 ◆ Warehouse CPM schedule, backward pass (LS-LF).

MODULE 44108-08 ◆ SCHEDULING 8.19

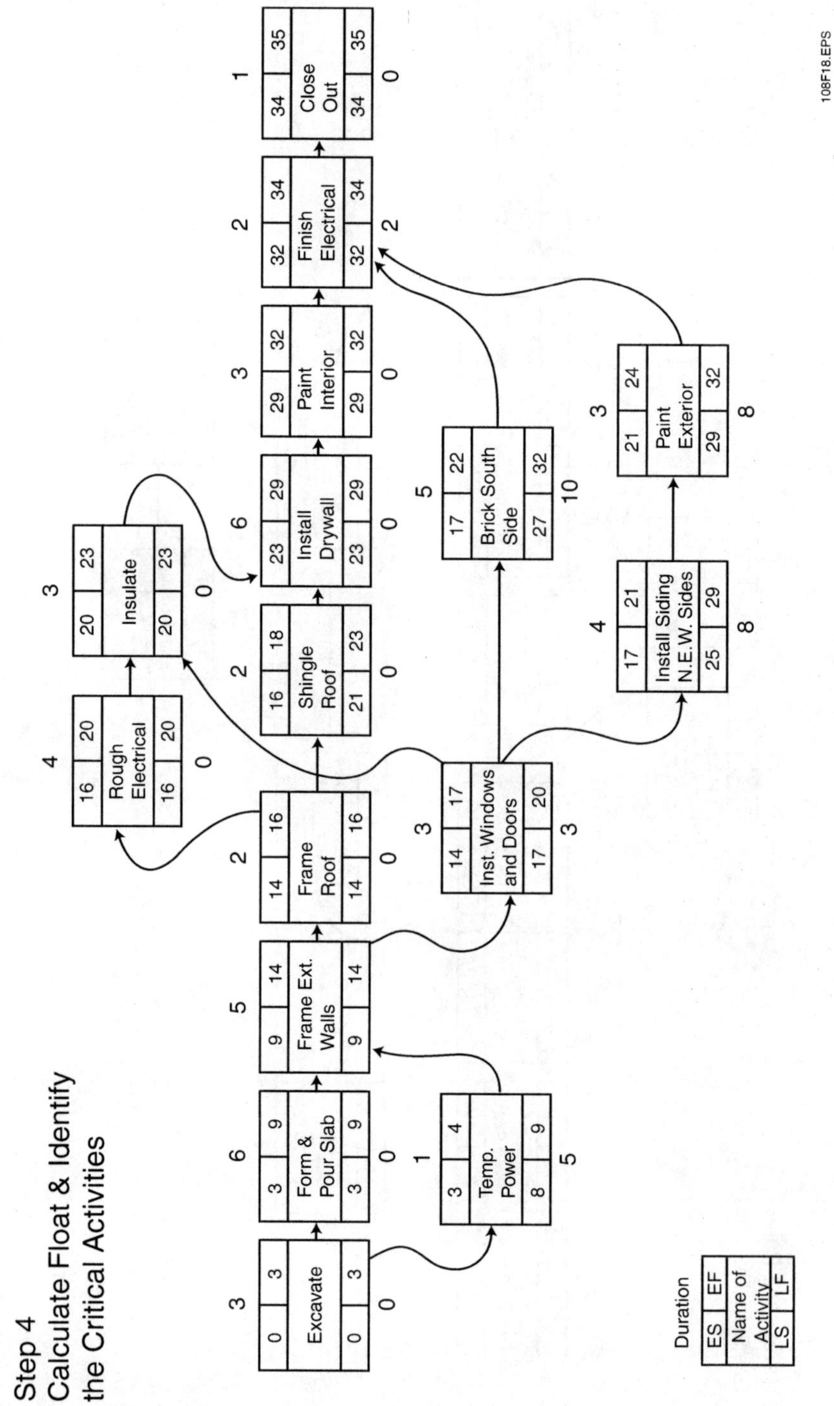

Figure 18 ◆ Warehouse CPM schedule, calculate float and identify the critical activities.

Instructor's Notes:

8.20 PROJECT MANAGEMENT

installed until certain parts of the job are completed—and they arrive before that happens." Chuck Greco, CEO of Linbeck Construction in Houston, Texas, indicates, "on the typical project, the rate of completion, meaning parts of the job being finished by the time the contractor said they would be finished, is usually only around 40 percent."

Implementing a new methodology requires critical change in an organization. Companies that embrace significantly different practices, such as lean construction with its systems thinking approach, must build learning organizations and create a culture and an environment that enable their people to continually expand their knowledge and work as teams.

5.0.0 ◆ FLOAT TYPES

Because construction schedules do not usually run like clockwork, you can expect to build some float into the schedule. Float is the amount of time that a project task can be delayed without causing a delay to either subsequent tasks or the project completion date.

- *Total float* is the amount of time before you start affecting the project. It is positive, zero, or negative.
- *Free float* is a measure to an affect on a project until you start affecting its successor activities. It does not affect early start of successor activities.
- *Independent float* does not affect early start or later activities. It cannot be used by other activities because the activity owns the float.
- *Path, string,* or *shared float* belongs to or is shared by a path or string of activities.
- *Negative float* is a type of total float seen when the backward pass is done with an early, forced late finish date.

Figures 19 and 20 illustrate float types.

6.0.0 ◆ LAGS

Lags are used when you want activities to overlap without breaking the activity into parts or percentages. A lag is a modification of a logical relationship that directs a delay in a successor activity.

Lags applied to finish-to-finish (FF) work well if the completion of the latter activity depends upon the completion of the prior activity in the event the preceding activity is delayed. Without lags, the schedule in *Figure 21* would take 21 days to complete. With lags it is completed in 17 days.

Some schedulers prefer to use lags on both start-to-start (SS) and finish-to-finish (FF) lag in order to properly tie the activities together. Another popular application of a lag is a finish-to-start (FS), where a time period is required between the end of the first activity and before the second activity can start. For example, a schedule may place an FS lag between placing concrete and before stripping or loading, to allow the concrete to cure.

Another technique many schedulers use is to not use lags at all, but rather to break the activities into greater detail, as shown in *Figure 21.* This gives more direction to the responsible subcontractors or craftworkers, so they know specifically when and where they are expected to do their work. This level of detail would be typical of a short-interval production schedule.

7.0.0 ◆ USING THE SCHEDULE

You will use the schedule in a number of ways, including:

- Balancing resources
- Determining the effects of a change order
- Enforcing a schedule with subcontractors
- Documenting problems and progress

7.1.0 Using the Schedule to Balance Resources

The schedule can be used to balance resources, including labor, equipment, money, tools, etc. It can also be used to forecast cash requirements. *Figures 22-24* provide examples of cash and equipment leveling and a cash flow forecast. The data in *Table 1* is used to calculate these charts.

This project is a large excavation project and XYZ Company has just been awarded the contract. The project manager needs to know if the company owns enough D-8 Cat bulldozers to do the job. Another need is to forecast the cash requirements for the contract.

7.2.0 Determining the Effects of a Change Order

Effectively managing change orders can keep a project on schedule. The schedule is updated at regular intervals and the effects of potential changes are analyzed in order to properly price and determine the effects of the change. If there is additional time needed that would affect the project completion date, this must be calculated immediately. Impacts are not just about time.

MODULE 44108-08 ◆ SCHEDULING 8.21

Assign reading of Sections 5.0.0-6.0.0

Show Transparency 31, Float Types. Define float and explain the various types of float.

Show Transparencies 32 and 33 (Figures 19 and 20) to illustrate how to calculate float.

Show Transparency 34 (Figure 21) and define lags. Explain that the bottom of the figure shows the activities broken into greater detail.

Assign reading of Sections 7.0.0-7.4.0.

Show Transparency 35, Use of the Schedule.

Show Transparencies 36–39 (Table 1 and Figures 22–24). Walk through the example with participants.

Explain how the schedule is used to balance resources.

Discuss how to manage change orders.

Ask participants to describe their experiences with change orders and resulting effects on the project cost and schedule.

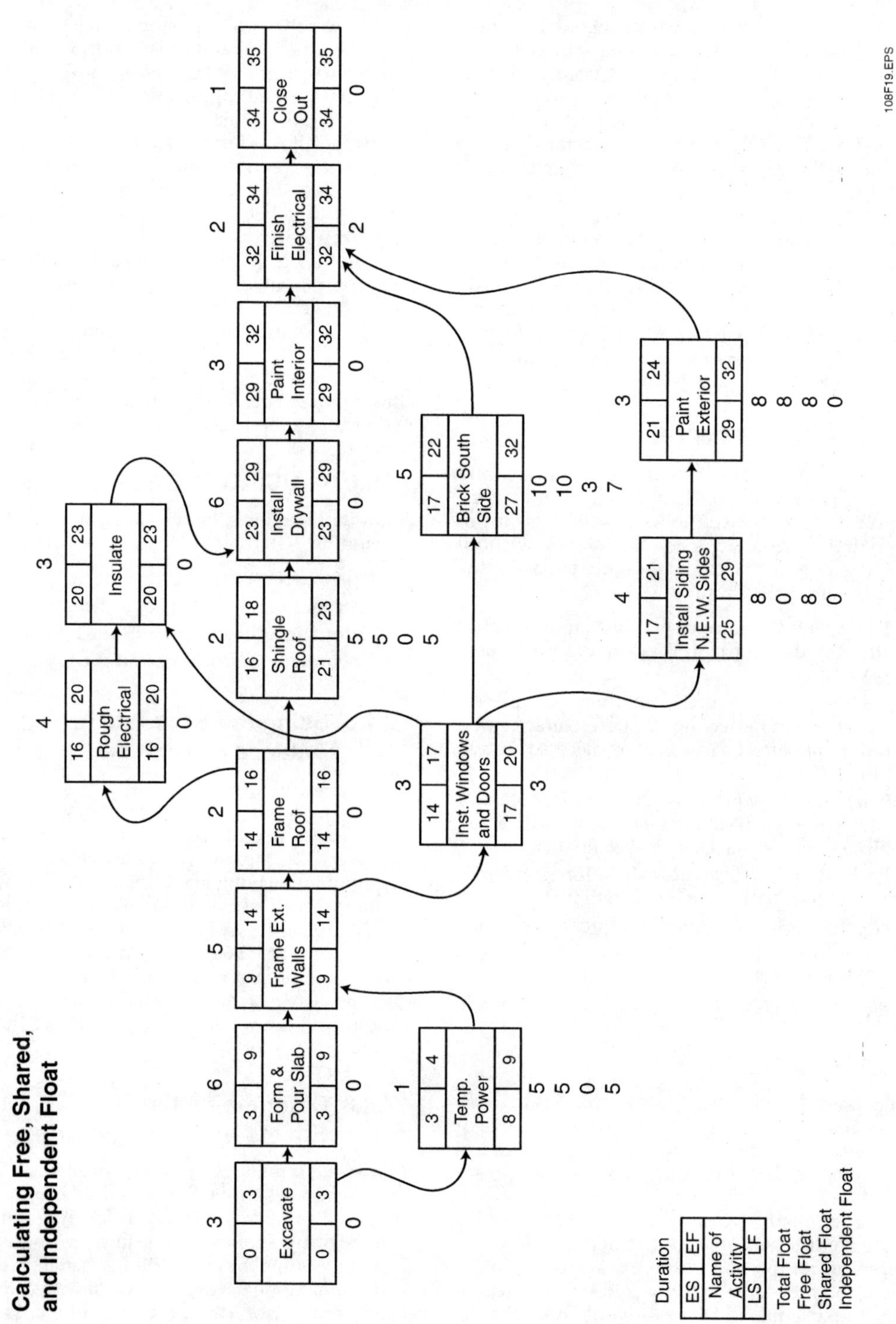

Figure 19 ◆ Warehouse CPM schedule, calculating free, shared, and independent float.

Instructor's Notes:

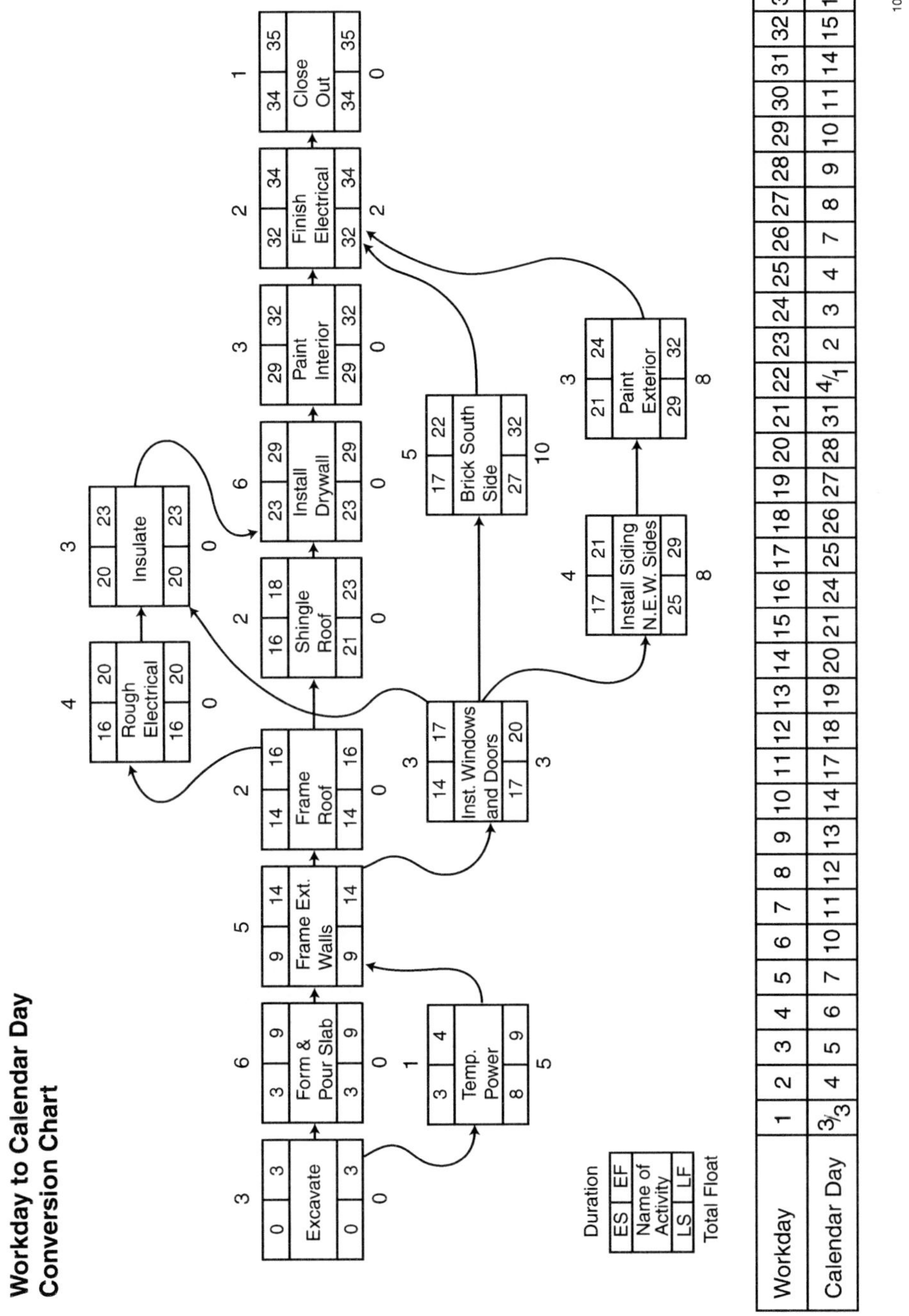

Workday	1	2	3	4	5	6	7	8	9	10	11	12	13	14	15	16	17	18	19	20	21	22	23	24	25	26	27	28	29	30	31	32	33	34	35
Calendar Day	3/3	4	5	6	7	10	11	12	13	14	17	18	19	20	21	24	25	26	27	28	31	4/1	2	3	4	7	8	9	10	11	14	15	16	17	18

Figure 20 ◆ Warehouse CPM schedule, workday to calendar day conversion chart.

MODULE 44108-08 ◆ SCHEDULING 8.23

Show Transparency 40, Enforcing a Schedule. Discuss the steps involved in enforcing a schedule with subcontractors.

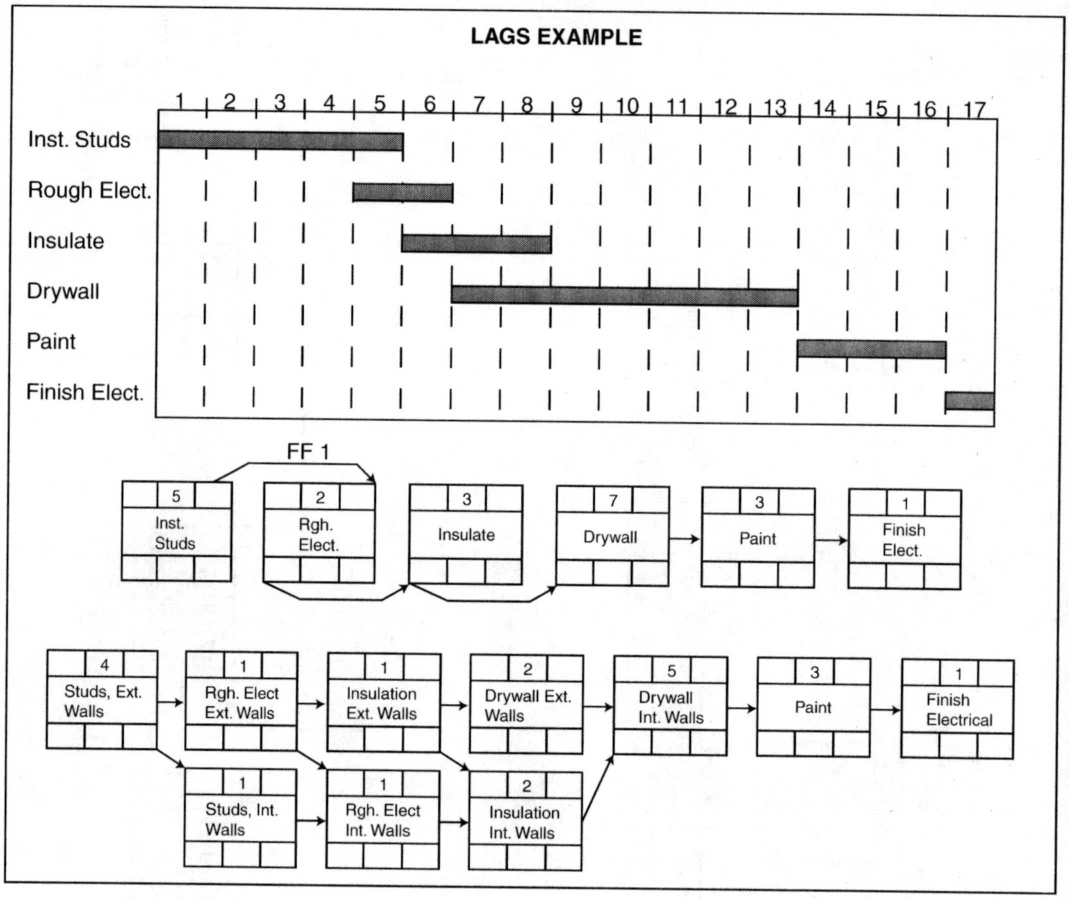

Figure 21 ◆ Lags example.

108F21.EPS

Change orders can affect other aspects of the schedule such as workforce diversity, availability and so forth. Look at the scope of the change order on the project when you are evaluating its implication.

7.3.0 Enforcing a Schedule

The health of a project depends on sticking to a schedule. It is up to you to enforce the schedule with your subcontractors. Use the following tips to keep your schedule on track:

Step 1 Hold a preliminary scheduling meeting, or pre-construction conference, with all of the subcontractors and suppliers involved. This will allow coordination between subcontractors and suppliers and eliminate potential bottlenecks.

Table 1 Cash and Equipment Loading Data

Activity	Duration (Months)	No. of D-8 Cats per Activity	Cost Estimate ($ 000s)	Cost per Month ($ 000s)
B	3	2	90	30
D	3	2	120	40
K	5	5	300	60
H	4	2	120	30
F	7	2	175	25
A	4	2	80	20
E	6	6	540	90
G	2	2	100	50
J	3	2	120	40
C	5	2	150	30

108T01.EPS

Instructor's Notes:

8.24 PROJECT MANAGEMENT

Cash Leveling

Activity	1	2	3	4	5	6	7	8	9	10	11	12	13	14	15	16	17	18	19	20	21	22	23	24
B	30	30	30																					
D				40	40	40																		
H				30	30	30	30																	
K				60	60	60	60	60																
E								90	90	90	90	90	90											
A									20	20	20	20												
F									25	25	25	25	25	25	25									
J														40	40	40								
G																50	50							
C																		30	30	30	30	30		
Total Cost/Month	30	30	30	130	130	130	90	150	135	135	135	135	115	65	65	90	50	30	30	30	30	30		

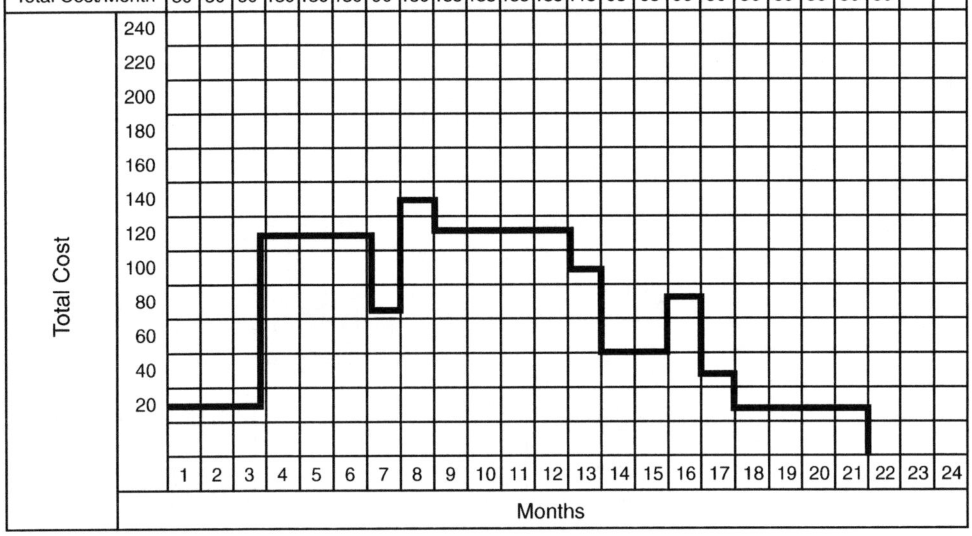

Since Activity E has one day of float, it could start on the late start day and thereby reduce the maximum cash flow from $150 to $135 per day

108F22.EPS

Figure 22 ◆ Example of cash leveling.

MODULE 44108-08 ◆ SCHEDULING 8.25

Equipment Leveling

Activity	1	2	3	4	5	6	7	8	9	10	11	12	13	14	15	16	17	18	19	20	21	22	23	24
B	2	2	2																					
D				2	2	2																		
H				2	2	2	2																	
K				5	5	5	5	5																
E								6	6	6	6	6	6											
A									2	2	2	2												
F									2	2	2	2	2	2	2									
J														2	2	2								
G																2	2							
C																		2	2	2	2	2		
Total Cost/Month	2	2	2	9	9	9	7	11	10	10	10	10	18	4	4	4	2	2	2	2	2	2		

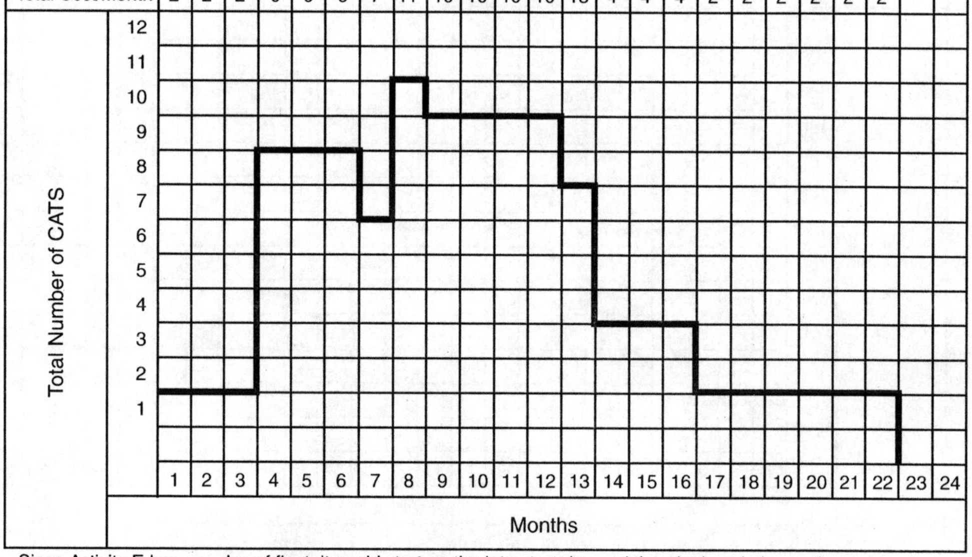

Since Activity E has one day of float, it could start on the late start day and thereby knock the peak number of men/day down to 10.

108F23.EPS

Figure 23 ◆ Example of equipment leveling.

8.26 PROJECT MANAGEMENT

Instructor's Notes:

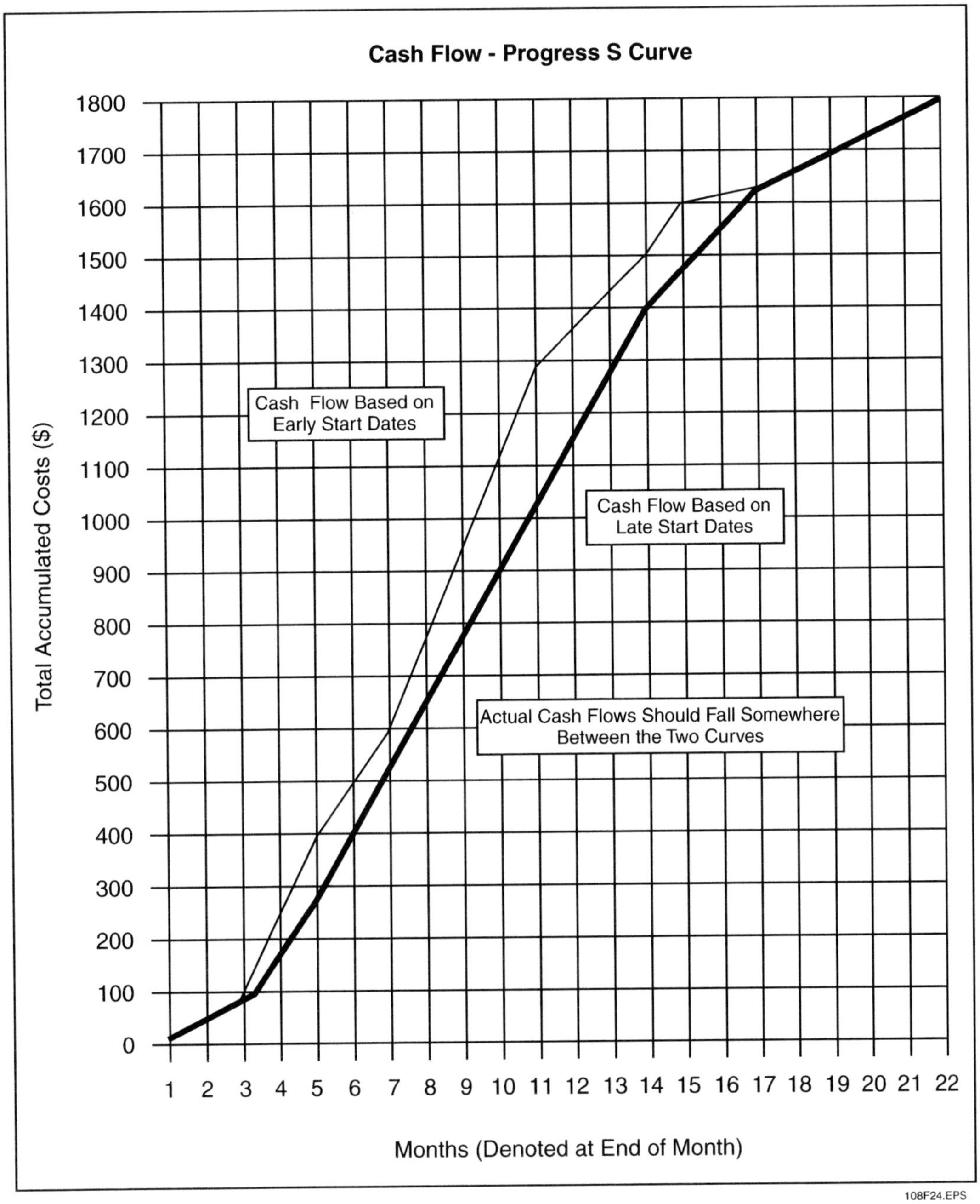

Figure 24 ◆ Cash flow – progress S curve.

MODULE 44108-08 ◆ SCHEDULING 8.27

Show Transparency 41, Documenting Problems and Progress. Emphasize the need to keep copies of all schedules and updates.

Step 2 Don't schedule two subcontractors on the same job, in the same place, and at the same time.

Step 3 Give ample notice and have the job on schedule for the subcontractor.

Step 4 Email or fax reminders to contractors and subcontractors about dates when they are scheduled for a particular job.

Step 5 Encourage cooperation and improvement between subcontractors. If they have a better idea about how to perform the job that will reduce the time or cost of the job, accept it willingly.

Step 6 Base payment on compliance with the schedule. If the subcontractor is on schedule, pay him on time; conversely, if the subcontractor delays the schedule, it may be appropriate to delay his payment. Be aware of and careful with the terms of the contract and especially progress reports.

Step 7 Whenever possible, reward superior performance. If the subcontractor is responsible for substantial savings as a result of excellent performance, reward that effort with a bonus and a letter of commendation.

Step 8 Treat subcontractors as if they were your own employees because, in a sense, they are.

Step 9 Tie the schedule to the contract.

7.4.0 Using the Schedule to Document Problems and Progress

Keep copies of all schedules including as planned, in addition to all updates. Make sure the schedule agrees with the daily log. A complete list of items to track or maintain might include the following:

- The preliminary (or 30- or 60-day) schedule if this was required or was prepared
- Records of input provided by consultants, subcontractors, or other prime contractors relating to the duration of activities or the sequencing thereof
- Copies of the initial schedules submitted to the owner or subcontractors for review or comment
- Any hand-drawn schedules utilized by personnel on an informal basis in the field to explain how a particular portion of the work is to be constructed or within what time frames or sequence of events
- Any marked-up schedules used in the field in the updating process
- Approved schedule updates
- Copies of reports of any CPM or other scheduling consultant on the project
- Any supplementary schedules used on the project, such as pour schedules, finish schedules or detailed daily or weekly schedules
- Any in-house schedules prepared for the use of trades, subcontractors, or suppliers

Instructor's Notes:

8.28 PROJECT MANAGEMENT

Review Questions

1. Give three reasons for using formal schedules:
   1. ______
   2. ______
   3. ______

2. List the four elements of an effective scheduling system:
   1. ______
   2. ______
   3. ______
   4. ______

3. List three types of scheduling systems:
   1. ______
   2. ______
   3. ______

4. For the project activity bar chart in *Figure 1*, fill in the blanks with the corresponding activities based on the following information: Excavation (10 days) must be 50 percent completed before foundations (15 days) can begin, while structural steel (20 days) cannot commence until foundations are at least 50 percent complete.

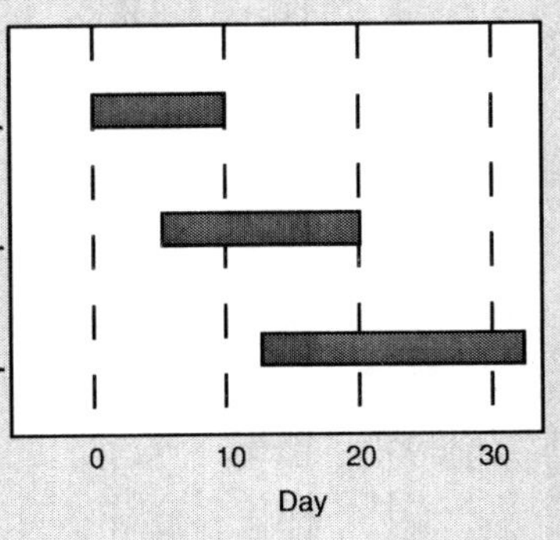

Figure 1

Use *Figure 2* to answer Questions 5-8.

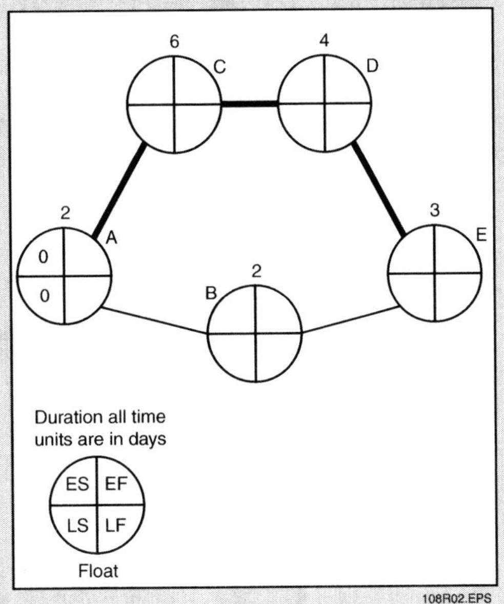

Figure 2

5. Outline the activities on the critical path diagram in *Figure 2*.

6. Calculate the float for Activity B in *Figure 2*. ______

7. Calculate the earliest start for Activity D in *Figure 2*. ______

8. Calculate the latest finish for Activity E in *Figure 2*. ______

9. Tying a schedule to a contract is ______ a schedule.

10. Describe a short-interval production schedule (SIPS).

______

______

Have participants complete the Review Questions, and go over the answers prior to administering the Module Examination.

Show Transparencies 42 and 43 to provide the answers to Review Questions 4–8.

MODULE 44108-08 ◆ SCHEDULING 8.29

Summarize the major concepts presented in this module.

Summary

This module discussed the importance of scheduling projects and resources. You learned the types of scheduling systems that are used in the industry, lean construction methods, and how to construct and interpret schedule diagrams.

Notes

Instructor's Notes:

8.30 PROJECT MANAGEMENT

Trade Terms Quiz

1. When using the ______________, schedule activities are graphically linked by one or more logical relationships to show the sequence in which the activities are to be performed.
2. The ______________ is a technique for planning the most efficient way to achieve a given objective by determining the activities and events required and showing how they relate to each other in time.
3. The distances to the previous elements of the sequence that are used to generate the next element are called ______________.
4. Planning construction in such a way that leads to the continuous process of eliminating waste, meeting or exceeding all customer requirements, focusing on the entire value stream and the pursuit of perfection in the execution of a constructed project is called ______________.

Trade Terms List

Critical path method (CPM)
Lags
Lean scheduling
Precedence diagramming method (PDM)

Have participants complete the Trade Terms Quiz, and go over the answers.

Administer the Module Examination. Record the results on Craft Training Report Form 200, and submit the results to the Training Program Sponsor.

Trade Terms Introduced in This Module

Critical path method (CPM): A technique for planning the most efficient way to achieve a given objective by determining the activities and events required and showing how they relate to each other in time.

Lags: Activities in schedules having successors that are scheduled later than activities that have no successors.

Lean scheduling: Planning construction in such a way that leads to the continuous process of eliminating waste and promoting efficiency through techniques that include team-based practices, efficient scheduling of resources (including workforce and just-in-time materials inventory and deliverables), and adjusting schedules in real time, with the goal of meeting or exceeding all customer requirements.

Precedence diagramming method (PDM): A schedule network diagramming technique in which schedule activities are represented by boxes; schedule activities are graphically linked by one or more logical relationships to show the sequence in which the activities are to be performed.

Instructor's Notes:

Resources & Acknowledgments

References

Blair, Dr. Gerard, "Starting to Manage: The Essential Skills," published by The Institute of Electrical Engineers (IEEE).

Ellwood, J. Mark, President, Pace Productivity, Inc., Toronto, Canada.

Howell, Greg and Glenn Ballard, Lean Construction Institute, Louisville, Colorado, www.leanconstruction.org.

International Group for Lean Construction, www.iglc.net.

VIP Quality Software, Odessa, Ukraine, www.vip-qualitysoft.com/timemanagement.

Womack, James P. and Daniel T. Jones, *Lean Thinking: Banish Waste and Create Wealth in Your Corporation*. New York: Simon & Schuster, 1996.

Acknowledgment

Jacobs Engineering Group, Inc.

MODULE 44108-08 ◆ SCHEDULING 8.33

MODULE 44108-08 — TEACHING TIPS

The following are suggested activities or instructional methods to help you teach the material in this module

General

When you call on someone to answer a question, the rest of the class relaxes or eve tunes out because they expect that the question and answer will take place onl between you and the participant you called on. Instead, use this technique to involv more participants in answering questions and to keep them on their toes.

1. Ask participants to define a term or explain a concept.
2. After one participant has answered, ask a participant seated nearby if the answer is right. Then ask whether a participant in the back of the room agrees.
3. Ask participants to explain why they think an answer is right or wrong.
4. Use the session to clear up incorrect ideas and encourage participants to learn from their mistakes.

Sections 4.3.0 – 4.3.3 AOA Diagrams

Assign Problems 1 through 6 (below) one at a time. Have the participants drav them, and then draw the correct relationships on the board. Allow 40 minutes fo this exercise.

Draw an AOA diagram for the following projects:

1. Activity A precedes Activity B
2. A precedes B & C
   B & C are concurrent
3. A & B precede C & D
   C & D precede K
4. A precedes C
   B precedes C & D
   A & B are concurrent
   C & D precede K
5. A precedes B & D
   B precedes E
   D precedes E & C
   E & C precede M
6. A, B, & E are beginning activities
   A precedes C
   B precedes C & D
   E precedes D & F
   C, D, & F precede G

MODULE 44108-08 — ANSWERS TO REVIEW QUESTIONS

Answer **Section**

1. Any three of the following:
   1. Reduce construction time (3.0.0)
   2. Reduce costs of labor, overhead, interest on loans and capital
   3. Provide a more continuous flow of work
   4. Increases productivity
   5. Provides employees and subcontractors specific milestones to work toward and indications of progress toward those goals
   6. Forces detailed thinking
   7. Improves communication
   8. Improves the company's image and shows professionalism
   9. Provides your company with a competitive edge
   10. Provides control and management of material, labor, money, equipment, and subcontractors
   11 Provides owners with relevant project data, which is increasingly important as owners are becoming more sophisticated and are requiring tighter controls

2. 1. planning (4.0.0)
   2. work breakdown structure (WBS)
   3. scheduling methods (bar charts, CPM, SIPS)
   4. periodic updates

3. 1. Check sheets and magnetic boards (4.2.0)
   2. Bar chart
   3. Network systems (CPM/PERT)

4. (4.3.0)

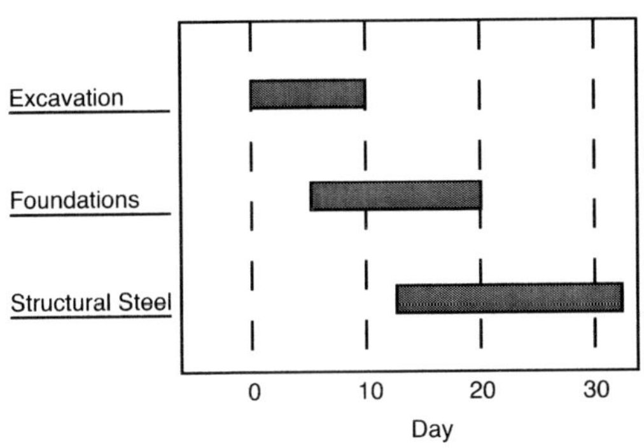

108R03.EPS

5. The bold lines linking Activities A to C, C to D, and D to E indicate the critical path. Activity B is not in the critical path.

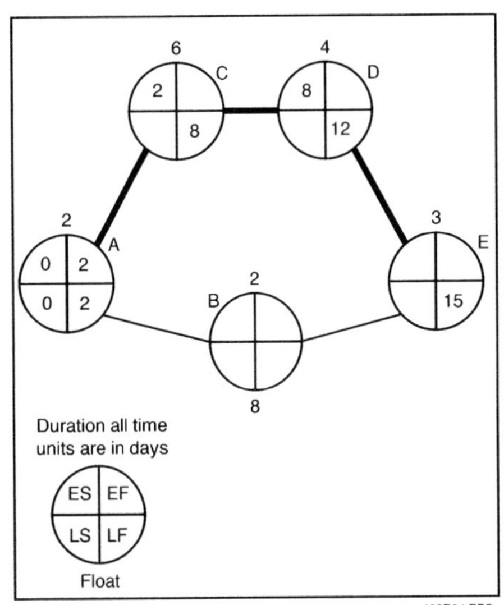

108R04.EPS

6. 8 days
7. Day 8
8. Day 15
9. Enforcing (7.3.0)
10. A short interval schedule is a three-to-four-week "look-ahead" schedule (4.4.0)

MODULE 44108-08 — ANSWERS TO TRADE TERMS QUIZ

1. precedence diagramming method (PDM)
2. critical path method (CPM)
3. lags
4. lean scheduling

CONTREN® LEARNING SERIES — USER UPDATE

NCCER makes every effort to keep these textbooks up-to-date and free of technical errors. We appreciate your help in this process. If you have an idea for improving this textbook, or if you find an error, a typographical mistake, or an inaccuracy in NCCER's Contren® textbooks, please write us, using this form or a photocopy. Be sure to include the exact module number, page number, a detailed description, and the correction, if applicable. Your input will be brought to the attention of the Technical Review Committee. Thank you for your assistance.

Instructors – If you found that additional materials were necessary in order to teach this module effectively, please let us know so that we may include them in the Equipment/Materials list in the Annotated Instructor's Guide.

Write: Product Development and Revision
National Center for Construction Education and Research
3600 NW 43rd St, Bldg G, Gainesville, FL 32606

Fax: 352-334-0932

E-mail: curriculum@nccer.org

Craft ______________ Module Name ______________

Copyright Date ______________ Module Number ______________ Page Number(s) ______________

Description

____

____

____

(Optional) Correction

____

____

(Optional) Your Name and Address

____

____

Module 44109-08

Resource Control

NCCER STANDARDIZED CRAFT TRAINING PROGRAM

The National Center for Construction Education and Research (NCCER) provides a standardized national program of accredited craft training. Key features of the program include instructor certification, competency-based training, and performance testing. The program provides trainees, instructors, and companies with a standard form of recognition through a National Craft Training Registry. The program is described in full in the *Guidelines for Accreditation*, published by NCCER. For more information on standardized craft training, contact the NCCER by writing us at 3600 NW 43rd St., Bldg. G, Gainesville, FL 32606; calling 352-334-0911; or emailing info@nccer.org. More information may be found at our website, www.nccer.org.

HOW TO USE THIS ANNOTATED INSTRUCTOR'S GUIDE

Each page presents two sections of information. The larger section displays each page exactly as it appears in the Trainee Module. The narrow column ties suggested trainee and instructor actions to each page and provides icons (detailed below) to call your attention to material, safety, audiovisual, or testing requirements. The bottom of each page includes space for your notes.

The **Audiovisual** icon indicates an appropriate time to show a transparency or other audiovisual aid.

The **DVD** icon indicates an appropriate time to discuss or show segments of the video.

The **Classroom** icon prompts you to define a term, stress a point, ask trainees to explain a concept, or give examples.

The **Demonstration** icon directs you to show trainees how to perform tasks.

The **Examination** icon tells you to administer the written module examination.

The **Homework** icon is placed where you may wish to assign reading for the next class, assign a project, or advise trainees to prepare for an examination.

The **Laboratory** icon is used when trainees are to practice performing tasks.

The **Materials** icon is a reminder for you to gather materials needed for classes, labs, and testing.

The **Performance Testing** icon tells you to administer a performance test or a portion thereof.

The **Safety** icon is used to emphasize safety issues. It is often keyed to *Caution* and *Warning!* statements in the Trainee Module.

The **Teaching Tip** icon indicates additional guidance is available, such as how to conduct an exercise, get the most educational value from a field trip, or encourage class participation. Teaching Tips may expand on a feature (*Think About It, Did You Know?*) or provide *Quick Quizzes* or similar exercises. You will be referred to the Teaching Tips section at the back of the module if there is additional material.

The **Combination** icon indicates that the laboratory listed corresponds with a performance task. If desired, you can note the proficiency of the trainees during the laboratory, and use it to satisfy performance testing requirements.

PREPARATION

Before teaching this module, you should review the Objectives, Performance Tasks, Materials and Equipment List, and Module Outline. Be sure to allow ample time to prepare your own training or lesson plan and gather all required materials and equipment.

MODULE OVERVIEW

This module discusses production, production control, and production control standards. It explains the manager's role in the production process and factors that can affect productivity. Methods for resource control and ways to evaluate and improve productivity are also covered.

PREREQUISITES

Please refer to the Course Map in the Participant Module. Prior to training with this module, it is recommended that the participant shall have successfully completed *Project Management,* Modules 44101-08 through 44108-08.

OBJECTIVES

Upon completion of this module, the participant will be able to do the following:

1. List the five elements of production control.
2. Recognize when production is under control.
3. Describe the role of reports in production control.
4. Identify and explain the major factors which affect production control.
5. Describe methods for alleviating the negative effects of the major production control factors.
6. List the three production standards and specify when they are to be used.
7. Explain the three methods for evaluating productivity.
8. Explain and give examples of production control alternatives.
9. Identify the resources that must be controlled and the project manager's role in the process.
10. Describe the role of the project manager in evaluating production both during and after a project.
11. Define debriefing and describe how it is accomplished and its value to production control.

MATERIALS AND EQUIPMENT LIST

Desktop or laptop computer or DVD player
Multimedia projector and screen
Minor Decisions: Major Impact – How to Deal with Real Issues in Project Management DVD
Transparencies
Blank acetate sheets
Overhead projector and screen
Transparency pens
Whiteboard/chalkboard
Markers/chalk
Pencils and scratch paper
Masking or painter's tape
Measuring tape
Flip chart
Copies of Caisson Case Study handout
Module Examinations (located in the Test Booklet)

TEACHING TIME FOR THIS MODULE

An outline for use in developing your lesson plan is presented below. Note that each Roman numeral in the outline equates to one session of instruction. Each session has a suggested time period of 2½ hours. This includes 10 minutes at the beginning of each session for administrative tasks and one 10-minute break during the session. Approximately 10 hours are suggested to cover *Resource Control*.

Topic	Planned Time
Session I. Production Control and Productivity Improvement	
A. Production Control Reports and Comparisons	________
B. Production Control Factors	________
C. Productivity Study	________
D. Production Control Standards	________
E. Basic Improvement Methods	________
F. Evaluating Productivity	________
G. More Detailed Activity/Productivity-Based Methods	________
H. Improving Work Methods	________
I. Production Control Alternatives	________
Session II. Controlling Resources	
A. The Project Manager's Role	________
B. Material Control	________
C. Equipment Control	________
D. Tool Control	________
E. Labor Control	________
F. Increasing Labor Efficiency	________
G. Evaluation and Debriefing	________
Session III. Resource Control Exercise	
A. Teaching Tip, Resource Control	________
Session IV. Caisson Case Study; Review and Testing	
A. Teaching Tip, Caisson Case Study	________
B. Review	________
C. Module Examination	________

1. Trainees must score 70% or higher to receive recognition from NCCER.
2. Record the testing results on Craft Training Report Form 200, and submit the results to the Training Program Sponsor.

USING THE PM DVD WITH *RESOURCE CONTROL*

Samantha's crew is suffering a productivity loss because of the 'missing tool' and the time wasted on arguments and resolutions.

As presented in Section 2.0.0, discuss with the class which production control factors have contributed to the scheduling and productivity problems.

Suggested response:** **Some of the factors that could apply include poor worker attitude; inadequate crew size and composition; lack of trained personnel; inefficient work activities or methods; disorganized job site; ineffective supervision; poor coordination of trades; down time; and insufficient tool resources.

Section 4.4.0, Tool Control, presents an opportunity to discuss some tool control systems and processes that Samantha and her team might put into place to prevent the loss of tools in the future.

Suggested response:** **Samantha needs to ensure that there is an inventory control system, adequate inventory of tools to do the job, and adequate control over the issuance and return of tools. Some of the tool control systems she may need to implement for her workforce and for her contractors include crew gang box checked out to the supervisor, master-keyed gang boxes, tool audits, labeled tools, and a receipt system for checking out and returning tools.

Referencing Section 4.5.0, Labor Control, ask the students what factors of labor control might have contributed to the problem in the video and how might they be improved in the future?

Suggested response:** **Searching or waiting for tools; idle time; lack of motivation (e.g., Anthony's attitude); lack of effective supervision (e.g., Keith not monitoring use and return of tools); poorly scheduled work; coordination issues with other trades; overtime; and time required to train new workers.

In Section 6.0.0, Evaluation and Debriefing, ask the students whether conducting a team debriefing might have helped prevent these problems, and others, from recurring.

Suggested response:** **Regularly scheduled debriefings would have enabled the crew leaders to point out the problems on the job site to Samantha, enabling her to act before the situation deteriorated and created a crisis.

(continued)

USING THE PM DVD WITH *RESOURCE CONTROL* *(Continued)*

Class Exercise: Divide the class into groups and instruct each group to develop a scenario to calculate the impact of the missing tool problem on productivity. Develop your own scenario or leave it up to the students to develop the parameters, or use the following example as a guideline. Refer the class to Labor Control in the Resource Control module for further guidance. Have each group present its scenario and the resulting cost to the project.

Sample Exercise: Samantha's project is behind schedule and has been incurring a million-dollar-a-day loss. She instituted productivity quotas in an attempt to stem the losses. The recent missing tool problem has created further delays and loss in productivity for this job. Acting as project manager, determine the productivity loss that has occurred because of this incident. Determine the time involved by Tony, Keith, and Craig and their crews, and consider Samantha's time. To simplify the calculation, you may want to use an average hourly rate for each crew. Factors you should consider include:

- Loaded labor rates
- Time lost by each crew member for each identifiable incident associated with the problem
- Idle time for any crew member
- Any labor added to compensate for the delays
- Any overtime incurred

Tasks can include:

- Searching for the tool
- Lack of tools to perform the work
- The confrontation
- Discussions with Samantha
- Samantha's time to resolve the issue
- Clean-up (housekeeping) to get the site in order.

Project Management

44109-08

Resource Control

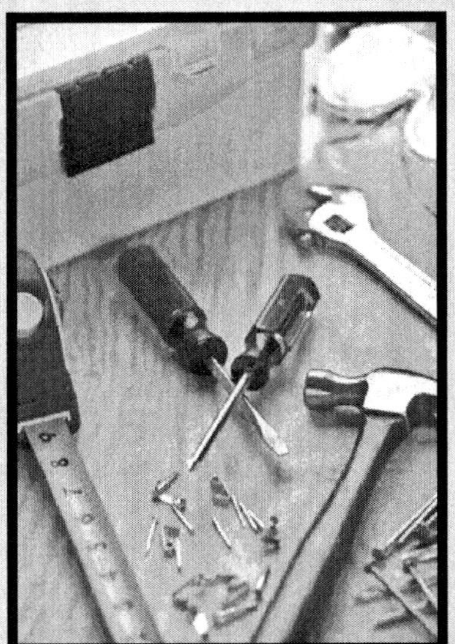

Assign reading of Module 44109-09.

44109-08
Resource Control

Topics to be presented in this module include:

Overview

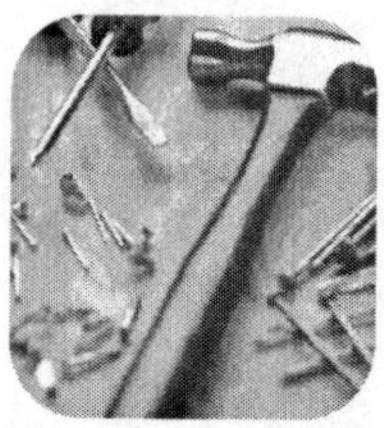

The project manager is responsible for controlling production so that the project is built within the specified time and allocated budget. This means controlling the use of resources, maintaining a safe work environment, being constantly aware of productivity rates, and, when necessary, reviewing and revising work methods. The resources that must be controlled are material, tools and equipment, labor, time, space, and money. This module examines the topic of resource control and the various parameters within it.

Instructor's Notes:

Objectives

When you have completed this module, you will be able to do the following:

1. List the five elements of production control.
2. Recognize when production is in control.
3. Describe the role of reports in production control.
4. Identify and explain the major factors which affect production control.
5. Describe methods for alleviating the negative effects of the major production control factors.
6. List the three production standards and specify when they are to be used.
7. Explain the three methods for evaluating productivity.
8. Explain and give examples of production control alternatives.
9. Identify the resources that must be controlled and the project manager's role in the process.
10. Describe the role of the project manager in evaluating production both during and after a project.
11. Define debriefing and describe how it is accomplished and its value to production control.

Trade Terms

Five-minute rating
Production
Productivity
Production quality standard
Production resource standard
Production time standard

Prerequisites

Before you begin this module, it is recommended that you successfully complete *Project Management*, Module 44101-08 through 44108-08.

This course map shows all of the modules in the *Project Management* curriculum. The suggested training order begins at the bottom and proceeds up. Skill levels increase as you advance on the course map. The local training program sponsor may adjust the training order.

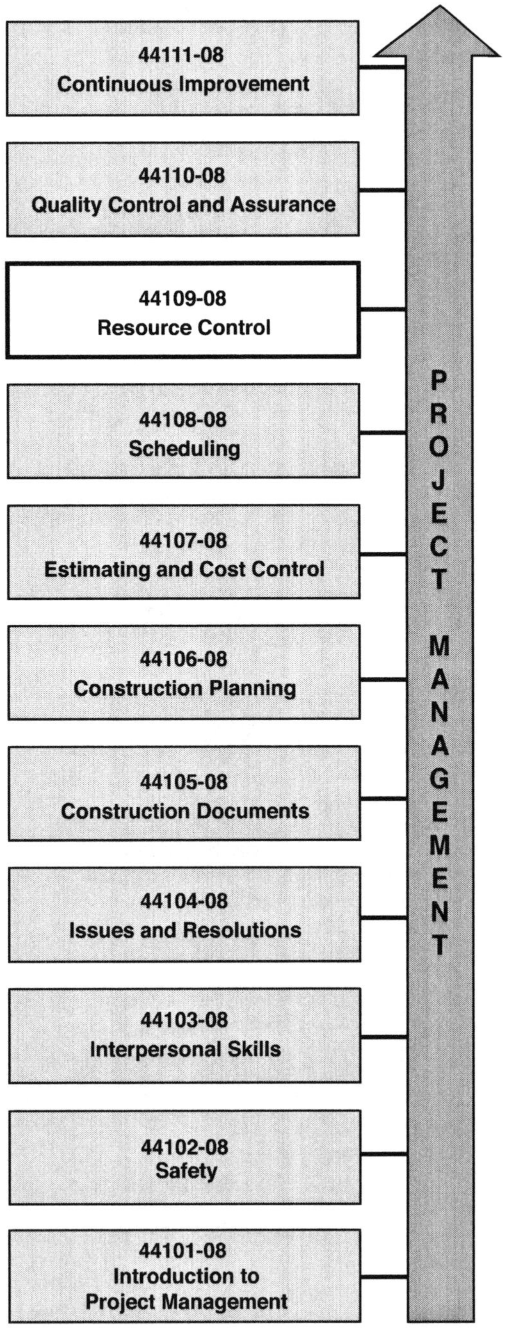

109CMAP.EPS

Ensure that you have everything required to teach the course. Check the Materials and Equipment list at the front of this module.

See the General Teaching Tip at the end of this module.

Explain that terms shown in bold are defined in the Glossary at the back of the module.

Show Transparency 1, Objectives. Review the goals of the module, and explain what will be expected of the participant.

MODULE 44109-08 ◆ RESOURCE CONTROL 9.1

Define the terms production and productivity and give examples of each.

Show Transparency 2, Production and Productivity.

List the resources required for effective production control.

Discuss the objective of production control. Ask participants to list the conditions that indicate that this objective has been met.

Show Transparency 3, Conditions for Successful Projection Completion.

Show Transparency 4, Tools for Controlling Job Production. Discuss the ways that a project manager can control job production.

Show Transparency 5, Production Control Reports.

Explain the purpose of reports and stress the need for accurate reporting.

Discuss production control factors and use examples as appropriate.

1.0.0 ◆ INTRODUCTION

In order to effectively control resources, the project manager must be able to distinguish between **production** and **productivity**. Production is a term describing the total quantity of work accomplished. Productivity describes the amount of work performed, using a given level of effort within a given unit of time. The difference can best be illustrated through an example.

A job requires that 105 doors must be hung in an office building. The production goal, then, is 105 doors. In estimating the cost of the job, the company consults its records of past performance to find out how many doors per hour (or workday) a person or crew can install. This figure is the productivity (sometimes referred to as the production rate) of the labor involved.

If a crew's productivity is three doors per hour, the production goal of 105 doors can be met in 35 crew hours. The same methods can also be applied to equipment usage.

2.0.0 ◆ PRODUCTION CONTROL

The estimate and the estimate breakdown provide basic information for planning and managing the project. The job plan and schedule provide a time frame and a sequence of action. The reporting system provides data for comparing unit actual costs with unit estimated costs. And all of these resources provide information for effective production control.

The objective of production control is to manage production in such a way that construction is completed successfully. This objective is met when at least three conditions are satisfied:

- Work meets standards of quality
- Work is completed on time
- Work is accomplished within the budget

The project manager has a set of tools available for controlling job production, including:

- Reviewing reports on labor, production, materials, and equipment
- Predicting and correcting avoidable conditions that slow production
- Monitoring quality, time, and resource standards
- Preparing alternate plans and leveling resources
- Conducting evaluations and debriefings

2.1.0 Production Control Reports and Comparisons

Reports provide information on actual job costs. and actual labor and equipment production rates. The project manager can compare the data from these reports with the project's original and revised estimates and production rates to determine job status. Major variances between report data and the estimate are an important indication that corrective action is required.

The three types of reports most important for production control are as follows:

- *Weekly labor and production reports* – These reports give up-to-date information on the job's human resources—information that can be compared with estimated labor costs and production rates.
- *Material quantity reports* – These reports tell the project manager how much work has been completed, providing quantities to compare with the estimate.
- *Equipment reports* – Equipment reports list costs being charged against company-owned or rented equipment on the job. They also include down time records. This information can be compared against the project's estimated equipment costs and productivity.

Your company may refer to these reports by different names.

Obviously, a project manager cannot manage effectively without adequate information from these reports. Every project manager is handicapped when the information is incomplete, inaccurate, or late. And because a company's estimates for future jobs are based on reports of current work, the entire company suffers when reports are deficient, inaccurate, or outdated.

Consequently, the project manager must insist that all job reports be complete, accurate, and timely. With these reports in hand, the manager must perform whatever analyses are necessary to determine whether the job is on track or not.

2.2.0 Production Control Factors

Many factors affect productivity on a construction job, including the following:

- Material shortage
- Undocumented change orders
- Ineffective supervision
- Poor coordination of trades
- Excessive use of casual and scheduled overtime

Instructor's Notes:

- Errors and omissions in drawings and specifications
- Multiple change orders
- Design complexity
- Design omissions
- Working in congested areas, high-rise buildings, and existing facilities
- Excessive reassignment of personnel from job to job on the same project
- Adverse weather
- Inadequate lighting
- Unanticipated subsurface conditions (e.g., water)
- Governmental regulations
- High rate of absenteeism
- High job turnover rate
- High accident rate
- Lack of trained personnel
- Poor worker attitude
- Inadequate crew size and composition
- Size and duration of project
- Timeliness of decisions
- Uncontrolled coffee and meal breaks
- Impractical quality control tolerances
- Time of day and day of week
- Inadequate temporary facilities (e.g., parking, change rooms, restrooms)
- Inefficient work activities
- Disorganized job site

The project manager must be aware of such factors on the jobs and take action to correct them or to change plans to negate their effects. The factors which seem to cause project managers the most problems are discussed in the following sections.

2.2.1 Change Orders

Change orders are orders for work that is not shown on the original drawings or in the specifications. Examples are:

- Changing the location of a door
- Relocating a heat duct after it has been installed
- Increasing the size of a pipe before materials are ordered

Change orders disrupt production because they require preparation, negotiation, and the processing of paperwork. Also, they may require work that has already been installed to be moved or replaced, or call for major changes in job sequence or schedule.

Research shows that change orders affect worker productivity in the following ways:

- The need for increased work output results in physical fatigue.
- If the change order is for rework, the motivation of both workers and supervisors deteriorates.
- The project schedule and the momentum of work are disrupted.
- Supervisors must take on added responsibility and their thought patterns are disrupted.
- Additional work may require learning new skills.
- Untimely procurement and delivery of additional materials and equipment can hinder work.
- Changes often require reassignment of labor and changes in crew makeup.

Because of these negative effects, it is very difficult, if not impossible, to accurately estimate the actual cost of a change. The negative effects counter the popular notion that change orders are a way to increase profit. There is no direct relationship between the size of a change and its effect on productivity; however, the more complex the change, the greater its negative effect. The project manager should be aware that changes are inevitable and must be accomplished as efficiently as possible.

2.2.2 Schedule Changes

Schedule changes always disrupt production, and are seemingly inevitable because of either errors in planning or unexpected on-the-job factors. The estimate may not have allowed enough production time per unit, the original schedule may have been unrealistic, or a subcontractor may not have been able to meet the schedule. These disruptions occur on construction jobs of all sizes, and the project manager must be able to overcome them by developing alternate plans and providing decisive leadership.

2.2.3 Construction Mistakes

Construction mistakes usually result from a failure to understand drawings or specifications before starting the job, from employing improper installation methods, or from sacrificing quality to speed production. The project manager must make it clear that such costly and avoidable errors are unacceptable on any project.

Show Transparency 6, Problematic Production Control Factors. Review the items and ask participants to discuss how these factors create problems.

Review and discuss how change orders affect productivity.

Explain how a project manager can best overcome the disruptions caused by schedule changes.

Ask participants to list reasons why construction mistakes occur and stress that mistakes are never acceptable.

Discuss which production control factors have contributed to the scheduling and productivity problems in the scenarios.

MODULE 44109-08 ◆ RESOURCE CONTROL 9.3

Discuss productivity issues resulting from untimely deliveries, shortages, and waste.

Point out that administrative problems are primarily due to poor communication.

Review and discuss how labor turnover affects productivity.

Show Transparencies 7–9 (Tables 1–3). Discuss how temperature and humidity levels can result in lower productivity.

Discuss the impact of accidents on production. Ask participants to suggest ways to reduce accident rates.

Show Transparencies 10 and 11 (Figures 1 and 2). Discuss how overtime can adversely affect productivity.

Explain why working in congested areas can be unproductive.

2.2.4 Untimely Deliveries, Shortages, and Waste

Late deliveries are serious if they leave crews idle. The project manager should be sure that job supervision monitors deliveries and moves crews to other work when materials are not available on time for a particular job. Early deliveries of materials can also disrupt the job plan and, of course, production.

Shortages and waste also add to job costs. It takes time to inspect deliveries to make sure all materials are received and to take steps to avoid waste. However, this is time well spent.

2.2.5 Administrative Problems

Administrative problems are almost always communication problems. Lack of communication between the construction company and the client—or the project manager and the subcontractor—lead to misunderstandings, friction, and delays until the problems are worked out.

Among the most common problems in this category is a lack of clarification on drawings and specifications as well as improper handling of contract administration items. Back charges are particularly disruptive because they take time for all parties to process. Therefore, they should be avoided whenever possible.

2.2.6 Labor Turnover

Turnover is defined as individuals quitting on a voluntary basis or being terminated for cause. It does not include reductions in force. Every 10 percent rate of craftworker turnover increases payroll by 2.5 percent. Labor turnover can result from poor supervision by the field supervisor or disputes and general unrest that may not be the fault of the supervisor. In either case, all levels of field supervision are responsible for avoiding excessive turnover. The project manager should monitor labor turnover and give the field supervisor all necessary assistance to avoid problems.

2.2.7 Weather

Bad weather interrupts the smooth pace of production and can damage work that has already been installed. It can change site conditions and certainly affects the work schedule. Keeping an eye on the weather is an important part of the job for both the project manager and field supervisor. Both should have alternate plans for redirecting labor and protecting installed work when weather turns bad.

Numerous studies have been conducted on the effects of temperature and humidity and have shown that workers are most efficient when the temperature is between 40°F and 80°F and the relative humidity is below 80 percent. *Tables 1, 2,* and *3* illustrate various temperatures and humidity levels and their effect on productivity.

2.2.8 Accidents

It takes little imagination to think of the many ways that accidents disrupt production. The project manager must take steps to identify and correct unsafe conditions before they result in costly accidents. Refer to the *Safety* module for assistance in ensuring a safe project.

2.2.9 Overtime

Overtime is any time worked in excess of a regular workday of 8 hours or a workweek of 40 hours. Frequently scheduled overtime reduces productivity because workers expend energy at a pace established over long periods of adaptation. An increase in hours of work forces an adjustment to a slower pace—a natural biological reaction for avoiding fatigue. Prolonged overtime results in a loss of productivity and an increase in injuries and absenteeism.

Studies have shown that if overtime continues beyond six to eight weeks there tends to be a negative impact on productivity. Refer to *Figure 1* for a graph showing the impact as well as the example in *Figure 2* in the presentation of this graph.

As overtime efficiency decreases, research has found there is an increase in disruptions. The major disruptions are caused by:

- Material availability
- Tool availability
- Equipment availability
- Information availability
- Rework
- Congestion

In light of these facts, the project manager should avoid scheduled overtime for relatively long periods.

2.2.10 Working in Congested Areas, High-Rise Buildings, and Existing Facilities

Working in congested areas always hinders productivity, due primarily to the limited amount of space available for movement of labor, equipment, and receipt and storage of materials. Trade coordination also becomes especially critical in congested areas.

Instructor's Notes:

9.4 PROJECT MANAGEMENT

Table 1 Temperature and 30% Relative Humidity Effect on Productivity

BASED ON RELATIVE HUMIDITY OF 30%		
Effective Temperature	Productivity	Loss of Productivity
-10° F	65%	35%
0° F	78%	22%
10° F	88%	12%
20° F	94%	6%
30° F	99%	1%
40° F	100%	0%
50° F	100%	0%
60° F	100%	0%
70° F	100%	0%
80° F	99%	1%
90° F	93%	7%
100° F	83%	17%
110° F	64%	36%

109T01.EPS

Table 2 Temperature and 60% Relative Humidity Effect on Productivity

BASED ON RELATIVE HUMIDITY OF 60%		
Effective Temperature	Productivity	Loss of Productivity
-10° F	60%	40%
0° F	75%	25%
10° F	87%	13%
20° F	94%	6%
30° F	99%	1%
40° F	100%	0%
50° F	100%	0%
60° F	100%	0%
70° F	100%	0%
80° F	98%	2%
90° F	93%	7%
100° F	80%	20%
110° F	58%	42%

109T02.EPS

Show Transparencies 12 and 13 (Figures 3 and 4). Review and discuss how the number of workers on a project can affect productivity.

See the Teaching Tip for Section 2.2.10 at the end of the module.

Table 3 Temperature and 90% Relative Humidity Effect on Workers' Productivity

BASED ON RELATIVE HUMIDITY OF 90%		
Effective Temperature	Productivity	Loss of Productivity
-10° F	56%	44%
0° F	72%	28%
10° F	82%	18%
20° F	89%	11%
30° F	92%	8%
40° F	96%	4%
50° F	98%	2%
60° F	98%	2%
70° F	96%	4%
80° F	93%	7%
90° F	85%	15%
100° F	57%	43%

109T03.EPS

Research has shown that increasing the number of workers on the project can result in a decrease of productivity. Refer to *Figure 3* for a graph illustrating the impact as well as the example in *Figure 4* in the presentation of this graph.

Working in high-rise buildings also lowers productivity, because moving personnel and materials vertically takes considerably more time than moving them horizontally. In addition, the location of most high-rise buildings (usually in metropolitan centers) and the working conditions involved also take their toll. Studies show that productivity decreases when construction gets above the third level. The project manager needs to keep this in mind when evaluating the progress of a high-rise building project.

When working in existing facilities, the project manager must take special precautions. To protect the existing building structure, finishes, systems, and furnishings from potential damage that may result from new construction. Furthermore, take care not to interrupt ongoing activities in the facility.

MODULE 44109-08 ◆ RESOURCE CONTROL 9.5

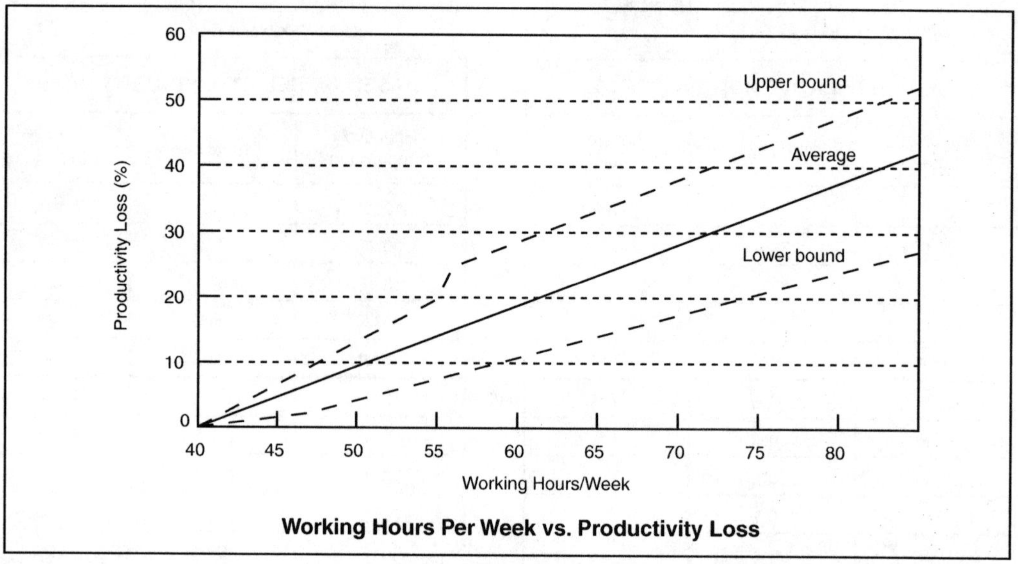

Figure 1 ◆ Working hours per week versus productivity loss.

EFFECT ON PRODUCTIVITY BY WORKING OVERTIME

Given: The job entails pulling 420,000 lf of cable. It is planned to work 40 hours per week with a labor force of 20 craftworkers pulling an average of 35 lf of cable per hour.

Total duration of activity:

$$\text{Activity Duration} = \frac{420{,}000}{35 \times 20 \times 40} = 15 \text{ weeks}$$

Prior to the start of the installation of the cable, the contractor is instructed by the client to accelerate the work and complete the job in 9 weeks.

To accomplish the new goal, management decided to go into a 60-hour week overtime schedule. When calculations show this strategy to be inadequate, it is decided to also double the number of workers on the site. In each case, it is assumed that the level of management control is increased to reflect the changes in hours worked and workers employed. No specific measures are taken to change any of the other variables.

Effect of only working overtime:

The average loss of productivity associated with a 60-hour week is 20%, as shown in Figure 1, "Working Hours Per Week vs. Productivity Loss."

$$\text{Reduced Productivity Achieved} = 35 - \frac{(35 \times 20)}{100} = 28 \text{ lf/hr}$$

$$\text{Workhours to complete job} = \frac{420{,}000}{28} = 15{,}000 \text{ hours}$$

$$\text{Duration of activity} = \frac{15{,}000}{20 \times 60} = 12.5 \text{ weeks}$$

$$\text{Original estimated worker hours} = \frac{420{,}000}{35} = 12{,}000 \text{ hours}$$

Therefore, working overtime would result in a loss of 15,000 – 12,000 = 3,000 hours

This represents 25% of the estimated hours of 20% of the actual hours that would be worked in overtime.

109F02.EPS

Figure 2 ◆ Effect on productivity by working overtime.

Instructor's Notes:

9.6 PROJECT MANAGEMENT

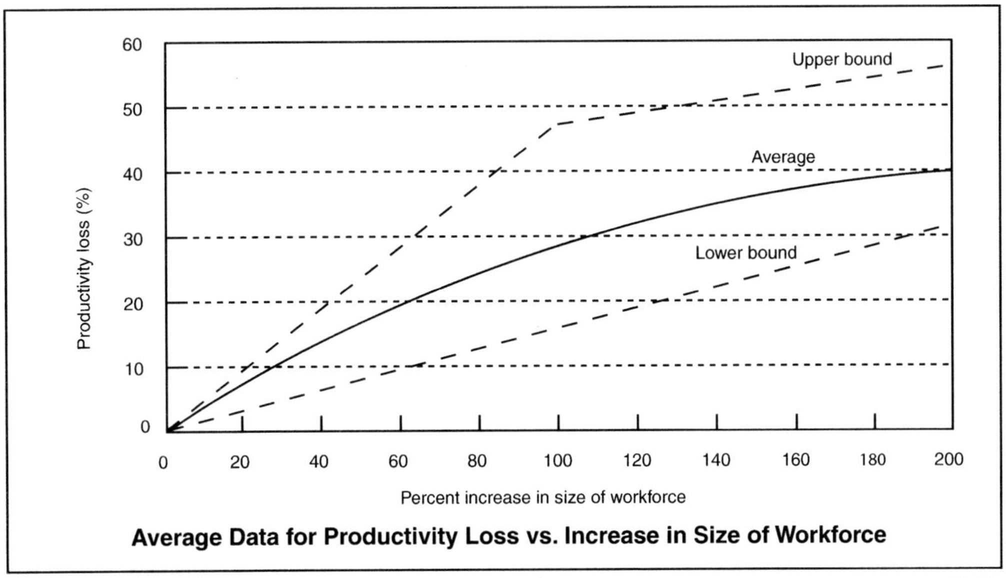

Figure 3 ◆ Average data for productivity loss versus increase in size of workforce.

EFFECT ON PRODUCTIVITY BY INCREASING THE SIZE OF THE WORKFORCE

Refer to *Figure 3*, "Average Data For Productivity Loss vs. Increase in Size of Workforce". The loss of productivity resulting from doubling the number of workers is 27%.

Thus, for a 60-hour week, productivity is reduced to:

$$\frac{28\,(100 - 27)}{100} = 20.44 \text{ lf/hr}$$

$$\text{Worker hours to complete activity} = \frac{420{,}000}{20.44} = 20{,}548 \text{ hours}$$

$$\text{Duration of activity} = \frac{20{,}548}{2 \times 20 \times 60} = 8.56 \text{ weeks}$$

This equated to a reduction from the original estimate of

$$\frac{15 - 8.56}{15} \times 100 = 43\%$$

Therefore, the effect of working overtime and doubling the size of the workforce is 20,548 – 12,000 = 8,548 worker hours. This represents about 71% of the original estimated worker hours or 42% of the actual hours planned to be worked.

Conclusion:

Add 71% to estimated labor hours to compensate for losses in productivity caused by working overtime and doubling the workforce.

Questions:

1. If the wage rate of the workers is $18.50 per hour per worker what would be the dollar loss if the effects of overtime and doubling the workforce were not accounted for?
2. Based on the same activity and original total quantity to place, original worker productivity and original number of workers, what would be the effect on productivity if management decided to work 50 hours a week?

109F04.EPS

Figure 4 ◆ Effect on productivity by increasing the size of the workforce.

MODULE 44109-08 ◆ RESOURCE CONTROL 9.7

Ask participants to give examples of inefficient work methods.

Point out that it is the project manager's responsibility to determine the most efficient process for a project.

Review and discuss reasons for unproductive time.

Show Transparency 14, Categories of Unproductive Time.

Show Transparency 15, Factors Affecting Production Time Standard.

Explain that production control standards are used to measure whether construction is acceptable in terms of quality, time, and resources.

Discuss the production time standard and explain why a realistic schedule is important.

Ask participants to list indications that a project is not on schedule.

2.2.11 Inefficient Work Methods

Many work processes used in the construction industry today can be improved resulting in a more cost-effective project. For example, preassembling mechanical units and installing them rather than installing all of the individual components in place will save a contractor both time and money. Also, placing concrete by pumping it rather than by using a crane and bucket may also result in substantial savings in time and money for the project.

The project manager, during the project planning phase, needs to assess each of the processes that are planned to be used to find out if possible improvements can be made. This information is in the *Continuous Improvement* module.

2.3.0 Productivity Study

A recent study by Clemson University's Construction Science and Management department concluded that the typical craftworker is only 40-percent productive—that is, 3.2 hours out of an 8-hour day. The same study concluded that the remaining unproductive time could be categorized into four areas. These areas, with the corresponding percentages, were as follows:

- Administrative delays caused by management, resulting in workers waiting for resources, instructions, etc.—20%
- Inefficient work methods such as described above—20%
- Restrictions in the workplace such as overstaffing, working in confined spaces, etc.—15%
- Personnel delays not including lunch and planned breaks—5%

It is critical that project managers be aware of the various production factors that can negatively impact their projects. When project management carefully estimates productivity and plans its jobs, pays close attention to scheduling, maintains close communication with everyone associated with the job, and acts quickly and decisively to head off problems, negative effects can be minimized.

Many factors affecting productivity are beyond the control of the contractor. Because of this, it may be possible to file claims for additional compensation or time when such unanticipated factors adversely affect productivity and disrupt the job schedule. The project manager should know the contract documents well enough to know when filing a claim may be possible. In cases where a claim can be filed, the project manager must make sure that accurate records are kept to document all problems. Complete and accurate documentation is absolutely necessary to substantiate any claim.

2.4.0 Production Control Standards

Standards are benchmarks against which construction is measured to determine whether or not it is acceptable. The three most important standards in construction are quality, time, and resources.

2.4.1 The Production Quality Standard

The **production quality standard** is set by the specifications and must be followed by all construction managers. The field supervisors are directly responsible for the quality of work performed by their crews and the subcontractors, but they cannot do the job without help. Their efforts require the support of the project manager.

The project manager's role in controlling quality is critical—clients should receive the quality they are paying for. You will cover quality in greater detail in the modules *Quality Control and Assurance* and *Continuous Improvement*.

2.4.2 The Production Time Standard

The overall **production time standard** is set by the owner, agreed to by the contractor, and incorporated into the job plan and schedule. As a result, the project manager knows whether the time standard is being met by knowing how well the job is meeting the schedule.

What signs indicate that the job is not on schedule? There are many. Some of the major ones include:

- Materials constantly arriving late
- Subcontractors missing deadlines
- Excessive overtime labor costs
- Project activities exceeding their allotted times
- Key milestones being missed

These are indications that the project manager must be aware of and be prepared to deal with.

Of course, it is not enough simply to have a schedule. The schedule must be realistic. A schedule that pushes production faster than necessary can be just as bad as one that paces production slower than necessary. Both waste resources. Scheduling is covered in greater detail in its own module.

Instructor's Notes:

2.4.3 The Production Resource Standard

The estimate establishes the **production resource standard**. Therefore, the resource standard is met by accomplishing all the work within the estimate.

The problem facing management is that resources are usually limited. For example:

- The right amount or type of labor is not always available.
- The materials and equipment needed to carry out the best—the optimum—schedule are sometimes lacking.
- The field supervisor may be unable to organize and supervise the required number of simultaneous operations because resources are limited.

Because resources are limited, management must pay particularly close attention to reports, particularly the cost report, for such production indicators as excessive overtime, unusual equipment maintenance and repair costs, equipment that is idle for long periods, and unusual labor turnover. These signal that production needs to be corrected, and it is important that the management team (both the project manager and the job supervisor) make these corrections immediately to avoid further resource waste.

The project manager can evaluate job performance in light of production control standards by frequently asking key questions during the course of each project:

- Is the quality of work acceptable?
- Is the project meeting the schedule?
- Is the work being done within the estimate?

3.0.0 ◆ PRODUCTIVITY IMPROVEMENT

There are two basic means that a project manager has for improving productivity without increasing costs: increase production without increasing installation time and decrease installation time without reducing production.

In one example, a crew of three electricians is hanging lighting fixtures at a rate of 3 per hour. A new person is added to the crew and the crew's production goes up to 3½ fixtures per hour. What is the result? Production has increased, but productivity has decreased and cost has increased. The 33-percent increase in labor has increased production by only a half-fixture per hour, 17 percent. If, however, the work method is changed so that the three electricians can install 4 fixtures per hour, productivity has been increased by 33 percent without increasing costs. The result is an overall decrease in installed unit cost.

Improving productivity, then, is a matter of finding ways to use labor, material, and equipment as efficiently as possible. Studies show that major increases in productivity may be realized by the following methods:

- Improving supervision
- Using innovative materials, equipment, and tools
- Prefabricating and installing as many components and units as possible
- Motivating workers
- Using a material-handling crew to receive, inventory, store, and distribute resources to crews
- Standardizing as many processes as possible
- Pre-planning as many activities as possible
- Assuring availability of adequate tools, equipment, and materials when needed
- Implementing job-site safety programs
- Providing on-the-job training
- Studying job activities and altering work methods as needed
- Improving the crew makeup

3.1.0 Basic Improvement Methods

The project manager can reduce or eliminate the impact of the major factors that affect productivity in a negative way by considering the suggestions in the topics that follow.

3.1.1 Change Orders

Reduce the effects of change orders by anticipating changes and making alternate plans. Instruct field personnel about the overall effects of change orders and get their commitment to work on changes as a normal part of the construction process. If at all possible, work with a complete set of construction documents. Become involved in the design process and perform a feasibility review.

The project manager should establish a formal method for documenting, monitoring, and handling change order requests. Instructions for dealing with change orders are often in the contract documents, but are frequently not adhered to. Common examples of where change orders are not formally derived and documented are:

- Late or defective owner-furnished materials
- Late or defective design
- Interference by other contractors

Show Transparency 16, Production Resource Standard. Discuss the problem of limited resources.

Explain the two ways a project manager can improve productivity without increasing costs. Review and discuss the example.

Point out that increasing productivity is a matter of finding ways to use labor, material, and equipment as efficiently as possible.

Show Transparency 17, Achieving Major Increases in Productivity. Review the items and ask participants for their input.

Discuss ways to minimize the effects of change orders.

MODULE 44109-08 ◆ RESOURCE CONTROL 9.9

Show Transparency 18 (Figure 5). Discuss how this form can be used to track turnover on a project.

Discuss the following ways to manage turnover: track it, conduct exit interviews, and develop an employee motivation program.

3.1.2 Schedule Changes

Schedule changes can disrupt work flow and lead to extra expenses. Anticipate changes as much as possible and have alternate plans available.

3.1.3 Construction Mistakes, Late Deliveries, Shortages, and Waste

Institute effective field supervision and a thorough expediting process. Stay aware of the progress of the job and any variances between resources used and resources budgeted.

3.1.4 Administrative Problems

Keep everyone associated with the project fully informed about matters that concern them. Anticipate problems as much as possible and prepare alternate plans to deal with them when they occur.

3.1.5 Labor Turnover

Because turnover can have a negative impact on profit, it is critical that project management monitor labor turnover through a formal process. *Figure 5* is a form that can be used to track turnover by project. It is recommended that the project manager working with others document the required information and calculate their percent turnover each week. Once the rates of turnover are known, the economic impact of it can be developed and steps can be taken to make any needed improvements. Remember that a turnover rate of 10 percent increases payroll costs by 2.5 percent.

In addition to monitoring the rate of turnover, project management should also conduct exit interviews with those people who voluntarily have left the company whenever possible. This exercise will result in the most frequent reasons why employees leave from which plans can be made and implemented to reduce turnover. In

RATE OF RETENTION LOG

PROJECT NAME ____________________

WEEK OF ____________________ ROR WEEKLY GOAL ________ %

COMPLETED BY ____________________ TITLE ____________________

	SUNDAY	MONDAY	TUESDAY	WEDNESDAY	THURSDAY	FRIDAY	SATURDAY	TOTAL FOR WEEK	AVERAGE PER WEEK
DATE									
TOTAL NUMBER OF CRAFTWORKERS ON SITE									
TOTAL NUMBER OF QUITS AND TERMINATIONS FOR CAUSE									

$$\text{WEEKLY RATE OF RETENTION} = \left(\frac{\text{TOTAL WEEKLY AVERAGE NUMBER OF WORKERS} - \text{TOTAL WEEKLY AVERAGE NUMBER OF QUITS/TERMINATIONS}}{\text{TOTAL WEEKLY AVERAGE NUMBER OF WORKERS}}\right) \times 100$$

109F05.EPS

Figure 5 ◆ Rate of retention log.

Instructor's Notes:

addition, develop and implement an effective employee motivation program that uses the concepts discussed in the *Interpersonal Skills* module to address the specific needs and policies of the company.

3.1.6 Weather

Review weather forecasts daily. Have alternate plans ready so that when inclement weather strikes workers are not forced to work in adverse conditions. Increase the number of breaks during periods of extreme temperature or humidity. Maintain good temperature and humidity conditions at the work site whenever possible. Build temporary enclosures when the temperature drops below 40°F. Consider working four 10-hour days to provide make-up time necessitated by adverse weather. If, for example, you normally work 10-hour days Monday through Thursday, and it rains Tuesday, causing all workers to be sent home, you can make up for the lost day by working Friday.

3.1.7 Accidents

Develop and implement a company safety program. Refer to the *Safety* module for additional information.

3.1.8 Overtime

Avoid overtime as much as possible, especially over long periods of time. When required, schedule it on alternate weeks. If using a network type of schedule such as critical path method (CPM) or precedence, carefully consider the effects of overtime before implementing it. Refer to *Figure 6* for an example of how to plan for an overtime schedule.

Ask participants to suggest ways to minimize the impact of weather on a project.

Review and discuss how to develop and implement a company safety program.

Show Transparency 19 (Figure 6). Discuss why overtime may not be the best way to increase production over an extended period of time.

PLANNING AN OVERTIME SCHEDULE

MATERIALS

1. Organize a cleanup crew.
2. Organize a material delivery crew.
3. Reassess material storage areas.
4. Identify long lead time items.

EQUIPMENT AND TOOLS

1. Increase equipment and tool maintenance efforts.
2. Increase the inventory of spare parts and hand tools.

INFORMATION

1. Apply additional supervision to identify design errors and omissions.

ORGANIZATION

1. Rotate the crews working overtime so that no single crew or individual works more than 3 to 5 consecutive weeks of overtime.
2. Shift to smaller crews or work teams within a crew.
3. Monitor progress weekly instead of biweekly or monthly.
4. Improve communications with crews.
5. Improve communications with other contractors.
6. Organize a special crew to handle changes.
7. Organize a setup/preparation crew.
8. Insist on greater backlogs before beginning work.
9. Perform detailed planning using short-interval scheduling (SIS).
10. Stress the crew performing 100 percent of their work.
11. Use rolling 4-10 (4 days a week, 10 hours a day) work schedules.

METHODS

1. Use modular or preassembled components.
2. Shift to system or package concept.

109F06.EPS

Figure 6 ◆ Planning an overtime schedule.

MODULE 44109-08 ◆ RESOURCE CONTROL 9.11

Discuss ways that a project manager can minimize the problem of working in a congested area.

Explain how productivity field measurements and activity sampling may be used if there is a discrepancy between estimated and actual productivity.

Review the masonry example for evaluating productivity.

Show Transparency 20, Methods for Evaluating Productivity.

Explain the difference between efficiency and productivity.

Explain the purpose of the five-minute rating and describe how it is conducted.

3.1.9 Working in Congested Areas and High-Rise Buildings

Conduct careful pre-construction planning to allow for the effects of these special work environments on the estimate and the schedule. The project manager should perform a personnel density analysis (i.e., square feet per person) to determine the effective number of people working in a specific space.

A good rule is to plan for no less than 200 square feet per worker. If it is clear that production targets cannot be met, the project manager should adapt the strategy by instituting actions such as:

- Double shifting
- Changing crew composition
- Changing the work method

3.2.0 Evaluating Productivity

The project manager can determine whether the job is being constructed for the estimated cost and within the scheduled time by comparing actual productivity (or rate of production) to estimated productivity. Discrepancies indicate potential problems that can have devastating effects if allowed to continue.

For the estimating stage, crew sizes and output are determined from historical company records of similar jobs. Usually, the historical data must be adjusted to fit the new job. For example, the company's records show that for a specific masonry job a crew of two masons and two helpers can place 1,800 bricks per 8-hour day (225 bricks per hour). The project's quantity survey shows that 27,000 bricks must be laid. At the crew rate estimated from the company's records, the job will take 15 days to complete.

As the job progresses, field reports show that average crew production is only 200 bricks per hour (1,600 per day). This should signal the project manager to immediately investigate the situation and implement whatever changes are necessary to bring productivity in line with the estimate.

When the project manager recognizes a problem, it's important to take steps to improve productivity and to ensure that those steps are effectively implemented. This often requires performing productivity field measurements. These measurements not only allow making corrections to work methods, but provide company estimators with updated productivity information for refining their own estimates for future work.

While one could generally assess the efficiency of all workers on a project, the most effective way is to concentrate on the activities and associated crews where problems exist. Crew assessment techniques are based on what is known as activity sampling.

3.2.1 Activity Sampling

The project manager must visit the project regularly to assess the efficiency of activities which the weekly activity report indicates have fallen behind schedule. When visiting the projects, there must be a reliable method for identifying productivity problems.

One group of such methods is known as activity sampling, which is formally defined as assessing the work in a specific activity during a specific period of time for the purpose of determining its efficiency. It should be noted that efficiency is not the same as productivity in that activity sampling analyses assess the percent of time craftworkers are engaged in productive work (efficiency) as compared to the amount of work they install within a specific time period (productivity). Research has shown that to be 90 to 95 percent sure that the workers are attaining estimated productivity, they should be 60 percent or more efficient. It is nothing more than observing and classifying a small part of a total activity. The most common methods of activity sampling for crews are the **five-minute rating** and the crew balance chart. An additional method, based on the activity sampling, is known as the delay survey.

3.2.2 The Five-Minute Rating

The purpose of the five-minute rating is to:

- Measure the efficiency of a crew
- Identifies possible causes for activity delays
- Indicate where more thorough, detailed planning could result in cost and savings

The five-minute rating uncovers construction delays such as crews or crafts interfering with each other, poor work methods, and shortages of equipment and materials. In addition, it can be used to measure the efficiency of a work method.

For example, a five-minute rating shows where using two trucks instead of three would be adequate or where a twelve-person crew can be reduced to ten. To conduct a five-minute rating, the project manager must observe an activity for at least five minutes for a continuous operation such as grading of soil for a roadbed or painting of walls or for one cycle for cyclic operations such

Instructor's Notes:

9.12 PROJECT MANAGEMENT

as loading a dump truck or installing a prefabricated reinforced concrete wall panel. In addition, to obtain more reliable assessment, a minimum of ten observations should be made for the same crew performing the same activity and these observations should be made on different days and at different times during the day. While a project manager could observe a crew without their knowledge, it would be more effective to involve the crew in the process as a joint effort to improve crew efficiency and thus productivity.

Using a form such as the one in *Figure 7*, the project manager notes whether or not each worker is in fact working on the minute—not approaching the minute or after the minute. If he or she is working, a check mark is placed opposite the minute observation under the worker's designation. The project manager continues to take readings every minute on the minute until at least five minutes has elapsed or one cycle has been completed. From this information the project manager can calculate a five-minute rating.

Figure 7 shows a five-minute rating analysis for erecting a precast concrete panel. The observations are made every minute and are listed on the left. Across the top of the form are the individual skill classifications of the members of the crew. The right-hand side, opposite each time designation, describes the activity or tasks going on during that specific time. A check mark indicates that the crew member was working on the particular task when the observation was made. A blank space indicates that the crew member was not working or otherwise being productive at the time the observation was made.

The example shows that at 2:14 p.m. ironworker number one was involved in loading the panel, but at 2:18 p.m. that ironworker was not involved in aligning the panel. In fact, at 2:18 p.m. that ironworker was doing no productive work.

For the purpose of the five-minute rating, working is defined as:

- Carrying, holding, or supporting
- Participating in such active physical work as:
  - Measuring
  - Laying out
  - Reading drawings
  - Filling in time cards
  - Writing orders
  - Giving instructions
  - Operating a machine or piece of equipment
  - Discussing the work, provided it can be determined that such is the case

Non-working personnel are those who are:

- Waiting for another person to finish work
- Talking while not actively working
- Attending self-operating machines, unless engaged in useful task
- Walking empty-handed

After the project manager's observations are completed, the total number of check marks and the total number of boxes are each added up. In *Figure 7*, the total number of check marks is 36 and the total number of boxes, whether they contain check marks or not, is 78. The check mark total (36) is divided by the box total (78), resulting in 0.46 or a 46-percent five-minute rating. A five-minute rating of less than 60 percent is generally considered unsatisfactory.

Five-minute ratings are quick, economical, and relatively easy to use. They give an overview of the work situation and guide the project manager in solving problems. However, the results are only indicators, not conclusive evidence, of ineffective performance and should not be used as a basis for discipline or discharge.

The way to increase effectiveness is to have as many members of the crew working as much of the time as possible. *Figure 7* shows that, during erection of precast concrete panels, some crew members are busy only for relatively short periods. This should suggest to the project manager that perhaps not all of the people in the crew are needed or that the work method could be changed to result in more of the crew working more of the time. In this particular case, perhaps the project manager could increase productivity by removing one ironworker from the crew and having the welder perform the ironworker's tasks, assuming, of course, that the welder has been trained in ironwork. Making such a change in the crew would change the five-minute rating results: 36 effective time units divided by 65 (78 – 13) total time units equals 0.55 or 55 percent efficiency—a 19 percent improvement.

3.2.3 Crew Balance Chart

Determination of proper crew balance is critical. Crew balance is defined as the proper deployment and efficient use of personnel involved in the work process. In other words, it's having the right number and type of personnel assigned to each job. The interrelationships among members of the crew and their equipment can be displayed graphically on a crew balance chart.

A crew balance chart is a graphical depiction of the efficiency of each person in the crew and

Show Transparency 21 (Figure 7). Review the chart, using the example provided.

Show Transparency 22, Definition of Working. Explain that these are the criteria for performing the five-minute rating.

Define non-working personnel.

Explain how the five-minute rating is tallied.

Point out that the goal is to have as many members of a crew working as much of the time as possible.

Using the example, review and discuss how a change in the crew would change the five-minute rating results.

Define crew balance.

MODULE 44109-08 ◆ RESOURCE CONTROL 9.13

Job Name XYZ Fast Food

Supervisor S. Lord

Start	1 Ironworker #1	2 Ironworker #2	3 Carpenter #1	4 Carpenter #2	5 Carpenter #3	6 Welder	
2:13	✓						Crew waiting for panel to be placed
:14	✓	✓	✓	✓	✓		Installing panel / welding waiting
:15	✓	✓	✓	✓	✓		Tack rebar
:16		✓	✓	✓	✓		Install upper brace bolts
:17		✓		✓	✓		Install braces
:18			✓	✓	✓		Align panels
:19			✓	✓	✓		Ditto
:20			✓	✓	✓		Ditto
:21	✓	✓	✓				Unhook crane rigging
:22	✓	✓	✓				Ditto
:23						✓	Welder tacks rebar/crew waits for next panel
:24						✓	
:25						✓	

Total Time Units: 78 Effective Time Units: 36 Five-Minute Rating: 46%

109F07.EPS

Figure 7 ◆ Five-minute rating.

Instructor's Notes:

9.14 PROJECT MANAGEMENT

thus the activity itself. As *Figure 8* shows, a vertical bar is drawn for each worker on the crew. The shaded areas show the times that the crew member was generally working; the blank areas show the unproductive or nonworking times. The crew balance chart tells the supervisor at a glance which workers were most productive (those with the most shaded areas) and which were less productive (those with the most blank areas).

The left side of *Figure* 8 shows an example of a three-person mechanical crew preassembling sections of an air supply duct. The process involves bringing raw materials to the fabrication site, fastening them up, fitting and aligning them, and then carrying the assembled sections away. This cycle is repeated continuously.

A close look at the chart shows that one of the journeymen was not needed and could have been assigned to a different task. Properly balancing the crew and the work would produce a crew balance chart like that on the right side of *Figure 8*.

Of course, each task must be analyzed separately. If, for instance, removing completed sections of duct work and bringing raw materials to the work area requires 30 minutes instead of 15, the supervisor might consider keeping a third person on the crew to reduce the waiting time of the worker assembling the duct. Unless time data can be accurately predicted before an operation begins, the formulation of a crew balance chart is usually a result of close observation of several cycles of the same work process.

3.2.4 Delay Survey

The delay survey shown in *Figure 9* is a tool for identifying delay problems in the field. The survey form lists probable causes of job delays and gives the supervisor space to record the number of hours involved in the delay and the number of workers affected.

Show Transparency 23 (Figure 8). Review the crew balance chart and explain how it is used. Refer to the example.

Show Transparency 24 (Figure 9). Explain that a delay survey lists probable causes of delays in the field.

Explain how a delay survey is used to determine lost work hours.

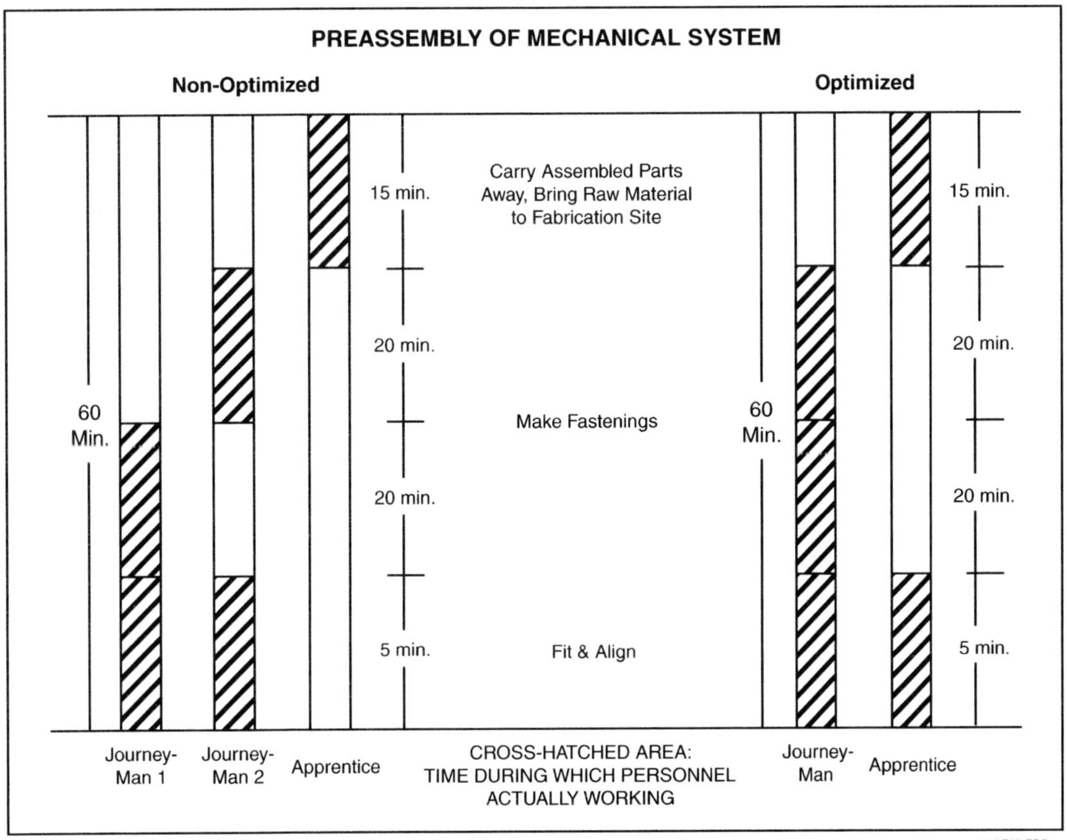

Figure 8 ◆ Crew balance chart.

MODULE 44109-08 ◆ RESOURCE CONTROL 9.15

List other methods used to analyze productivity: the stop-watch method, time-lapse motion pictures, and videotaping.

Ask participants when these methods might be used.

DELAY SURVEY

Supervisor's Name ____________________ Craft/Crew Designation ____________________

Date of Evaluation ____________________ Number in Crew(s) ____________________

PROBLEMS CAUSING DELAYS

	Work Hours Lost		
	Number of Hours ×	Number of Workers =	Work Hours
1. Waiting for materials at warehouse			
2. Waiting for materials not received or not ordered			
3. Waiting for tools or tools not available			
4. Waiting for equipment			
5. Equipment breakdowns			
6. Changes/rework-design errors			
7. Changes/rework-prefabrication errors			
8. Changes/rework-field errors			
9. Move to other work			
10. Waiting for information			
11. Interference with other crews			
12. Overcrowded working areas			
13. Plant coordination/authorizations			
14. Other			
TOTALS			

109F09.EPS

Figure 9 ◆ Delay survey.

Once the supervisor has completed the form with information from observations, the number of delay hours is multiplied by the number of delayed workers to determine the number of worker hours lost in connection with each type of delay. The columns are then totaled and the sums entered at the bottom. This form can be used by supervisors to compare the whole crew, or supervisors may elect to have the crew fill it out themselves.

A delay survey is not an exact method for determining the impact of delays on job progress, but it does provide some helpful information on delays and the magnitude of their effect on productivity.

3.3.0 More Detailed Activity/Productivity-Based Methods

Although the described methods offer the advantage of immediate results, their accuracy is limited. For situations where a more accurate or detailed analysis is required, other methods must be used. These include stop-watch studies, time-lapse motion pictures, and videotaping. Each is based on four general aims, which are to:

- Document in detail how the job is presently being done.
- Analyze current work methods in light of how they achieve job objectives. Each method pays close attention to movement of workers and materials, mobilization and demobilization of tools and equipment, material handling tasks, delays, transportation activities, and cleanup.
- Develop a new work method incorporating only those elements of movement, material, and time that are needed to meet job objectives, and do away with any unnecessary aspects of accomplishing the task.
- Implement the new method and follow up to see if, in fact, it achieves desired results more efficiently.

Instructor's Notes:

9.16 PROJECT MANAGEMENT

3.4.0 Improving Work Methods

Changing work methods is a means of putting a project back on schedule and for saving money on jobs that are on schedule. Work improvement methodology is also useful during pre-construction planning for developing more efficient ways to complete specific tasks.

Since project managers are responsible for the construction of their projects, and thus the various related jobs and activities, they must always be on the lookout for work methods that can increase productivity. They should also be familiar with methods of installation for new types of equipment.

The basis of any method for program improvement is a systematic approach to solving a problem. Many of the tools needed for this task can be found in the *Interpersonal Skills* module.

The first step is to identify the job that needs to be improved. It might be a job that is behind schedule, over budget, or one that is critical to other jobs. It is most likely one that can net substantial savings if a more efficient method of doing it is implemented, frequently one that will be repeated many times in the course of the project.

The second step is to gather the facts related to the situation. Changes in methods cannot be based on hunches or gut reactions. This is where the productivity measuring methods described earlier come into play. It is helpful if the project manager involves supervisors (and possibly others) who may be affected by the resulting changes. They should be informed about what is happening, how they will be involved, and how the outcome of the process will be used.

Once facts are gathered, the project manager must analyze all relevant factors using the job analysis methods detailed in the *Construction Planning* module. How can the job be done differently? How can operations be simplified, unnecessary activities eliminated, task sequences improved, and new tools, equipment and materials used? This part of the process must involve supervisors and workers, who are often best equipped to define and establish work methods that will increase the production rate without increasing the cost of the job.

Finally, the project manager must get a commitment from the field employees that they will implement and follow the plan. He must follow up to see if, in fact, productivity improves; if not, he should revise the plan accordingly.

Regardless of which method is used, the project manager must first commit resources, mainly time and money, to the effort. The cost of implementing the evaluation could increase the overall cost of the job. Note also that these techniques are most applicable to evaluating repetitive operations or those which occur over a relatively long period of time where the results of any corrective action can be realized before the project ends.

3.5.0 Production Control Alternatives

To improve production and keep it on track—and to react quickly and decisively to disruptions—having alternate plans and leveling resources are the most useful alternatives for improving productivity.

3.5.1 Alternate Plans

Sharp managers are always looking for better ways to do the job, alternate plans that will get the job done for less money. But they don't chase after every new idea; they carefully weigh the advantages and disadvantages of a possible alternative before they change their original plans.

3.5.2 Leveling Resources

One of the first changes management is likely to make in the plans is to level (adjust) the resources. This is done as follows:

- Moving excess labor from one job to another that has a shortage of labor
- Moving idle equipment from one job site to another that is in need of equipment
- Changing equipment that is the wrong size, or is otherwise inappropriate for one job to another job where it is more appropriate
- Redistributing supervisors so that leadership is used effectively where it is needed
- Adjusting rates of activity on non-critical activities to smooth out staffing peaks

4.0.0 ◆ CONTROLLING RESOURCES

The control of resources is the responsibility of everyone within the construction company. Each person directly connected with the project has responsibility in the control process. The field supervisor is responsible for the day-to-day control activities of material receipt, testing, inspection and use, productivity of the labor, use and care of equipment and tools, and the productivity of the equipment. The *Project Supervision* curriculum describes the field supervisor's role and each project manager should refer to it.

Review and discuss the steps for improving work methods.

Explain the benefit of having alternate plans.

Discuss ways of leveling resources and how this can improve productivity.

Assign reading of Sections 4.0.0–6.0.0.

Explain that everyone directly involved in the project is responsible for controlling resources.

See the Teaching Tip for Sections 4.0.0–5.0.0 at the end of this module.

Review the responsibilities of the field supervisor in the control process.

MODULE 44109-08 ◆ RESOURCE CONTROL 9.17

Review the responsibilities of the project manager in the control process.

Show Transparency 25, Information Sources for Material Control. Generally explain each document and point out that they contain information about what is needed on the job site and when.

Review and discuss the project manager's responsibilities for material, equipment, and tool control.

Discuss various tool control systems and the advantages and disadvantages of each.

Show Transparencies 26 and 27 (Figures 10 and 11).

Show Transparency 28, Equipment Control.

Show Transparency 29 (Figure 12).

Discuss how a tool control system may have prevented tools getting lost on Samantha's job site.

4.1.0 The Project Manager's Role

The project manager's responsibility for the control of the job resources is broader than that of the field supervisor; however, the project manager must work closely with the field supervisor to keep job resources under control. Keep in mind that all project resources are under control (well-managed) when they are what you want (quantity, quality), when you want them (date, time), and where you want them (location).

4.2.0 Material Control

The project manager is responsible for having all required materials on the job site when they are needed. The following are information sources used to determine what is needed on the job site and when:

- The estimate
- The schedule
- Bills of material
- Purchase orders
- Expediting forms
- Company procedures

Computer software that tracks material ordering and delivery helps to keep control of the job materials. The project manager must also be sure that the price being paid for the material (including taxes, delivery, and related costs) either matches or is below that quoted in the estimate. If unit costs exceed those in the estimate, the project manager must find out why.

The project manager must also develop, or at least implement, a procedure for handling additional purchases needed to cover shortages and acquire miscellaneous items needed during the course of the job. When such purchases are required, the project manager is responsible for seeing that the materials are properly ordered, delivered on time, and charged against the correct cost code, making sure that the additional charges do not bring the final cost of a specific material over the estimate.

Figure 10 shows a flow chart that summarizes the material organization process. *Figure 11* represents general guidelines for controlling materials.

4.3.0 Equipment Control

As project manager, it's your responsibility to see that required construction equipment is on the job when it is needed. The job schedule provides the information you need to determine delivery dates and the job estimate and plan provide data on the types and sizes of equipment needed. Once equipment is on site, work closely with the field supervisor to ensure that:

- The equipment is being used correctly.
- The equipment is being maintained according to its preventive maintenance schedule.
- The productivity of the equipment is at or above the estimated rate.
- The equipment is returned or sent to another site when it is no longer needed.

Figure 12 represents basic information for controlling equipment costs and planning efficient equipment usage.

4.4.0 Tool Control

The project manager is also responsible for ensuring that the labor force has the tools it needs to do the job and is trained in proper tool maintenance and usage. Day-to-day responsibility for tool use and care belongs to the field supervisor. However, the project manager must work closely with supervisors to keep tool usage under control.

If workers provide their own tools, the project manager needs to work closely with the field supervisor to see that the tools are the correct type and are being used and cared for properly. Using the wrong tool, a poorly maintained tool, or incorrect use of a tool can contribute to poor productivity.

If the company is providing hand and power tools, you must make sure that they have been ordered and are on the job site when needed so construction is not delayed. In addition, you must be sure systems have been established for:

- Parts ordering and inventory organization
- Maintenance and repair
- Replacement
- Control

Among the most successful tool control systems are the following:

- *Crew Gang Box* – Tool packages are checked out to the crew supervisor.
- *Master-Keyed Gang Boxes* – Each crew is given a gang box with a key lock. Each crew has a key for its gang box lock, but only supervisors have master keys for all gang boxes.
- *Tool Audits* – All gang boxes, whether crew or master-keyed, are audited on a frequent but irregular basis to determine the number of tools in each crew's possession. Results of the audit are compared to the crew's original inventory.

Instructor's Notes:

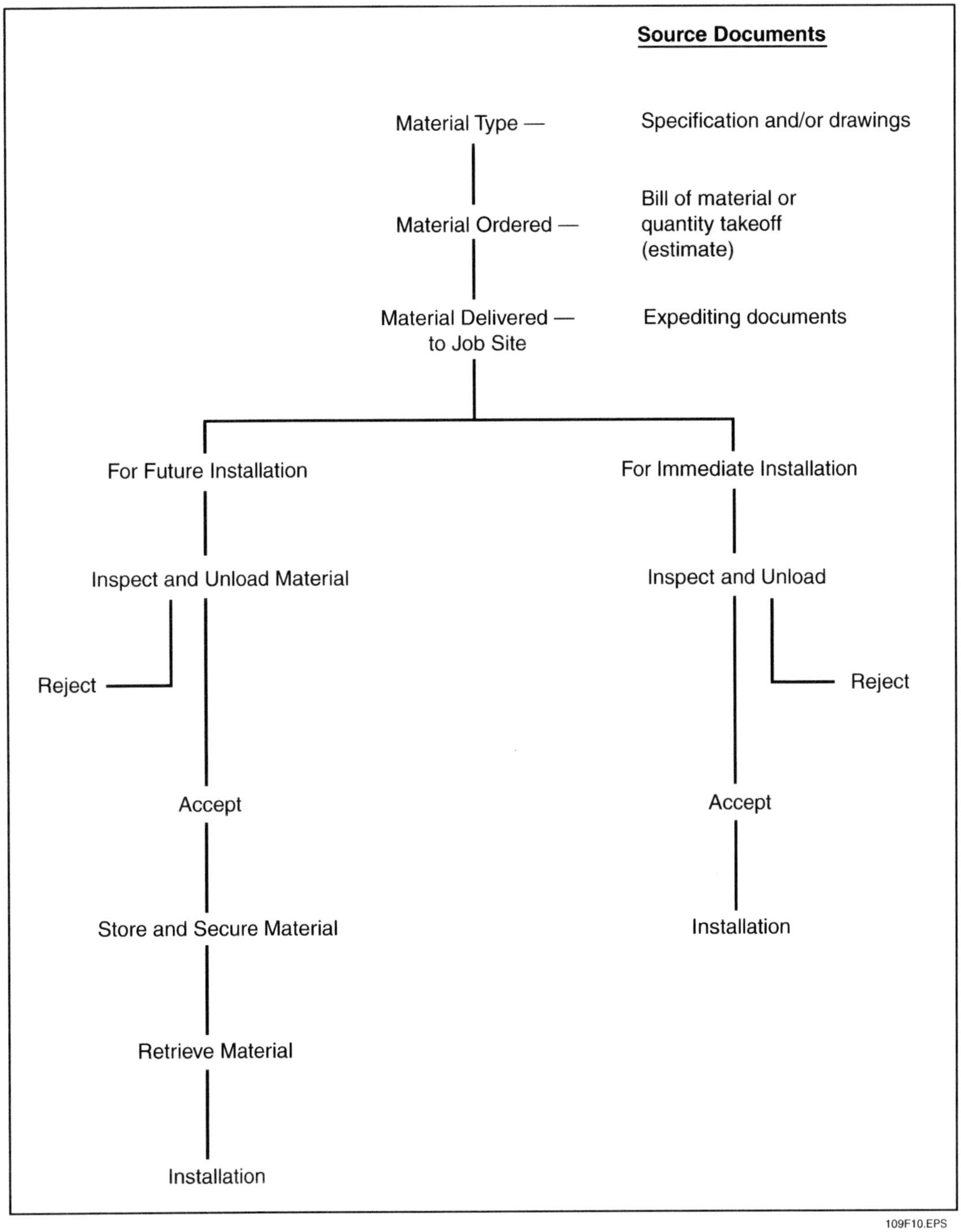

Figure 10 ◆ Material organization.

MODULE 44109-08 ◆ RESOURCE CONTROL 9.19

MATERIAL CONTROL GUIDELINES

GENERAL RULE OF THUMB

1. Check to see that all needed materials are on hand or that the delivery is scheduled.
2. Place orders for material needed late on the job.

PHYSICAL MATERIAL CONTROL AT THE JOB SITE – QUESTIONS TO ASK

1. What can be done to make sure we receive everything we sign for?
2. Is the material easy to steal?
3. Is it worth stealing?
4. Can the weather hurt it?
5. Does the risk of loss justify the cost of production?
6. Will the paperwork cause the proper cost account to be charged with correct amount?

POSSIBLE SOLUTIONS

1. When you receive a shipment, make sure you are getting what you ordered and that it is all there and in good condition.
2. After you have received it, be sure you know where you put it and that it is as well protected from loss, damage, or theft as is reasonably possible.
3. Whether it is delivered directly to the job for immediate use or is stored in a yard, warehouse or stockroom, be sure you have a written record of who is getting it and for what before you release it. The person to whom you deliver it should do the same.

109F11.EPS

Figure 11 ◆ Material control guidelines.

EQUIPMENT COSTS AND PLANNING CONSIDERATION COSTS

1. Operation Costs
   a. Fuel
   b. Oil
2. Repairs
   a. Preventive Maintenance
   b. Crash Maintenance
3. Operator
4. Downtime due to equipment itself or other equipment
5. Transportation in and out
6. Set-up and take-down time
7. Inspection
8. Housing and/or storage of equipment, equipment parts and maintenance facilities
9. Other minor costs

PLANNING CONSIDERATIONS

1. Various pieces of major equipment needed
2. Kinds of work they are to do
3. Date and time each piece of equipment is to be available for use on each kind of work
4. Equipment available
5. Equipment layout that coordinated the various pieces most efficiently
6. Timely delivery to ensure productivity
7. Properly trained operators

109F12.EPS

Figure 12 ◆ Equipment costs and planning considerations.

Instructor's Notes:

- *Brass System* – Each worker is given a numbered and recorded brass token that is handed in to check out a tool from the stock room. When the tool is returned, the token is given back. The stockroom clerk records the tool, its serial number, the date it is checked out, the date it is expected back, and the date it is actually returned.
- *Permanently Labeled Tools* – Tools are permanently marked with the company name and perhaps a serial number. In some such systems, tools are color-coded.
- *Receipt Book System* – Everything that goes out is recorded by a stockroom clerk in a receipt book. A copy of the receipt is kept in the book and a copy given to the employee. When the tool is returned, the copy that was in the book is given to the employee.
- *Bar Coding* – In this method, every worker on the project site is issued an identification card or name badge with a unique bar code imprinted on it. To check out a company tool, a worker presents the card to the person in charge of the tool storage area and requests the equipment needed. The tool room clerk logs the person's name into the tool inventory computer system by scanning the bar code. Then the tool room clerk does the same for each tool by scanning the bar code.

  The bar code description can be placed on each tool or placed on paper, along with all other tool bar codes, and collected in a three-ring binder. The binder method is more manageable, because you don't have to be concerned about having the bar code on the tool disfigured.

  When the task is completed, the craftworker returns the tools to the checkout point. At this time, the individual checking in the tools reverses the procedure, adding a step in the computer to indicate that it is a tool check-in as opposed to check-out.

  Once the individual's bar codes and those for the tools have been entered into the computer, you can generate an immediate inventory of tools checked out and to whom.

Studies have shown that the key to an effective tool control system lies in limiting:

- The number of people allowed access to stored tools
- The number of people held responsible for tools
- The number of crews using the same tools
- The ways in which a tool can be returned to storage

Figure 13 provides guidelines for controlling tools on the job.

4.5.0 Labor Control

Labor is the most difficult resource to control. Many studies show that construction personnel are only 35 to 40 percent efficient; that is, in any typical hour of any typical day only 35 to 40 percent of the time is spent performing the task assigned. The balance (60 to 65 percent) is devoted to non-productive or indirect tasks and is thus a waste of money.

A typical worker's day includes about a dozen unproductive activities, including:

- Searching for materials, tools, and equipment
- Walking with materials, tools, and equipment
- Working on jobs outside the worker's trade
- Being delayed by other trades' people
- Waiting for materials, tools, and equipment
- Walking and loading
- Being idle for no apparent reason
- Talking about topics other than work
- Personal delays

To improve productivity, the project manager must work with field supervisors to minimize these unproductive activities and the time spent on them. This means understanding why a worker does not do a full day's work. Workers fail to do a reasonable day's work due to one or more of the following reasons:

- Looking for or waiting for materials, tools, equipment, and/or labor
- Lack of information
- Lack of assignments
- Interference
- Lack of effective supervision
- Lack of motivation
- Lack of ability

Lack of information, assignments, and tools, as well as interference, can be eliminated through effective planning, scheduling, and supervision. Direction, motivation, and ability can be reinforced through effective leadership and training that instills incentive, integrity, initiative, and loyalty.

Labor costs can be minimized by careful planning of work so that each person knows what to do, and proper scheduling of workers so that they know who is to do what, how many it takes to do it, and when it should be done. *Figure 14* provides guidelines for controlling and increasing the productivity of labor.

Show Transparency 30, Successful Tool Control Systems.

Show Transparency 31 (Figure 13). Review and discuss the guidelines for controlling tools on the job.

Discuss the problem of inefficiency.

Show Transparency 32, Unproductive Activities. Review and discuss the items on the list.

Show Transparency 33, Causes of Unproductive Activities. Ask participants to give examples from their experiences.

Discuss ways that productivity can be maximized and labor costs minimized.

Discuss how labor control factors also could have impacted Samantha's schedule.

MODULE 44109-08 ◆ RESOURCE CONTROL 9.21

Show Transparency 34 (Figure 14).

Show Transparency 35 (Figure 15). Walk participants through the the case study.

Show Transparency 36, Achieving Minimum Labor Costs. Discuss how these actions can help workers to be more efficient.

Discuss how debriefings may have prevented the production delays on the site.

TOOL CONTROL GUIDELINES
• Establish a complete list of tools for the job. • Establish who provides the tools. • Establish standards for tools purchased by the contractor. • Establish controlled, organized, and accessible tool storage areas. • Establish and maintain tool inventory. • Instruct workers in the safe use of tools. • Stamp and identify tools. • Maintain spare parts. • Use proper tools for the job.

109F13.EPS

Figure 13 ◆ Tool control guidelines.

LABOR CONTROL GUIDELINES
• Types and amounts of labor will vary for each job. • Determine skilled and unskilled labor requirements. • Consideration should be given to the contractor's permanent organization as well as local and imported labor from other locations. • Watch for construction, labor and wage trends. Large construction activity in a locality will mean labor pool will be low, wages will be high, and efficiency will be low. • Wages depend on the amount of work and labor pool available. Other factors include union rules and regulations, government regulations, economic conditions, and work available in other fields. • Estimate the time and type of work to be done. Schedule the date, time and type of work to be performed. • Make required reassignments. • Be sure everyone knows what to do and where to go. • Check subcontractors to observe progress. • Make sure your people have properly completed their work in preparation for a subcontractor or in conjunction with the subcontractor.

109F14.EPS

Figure 14 ◆ Labor control guidelines.

Figure 15 is a productivity case study of a construction company showing the actual cost per 100 worker average of unproductive activities and possible savings through the implementation of the noted action.

5.0.0 ◆ INCREASING LABOR EFFICIENCY

For years, a manufacturing plant had used a process that yielded only 70 percent good product, despite sporadic efforts to improve it, and plant staff eventually treated the 30-percent loss as normal. Cost standards reflected this and machine capacity was adjusted to allow for it. The control system sounded an alert only when the 30-percent loss was exceeded; as a result, the control procedure actually perpetuated the poor performance. The fact that their performance was chronically inferior escaped company personnel. This example could apply to any organization.

A top construction organization is not made up of one or two stars that can make a spectacular showing. It's a hard-hitting, unified team with each person doing all the things necessary to get the job done efficiently and on time.

Minimum labor costs are attained through the following actions:

- Carefully plan the work—know what the craftworkers should do.
- Properly schedule workers so that you know who is doing the work and when it should be done.
- Ensure that the right tools and materials are available so that workers do not stand by waiting for something to work with.
- Educate or coach the workers about the easiest and best ways to do the job and invite their input.
- Know and control how much labor is being used while the job is going on.
- Get the workers to do a full day's work.

6.0.0 ◆ EVALUATION AND DEBRIEFING

The techniques discussed in this module are aimed at improving the performance of the construction effort. Improvement is always appropriate, but it is sometimes urgently necessary, such as when there is a question concerning whether the correct method is being used, or whether the project is going according to plan.

Instructor's Notes:

9.22 PROJECT MANAGEMENT

A SUMMARY OF UNPRODUCTIVE TIME STUDY	Cost per Year 100 Worker Avg.	Possible Savings	$ Savings Per Year
1. Stopped work to get information. When one worker in a work crew stops to look at a drawing, everyone in the crew also looks at the drawing. Closer supervision can reduce this.	$ 265,018	25%	$ 66,255
2. Going for material to use at work site. Much time was spent on trips to get material. Proper planning of the complete material requirements at least a day in advance would eliminate this. This also requires adequate material store facilities at the work site, and a crew soley devoted to material handling.	$ 265,018	70%	$ 185,513
3. Crew members not necessary to work being done. Closer attention to the proper size work crew is necessary to reduce this.	$ 426,305	10%	$ 42,631
4. Stopped for lack of tools. Proper advance planning is necessary.	$ 5,060	80%	$ 4,048
5. Stopped for lack of material. Having all the material required for the day at the work site is necessary.	$ 186,588	80%	$ 149,271
6. Idle time. Even though it is not possible to have an employee work 100% of the time, better supervision will reduce this.	$ 509,795	50%	$ 254,898
7. Start-up, clean-up, breaks, etc. There are 8 periods during the day when production is down. They are: a. First 15 minutes in the morning b. 15 minutes before the coffee break c. 15 minute coffee break d. 15 minutes after coffee break e. 15 minutes before lunch f. 15 minutes after lunch g. 15 minute afternoon rest period h. 15 minutes prior to clean-up Closer supervision is needed during these times to manage this.	$ 755,205	50%	$ 377,603
TOTAL SAVINGS			**$ 1,080,219**

109F15.EPS

Figure 15 ◆ A summary of unproductive time study.

MODULE 44109-08 ◆ RESOURCE CONTROL 9.23

Explain evaluation as it relates to resource control.

Describe the purpose and process of debriefing.

Point out that evaluation and debriefing should be fair and constructive.

See the Teaching Tip at the end of this module: Caisson Case Study.

Several other modules in this curriculum offer tools for evaluating performance; however, it is important to mention two in the context of resource control. They can help point out when an improvement technique may be necessary to achieve the goals of the project. These two tools are the evaluation and the debriefing.

Most of us think of an evaluation as something done at the end of an activity or project. Actually, effective evaluation is ongoing and is done at any time. An evaluation is a judgment based on measurement of performance or conditions. The measurement might be in the form of answers to questions such as, "Did I make the right decision about how to handle this problem?" or "Is this job proceeding as planned?" Evaluations can be conducted to judge overall company performance, performance of an entire crew, or performance of an individual.

A debriefing is simply a discussion of a job or project in which both positive and negative aspects of planning, execution, and completion are reviewed. The sole purpose of a debriefing is to obtain an individual's or a group's observations and opinions so that, in the future, the company can avoid repeating errors and can duplicate behavior that was most productive. Debriefings are not conducted in order to lay blame or develop grounds for disciplinary action.

When conducted regularly during the construction process, evaluations and debriefings help the project manager solve problems before they get out of hand. Ongoing evaluation flags potential problems and gives managers time to make changes before production gets too far off track.

Because it is part of the job, you must understand the purpose of evaluations. In evaluating supervisors, subcontractors, and others associated with a construction project, you are not simply trying to identify those who are doing less than satisfactory work, but also trying to help everyone on the team improve their performance.

To do this, you must be certain that evaluations are fair and that criticisms are constructive. Always aim at offering helpful advice in a way that it will be accepted and used to improve performance. In doing each evaluation, keep in mind personal experiences and base your own evaluation and debriefing techniques on those which you found most helpful in your own life.

Instructor's Notes:

9.24 PROJECT MANAGEMENT

Review Questions

1. List the five elements of production control.
   1. ______________________
   2. ______________________
   3. ______________________
   4. ______________________
   5. ______________________

2. List three production control standards.
   1. ______________________
   2. ______________________
   3. ______________________

3. List five major factors which affect productivity.
   1. ______________________
   2. ______________________
   3. ______________________
   4. ______________________
   5. ______________________

4. To be sure that workers are obtaining estimated productivity, they should be ________ percent or more efficient.
   a. 40
   b. 50
   c. 60
   d. 70

5. Besides activity sampling, three more accurate, although more complex, methods for evaluating job-site productivity are:
   1. ______________________
   2. ______________________
   3. ______________________

6. Two production control alternatives are _____ and _____.
   a. alternate plans; labor study
   b. leveling resources; alternate plans
   c. leveling resources; work analysis
   d. work analysis; labor study

7. List the six major information sources used to determine job site needs.
   1. ______________________
   2. ______________________
   3. ______________________
   4. ______________________
   5. ______________________
   6. ______________________

8. ________ is the most difficult resource to control on any job.
   a. Material
   b. Labor
   c. Environment
   d. Equipment

9. List the six steps for attaining minimum labor costs.
   1. ______________________
   2. ______________________
   3. ______________________
   4. ______________________
   5. ______________________
   6. ______________________

10. What is the main purpose of a debriefing?
    a. to lay blame and initiate disciplinary action
    b. to acknowledge and reward quality workmanship
    c. to gather opinions on what people like/dislike about their jobs
    d. to gather opinions on what went right and wrong on project

Have participants complete the Review Questions, and go over the answers prior to administering the Module Examination.

MODULE 44109-08 ◆ RESOURCE CONTROL 9.25

Summarize the major concepts presented in this module.

Summary

As a project manager, it is your responsibility to control production so that the project is constructed for the estimated cost, within the scheduled time, and to the quality standards designated in the specifications and drawings. This means that you must continually monitor productivity on the job site and work to improve it.

This module explained the necessity of being involved on a daily basis in controlling the use of resources and reducing costs by such means as:

- Implementing more efficient work methods
- Using more effective tools and equipment
- Providing training and re-training of personnel
- Reviewing and following up on production records

Notes

Instructor's Notes:

9.26 PROJECT MANAGEMENT

Trade Terms Quiz

1. The total quantity of work accomplished is called _________.
2. A _________ is the project manager's record of whether each worker is working.
3. Work accomplished within the estimate is the _________.
4. _________ are set by the specifications and must be followed.
5. The typical measure of output per worker or output per labor hour is known as _________.
6. _________ is set by the owner and incorporated into the job plan and schedule.

Trade Terms List

Five-minute rating
Production
Productivity
Production quality standard
Production resource standard
Production time standard

Have participants complete the Trade Terms Quiz, and go over the answers.

Administer the Module Examination. Record the results on Craft Training Report Form 200, and submit the results to the Training Program Sponsor.

MODULE 44109-08 ◆ RESOURCE CONTROL 9.27

Trade Terms Introduced in This Module

Five-minute rating: The project manager observes the task area for five minutes and records whether each worker is working or not working.

Production: The total quantity of work accomplished.

Productivity: Output per unit of input: a measure of efficiency.

Production quality standard: Set by the specifications and must be followed by all construction managers.

Production time standard: Set by the owner, agreed to by the contractor, and incorporated into the job plan and schedule.

Production resource standard: Work accomplished within the estimate.

Instructor's Notes:

Resources & Acknowledgments

Acknowledgment

This module was developed in cooperation with Roger Liska, Ph.D., Department Chair and Professor, Department of Construction Science and Management, Clemson University, Clemson, South Carolina.

Figure Credits

R.M.W. Horner and B.T. Talhouni, *Effects of Accelerated Working, Delays, and Disruption of Labor Productivity*, Chartered Institute of Building, Ascot, UK, 109F01, 109F03

MODULE 44109-08 ◆ RESOURCE CONTROL 9.29

The following are suggested activities or instructional methods to help you teach the material in this module

General

When you call on someone to answer a question, the rest of the class relaxes or even tunes out because they expect that the question and answer will take place only between you and the participant you called on. Instead, use this technique to involve more participants in answering questions and to keep them on their toes.

1. Ask participants to define a term or explain a concept.
2. After one participant has answered, ask a participant seated nearby if the answer is right. Then ask whether a participant in the back of the room agrees.
3. Ask participants to explain why they think an answer is right or wrong.
4. Use the session to clear up incorrect ideas and encourage participants to learn from their mistakes.

Section 2.2.10 *Crowded Work Areas*

This activity will enable participants to experience the difficulty of working in small spaces or overcrowded conditions. The instructor will need masking tape or painter's tape, and a measuring tape. Allow 10 to 15 minutes for this exercise.

1. Start by measuring and taping out a 6' × 6' square on the floor. (Dimensions can change to accommodate larger or smaller class sizes.)
2. Ask one participant to enter the square and perform three jumping jacks. Observe that this can be done easily without any problems or interference.
3. Ask other participants to join the square, one by one, and as each one joins, the group in the square should perform three jumping jacks simultaneously without interfering with the others. How many can fit in the square and still effectively perform jumping jacks?
4. Point out that as this becomes difficult when there are just two or three participants, the same is true when adding more workers to a small work space. Emphasize that each worker needs to have enough space to effectively perform the job free from interference.

Session III *Resource Control*

This activity will give participants the opportunity to fully realize the importance of resource control. This exercise is divided into two parts. The objective of the first part is to identify the problems project managers may face when managing the four major resources. The objective of the second part is to solve the major problems identified in the first part. Participants will need their Participant Guides and pencil and paper to takes notes for discussion and presentation. The instructor will need a flip chart and markers. Allow 2 to 3 hours for this exercise.

1. Note that quality is one of the three construction performance standards. Ask participants to list the other two (cost and timeliness). The Instructor should create a list of these three standards on the flip chart.
2. Ask participants to list the four types of project resources that are needed to complete all of the work (tools, materials, equipment, and labor). The Instructor should create a list of the four project resources on the flip chart.
3. Break the class into groups of three to five individuals each, depending on the total size of the class. Provide the groups with the following instructions:
   - Each group should select a team leader. The responsibility of the leader is to facilitate the team's activities and to take notes.

- Using the brainstorming technique, choose a project resource, such as materials, and identify all of the problems project managers may having in managing this resource. The group leader should keep a list of all the problems the group identifies. The purpose of this brainstorming activity is only to compile a list of the problems. Emphasize to the groups that they should not attempt to solve the problems yet. Repeat this step until the group has identified the problems for each resource. Allow 15 minutes to 20 minutes to create lists for all four resources.
- After the lists for all four resources have been compiled, each group should review the lists and by consensus identify the top three problems in each list, based on the opinions of the group members. Allow 10 minutes for this part of the exercise.
- The leader of each group should list the top three problems for each resource on one sheet of flip chart paper. Place the sheets for each group in front of the room on the wall.
- Once all of the groups have placed their sheets on the wall, instruct the class to take a 15-minute break. During the break, the instructor should identify the top two problems for each resource based on the frequency of their listing on all the individual flip chart pages. List the top two problems for each resource on a new flip chart page and display it for the class.
- After the break, review what was determined to be the top two problems in managing each resource and answer any questions.
- Assign each group two problems associated with one resource (for example Team 1 would address problems with tools, Team 2 would address problems with materials, etc.). The objective is for each group to come up with a solution to the assigned problems (This could be two separate plans or one combined plan, depending on the problems).

  The solution should include the following:

  a. A step-by-step procedure to solve the problem.

  b. For each step, list the person who would be responsible for implementing it.

  c. When implemented, list how should the solution should be monitored to be sure it will work and any changes needed to make it work.
- The team leader for each group must record the above steps and present the group's solutions to the class. Allow 20 to 30 minutes for this part of the exercise.
- Bring the class together and allow the leader of each group to make their presentation relative to the first resource. Allow all groups assigned to the first resource to present before moving to the solutions for the second resource. Allow time for any questions or discussion from the class after each presentation. Allow 5 to 10 minutes for each group presentation with questions and answers.
- After all groups assigned a specific resource have presented, the instructor should review the material concerning that resource contained in the participant's manual and engage the students in a discussion to ensure all of their individual problems in managing that resource have been discussed. Explain any new information that has not been presented on how to effectively and efficiently manage the respective resource. The instructor's presentation for each resource should be limited to 10 minutes.
- Once the instructor has completed the presentation, move on to the next resource and continue the same presentation steps above until all four resources have been covered. The instructor should take appropriate breaks so as not to have participants sitting for more than 60 minutes at any one time.
- To conclude this activity (after all group and instructor presentations) ask the class to identity the common elements of all of the solutions such as needing to document, needing management commitment, needing resources, following up, holding individuals accountable, etc. Allow 5 to 10 minutes for this part of the exercise.

Session IV *Caisson Case Study*

This activity will give participants the opportunity to develop the most efficient plan to prepare for and drill caissons, estimate the cost of performing the work, and develop a CPM network diagram. Participants will need their Participant Guides, pencil, and paper. The Instructor will need a flipchart, markers, overhead projector, screen, blank transparency sheets, Blank Work Analysis Sheet; and copies of the Caisson Case Study, including two copies of the Blank Work Analysis Sheet, for each participant. Allow 1½ to 2 hours for this exercise.

1. Explain to participants that this exercise includes information from *Estimating and Cost Control*, *Scheduling*, and *Resource Control*. The participants will have the opportunity to use what they learned in these three modules to complete the case study. If the instructor prefers, the participants may work in teams of two to complete the case study.
2. Pass out the Caisson Case Study to participants (see the following pages). Read through the case study as a class, review the figures, and answer any questions and clarify any issues the participants may have at this time.
3. Have participants fill out the Work Analysis Sheets and develop a CPM network diagram. Point out that the total cost to perform the job must be noted on the last Work Analysis Sheet, and the critical path must be highlighted on the CPM network diagram. Allow 1 hour for this part of the exercise. The instructor may walk around the class to offer assistance and answer questions if he or she so chooses.
4. Once the participants have completed the Work Analysis Sheets and CPM, ask participants to share their answers. Show a Blank Work Analysis Sheet on the overhead projector, and fill it out with the correct information (see completed Work Analysis Sheet answer key). Use blank transparency sheets to draw a CPM network diagram for the class, and highlight the critical path.

Caisson Case Study Handout

Based on the information provided below, develop the most efficient plan to prepare for and drill caissons, estimate the cost of performing the work and develop a CPM network diagram. The plan must be formalized and placed on the Work Analysis Sheets provided with this case study. In addition, a CPM network diagram must be developed. The critical path must be highlighted on the diagram. The total cost to perform the job must appear on the last work analysis sheet.

A. The Job

Your company has been awarded a contract to:

- layout site
- strip site
- stockpile topsoil
- layout caissons
- drill caissons
- inspect caissons
- cover caissons
- leave site (no cleanup needed)

B. Planning Parameters

The following planning parameters are in effect for this job and cannot be changed:

1. Production rates, wage rates, and equipment costs are presented below.
2. The work week is typically five 8-hour days. The typical 8-hour day includes one-half hour for lunch and the appropriate breaks. The production rates take lunch and rest periods into consideration. Overtime is paid after 8 hours in any one work day. Overtime must also be paid for working on weekends and holidays. The overtime rate is 1½ times straight time. The typical work day begins at 8:00 AM but this is not mandatory.
3. All activities must fall in sequential order (follow one another), except the following which can happen with some overlap (1 hour):
   - Transportation activities
   - Layout of site and stripping and stockpiling
   - Stripping and stockpiling and layout of caissons
4. The sequence of the drilling and inspection operation must follow one another exactly as:
   - Set up drill rig
   - Drill
   - Inspect

   None of these activities can happen concurrently. Also, for any one caisson, all of these must be completed before moving to the next caisson, or prior to leaving the job site for lunch or for the day.
5. Lunch must be allowed after six or more hours of work. Lunch time is ½ hour, paid.
6. Must pay all workers for four hours minimum if they show up on the job site.
7. Daylight hours are 7:00 AM to 6:00 PM. If you work other than these hours, you must add another $100/day for lighting rental charges. This does not apply for transportation activities.
8. All equipment for the work is transported on two trucks (one having the drilling unit as a permanent attachment and the other the flatbed truck). There is also a pickup truck that the foreman drives.
9. The bulldozer used for stripping and stockpiling is carried to the job by a flatbed truck. Bulldozer can be returned after it has completed work. The flatbed truck will stay on site until the bulldozer is finished with activity. The same is true for the flatbed truck driver.
10. Truck drivers are:
    - Flatbed – special driver
    - Pickup – foreman
    - Caisson Drill Rig – one of the operators

Copyright © 2008 National Center for Construction Education and Research. Permission is granted to reproduce this page provided that copies are for local use only and that each copy contains this notice.

11. Must pay all equipment and drivers one hour travel time to the job at the very beginning of the job and one hour travel time upon completion of the specific job. No other travel time is paid.
12. Assume no delay.

C. Labor Requirements

Overall labor requirements, based on effective equipment operation:

- One foreman
- Two laborers
- Two operators

Craftworkers are not trained nor experienced to assume the duties of different craftworkers (i.e., operator cannot perform labor work).

D. Labor Productivity (Time Involved)

Based on company records and experience for the type of soil, the following average times will be used:

- Layout of general area – 2 employees – 2 hours (1 foreman and 1 laborer)
- Layout of caissons center – 2 employees – 2 hours (1 foreman and 1 laborer)
- Stripping and stockpiling – 1 employee – 2 hours (1 operator)
- Set-up of drill rig:
  First setup and final takedown – 1 hour (1 foreman, 1 operator, 2 laborers)
  Intermediate setup – ½ hour (entire crew: 1 foreman, 2 operators, 2 laborers)
  – Drilling holes – 1 hour per hole (entire crew)
  – Inspection – ½ hour per hole (entire crew and engineer)

E. Pay Rates

- Operator - $12.00/hour
- Laborers - $8.50/hour
- Foreman - $18.00/hour
- Engineer - $20.00/hour

F. Equipment Costs

- Pickup Truck - $150.00/24-hour day
- Drill Rig - $75.00/hour
- Bulldozer - $50.00/hour
- Flatbed Truck - $65.00/hour

G. Material Cost – Negligible

Copyright © 2008 National Center for Construction Education and Research. Permission is granted to reproduce this page provided that copies are for local use only and that each copy contains this notice.

CAISSON FOUNDATION PLAN (N.T.S.)

SIX 30" CAISSONS EQUALLY SPACED

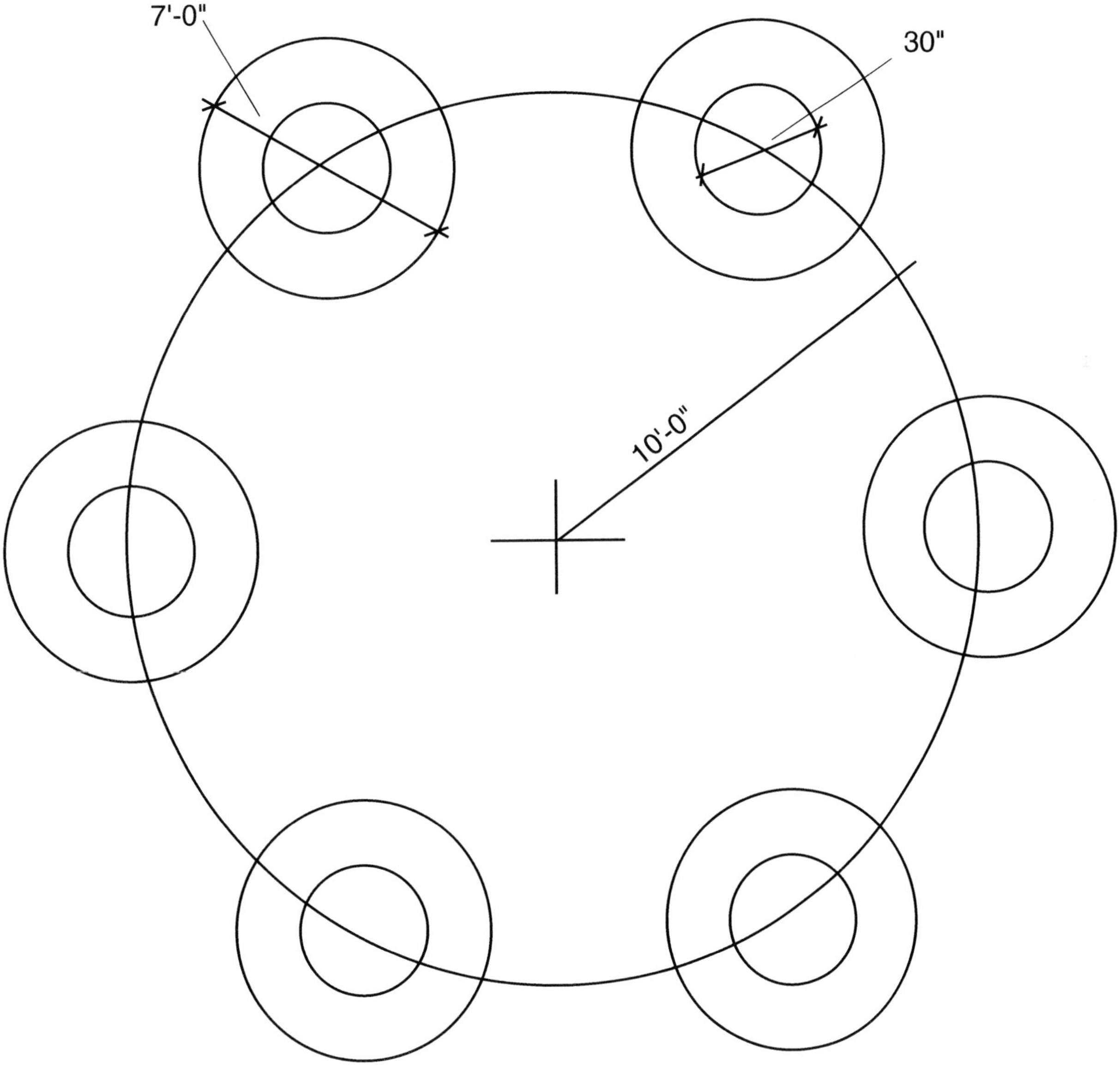

109H01.EPS

Copyright © 2008 National Center for Construction Education and Research. Permission is granted to reproduce this page provided that copies are for local use only and that each copy contains this notice.

TYPICAL CROSS SECTION (N.T.S.)

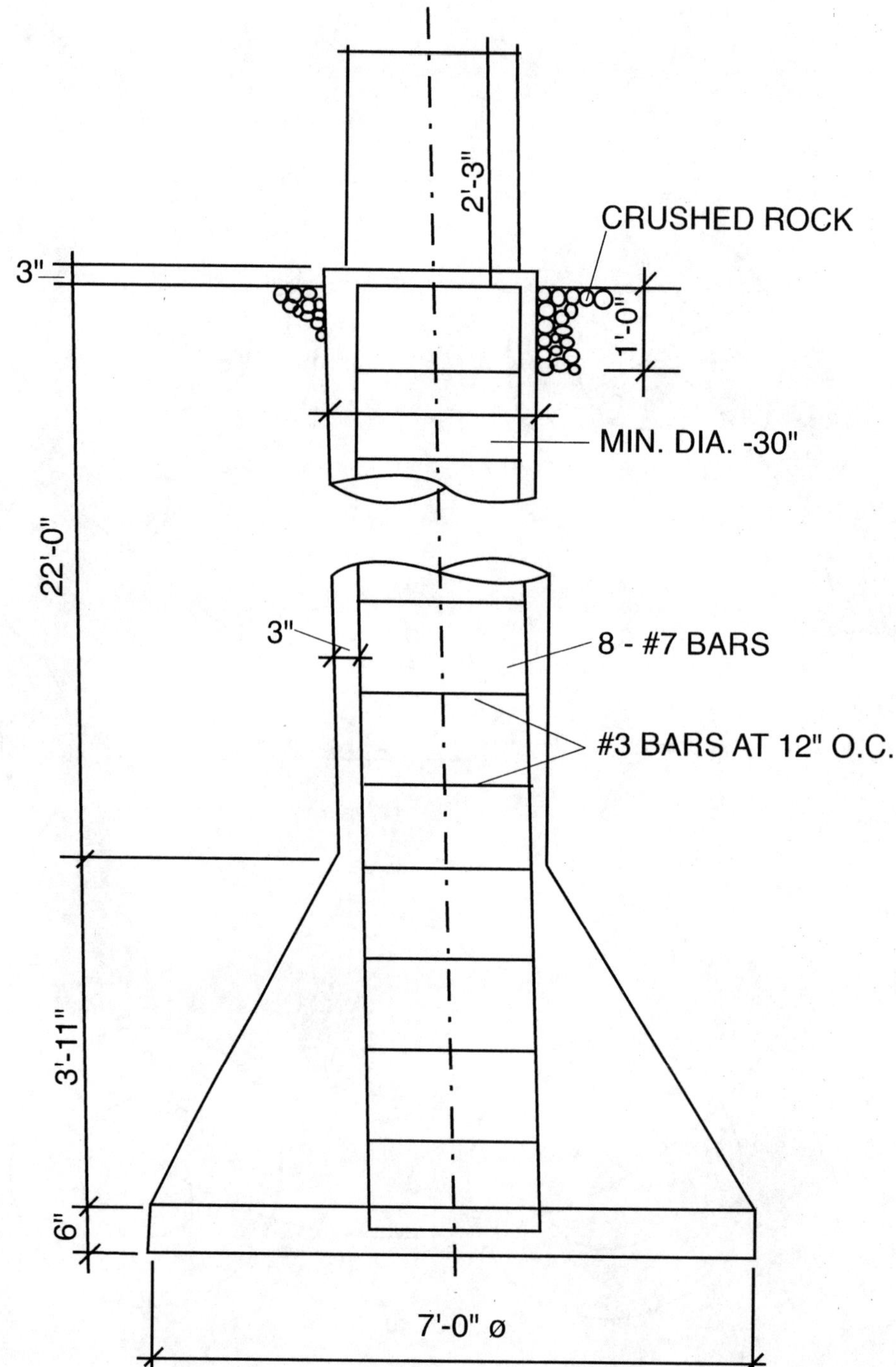

109H02.EPS

Copyright © 2008 National Center for Construction Education and Research. Permission is granted to reproduce this page provided that copies are for local use only and that each copy contains this notice.

SEQUENCE OF DRILLING OPERATION

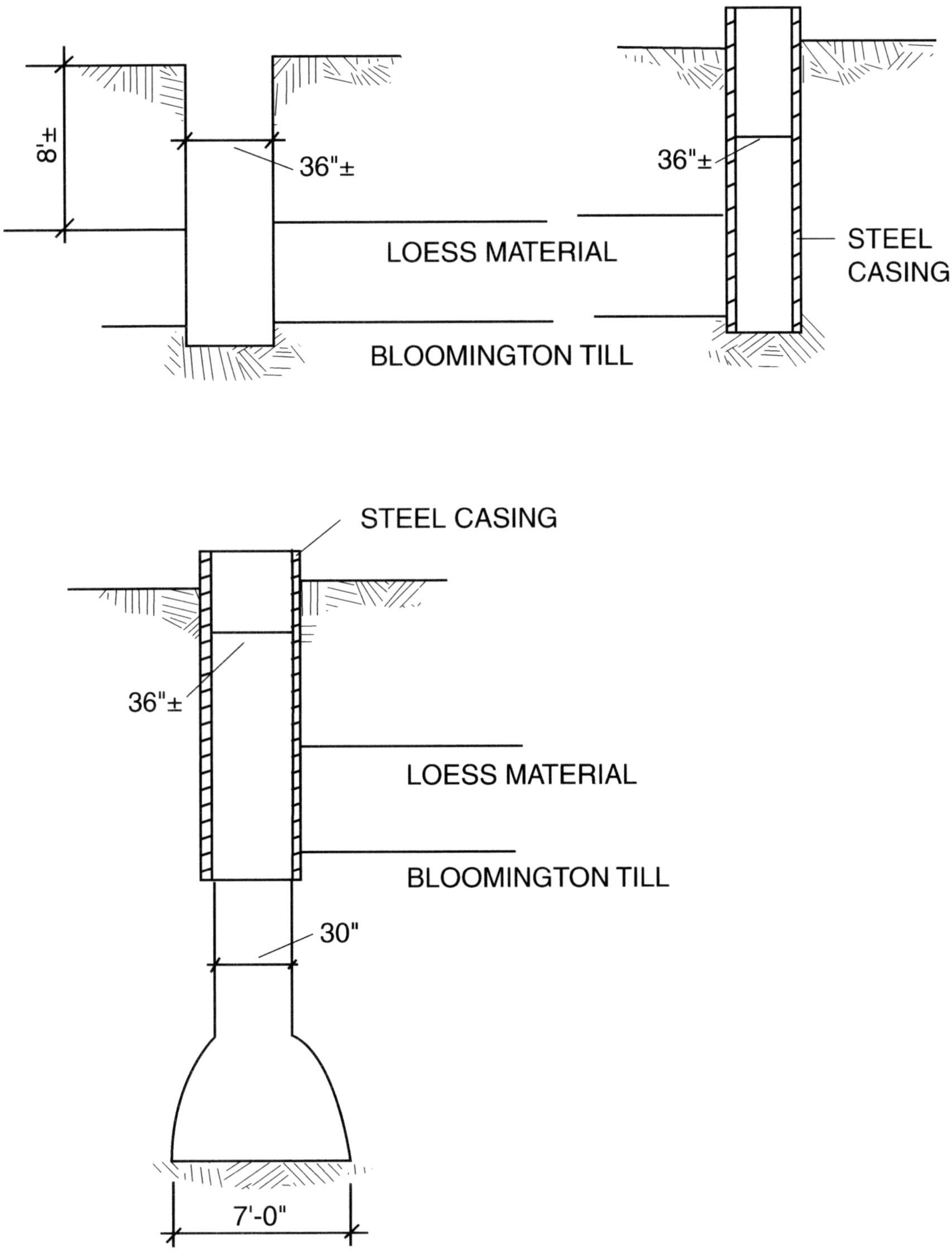

109H03.EPS

Copyright © 2008 National Center for Construction Education and Research. Permission is granted to reproduce this page provided that copies are for local use only and that each copy contains this notice.

SCHEMATIC OF TRUCK DRILLING RIG

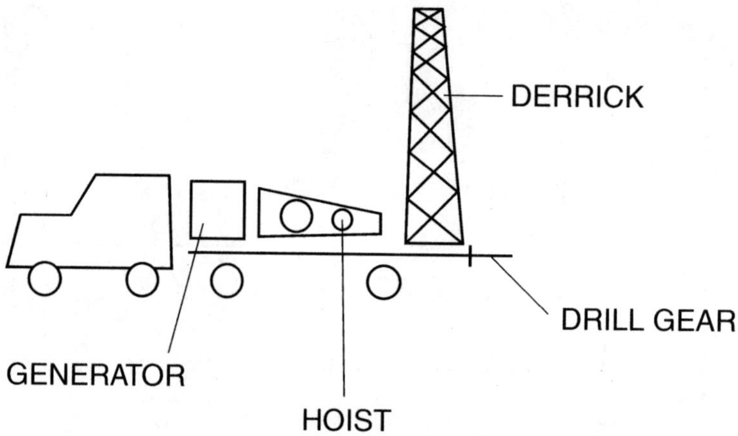

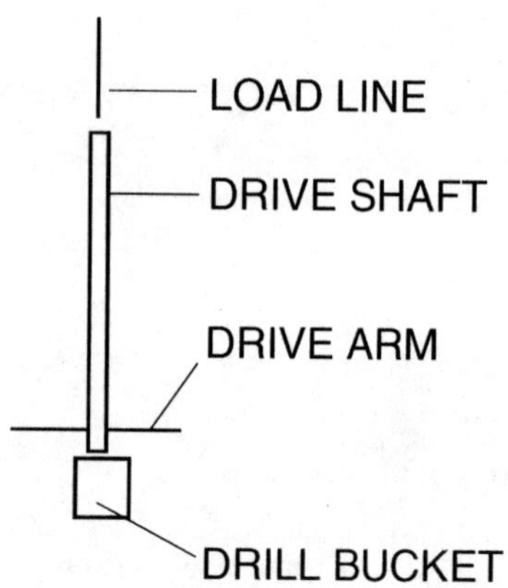

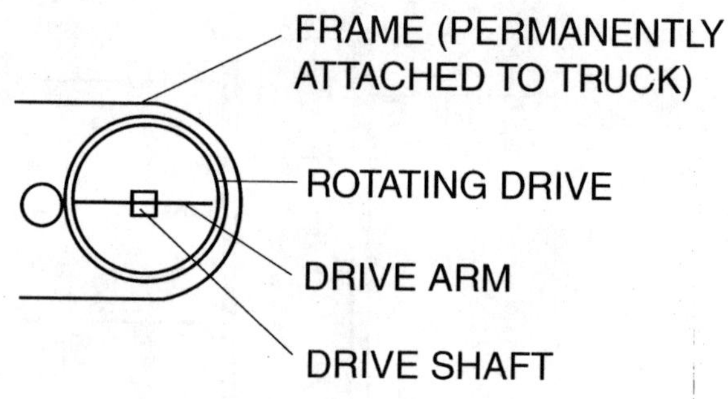

DRIVE ARM	IS SEPARATE INDIVIDUAL PIECE CAN BE REMOVED LUGS ON ROTATING DRIVE TURN DRIVE ARM ROTATION IN EITHER DIRECTION
DRIVE SHAFT	IS SEPARATE INDIVIDUAL PIECE RAISED AND LOWERED THROUGH HOLE IN DRIVE ARM BY LOAD LINE LOWER END ATTACHED TO DRILL BUCKET VARIETY OF BUCKETS WHICH CAN BE INTERCHANGED TO DRIVE SHAFT
DRIVE GEAR	OPERATED THROUGH GENERATOR ATTACHED TO A GEAR BOX

109H04.EPS

Copyright © 2008 National Center for Construction Education and Research. Permission is granted to reproduce this page provided that copies are for local use only and that each copy contains this notice.

Date:

Job:

Date/ Time	Activity	Labor				Equipment				Total L&E	Comments
		Class	Hours	Rate	Total	Class	Hours	Rate	Total		

109H05.EPS

Date:

Job:

Date/ Time	Activity	Labor				Equipment				Total L&E	Comments
		Class	Hours	Rate	Total	Class	Hours	Rate	Total		

109H05.EPS

Copyright © 2008 National Center for Construction Education and Research. Permission is granted to reproduce this page provided that copies are for local use only and that each copy contains this notice.

MODULE 44109-08 — ANSWERS TO CAISSON CASE STUDY

Date: 15-Jul-08
Job: 2008-145

Date/ Time	Activity	Labor				Equipment				Total L&E	Comments
		Class	Hours	Rate	Totals	Class	Hours	Rate	Total		
Day # 1											
7 AM - 8 AM	Move foreman and truck on site	F	1	$18.00	**$18.00**	Pickup Truck	1 day	$150.00	**$150.00**	**$168.00**	No travel for laborers
8 AM - 9 AM	Layout 1st half General Site	F	1	$18.00							
		L	1	$8.50	**$26.50**					**$26.50**	
8 AM - 9 AM	Move flatbed & dozer to site					Flatbed & dozer	1	$65.00			No travel for operator
								$50.00	**$115.00**	**$115.00**	
9 AM - 10 AM	Layout 2nd half General Site	F	1	$18.00							
		L	1	$8.50	**$26.50**					**$26.50**	
9 AM - 10 AM	Strip & stockpile 1st half	O	1	$12.00		Flatbed & dozer	1	$65.00			
					$12.00			$50.00	**$115.00**	**$127.00**	
10 AM - 11 AM	Layout 1st half Caissons	F	1	$18.00							
		L	1	$8.50	**$26.50**					**$26.50**	
10 AM - 11 AM	Strip & stockpile 2nd half	O	1	$12.00		Flatbed & dozer	1	$65.00			
					$12.00			$50.00	**$115.00**	**$127.00**	
11 Am - 12 PM	Layout 2nd half Caissons	F	1	$18.00							
		L	1	$8.50	**$26.50**					**$26.50**	
11 Am - 12 PM	Drill rig to site w/operator	0	1	$12.00	**$12.00**	Drill rig	1 day	**$1,800.00**	**$1,800.00**	**$1,812.00**	
11 Am - 12 PM	Return flatbed and dozer	0	1	$12.00		Flatbed & dozer	1	$65.00			Operator idle on site
					$12.00			$50.00	**$115.00**	**$127.00**	
12 PM - 12:30 PM	Lunch	2 - O	0.5	$24.00							
		1 - L	0.5	$8.50							
		1 - F	0.5	$18.00	**$25.25**					**$25.25**	
12:30 PM - 1:30 PM	Set up drill rig	2 - O	1	$24.00							
		1 - L	1	$8.50							
		1 - F	1	$18.00	**$50.00**					**$50.50**	
Totals This Page					**$247.75**				**$2,410.00**	**$2,657.75**	$2,657.75

MODULE 44109-08 — ANSWERS TO REVIEW QUESTIONS

	Answer	Section
1.	1. reports 2. factors 3. standards 4. alternatives 5. evaluation	(2.0.0)
2.	1. acceptable quality of work 2. project on schedule 3. work accomplished within estimates	(2.0.0)
3.	Any five of the following: 1. change orders 2. schedule changes 3. construction mistakes and rework 4. administrative problems 5. labor turnover 6. weather 7. accidents 8. overtime 9. difficult conditions 10. inefficient methods	(2.2.0)
4.	c.	(3.2.1)
5.	1. stopwatch studies 2. time-lapse photography 3. videotape	(3.3.0)
6.	b.	(3.5.0)
7.	1. estimate 2. schedule 3. bills of material 4. purchase orders 5. expediting forms 6. company procedures	(4.2.0)
8.	b.	(4.5.0)
9.	1. carefully planning work 2. properly schedule workers 3. have the right tools and materials 4. educate, coach and obtain worker input 5. control the labor being used 6. get a full day's work from personnel	(5.0.0)
10.	d.	(6.0.0)

MODULE 44109-08 — ANSWERS TO TRADE TERMS QUIZ

1. production
2. five-minute rating
3. production resource standard
4. Production quality standard
5. productivity
6. Production time standard

CONTREN® LEARNING SERIES — USER UPDATE

NCCER makes every effort to keep these textbooks up-to-date and free of technical errors. We appreciate your help in this process. If you have an idea for improving this textbook, or if you find an error, a typographical mistake, or an inaccuracy in NCCER's Contren® textbooks, please write us, using this form or a photocopy. Be sure to include the exact module number, page number, a detailed description, and the correction, if applicable. Your input will be brought to the attention of the Technical Review Committee. Thank you for your assistance.

Instructors – If you found that additional materials were necessary in order to teach this module effectively, please let us know so that we may include them in the Equipment/Materials list in the Annotated Instructor's Guide.

Write: Product Development and Revision
National Center for Construction Education and Research
3600 NW 43rd St, Bldg G, Gainesville, FL 32606

Fax: 352-334-0932

E-mail: curriculum@nccer.org

Craft ______ Module Name ______

Copyright Date ______ Module Number ______ Page Number(s) ______

Description ______

______

______

______

(Optional) Correction ______

______

______

(Optional) Your Name and Address ______

______

______

Module 44110-08

Quality Control and Assurance

NCCER STANDARDIZED CRAFT TRAINING PROGRAM

The National Center for Construction Education and Research (NCCER) provides a standardized national program of accredited craft training. Key features of the program include instructor certification, competency-based training, and performance testing. The program provides trainees, instructors, and companies with a standard form of recognition through a National Craft Training Registry. The program is described in full in the *Guidelines for Accreditation*, published by NCCER. For more information on standardized craft training, contact the NCCER by writing us at 3600 NW 43rd St., Bldg. G, Gainesville, FL 32606; calling 352-334-0911; or emailing info@nccer.org. More information may be found at our website, www.nccer.org.

HOW TO USE THIS ANNOTATED INSTRUCTOR'S GUIDE

Each page presents two sections of information. The larger section displays each page exactly as it appears in the Trainee Module. The narrow column ties suggested trainee and instructor actions to each page and provides icons (detailed below) to call your attention to material, safety, audiovisual, or testing requirements. The bottom of each page includes space for your notes.

The **Audiovisual** icon indicates an appropriate time to show a transparency or other audiovisual aid.

The **DVD** icon indicates an appropriate time to discuss or show segments of the video.

The **Classroom** icon prompts you to define a term, stress a point, ask trainees to explain a concept, or give examples.

The **Demonstration** icon directs you to show trainees how to perform tasks.

The **Examination** icon tells you to administer the written module examination.

The **Homework** icon is placed where you may wish to assign reading for the next class, assign a project, or advise trainees to prepare for an examination.

The **Laboratory** icon is used when trainees are to practice performing tasks.

The **Materials** icon is a reminder for you to gather materials needed for classes, labs, and testing.

The **Performance Testing** icon tells you to administer a performance test or a portion thereof.

The **Safety** icon is used to emphasize safety issues. It is often keyed to *Caution* and *Warning!* statements in the Trainee Module.

The **Teaching Tip** icon indicates additional guidance is available, such as how to conduct an exercise, get the most educational value from a field trip, or encourage class participation. Teaching Tips may expand on a feature (*Think About It, Did You Know?*) or provide *Quick Quizzes* or similar exercises. You will be referred to the Teaching Tips section at the back of the module if there is additional material.

The **Combination** icon indicates that the laboratory listed corresponds with a performance task. If desired, you can note the proficiency of the trainees during the laboratory, and use it to satisfy performance testing requirements.

PREPARATION

Before teaching this module, you should review the Objectives, Performance Tasks, Materials and Equipment List, and Module Outline. Be sure to allow ample time to prepare your own training or lesson plan and gather all required materials and equipment.

MODULE OVERVIEW

This module defines the role of project managers in relation to quality control and quality assurance. It emphasizes management's concerns about quality, explains project quality management, and presents guidelines on how to develop an effective quality control plan. It also discusses how to identify, assess, and measure weaknesses to avoid rework.

PREREQUISITES

Please refer to the Course Map in the Participant Module. Prior to training with this module, it is recommended that the participant shall have successfully completed *Project Management,* Modules 44101-08 through 44109-08.

OBJECTIVES

Upon completion of this module, the participant will be able to do the following:

1. Define quality control and quality assurance.
2. Describe the essential components of an effective quality control and assurance program (or process).
3. Explain how to develop an effective quality control and assurance process.
4. Explain how to monitor the causes and costs of rework.

MATERIALS AND EQUIPMENT LIST

Desktop or laptop computer or DVD player
Multimedia projector and screen
Minor Decisions: Major Impact – How to Deal with Real Issues in Project Management DVD
Transparencies
Blank acetate sheets
Overhead projector and screen
Transparency pens
Whiteboard/chalkboard
Markers/chalk
Pencils and scratch paper
Module Examinations (located in the Test Booklet)

ADDITIONAL RESOURCES

This module is intended to present thorough resources for task training. The following reference works are suggested for both instructors and motivated participants interested in further study. These are optional materials for continued education rather than for task training.

American Society for Quality, Milwaukee, Wisconsin, www.asq.org

American Concrete Institute, Farmington Hills, Michigan, www.concrete.org

Environmental Protection Agency, Washington, D.C., www.epa.gov/hudson/012607_cqap_ph1-fswc.pdf

Haas, Dustin J., QA Guidebook. Accessed at www.oregon.gov/ODOT/HWY/QA/docs/qa_guidebook.pdf

Harrison, Jon, CQE Senior Quality Engineer, Construction Quality Assurance White Paper February 2005, Page2, Performance Validation. Accessed at www.perfval.com/news/Construction-Quality Assurance _WhitePaper_2005.pdf

LA DPW Engineering, Bureau of Engineering, Project Manual, page 2 eng.lacity.org/techdocs/pdm/Chapter09/Procedure9_1.pdf

Unified Facilities Guide Specification (UFGS), 01/01/2007, PDF document www.wbdg.org/ccb/DOD/UFGS/UFGS%20COMPLETE.pdf

TEACHING TIME FOR THIS MODULE

An outline for use in developing your lesson plan is presented below. Note that each Roman numeral in the outline equates to one session of instruction. Each session has a suggested time period of 2½ hours. This includes 10 minutes at the beginning of each session for administrative tasks and one 10-minute break during the session. Approximately 5 hours are suggested to cover *Quality Control and Assurance.*

Topic	Planned Time
Session I. The Fundamentals and Components of Quality Control and Assurance	
A. An Introduction to Quality Control and Assurance	________
B. Fundamentals of Quality Control and Assurance	________
C. Components of an Effective Quality Control and Assurance Process	________
Session II. Developing an Effecting Quality Control and Assurance Process, Monitoring Rework, and Project- and Job-Specific Quality Checklists; Review and Testing	
A. Developing an Effective Quality Control and Assurance Process	________
B. Monitoring Rework	________
C. Project- and Job-Specific Quality Checklists	________
D. Review	________
E. Module Examination	________
1. Participants must score 70% or higher to receive recognition from NCCER.	
2. Record the testing results on Craft Training Report Form 200, and submit the results to the Training Program Sponsor.	

USING THE PM DVD WITH *QUALITY CONTROL AND ASSURANCE*

Use Section 4.4.0 to discuss with the class the impact the problems presented throughout the video are likely to have on the quality of this job.

Suggested response: To meet the productivity quotas, workmanship may suffer and shortcuts may be taken that impair the quality of the job. The need to compensate for delays in the project may lead to the reduction or elimination of inspections that could catch quality issues at an early stage. Rework may be required to resolve quality problems, leading to further delays and cost impacts.

Under the topic of Quality Control, ask your students to identify some specific processes and controls that could prevent the problems in the scenarios from recurring.

Suggested responses:

- *Involve the team in the day-to-day quality improvement process.*
- *Implement the tool and labor control processes that are a critical part of resource control.*
- *Schedule regular crew debriefings to identify problems before they get out of hand.*
- *Ensure that the team understands the schedule and importance of meeting deadlines.*
- *Implement inspections for specific job requirements, as well as for areas that can lead to serious problems, such as poor housekeeping and safety hazards.*
- *Implement an effective communication process between Samantha and her team leaders, as well as among the team leaders as a group.*
- *Anticipate any need for additional workers as the job progresses and ensure that new or replacement workers are trained and able to perform as soon as they are called to the job.*

Project Management

44110-08

Quality Control and Assurance

Assign reading of Module 44110-08.

44110-08
Quality Control and Assurance

Topics to be presented in this module include:

Overview

In any construction company, quality should be as much the responsibility of the president as the project manager and craftworker. As project manager, you must be aware of the fundamental concepts of an effective quality control and assurance process and apply them in day-to-day management activities.

Many companies have established quality standards covering all aspects of quality control and assurance, from pre-planning to final inspection. Project managers in such companies should become familiar with these standards and incorporate them into their daily activities.

The Project Management Institute (PMI) has established standards (the *Project Management Body of Knowledge* or the *PMBOK® Guide*) to ensure that a construction project takes into account quality planning, assurance, and control. PMI's guidelines for project quality management are tailored to be compatible with the International Organization for Standardization's ISO 9000 and 10000 series. Also influential are guidelines established in total project quality management (TQM) processes. Together, these proprietary and non-proprietary systems provide the foundation for today's project quality management process.

Instructor's Notes:

Objectives

When you have completed this module, you will be able to do the following:

1. Define quality control and quality assurance.
2. Describe the essential components of an effective quality control and assurance program (or process).
3. Explain how to develop an effective quality control and assurance process.
4. Explain how to monitor the causes and costs of rework.

Trade Terms

Project quality management (PQM)	Quality assurance (QA)
Quality	Quality control (QC)

Prerequisites

Before you begin this module, it is recommended that you successfully complete *Project Management,* Modules 44101-08 through 44109-08.

This course map shows all of the modules in the *Project Management* curriculum. The suggested training order begins at the bottom and proceeds up. Skill levels increase as you advance on the course map. The local Training Program Sponsor may adjust the training order.

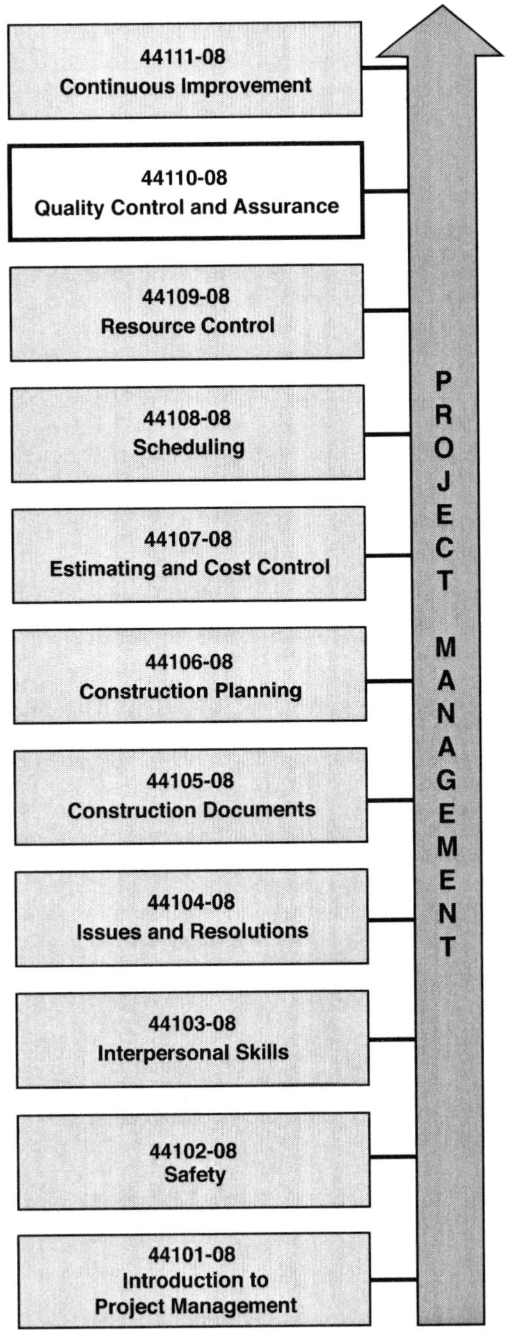

110CMAP.EPS

Ensure that you have everything required to teach the course. Check the Materials and Equipment list at the front of this module.

See the General Teaching Tip at the end of this module.

Explain that terms shown in bold are defined in the Glossary at the back of the module.

Show Transparency 1, Objectives. Review the goals of the module, and explain what will be expected of the participant.

Introduce the concept of quality standards and the importance of meeting the contractual obligations of time, cost, and quality.

Define quality control and quality assurance.

Explain project quality management (PQM) and note the three phases of an effective PQM plan.

Point out the distinction between quality control and quality assurance and who is responsible for each.

1.0.0 ◆ INTRODUCTION

Every project has specific goals and all contracts essentially focus on these three in particular:

- Construction should be completed within a specific time.
- Construction should be done for a specific cost.
- Construction should be performed in a quality manner.

All construction must be performed to meet the quality standards specified in the contract documents. As a project manager, you need to be aware of differences between quality control and quality assurance because both are part of the project quality process. Keep in mind that all three areas—time, cost, and quality—are equally important, and you must not sacrifice one to attain the others.

Quality is not a new concept to U.S. businesses. Since the 1980s, with the advent of increased global competition, organizations have seen a need to improve the quality of their products and services to remain competitive in the world marketplace. More effective quality control and assurance programs are needed throughout the construction industry. Research conducted by the Construction Industry Institute (CII) noted that when quality management systems are in place, as much as 12 percent of overall project costs can be saved when rework is substantially reduced or eliminated. If indirect cost offsets of rework are added to the direct costs, this percentage could exceed 25 percent of the total installed project costs. Including both direct and indirect costs, the average contractor has an opportunity to increase his profit margin by 5 percent (0.20 × 25 percent) by eliminating rework.

Project quality management is nothing more than controlling the day-to-day use of job resources to ensure that a project is completed in accordance with drawings and specifications. It also may mean executing a project according to company standards that exceed those in the contract documents. Many people feel that quality control lies only in the work being accepted by the designated contracting party (i.e., owner, architect/engineer, construction manager). They feel that acceptance magically makes the completed work a quality product. This is erroneous thinking.

Quality involves more than acceptance of the final product. It is an ongoing process of performing in the best and most cost-effective way possible. **Quality control** is carrying out the day-to-day activities of the company to ensure a coordinated flow of events that meet all specifications. Assurance is ongoing evaluation of the quality control process to make sure it is achieving its purpose. The acceptance of the final product is merely a benefit of effective quality control. Project quality management is doing the job right the first time, thus saving rework and warranty costs.

To consistently provide quality construction, a contractor must first realize how profoundly quality affects the health of his company. He must then formulate a well-documented, understandable, practical project quality management process for the company—one that incorporates input from all levels of the organization, not only top management.

2.0.0 ◆ FUNDAMENTALS OF QUALITY CONTROL AND ASSURANCE

The *PMBOK® Guide* identifies the general phases that make up any project quality management plan. Each phase has unique inputs, tools, and techniques that a company can modify and integrate with its project quality management plan. The phases are:

- Quality planning
- Quality control
- Quality assurance

No matter how large or small the company may be, quality control and assurance processes must be in place and incorporated into all phases of a project from pre-construction to completion. The systems of quality control and assurance provide checks and balances to ensure product quality. In both instances, the project manager has to be knowledgeable about what needs to happen and who is responsible for making it happen.

To understand the difference between quality control and quality assurance, let's look at the concepts in the following way. Quality control (QC) is an internal function performed by the contractor. It is similar in concept to your company controller or chief financial officer checking financial records to see that they conform to generally accepted accounting principles (GAAP). The company's financial department also has a system of internal controls that dictate who can perform specific functions to maintain the integrity of the financial process and eliminate mistakes in accounting.

Quality assurance (QA) is performed by the owner of the project or an outside assurance auditing firm hired by the owner. Again, we can compare this to our financial model. The company hires an outside financial firm to audit its

Instructor's Notes:

10.2 PROJECT MANAGEMENT

financial records and financial processes to ensure that they comply with GAAP. This outside auditing firm will then report to the company's chief executive any deficiencies that occur in the processes and any specific items in the financial records that need to be changed or corrected. Once these flaws are identified to the company, it is the company's responsibility to correct them. The results of these audits can affect the company's ability to obtain new business and bank financing. Smart business people make sure their companies are adhering to the proper processes and standards.

Quality control and quality assurance work the same way. As the contractor, your company is responsible for quality control; that is, doing the job to meet the specifications and conditions of the project owner (your customer) and any industry standards that apply. When you perform quality control, you audit the products and workmanship that go into the project you will deliver to your customer. You define the processes for testing and inspection of products and workmanship, and you specify precisely who performs the functions and the skills and certification that they need.

The project owner is responsible for quality assurance; that is, assessing and then letting your company know the deficiencies or defects they have detected in your work. Just as your company hires outside auditors to perform the audit of the company's financials, the project owner may hire a professional quality assurance audit firm to perform the QA audit. Large development firms may have their own QA teams. Either way, this QA process is important to ensure that the contractors' deliverables are to the specifications and standards.

2.1.0 Quality Planning

Quality planning identifies standards that are relevant to completing the project. This includes the company's administrative procedures, policies, and shop standards, if applicable. Other planning considerations include reviewing all phases of the work to be completed such as:

- Quality and safety issues
- Project correspondence procedures
- Purchasing
- Staging equipment and materials
- Subcontractor agreements
- Inspecting and testing requirements and procedures
- Other relevant policies and processes

Planning also includes a focus on the customer. Meeting the customer's needs is essential to a successful project, but who is the customer? There are two major customer classifications—internal and external. Internal customers are company employees. For example, when a project manager gives instructions to a first-line field supervisor, the project manager is the supplier and the supervisor a customer. Likewise, when a supervisor delegates work to his crew, he is the supplier of information and his crew the customers. If the information provided to the customer is incomplete or incorrect, the result will likely be poor-quality work through no fault of those receiving the information.

External customers can include other contractors, designers, vendors, government officials, and the client or owner for whom the facility is being constructed. The company and contractor must work in sync to achieve the project goals. To meet customer expectations, you need a set of tools and techniques that may include the following:

- Adhering to established construction standards
- Setting quality requirements
- Reporting corrected defects
- Using industry-accepted processes and standards
- Having knowledge of plans and specifications
- Ensuring prevention through tests and inspections
- Planning for and providing necessary resources
- Implementing an effective communication process
- Using internal benchmarks (review past and present project data)
- Using internal cost and price models
- Using internal workflow or cause and effect diagrams

A major result of quality planning is the project quality control and assurance plan, which defines the standards and describes the methods to measure success.

2.2.0 Quality Control

Quality control is about monitoring results to ensure they meet set standards. Quality control also identifies ways to reduce inefficiency and poor results. The project manager has two overarching goals regarding quality control: product results and management results. These two goals are the major forces behind quality control because

Show Transparency 2, Quality Planning Considerations, and Transparency 3, Tools and Techniques to Meet Customer Expectations.

Explain quality planning and the components included in the planning process.

Identify customer classifications and emphasize that quality includes a focus on the customer.

Review tools and techniques that lead to customer satisfaction.

Explain that a major outcome of quality planning is the project quality control and assurance plan.

MODULE 44110-08 ◆ QUALITY CONTROL AND ASSURANCE 10.3

Show Transparency 4, The Quality Process.

they focus on prevention and inspection. A product-results focus includes attention to the day-to-day job activities such as project correspondence, scheduling, deliverables, standard operating procedures (SOP), checklists, and other items identified in the project quality management plan. The management-results focus is on time (schedule), meetings (bid review, coordination, post-project), and other managerial-level policies and planning.

Review the definition and goals of quality control.

List the components of the quality control process.

The core of quality control rests on the implementation of testing and inspection processes to ensure the materials and systems meet the technical specifications. While this may sound simple, it requires a process to ensure that the results are valid and documented. The basic components of a quality control process generally include the following:

- A quality plan that specifies the tasks and standards
- A quality organization for the specific project
- Technical specifications
- Installation and testing programs, including the inspection, testing, and acceptance criteria and procedures
- Documentation

Show Transparency 5 (Figure 1). Review the sections of a typical quality control plan.

Show Transparency 6 (Figure 2). Describe the structure of a typical quality control organization.

2.2.1 Quality Control Plan

All important decisions and inputs regarding quality are made during the design and planning stages. It is at these pre-construction stages that the project plan is reviewed, and material specifications and functional performance are identified. During construction quality control focuses on conformance to these pre-construction decisions. If the specifications and performance criteria are wrong or not well-designed, the quality control process will be flawed and significant rework may be required. Nonetheless, these drawings and specifications are the criteria that govern your quality control process.

Another important component of quality control that is identified during pre-construction is safety. Project managers know that during the day-to-day quality control process of monitoring construction activities, safety is a concern—especially if defects are detected and rework is required. Therefore, not only are there direct costs for rework, but also the possibility of accidents and personal injuries. These indirect costs of insurance, inspection, and regulation can add additional costs. Project managers try to ensure that work is completed correctly and that there are no accidents.

The quality control plan is a comprehensive, well-defined, written set of procedures and activities aimed at delivering products that meet or exceed a customer's expectations, as expressed in the contract documents and other published sources. A quality control plan will identify the organization, or individuals, responsible for quality control and the specific procedures used to ensure delivery of a quality product. A quality control plan will also detail the method of accountability and required documentation. An outline of a sample quality control manual is shown in *Figure 1*.

2.2.2 Quality Control Organization

The quality control organization for each project identifies the roles and responsibilities of the individuals performing the quality control functions. At a minimum, it includes management (QC manager), experts and technicians in specific disciplines related to the project (engineers, technicians, and qualified craftworkers), and qualified inspectors. *Figure 2* represents a sample project organization chart for the quality control function.

To establish the QC organization, you will need to lay out the tasks to be performed by each of the groups and define the rules and requirements. Any technical certification or specific training requirements for persons reviewing the construction needs to be defined.

Briefly describe technical specifications.

2.2.3 Technical Specifications

The technical specifications for the products to be inspected and tested must be identified and referenced or included in the project quality manual. They should be sufficiently detailed to cover the necessary elements that ensure conformance to the specifications, but not too difficult to follow. Since the purpose of quality control is to adhere to the contract specifications and general conditions, the technical specifications may refer back to specific items in the contract documents and in the drawings and specifications. In other cases, the technical specifications may be manufacturer or industry specifications.

2.2.4 Inspection and Testing Program

The heart of the quality control process is the inspection, testing, and verification program. This program consists of the following:

- Scheduled inspections—job site inspections, audits, and reviews
- Testing requirements and procedures
- Acceptance criteria
- Production and management checklists and charts

Instructor's Notes:

Explain the role of inspection, testing, and verification as part of the quality control process.

Emphasize that documentation is critical and becomes part of the project's historical record.

Quality Control Plan

Section 1	Introduction	Describes the project, the contracts, and the quality program overview.
Section 2	Project QC Organization	Defines the QC structural hierarchy, key personnel and others (including subcontractors) in the QC organization, responsibilities and authorities, and training, experience, and certifications required of QC personnel.
Section 3	Technical Specifications and References	Describes the technical specifications that require testing and inspection. Cites references to relevant specifications for materials and workmanship and comments, such as specifications in the contract, manufacturers' specifications, regulatory and permitting requirements.
Section 4	Performance Monitoring Requirements	Summarizes the monitoring requirements.
Section 5	Inspection and Verification Activities	Details procedures for tracking inspection and acceptance activities.
Section 6	Construction Deficiencies	Details procedures for tracking deficiencies.
Section 7	Documentation	Details requirements for project documentation and procedures for document tracking and storage. Describes the final report to the customer at conclusion of the project.

110F01.EPS

Figure 1 ◆ Quality control manual outline.

Detailed inspection procedures must be set up to cover materials, records, and workmanship. Inspections should be routinely scheduled and include established inspection and test procedures for products (materials, plant, and machinery), workmanship (installation procedures), and processes (deliveries and scheduling). Acceptance criteria are established to measure conformance to requirements and ensure that deficiencies cited during the testing and acceptance process are documented. Production and management checklists are among the audit tools that can be used to record observations, test results, and defects and variances.

2.2.5 Documentation

Documentation is vital to the quality control process. The test results, checklists, acceptance, and variance reports all become part of the project's historical data. The amount of documentation and the level of detail required for the various

MODULE 44110-08 ◆ QUALITY CONTROL AND ASSURANCE 10.5

Discuss management's role in the review process.

List the components of a quality assurance plan.

Show Transparency 7 (Figure 3).

Describe the structure of a typical quality assurance organization.

Explain the role of QA teams in Review and Evaluation.

forms should be addressed in the QC Plan. A document control system must be set up to track documentation. Today, there are software systems to support tracking, version control, and electronic filing and storage of documents.

2.2.6 Management Reviews

The results of quality control include revising procedures and taking corrective actions. This includes the type and frequency of rework and changing processes and procedures that can be business- or field-related. The contractor should plan to hold regular management reviews throughout the project to ensure corrective actions can be taken in a timely manner. Management should look at cumulative performance data and quality metrics to assess the acceptability of the results and determine the longer range changes that need to be made to strive for zero defects.

2.3.0 Quality Assurance

Quality assurance consists of the activities the project owner takes to determine your QC process is functioning adequately and that the product will meet the requirements. In some case, regulatory bodies may also perform quality assurance functions, such as inspections required by law. The basic elements of a quality assurance plan include:

- A quality assurance organization
- Quality assurance review standards
- Review and evaluation of contractor's quality control plan
- Inspection and acceptance criteria
- Project sampling
- Documentation

2.3.1 The QA Organization

The QA organization is typically composed of a construction quality assurance officer (CQAO), a review coordinator, and a review team that will conduct the assurance process for the owner's project. The review team typically consists of representatives from various disciplines relevant to the project. In small organizations and on small jobs, the review coordinator may act as the CQAO. Quality assurance may also be an internal function of the owner. *Figure 3* represents a sample project quality assurance organization chart.

2.3.2 Review and Evaluation

The review teams develop work plans that guide reviews of their areas of expertise. These work plans also lay out the schedule for conducting reviews. For example, a lengthy and complex project may undergo quarterly reviews, while a smaller project may be subject to one or two reviews that conform to specific milestones in the

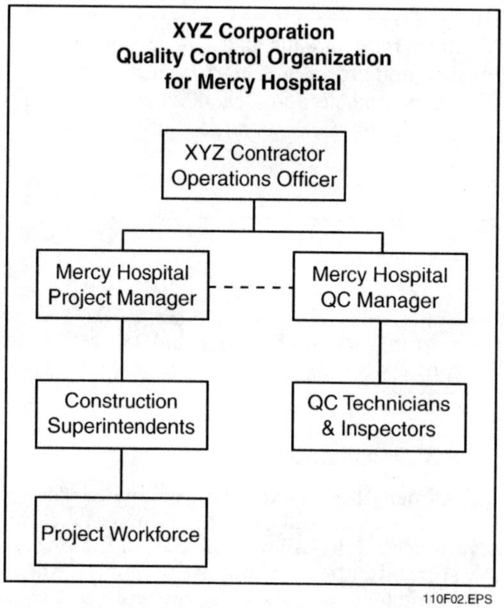

Figure 2 ◆ Sample quality control organization chart.

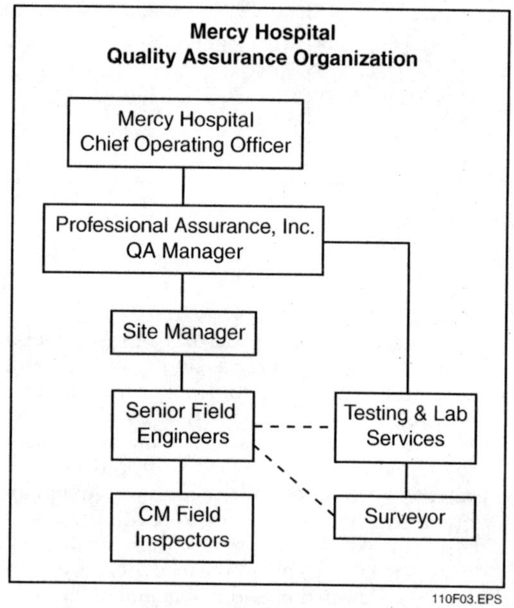

Figure 3 ◆ Sample quality assurance organization chart.

Instructor's Notes:

10.6 PROJECT MANAGEMENT

project schedule. The QA organization will audit the contractor's QC plan to ensure that it meets the standards required to deliver a quality product and that the plan provides for appropriate documentation, review, and corrective action.

2.3.3 Inspections

The QA organization also performs inspections and tests to verify the effectiveness of the quality control program. This is typically done at the discretion of the CQAO based upon the results of the contractor's QC tests and reports and the QA organization's audits. *Figure 4* shows an example of a typical quality inspection report.

2.3.4 Sampling

The QA organization may conduct on-site reviews of a sample of the activities on the project. These reviews will typically be conducted during the development phases of the project and at the conclusion of the project. The QA inspector must notify the contractor of a pending inspection. An inspection notification form is provided in *Figure 5*.

2.3.5 Documentation

The QA organization will design and document standards that will be used to review the contractor's quality control procedures. These standards are objective and based upon the contract specifications and general conditions.

The results of all reviews will be documented and feedback provided to the contractor. A deviation or failure in quality is noted on a non-conformance report. The documentation becomes a part of the contractor's and owner's historical archives for the project. *Figures 6* and *7* provide examples of a non-conformance report and a non-conformance letter that are provided to a contractor by the QA review team.

3.0.0 ◆ DEVELOPING AN EFFECTIVE QUALITY CONTROL AND ASSURANCE PROCESS

In a quality organization, everyone commits to preventing problems by focusing on doing the work correctly the first time. To instill the quality mindset in a company entails more than dictating quality policies and procedures. Craftworkers performing specific tasks must be knowledgeable, qualified, and current in their skills, and the company must provide appropriate up-to-date training. Management must properly plan so that the job processes flow on schedule. Implementing quality control and assurance programs takes cooperation and coordination among company executives, the project manager and his team, and the quality assurance organization.

3.1.0 Management Commitment

No quality program will be successful if top management is not 100 percent committed to it and provides the resources necessary to carry out the quality function. Management must develop a clear mission statement committing to quality while the company must invest in providing adequate resources to carry out quality functions. Management must also provide the project manager with the resources needed to complete the product to the specifications. The start-up costs of becoming a quality organization may seem high; however, there is documented evidence that quality produces profit and repeat business.

3.2.0 Project Manager's Commitment

As a project manager it is your responsibility to ensure your project is delivered to the specifications and customer's satisfaction. The following actions will help meet these goals.

- Infuse your team with ownership and personal pride in the quality work they produce. Involve them in the day-to-day process of improving the quality process.
- Select a qualified workforce to carry out project tasks. Make sure that workers receive training to keep their skills up-to-date.
- Ensure that the project requirements and their specific tasks are clearly understood by all workers and that any gaps are readily identified and corrected.
- Be certain that the schedule is understood and emphasize the importance of meeting deadlines and completing deliverables on schedule.
- To the extent you are responsible for testing and inspection, ensure that the specialists that monitor the quality procedures are highly qualified and hold the certifications required to perform the quality functions.

3.3.0 Implementing the Quality Processes

Implementing quality control in your organization requires that management and the workforce be committed and assume responsibility for their role in quality. It requires that the quality control procedures and processes are put into place to uncover

Show Transparency 8 (Figure 4). Discuss the QA organization's role in the inspection and testing process.

Show Transparency 9 (Figure 5). Explain on-site sampling.

Show Transparencies 10 and 11 (Figures 6 and 7). Explain the procedures for informing the contractor of non-conformance to the specifications.

Have participants review Sections 3.0.0–5.0.0

Review how to develop the QC and QA process.

Emphasize the commitment required from management and the project manager to the quality processes.

Professional Assurance, Inc. Quality Inspection Report	Job Number	Project Mercy Hospital	Page of 1/1

CONTRACTOR: ______________________ REPORT NUMBER: ______________________

TYPE OF INSPECTION:

☐ Preparatory Inspection. ☐ Initial Inspection. ☐ Follow-up Inspection. ☐ Completion Inspection.

DESCRIPTION/INTENT OF INSPECTION: ______________________

COMPONENTS/MATERIALS REVIEWED: ______________________

CONTRACTOR PERSONNEL CONTACTED: ______________________

APPLICABLE CONTRACTOR PROCEDURES, CHECK LISTS, INSTRUCTIONS: ______________________

RESULTS OF INSPECTION: ______________________

DEFICIENCIES NOTED: ______________________

RECOMMENDED CORRECTIVE ACTION: ______________________

NON-CONFORMANCES: ______________________

QA Inspector Signature: ______________________ Date: __________

110F04.EPS

Figure 4 ◆ Quality inspection report.

Instructor's Notes:

Professional Assurance, Inc. Inspection Notification Form	Job Number	Project Mercy Hospital	Page of 1/1

CONTRACTOR: ______________________ DATE: ______________

TYPE OF INSPECTION REQUESTED: ______________________

DATE AND TIME OF INSPECTION REQUESTED: ______________________

LOCATION OF INSPECTION REQUESTED: ______________________

OTHER COMMENTS: ______________________

CONTRACTOR SIGNATURE: ______________________ DATE: ______________

110F05.EPS

Figure 5 ◆ Inspection notification form.

MODULE 44110-08 ◆ QUALITY CONTROL AND ASSURANCE 10.9

Professional Assurance, Inc. (PAI) Non-Conformance Report	Job Number	Project Mercy Hospital	Page of 1/1
CONTRACTOR	REPORT NO.	DATE	

SPECIFICATION/DRAWING NO.

ITEM

PART I - To be completed by the inspector who detects a deviation.

DESCRIPTION OF NON-CONFORMANCE:

RECOMMENDED DISPOSITION: USE AS IS ☐ REWORK ☐ REPAIR ☐ SCRAP ☐

SIGNED ______________________ DATE ________

PAI FIELD ENGINEER

PART II - To be completed by the contractor who proposed the corrective action.

DESCRIPTION OF PROPOSED CORRECTIVE ACTION:

SIGNED ______________________ DATE ________

CONSTRUCTION ENGINEER

PART III - To be completed by the Engineer of Record.

RECOMMENDATION AND REMARKS:

Proposed corrective action status: Approved ☐ Rejected ☐

SIGNED ______________________ DATE ________

ENGINEER OF RECORD

PART IV - QUALITY CONTROL DISPOSITION To be determined by QC System Manager (Construction Engineer)

DECISION AND DISPOSITION INSTRUCTIONS: USE AS IS ☐ REWORK ☐ REPAIR ☐ SCRAP ☐

SIGNED ______________________ DATE ________

PAI FIELD ENGINEER

SIGNED ______________________ DATE ________

PAI CONSTRUCTION MANAGER

PART V - ENGINEERING DISPOSITION

METHOD OF APPROVALS:

☐ TELEPHONE ☐ MEMORANDUM ☐ FAX ☐ SPEC. CHANGE ☐ DRAWING CHANGE

CONVEYED BY

☐ PROJECT MANAGER ______________________ DATE ________

PART VI - DISPOSITION VERIFICATION

☐ CORRECTIVE ACTION WAS ACCOMPLISHED ON ________

SIGNED ______________________ DATE ________

PAI FIELD ENGINEER

SIGNED ______________________ DATE ________

PAI CONSTRUCTION MANAGER

110F06.EPS

Figure 6 ◆ Quality assurance non-conformance report.

Instructor's Notes:

Professional Assurance, Inc. Contractor Non-Conformance Letter (Sample)	Job Number	Project Mercy Hospital	Page of 1/1

Attention: ______________________________

Subject: Non-Conformance Report No. ____________

Gentlemen:

The attached Non-Conformance Report (NCR) details discrepancies on your contract.

Please review and take appropriate action to remedy this situation, including changing any procedures, methods and/or personnel necessary to preclude similar problems in the future. Your attention is specifically drawn to Item 12, disposition date.

We are available to discuss the attached with you.

Very truly yours,
Professional Assurance, Inc.

____

Joe Davenport
Construction Manager

cc: Program Manager
Project Manager
Construction Manager
Quality Assurance Department
Contract File

110F07.EPS

Figure 7 ◆ Quality assurance non-conformance notification letter.

MODULE 44110-08 ◆ QUALITY CONTROL AND ASSURANCE 10.11

Explain how to gain workforce commitment and the need for subcontractors and vendors to participate in the process.

Emphasize the need to adhere to standards and regulations to achieve quality.

Briefly describe the roles in the quality organization.

Show Transparency 12, Participant Exercise 1.

See the Teaching Tip for Sections 3.0.0–3.4.0 at the end of this module.

Show Transparency 13, Causes of Rework. Explain the elements of rework that the contractor can control.

and stamp out defects. Your company's role in the quality assurance process is to cooperate with the owner's QA team, participate in reviews, and implement the results of the QA findings for any defects and deficiencies. Together, these cooperative quality control and assurance processes will help ensure that the owner receives the product specified and expected.

3.3.1 Gaining Workforce Commitment

When top management commits to quality and you take responsibility for quality delivery of your project, it seems reasonable that the workforce will follow the lead. You still may face some challenges. At first, your workers may balk at more work as a result of quality procedures and inspections. They may feel threatened by the prospect of inspections and be intimidated by the new processes. To alleviate some of this anxiety, involve them in all levels and gain their input to the quality processes. Your workforce may need additional skills training to raise the level of quality to the standards required for the project. They will need to understand the new procedures that are part of the QC and QA process. Over time, the reduction in rework and personal pride in a job well done can make quality the number one priority of your workforce.

3.3.2 Subcontractors and Vendors

A comprehensive quality program needs to include your subcontractors or vendors. It is the contractor's responsibility to hire well-qualified suppliers. As the contractor responsible for successful completion of the project, you should require that they adhere to the quality processes in your quality control plan and that they participate in the testing, inspection, and reviews.

3.3.3 Adherence to Standards and Regulations

The contractor is required to adhere to codes and regulations dictated by law. You should develop good working relationships with inspection agencies and other regulatory agencies. This will help ensure that you comply with requirements as the job progresses, minimizing rework and schedule delays. You will be well-served to adopt industry best practices and standards that raise the level of quality in your company. This, too, may require additional training and continuing education for the workforce.

3.4.0 The Quality Organization

Management must assign a quality organization that is responsible for the quality process throughout the life of the project. The quality organization reports to top management with a dotted-line responsibility to the project manager. The quality manager and job site quality control teams that make up the quality organization must be independent of the workforce developing the project to ensure integrity of results.

Top management is responsible for developing the policies that govern quality in the organizations. The assigned quality organization will develop the quality manual for the project and design the inspection and test procedures, quality checklists, and other tools that will be used to measure quality on the project. It will also develop the reporting, documentation, and document control procedures. A good quality control plan will be built into the process during the preconstruction phase and take schedule and other impacts into consideration.

4.0.0 ◆ MONITORING REWORK

The contractor, unless involved in the design stage, has very little control over rework caused by the owner, designer, or vendor. Unless the contractor is also providing transportation, there is no control over that either. The causes you can control are construction errors, omissions, and changes.

4.1.0 Causes of Rework

There are hundreds, if not thousands, of specific causes for rework. Some of these are under the direct or indirect control of the contractor and some are not. For instance, a contractor who was not involved in the design phase of a project essentially has no control over any causes for rework due to design-related decisions. Studies indicate that approximately 80 percent of the causes of rework are related to changes, errors and omissions in design and 20 percent in construction.

All causes of rework can be grouped into five major categories:

- Owner changes
- Design errors, omissions, and changes
- Vendor errors, omissions, and changes
- Transportation errors
- Construction errors, omissions, and changes

Instructor's Notes:

4.2.0 Monitoring and Documenting Rework

In order to more effectively manage rework, it has to be formally monitored. Monitoring is a formal, structured way of noting when rework occurs, determining its cause, and assessing the cost impact.

The immediate responsibility of monitoring falls mainly on field supervisors, whether a foreman, superintendent or other supervisor, because this person is the first one to encounter the problem. Once a rework problem surfaces however, others such as the project manager, estimator, and CEO will become involved to some extent. No matter what kind of organizational structure exists in the company, the responsibility of monitoring rework must be placed in the appropriate person's job description. Furthermore, training needs to take place to provide the new knowledge and skills required to perform the monitoring activity in an effective and efficient manner.

The actual monitoring activity must include documenting all information related to it. Use a form, even if you have to design one, that can be used to capture the following information for each incidence of rework:

- Project identification
- Description of rework and when it was discovered
- Cause(s) of rework and person(s) responsible
- Names of personnel and their affiliation
- Any needed dates
- Description of how to perform rework
- Costs of rework
- Any other information needed to complete documentation, such as drawing numbers
- Signatures with dates are required

Whatever type of form is designed, it should meet company needs. An example of a rework form is shown in *Figure 8*.

The completed form should be kept in the project files. As part of a post-project review, all the forms applicable to a specific project should be discussed as part of that process. In addition, the completed forms for all projects should be reviewed from time to time by the appropriate person or persons to ascertain if patterns exist as to causes, person(s) responsible, times rework typically occurs, and other pertinent issues. Knowing this information can help the contractor prevent, or at least minimize, future problems pertaining to the same activities.

When implementing the use of any type of new system, employees who will be affected by it should be involved in its development. But it is important that, in the implementation stage, those using the system not be disciplined if the system identifies problems that include them. One of the best examples is related to rework.

In most cases, the first person to identify the need for rework is the front-line supervisor. This may also be the same person designated to initiate the completion of a form identifying the causes and other related information. In this case, you are essentially asking the person to report himself as a possible contributor to the problem. It is important that management not discipline the front-line supervisor, but work with him to take any corrective action to ensure that a similar type of rework does not occur again. Discipline should only be used in cases where the same problem recurs, and its prevention was within the control of the front-line supervisor who was previously supported by management to correct the unacceptable situation.

4.3.0 Cost of Rework

Project quality management must include the determination, documentation, and analysis of the costs of rework. Once the need for rework has been identified and documented, one of the next activities is to determine how much it will cost and to monitor and capture these costs in an appropriate manner.

Once the rework begins, all information related to it, such as labor hours worked, costs of material, and other factors, must be captured. This should be done in a consistent manner with the company's existing cost-control program. If a company does not have such a program, one must be developed if the project quality management is going to be effective.

There is one modification, however, that must be made to the existing cost-control system, and that is expanding the activity coding system to include rework as an added category to each activity category. If this is not done, all information captured will appear as if it is resulting as part of the original work and not rework. Since the original project estimate did not include quantities, time, and money for rework, it should be kept as a separate category in any cost control system.

For example, if XYZ Construction Company has an activity code of 0320 for placing concrete wall footings, then another code designation should be identified for rework or reworking concrete wall footings such as 0320.R. By having such a system, management, at any point during the project, can determine costs of rework sepa-

Show Transparency 14 (Figure 8). Explain who is responsible for monitoring rework and the documentation required.

Explain the critical need to capture the costs of rework and the need for job code revision to the cost-control system.

Remind the class that indirect costs must be included in the cost of rework.

MODULE 44110-08 ◆ QUALITY CONTROL AND ASSURANCE 10.13

FIELD QUALITY REPORT

Project: ______________ Contract No.: ______________

Contractor: ______________ Report No.: ________ Date: ________

Problem: ______________

Item Identification: ______________ Vendor: ______________

Drawing No.: ______________ Specification No.: ______________

Work Area: ______________

Date Found: ______________ Date Needed: ______________

Work Category: ______________

Demolition ☐ Earthwork ☐ Concrete ☐ Steel ☐ Sheet Metal ☐ HVAC ☐ Carpentry ☐ A/G Pipe ☐
Roofing ☐ Finishes ☐ Insulation ☐ Electrical ☐ Plumbing ☐ Specialties ☐ Other ☐

Possible Cause:

Design Change	☐	Shop Drawing Error	☐	Installation Problem	☐
Design Error	☐	Fabrication Error	☐	Other	☐
Design Omission	☐	Fabrication Omission	☐		
Specification Change	☐	Shipping Damage	☐		

Problem Found:

Receipt Inspection ☐ Has Vendor ________ Carrier ________ been notified? Yes ☐ No ☐

Before Installation ☐ Contact ______________ Date ________ N/A ☐

During Installation ☐ Correction to be made by Contractor ☐ Vendor ☐

After Installation ☐

Description of Problem: ______________

______________ Contractor ______________

Description of Solution: ______________

Estimated Cost: ______________

Schedule Impact: ______________

Material $ ______________ Contractor ______________ Date ________

Labor ER ______________ Date ________

Remarks ______________

Signed ______________ Date ________

110F08.EPS

Figure 8 ◆ Field quality report.

Instructor's Notes:

rate from regular work for one activity, a combination of activities, or the total project.

In terms of indirect costs of rework, it will be up to management to determine how and where to include these in the documentation. They can be either added as a percentage to the direct costs, or the indirect costs can be calculated for each individual project and appropriately documented. However accomplished, it is important not to forget to capture these costs.

4.4.0 Controlling and Avoiding Rework

Rework is a deviation of requirements caused by a lack of quality-oriented management and workers. Reducing and eliminating rework will increase the company's competitiveness and result in more work and a higher profit margin, not to mention more motivated and loyal employees. Rework can be reduced through project quality management. This program must include determining, monitoring, and documenting the causes and costs of rework.

With this information, management can take corrective steps in problem areas to ensure the quality of their processes and product. In the end, rework can be minimized by:

- Careful supervision
- Use of skilled workers
- Use of all tools and equipment in a correct and safe manner
- Thorough planning of all activities prior to their start
- Involvement of key workers and supervisors in the planning process (others outside the organization might have to be involved, such as a technical representative of a vendor)
- Involvement of key representatives of other organizations that are part of the project, such as subcontractors and vendors
- The development of a realistic schedule so undue pressure is not placed on the workforce
- Using job-specific quality checklists

5.0.0 ◆ PROJECT- AND JOB-SPECIFIC QUALITY CHECKLISTS

Before the start of any project, all quality-related requirements or standards must be identified to ensure that all work is carried out in accordance with them. This task is usually performed by the project manager, but could be done by anyone familiar with the specifications and standards, such as a QC manager, who would then submit findings to the project manager for review.

Because there are so many quality-related standards in a typical project specification, all the requirements must be documented. The most effective way to accomplish this is through an itemized list by project division. The resulting document will be a checklist of quality-related items.

5.1.0 Project Quality Checklist

The project quality checklist is quite simple to develop. To do so, review the technical specifications and drawings, beginning at the first relevant division, and abstract every item which is a quality-related requirement to be met and place it in the checklist format. This review should also include the establishment of the order of precedence of the contract documents. Many times, notes or drawings will override information found in technical specifications. An example of a checklist is shown in *Figure 9*.

If the contractor is a general contractor, and thus responsible to the owner for the entire project, a comprehensive checklist for all divisions should be developed. On the other hand, if the contractor is responsible for only a portion of the work (such as for a subcontractor), then it will be necessary to refer only to those relevant divisions of the specifications. Some recent sets of specifications have an added division which includes all quality-related requirements. However, this is an exception rather than the rule.

When the project quality checklist is completed, it can be used to:

- Schedule required tests and inspections and make sure they occur in a timely manner and in accordance with the contract documents
- Develop job-specific quality checklists to ensure that each job is prepared, performed, and evaluated in a quality manner
- Serve as a quick reference to the specifications, manufacturer's instructions, and other data that is needed in doing the work
- Identify those quality specifications (i.e., workmanlike manner) that may result in unclear assessment procedures for acceptance of the finished construction

When developing a project quality checklist, note that there are many standard or reference specifications referred to in the technical specs. Examples are building codes, such as the Uniform Building Code, and material design codes, such as the American Concrete Institute. These reference specifications either explain outcome performance requirements or how to perform a specific activity. In either case, it is important that

Explain how rework can be minimized.

Show Transparency 15, Participant Exercise 2.

See the Teaching Tip for Sections 4.0.0–4.4.0 at the end of this module.

Show Transparency 16 (Figure 9).

Introduce project- and job-specific quality checklists, and explain the need to document the quality requirements.

Demonstrate how the checklist is used.

Ask students how the problems presented in the video impact the quality of the job.

MODULE 44110-08 ◆ QUALITY CONTROL AND ASSURANCE 10.15

Project Quality Control Checklist

Project ______________________ **Location** ______________________

Completed by ______________________ **Date** ______________________

Specification Division Number	Specified Page Number	Quality Item
2-Earthwork	2-3	All filling and backfilling shall be deposited in not more than 8" layers.
	2-3	Each layer shall be compacted to a density equal to 95 percent Proctor.
	2-4	Field density tests shall be performed by XYZ Testing Laboratory, Inc., Washington, D.C.
	2-4	At least one test shall be conducted for each layer of fill for every 500 sq ft.
3-Concrete	3-1	Concrete shall have a minimum compressive strength of 2500 si at 28 days.
	3-2	Three test cylinders shall be made during each day's pour or for each 25 cubic yards of concrete.
	3-2	Speciments shall be taken by the Maresco Testing Laboratory.
	3-2	Speciments shall be made in accordance with current ASTM specifications C-30 and C-31.
4-Masonry	4-3	Masonry contractor is to lay all block in such a fashion as to not destroy the surface texture.
5-Metals	5-10	Contractor shall submit four copies of shop drawings of structural steel and iron for approval.
	5-12	Steel deck to be 22 gauge, galvanized as manufactured by PQR Company.
7-Thermal & Moisture Protection	7-3	Surfaces to which materials are to be applied should be dry and free from holes, cracks, projections and other conditions which would prevent adhesion of waterproofing material.
	7-6	Waterproofing with ABC GS206 System of Hydrocide, Mastic, and Fabric.
	7-9	Roofing contractor shall furnish owner with a written warranty that the roofing and flashing as installed will be watertight for a period of five (5) years from date of acceptance of the roof by the owner.
9-Finishes	9-9	No coat shall be applied until the preceding one is thoroughly dry and no paint shall be applied when temperature is 50°F or below or when surfaces are damp.
		Paint shall be applied according to manufacturer's directions.

110F09.EPS

Figure 9 ◆ Project quality control checklist.

Instructor's Notes:

the contractor obtain the reference specifications and have them available on the project site. These specifications will be used in the development of job-specific checklists.

5.2.0 Job-Specific Quality Checklists

The job-specific quality checklist is developed for or covers a specific job, such as placing concrete for a slab or installing electrical work. The job can be as general or specific as one wants to define it. The general rule is to develop one checklist for each type of work or job that is different from another in method, tools and equipment, and material used. Therefore, it is obvious that a checklist for placing lighting fixtures would be different than one for constructing wood stud walls. However, one checklist for placing concrete for footings may also suffice for piers.

The job-specific checklist contains two major pieces of information. First, the checklist should incorporate those items from the project quality checklist that pertain to the specific job, such as a brand name product, a specific method of installation, inspection or testing requirement, a particular grade of material, or final quality item.

The next major items are job-related tasks that have a direct effect on quality and need to be done in preparing for, performing, and cleaning up after the job. These items can be identified from:

- The experience of construction workforce
- Training
- Manufacturers and suppliers
- Trade associations
- Consultants

For instance, for the job of placing concrete for a slab, the concrete will have to be placed in accordance with American Concrete Institute standards and meet a specified strength (referencing items from the project quality checklist); the decking (or formwork) must be clean before placing the concrete; the installation of all screeds must be verified before concrete placement. These are all tasks that must be done in the total work method and which are normally not stated in the specifications.

All members of the project team affect quality. Each craftworker's and supervisor's attitude affects the quality process. The unique habits, styles, and personalities of team members often require customized personnel-specific checklists. Getting the craftworkers (especially those who will use the checklists) and supervisors involved in developing the checklists helps in making the checklists an effective tool and convincing personnel to share ownership of the company's concern for quality.

Contractors should prepare checklists for jobs for which they are responsible. General contractors can mandate the development of checklists by subcontractors through contract language or can develop them and make them available to subcontractors. It should be noted that these checklists include, where appropriate, certification documents, which can also be developed for vendors and included as part of the purchase order requirements. In this case, they would be used in receiving and inspection activities. Vendors and manufacturers can, however, develop similar checklists for their own in-house operations.

These checklists are usually used by field supervisors to prepare for, carry out, and evaluate each activity. Besides providing space for general information, such as project name and date, provide a signature space for the checklist user to sign. This signature will show commitment to the process. If other trade work is involved in the job, such as electrical and mechanical inserts in slabs, designated representatives of these trades should also sign the checklists, indicating they have checked and approved the installation of their work that needs to be completed before the completion of other work, such as placement of the concrete.

See *Figures 10, 11, 12,* and *13* for examples of job-specific checklists. At frequent intervals, the checklists should be reviewed by the project managers and field supervisors, and they should make appropriate corrections. Those using the checklists should be involved in this process. Keep in mind that no checklist can be used exactly as is for two similar projects. Modifications, additions, and deletions are always necessary, because jobs vary from one project to another. Nonetheless, a basic checklist is helpful. Using these checklists will not only help ensure that all tasks are complete in their work method, but also enable management to arrive at quicker decisions toward solutions for quality-related problems.

Show Transparencies 17–20 (Figures 10–13). Explain the two major components of job-specific checklists.

See the Teaching Tip for Sections 5.0.0–5.2.0 at the end of the module.

MODULE 44110-08 ◆ QUALITY CONTROL AND ASSURANCE 10.17

QUALITY CONTROL CHECKLIST
STRUCTURAL STEEL

Project ______________________ Location ______________________

Instructions: Perform the activities noted below and indicate date when completed. Also provide any general comments at the bottom of the form. If unable to perform any activity, notify your supervisor immediately. After completing the form, sign it and submit it to your immediate supervisor.

Item	Date Completed
1. Check for mill certificates.	________
2. Check major supporting members for heat numbers.	________
3. Check size and grade for compliance with plans and specifications.	________
4. Check for approval of welding procedures.	________
5. Verify that each welder is tested and qualified before working on the project.	________
6. Verify that stamp numbers are placed near full penetration welds.	________
7. Check preheating requirements.	________
8. Implement distortion controls.	________
9. Insure that temperature does not exceed 1100° F when straightening or cambering members.	________
10. Nondestructively test all lifting members and skid members.	________
11. Submit any approved field changes that differ from the construction drawings to Engineering for incorporation into "As-Builts" or record drawings.	________

____

Supervisor Date

110F10.EPS

Figure 10 ◆ Quality control checklist—structural steel.

Instructor's Notes:

____

____

____

QUALITY CONTROL CHECKLIST
FLASHING AND CAULKING

Project ______________________ Location ______________________

Instructions: Perform the activities noted below and indicate date when completed. Also provide any general comments at the bottom of the form. If unable to perform any activity, notify your supervisor immediately. After completing the form, sign it and submit it to your immediate supervisor.

Item	Date Completed
1. Verify that factory tests and inspections required for materials in this section have been certified by the manufacturer.	________
2. Check to see that surfaces to receive caulking are dry and clean.	________
3. Check upon completion to see that adjacent surfaces have been cleaned.	________
4. Check for proper application.	________
5. Conduct water test.	________
6. Ensure that caulking used will accept top coating and is compatible with the coating system.	________

____

Supervisor Date

110F11.EPS

Figure 11 ◆ Quality control checklist—flashing and caulking.

MODULE 44110-08 ◆ QUALITY CONTROL AND ASSURANCE 10.19

QUALITY CONTROL CHECKLIST
ARCHITECTURAL PAINTING

Project ______________________________ Location ______________________________

Instructions: Perform the activities noted below and indicate date when completed. Also provide any general comments at the bottom of the form. If unable to perform any activity, notify your supervisor immediately. After completing the form, sign it and submit it to your immediate supervisor.

Item	Date Completed
1. Check that all items that are not to be painted have been protected.	________
2. Check that interior surfaces are clean before painting.	________
3. Check that paint is stored on the site in sealed and labeled containers.	________
4. Check to see that sufficient time shall be allowed for drying between coats.	________
5. Check to see that sanding is done between coats.	________
6. Check to see that each coat of paint applied is thick enough to cover previous coat.	________
7. Check to see that all surfaces to be painted are clean of all dust, dirt, rust scale, loose particles, grease, oil, and other deleterious substances.	________
8. Check to see that wood doors and trim are primed immediately following delivery to site.	________
9. Verify sealing on tops and bottom of hung doors.	________
10. Check to see that all paint is cleaned up from all surfaces which are not to be painted.	________

____

Supervisor Date

110F12.EPS

Figure 12 ◆ Quality control checklist—architectural painting.

Instructor's Notes:

____

____

____

QUALITY CONTROL CHECKLIST
ELECTRICAL WORK

Project ______________________ Location ______________________

Instructions: Perform the activities noted below and indicate date when completed. Also provide any general comments at the bottom of the form. If unable to perform any activity, notify your supervisor immediately. After completing the form, sign it and submit it to your immediate supervisor.

Item	Date Completed
1. Check for compliance with the specifications. Check electrical subcontractor's submittal data for all materials and equipment included in the project.	________
2. Check the coordination of all conduit runs, air conditioning piping, and sheet metal work.	________
3. Conduct an operating test demonstrating that all equipment devices are performing in accordance with specification requirements.	________
4. Verify that all electrical work is in accordance with the National Electrical Code requirements.	________
5. Check installation of all exposed conduit runs to see that they are parallel and/or at right angles to the building.	________
6. Check and determine that all conduit is adequately supported.	________
7. Verify that all receptacles are grounded.	________
8. Check that all receptacles, ceiling and switch outlet boxes are installed to specification heights.	________
9. Check that all pull boxes are located in an accessible location.	________
10. Supervise all necessary tests to assure safety and performance of equipment.	________
11. Upon completion of the project, check that all circuits are as scheduled.	________
12. Implement a final operation of all electrical systems.	________
13. Verify light levels in all rooms.	________
14. Check ground, continuity, and polarization on all receptacles.	________
15. Submit any approved field changes that differ from construction drawings to Engineering for incorporation into "As-Builts" or record drawings.	________

____

Supervisor Date

110F13.EPS

Figure 13 ◆ Quality control checklist—electrical work.

MODULE 44110-08 ◆ QUALITY CONTROL AND ASSURANCE 10.21

Have participants complete the Review Questions, and go over the answers prior to administering the Module Examination.

Review Questions

1. The fundamentals of project quality management are quality _____, quality _____, and quality _____.

2. Quality control is about monitoring _____ to ensure they meet set standards.
   a. labor costs
   b. results
   c. equipment costs
   d. subcontractors

3. Two major classifications of customers are _____ and _____.
   a. doers; non doers
   b. internal; external
   c. contractors; owners
   d. vendors; suppliers

4. An important component of quality control identified during the pre-construction phase is _____.
   a. safety
   b. the delivery schedule
   c. a performance bond
   d. a change order

5. Identify two tools and techniques that are components of the quality control and assurance process.
   1. ______________________________
   2. ______________________________

6. One of the greatest opportunities for the application of a quality control and assurance process is the reduction of _____.
   a. time
   b. workforce
   c. rework
   d. absences

7. To be most effective, any procedure to assess a PQM process must involve the _____ to the construction project.
   a. general contractor
   b. subcontractor
   c. owner
   d. all of the above

8. Of all the causes of rework, the contractor has the greatest control over _____ errors, omissions, and changes.
   a. vendor
   b. designer
   c. construction
   d. client

9. The cost accounting system should have _____ codes for rework.

10. A project quality checklist can be used to:
    ____
    ____
    ____

Instructor's Notes:

____

____

____

10.22 PROJECT MANAGEMENT

Summary

Summarize the major concepts presented in this module.

An investment in quality will pay dividends to the construction firm. Once successfully implemented, quality work increases profits, brings in repeat business, and contributes to a proud and stable workforce.

All quality processes must be performed to meet the quality standards specified in the contract and to meet the project owner's expectations. As the project manager, your firm is responsible for quality control on the project, ensuring that materials and workmanship meet the requirements. Your goal is to minimize rework and strive for zero defects. The project owner performs the quality assurance role to ensure that your quality control processes are effective and that defects and deficiencies are corrected.

Implementing quality control and assurance requires commitment and involvement from everyone in the company. You must be prepared to provide the resources, skills training, and appropriate procedures and processes for your team in order for quality control measures to be accepted and applied.

Notes

MODULE 44110-08 ◆ QUALITY CONTROL AND ASSURANCE 10.23

Have participants complete the Trade Terms Quiz, and go over the answers.

Administer the Module Examination. Record the results on Craft Training Report Form 200, and submit the results to the Training Program Sponsor.

Trade Terms Quiz

1. The control of variation of workmanship, processes, and materials in order to produce a consistent, uniform product is called ____________.
2. The controlling of the day-to-day use of job resources to ensure that the project is completed in accordance with drawings and specifications is known as ____________.
3. The actions taken to ensure all standards and procedures are adhered to is known as ____________.
4. ____________ is the degree to which a product or service conforms with a given requirement.

Trade Terms List

Project quality management (PQM)
Quality
Quality assurance (QA)
Quality control (QC)

Instructor's Notes:

10.24 PROJECT MANAGEMENT

Trade Terms Introduced in This Module

Project quality management (PQM): Controlling the day-to-day use of job resources to ensure that a project is completed in accordance with drawings and specifications.

Quality: The degree of excellence of a product or service; the degree to which a product or service satisfies the needs of a specific customer; or the degree to which a product or service conforms with a given requirement.

Quality assurance: All actions taken to ensure that standards and procedures are adhered to and that delivered products or services meet performance requirements.

Quality control: The control of variation of workmanship, processes, and materials in order to produce a consistent, uniform product.

MODULE 44110-08 ◆ QUALITY CONTROL AND ASSURANCE 10.25

Resources & Acknowledgments

References

American Society for Quality, Milwaukee, Wisconsin, www.asq.org.

American Concrete Institute, Farmington Hills, Michigan, www.concrete.org

Environmental Protection Agency, Washington, D.C., epa.gov/hudson/012607_cqap_ph1-fswc.pdf

Haas, Dustin J., QA Guidebook. Accessed at www.oregon.gov/ODOT/HWY/QA/docs/qa_guidebook.pdf

Harrison, Jon, CQE Senior Quality Engineer, Construction Quality Assurance White Paper February 2005, Page2, Performance Validation. Accessed at www.perfval.com/news/ConstructionQualityAssurance_WhitePaper_2005.pdf

LA DPW Engineering, Bureau of Engineering, Project Manual, page 2 eng.lacity.org/techdocs/pdm/Chapter09/Procedure9_1.pdf

Unified Facilities Guide Specification (UFGS), 01/01/2007, PDF document www.wbdg.org/ccb/DOD/UFGS/UFGS%20COMPLETE.pdf.

Instructor's Notes:

10.26 PROJECT MANAGEMENT

MODULE 44110-08 — TEACHING TIPS

The following are suggested activities or instructional methods to help you teach the material in this module.

General

When you call on someone to answer a question, the rest of the class relaxes or even tunes out because they expect that the question and answer will take place only between you and the participant you called on. Instead, use this technique to involve more participants in answering questions and to keep them on their toes.

1. Ask participants to define a term or explain a concept.
2. After one participant has answered, ask a participant seated nearby if the answer is right. Then ask whether a participant in the back of the room agrees.
3. Ask participants to explain why they think an answer is right or wrong.
4. Use the session to clear up incorrect ideas and encourage participants to learn from their mistakes.

Section 3.0.0–3.4.0

Fundamentals of Quality Control

This exercise will give participants the opportunity to review the fundamentals of quality control and assurance, and learn more about how quality processes are implemented in various companies. They will need their Participant Guides and pencil and paper to take notes for discussion. The instructor will show Transparency 12, Participant Exercise 1. Allow 20 minutes for this exercise.

Divide the class into small groups and have them answer the questions below. After an appropriate amount of time, call on representatives of some of the groups to present their responses.

- Does your company have a project quality management process in place? If so, how are you involved?
- Why is involving those who report to you important to implementing any part of a total quality management process?
- Explain the steps in solving a quality-related problem by using a quality team.
- What are some of the questions that should be asked in weekly project progress meetings that may help prevent rework?
- Describe the components of an effective project quality management process.

Section 4.0.0–4.4.0

Causes of Rework

This exercise will familiarize participants with the causes of rework. Participants will need their Participant Guides and pencil and paper to takes notes for discussion. The instructor will show Transparency 15, Participant Excercise 2. The instructor will show Transparency 15, Participant Exercise 2. Allow 20 minutes for this exercise.

1. What are the major causes of rework?
2. What are some things the contractor can do to reduce rework?
3. If a field supervisor has the responsibility of reporting the need for rework caused by his own crew, why shouldn't he be immediately disciplined?
4. Why should rework be monitored?
5. What types of costs are associated with rework?
6. What is the difference between indirect and direct costs of rework? Give examples of each.
7. Why is it important to have a cost coding system that differentiates between normal work and rework?
8. Why should rework costs be monitored?

Sections 5.0.0–5.2.0

Developing Checklists

This exercise will help participants develop checklists for various tasks. Participants will need their Participant Guides and pencil and paper to takes notes for discussion. Allow 20-30 minutes for this exercise.

1. Ask participants to develop a job-specific quality control checklist for a process that their company would perform.
2. Call on two or three participants to present their checklists.
3. Ask other participants for feedback on the checklists presented.

MODULE 44110-08 — ANSWERS TO REVIEW QUESTIONS

	Answer	Section
1.	planning, assurance, control	(2.0.0)
2.	b.	(2.2.0)
3.	b.	(2.1.0)
4.	a.	(2.2.1)
5.	1. scheduled inspections 2. testing requirements and procedures 3. acceptance criteria 4. production management checklists and charts	(2.2.4)
6.	c.	(1.0.0)
7.	d.	(1.0.0)
8.	c.	(4.0.0)
9.	separate	(4.3.0)
10.	a. schedule required tests b. develop job-specific checklist c. serve as a quick reference to specifications and other information d. identify quality specs that may affect final acceptance	(5.1.0)

MODULE 44110-08 — ANSWERS TO TRADE TERMS QUIZ

1. quality control
2. project quality management (PQM)
3. quality assurance
4. Quality

CONTREN® LEARNING SERIES — USER UPDATE

NCCER makes every effort to keep these textbooks up-to-date and free of technical errors. We appreciate your help in this process. If you have an idea for improving this textbook, or if you find an error, a typographical mistake, or an inaccuracy in NCCER's Contren® textbooks, please write us, using this form or a photocopy. Be sure to include the exact module number, page number, a detailed description, and the correction, if applicable. Your input will be brought to the attention of the Technical Review Committee. Thank you for your assistance.

Instructors – If you found that additional materials were necessary in order to teach this module effectively, please let us know so that we may include them in the Equipment/Materials list in the Annotated Instructor's Guide.

Write: Product Development and Revision
National Center for Construction Education and Research
3600 NW 43rd St, Bldg G, Gainesville, FL 32606

Fax: 352-334-0932

E-mail: curriculum@nccer.org

Craft ______________________ Module Name ______________________

Copyright Date __________ Module Number __________ Page Number(s) __________

Description

____

____

____

(Optional) Correction

____

____

(Optional) Your Name and Address

____

____

Module 44111-08

Continuous Improvement

NCCER STANDARDIZED CRAFT TRAINING PROGRAM

The National Center for Construction Education and Research (NCCER) provides a standardized national program of accredited craft training. Key features of the program include instructor certification, competency-based training, and performance testing. The program provides trainees, instructors, and companies with a standard form of recognition through a National Craft Training Registry. The program is described in full in the *Guidelines for Accreditation*, published by NCCER. For more information on standardized craft training, contact the NCCER by writing us at 3600 NW 43rd St., Bldg. G, Gainesville, FL 32606; calling 352-334-0911; or emailing info@nccer.org. More information may be found at our website, www.nccer.org.

HOW TO USE THIS ANNOTATED INSTRUCTOR'S GUIDE

Each page presents two sections of information. The larger section displays each page exactly as it appears in the Trainee Module. The narrow column ties suggested trainee and instructor actions to each page and provides icons (detailed below) to call your attention to material, safety, audiovisual, or testing requirements. The bottom of each page includes space for your notes.

The **Audiovisual** icon indicates an appropriate time to show a transparency or other audiovisual aid.

The **DVD** icon indicates an appropriate time to discuss or show segments of the video.

The **Classroom** icon prompts you to define a term, stress a point, ask trainees to explain a concept, or give examples.

The **Demonstration** icon directs you to show trainees how to perform tasks.

The **Examination** icon tells you to administer the written module examination.

The **Homework** icon is placed where you may wish to assign reading for the next class, assign a project, or advise trainees to prepare for an examination.

The **Laboratory** icon is used when trainees are to practice performing tasks.

The **Materials** icon is a reminder for you to gather materials needed for classes, labs, and testing.

The **Performance Testing** icon tells you to administer a performance test or a portion thereof.

The **Safety** icon is used to emphasize safety issues. It is often keyed to *Caution* and *Warning!* statements in the Trainee Module.

The **Teaching Tip** icon indicates additional guidance is available, such as how to conduct an exercise, get the most educational value from a field trip, or encourage class participation. Teaching Tips may expand on a feature (*Think About It, Did You Know?*) or provide *Quick Quizzes* or similar exercises. You will be referred to the Teaching Tips section at the back of the module if there is additional material.

The **Combination** icon indicates that the laboratory listed corresponds with a performance task. If desired, you can note the proficiency of the trainees during the laboratory, and use it to satisfy performance testing requirements.

PREPARATION

Before teaching this module, you should review the Objectives, Performance Tasks, Materials and Equipment List, and Module Outline. Be sure to allow ample time to prepare your own training or lesson plan and gather all required materials and equipment.

MODULE OVERVIEW

This module describes the project manager's role in creating a culture of continuous improvement. It covers the fundamentals of a continuous improvement program and how to identify issues that require improvement. Participants will learn how to implement a continuous improvement process and measure results, with an emphasis on satisfying internal and external stakeholders.

PREREQUISITES

Please refer to the Course Map in the Participant Module. Prior to training with this module, it is recommended that the participant shall have successfully completed *Project Management,* Modules 44101-08 through 44110-08.

OBJECTIVES

Upon completion of this module, the participant will be able to do the following:

1. Describe the project manager's role in the culture of continuous improvement.
2. Explain the fundamentals of a comprehensive continuous improvement process as it relates to a project and company.
3. Present the objectives and explain the basic steps in implementing a continuous improvement process.
4. Describe some applications of continuous improvement.
5. Describe how to measure improvement.
6. Explain the importance of recognizing employees for embracing the continuous improvement process along with some of the major methods.

MATERIALS AND EQUIPMENT LIST

Desktop or laptop computer or DVD player
Multimedia projector and screen
Minor Decisions: Major Impact – How to Deal with Real Issues in Project Management DVD
Transparencies
Blank acetate sheets
Overhead projector and screen
Transparency pens
Whiteboard/chalkboard
Markers/chalk
Pencils and scratch paper
Module Examinations (located in the Test Booklet)

ADDITIONAL RESOURCES

This module is intended to present thorough resources for task training. The following reference works are suggested for both instructors and motivated participants interested in further study. These are optional materials for continued education rather than for task training.

1001 Ways to Reward Employees, Bob Nelson
Heathfield, Susan, available at http://humanresources.about.com/od/rewardrecognition/a/recognition_tip.htm
MACTEC, Atlanta, GA
Simon, Kerri, "80/20 Rule," accessed at www.isixsigma.com/library/content/c010527d.asp
Six Sigma Handbook, The, Thomas Pyzdek
Washington State University, Pullman, WA 99164

TEACHING TIME FOR THIS MODULE

An outline for use in developing your lesson plan is presented below. Note that each Roman numeral in the outline equates to one session of instruction. Each session has a suggested time period of 2½ hours. This includes 10 minutes at the beginning of each session for administrative tasks and one 10-minute break during the session. Approximately 5 hours are suggested to cover *Continuous Improvement.*

Topic	Planned Time
Session I. The Project Manager's Role in Continuous Improvement; Continuous Improvement Fundamentals	
A. Project Manager's Role in Continuous Improvement	________
B. Continuous Improvement Fundamentals	________
Session II. Implementing and Measuring Continuous Improvement; Employee Recognition; Review and Testing	
A. Implementing Continuous Improvement	________
B. Measuring Improvement	________
C. Employee Recognition	________
D. Review	________
E. Module Examination	________
1. Participants must score 70% or higher to receive recognition from NCCER.	
2. Record the testing results on Craft Training Report Form 200, and submit the results to the Training Program Sponsor.	

USING THE PM DVD WITH *CONTINUOUS IMPROVEMENT*

Referencing Section 3.3.3, Employee Involvement, ask the students if they think Samantha took advantage of employee feedback to resolve the situation.

Suggested response: Samantha asked for employee feedback and suggestions once the problem occurred. It appears she may not have sought employee feedback or recognized this particular problem prior to the situation getting out of control. By listening carefully to each team leader's concerns and explanations, she can improve employee satisfaction and enjoy the resulting increase in improved performance. She is also building a culture of teamwork.

Section 4.0.0, Implementing Continuous Improvement, presents an opportunity to discuss what methods Samantha could employ in the future to engage the team leaders in preventing other serious problems from cropping up.

Suggested response: Samantha can involve workers by encouraging them to take ownership for their work and solve problems on their own. She can engage them in ongoing project evaluations and debriefings to identify and resolve problems quickly. Engaging those closest to the problem is the most effective way to identify problem areas and devise improved processes to address the problems. Holding regular crew meetings is an important means of soliciting opinions and ideas from those closest to the tasks and processes. Conducting employee satisfaction surveys on a regular basis helps build a culture of teamwork that is necessary for a successful continuous improvement program. Samantha should analyze the process and events that occurred (find the root causes) and revise or eliminate processes and activities that do not work. To accomplish this in the most effective manner, she should engage her team leaders in this task. One method she and the team can use is process mapping.

Class Exercise: Ask the class to consider the roles of Samantha and her team leaders, Craig, Anthony, and Keith, and devise a plan for improvement using the principles of continuous improvement. Consider specific processes and functions that could improve as a result of the problems that surfaced on this job. If appropriate, process mapping can be used as a tool.

Note to Instructor: If the class is large enough, have the students form teams to discuss their approach and draft the improvement plan. Have the students present their plans to the entire class.

44111-08

Continuous Improvement

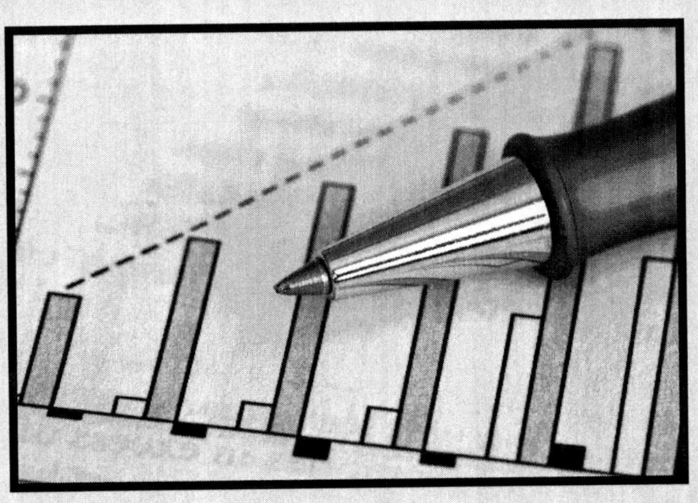

Assign reading of Module 44111-08.

44111-08
Continuous Improvement

Topics to be presented in this module include:

Overview

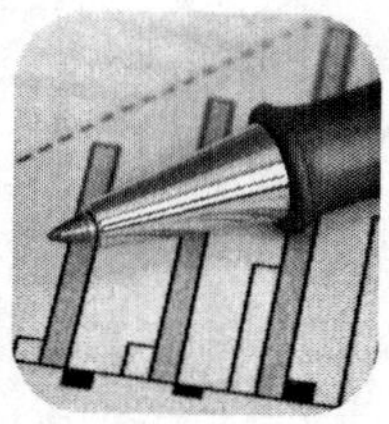

In today's construction industry, many companies have an improvement process that is integrated into their business model to better achieve their quality standards and client satisfaction. In this module, you will explore the process of continuous improvement. You will develop answers to key questions: What is continuous improvement? How is it performed? What makes it work?

Instructor's Notes:

Objectives

When you have completed this module, you will be able to do the following:

1. Describe the project manager's role in the culture of continuous improvement.
2. Explain the fundamentals of a comprehensive continuous improvement process as it relates to a project and company.
3. Present the objectives and explain the basic steps in implementing a continuous improvement process.
4. Describe some applications of continuous improvement.
5. Describe how to measure improvement.
6. Explain the importance of recognizing employees for embracing the continuous improvement process along with some of the major methods.

Trade Terms

Continuous improvement
Pareto analysis
Pareto Chart
Process map

Prerequisites

Before you begin this module, it is recommended that you successfully complete *Project Management*, Modules 44101-08 through 44110-08.

This course map shows all of the modules in the *Project Management* curriculum. The suggested training order begins at the bottom and proceeds up. Skill levels increase as you advance on the course map. The local Training Program Sponsor may adjust the training order.

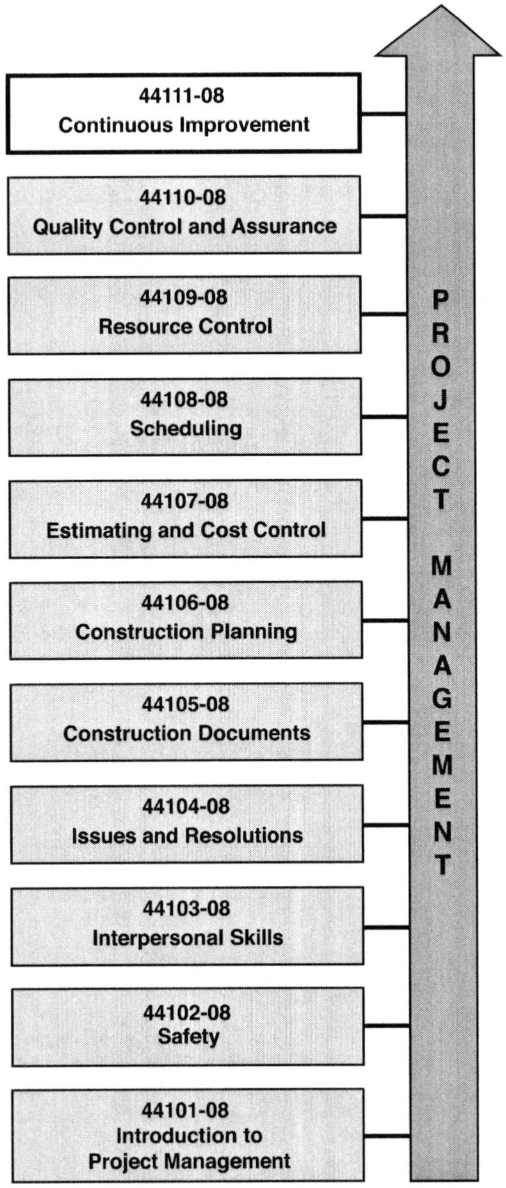

111CMAP.EPS

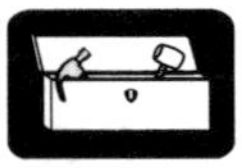

Ensure that you have everything required to teach the module. Check the Materials and Equipment list at the front of this module.

See the General Teaching Tip at the end of this module.

Show Transparency 1, Objectives. Review the goals of the module, and explain what will be expected of the participant.

Explain that terms shown in bold are defined in the Glossary at the back of this module.

MODULE 44111-08 ◆ CONTINUOUS IMPROVEMENT 11.1

Discuss the meaning and purpose of continuous improvement.

Review and discuss the project manager's role in the continuous improvement process.

Point out the direct relationship of employee satisfaction to business results and customer loyalty.

Show Transparency 2 (Figure 1). Review the employee satisfaction survey and discuss why this might be a helpful tool for project managers.

1.0.0 ◆ INTRODUCTION

To attain a sustained level of quality and client satisfaction, the contractor and employees must be knowledgeable in all areas of construction. Today's successful companies achieve a level of quality and improvement by using current knowledge about the entire construction process. Methodologies such as total quality management, partnering, lean management, and **continuous improvement** are all ways in which companies are taking an aggressive approach to meet or exceed the standards and specifications of the contract.

2.0.0 ◆ THE PROJECT MANAGER'S ROLE IN CONTINUOUS IMPROVEMENT

What do we mean by continuous improvement? The simple definition is something like this—a process that continues to achieve both acceptable and exceptional quality during a construction project, and a commitment by the company to provide a skilled and responsible workforce.

As a project manager, your company's reputation rides on how well you can do the following:

- Apply your company's quality control/assurance and continuous improvement processes
- Monitor projects
- Measure and analyze project and performance data
- Use your experience in construction
- Establish thresholds that identify trends indicating when a project is going outside allowed limits
- Apply the relationships between time, cost, quality, and productivity

It is important to remember that, along with monitoring schedules, specifications, and subcontractors, the project manager must know what goes on at the job site. For example, you must identify problems such as overlapping work, rework, interpersonal problems, resources, excessive scrap, communication gaps, and other situations that can possibly be solved through your company's continuous improvement process.

Continuous improvement means making things better by focusing on preventing problems before they occur. Coupled with your quality control/assurance program, it will enable your firm to develop improved business processes and create an environment of continuous learning and commitment to satisfying the interests of your stakeholders and your company.

Applying the principles of continuous improvement requires analyzing process and events that did not meet expectations or requirements, seeking solutions to prevent a recurrence, and formally implementing revised processes that work better. In doing so, you and your team must look at what caused the problem and why it occurred. Pinpointing the root causes of a process failure helps to:

- Reduce or eliminate redundancy and overlap
- Revise or eliminate activities
- Improve stakeholder relations and satisfaction

Another aspect of continuous improvement is monitoring employee satisfaction and continuing to build solid relationships based on employee feedback. Employee satisfaction has a direct relationship to business results (including revenues and profits) and customer loyalty. Satisfied employees perform better at their jobs. Employees with poor attitudes can have a negative affect on the project team and, in the worst case, the company's customers.

Ultimately, negative attitudes can affect various work processes on the job. Research indicates that processes account for 80 percent of all problems, while people account for the remaining 20 percent. This is referred to as the 80/20 rule, which is discussed in greater detail later in this module.

Employee satisfaction surveys are one way the project manager can formally obtain feedback from his team. There are several organizations that provide survey forms and other information related to employee satisfaction. Check with your human resources department for policies regarding employee satisfaction surveys and how they should be conducted in your company. A sample employee satisfaction survey form is shown in *Figure 1*.

As a project manager, you need to fully understand the company's process for doing business and how you can use it to improve quality. At the same time, you must build a culture of teamwork and engage the entire team in the improvement process.

Instructor's Notes:

11.2 PROJECT MANAGEMENT

EMPLOYEE QUALITY SATISFACTION SURVEY

Complete the three sections of this form and return it to Human Resources.

Section One – Quality of Work Environment

Rate each item below on a scale of 1 to 5 with 1 = strongly disagree; 2 = disagree; 3 = neutral; 4 = agree; 5 = strongly agree.

1. I feel loyal to the company. ________
2. The company provides opportunities for growth and advancement. ________
3. Management asks me for my ideas/input. ________
4. Management attempts to continually improve working conditions. ________
5. I frequently consider quitting. ________
6. I am less excited about my work than I was a year ago. ________
7. I can discuss job conditions openly with my supervisor. ________
8. There is a sense of team spirit in the company. ________
9. I care about the future of the company. ________

Section Two – Company Attitude Toward Quality

Check the answer which you feel is most appropriate for each statement.

1. Management is responsive to my suggestions to improve quality.
   Yes ________ No ________
2. Management is responsive to my requests for help with quality problems.
   Yes ________ No ________
3. Management provides me the tools and training I need to produce quality work.
   Yes ________ No ________
4. The help I receive with quality problems is adequate.
   Yes ________ No ________
5. Management recognizes and encourages good performance.
   Yes ________ No ________

Section Three – Improving Quality

1. List in a priority order five barriers to improving quality in the company. Begin the list with what you feel is the biggest barrier.

   a.
   b.
   c.
   d.
   e.

2. Relative to your work area, list below five causes of defects. Begin the list with the biggest cause.

   a.
   b.
   c.
   d.
   e.

111F01.EPS

Figure 1 ◆ Employee satisfaction survey.

MODULE 44111-08 ◆ CONTINUOUS IMPROVEMENT 11.3

Ask participants to list the four core principles of continuous improvement. Discuss ways that each principle might be implemented.

Show Transparency 3 (Figure 2) and explain the diagram.

Review and discuss the objectives of quality project management.

Explain that a primary objective of project quality management is to identify barriers to quality performance.

3.0.0 ◆ CONTINUOUS IMPROVEMENT FUNDAMENTALS

If a company does not strive to improve its processes, it will not be a credible competitor in today's construction industry. Continuous improvement of internal and external processes has become an integral part of many construction companies.

Successful organizations know that improved process increases productivity and ultimately contributes to satisfied clients. Gone are the days when quality was based solely on conforming to specifications and standards.

Continuous improvement within an organization focuses on four core principles:

- *Customer Focus* – Understand and respect the needs and requirements of customers. Strive to meet or exceed customer expectations.
- *Employee Involvement* – Provide employees (internal customers) the resources, training, and opportunity to influence decision-making through improvement efforts.
- *Results-Based Decision-Making* – Develop programs and services that are results-oriented and based on the use of reliable data.
- *Quality Improvement Focus* – Create a company environment and business processes that promote continuous improvement.

Continuous improvement is often described as a merger between the company's business process and workforce. Process is built into a company's business model. A process consists of a business process flow and data flow, whereas workforce procedures deal with material flow and specific job-related tasks (*Figure 2*).

3.1.0 Objectives of Project Quality Management

The following are some of the major objectives of project quality management:

- Improve efficiency and performance
- Provide an opportunity for all employees to participate in the implementation and attainment of the project's objectives
- Draw upon talents, experience, and knowledge of the employees to improve operations and customer satisfaction
- Encourage decision-making at the most appropriate level on the project
- Improve communication

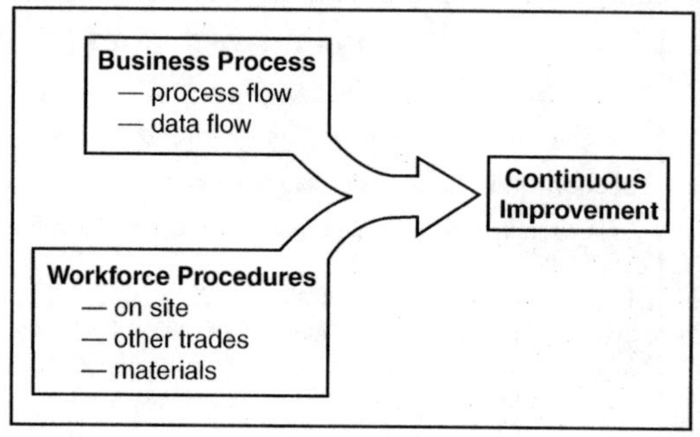

Figure 2 ◆ Continuous improvement cycle

- Promote teamwork
- Acknowledge the value of each employee by recognizing individual contributions to the project
- Provide an opportunity for personal growth and development so that each employee may realize his/her full potential
- Develop leadership skills necessary for effective management

Whether a company has a process in place or not, the project manager is responsible for meeting quality standards.

3.2.0 Main Purpose of Project Quality Management

The central purpose of a project quality management process is to identify barriers to quality performance and empower employees to remove them, within the limitations of the company, in order to meet the needs of both internal and external customers. Barriers to quality performance exist in any work environment and range from the specific way a job is performed to the relationship between a supervisor and his crew. Research has found that most barriers to quality stem from work processes and not from personal relationships.

If work processes occur without problems, the relationships among those involved are assumed to be good. For a project to meet quality standards, barriers must be identified and prevented or removed by those who are dealing most directly with those barriers. For example, if a specific activity often results in rework, those involved should be given the responsibility, authority, and resources to make the needed

Instructor's Notes:

11.4 PROJECT MANAGEMENT

changes to correct the situation. The resources needed will not only be physical, such as new tools or equipment, but other resources like the amount of time needed to solve the problem that is causing the rework.

The resources a company makes available will depend on such things as financial strength, upper management's level of commitment to the project quality management process, and the skills and knowledge company employees need to solve problems. Project managers are key players in this process, since they empower employees to identify and remove barriers in their work environment by applying the basic concepts discussed next.

3.3.0 Project Quality Management Process

Critical steps to implementing project quality management include project manager commitment, identifying barriers, involving employees, and solving problems.

3.3.1 Project Manager Commitment

Project quality management begins with your commitment as the project manager. Also important is being aware of the process's essential fundamentals and components through research and communicating with successful project managers who can guide and mentor your project quality management efforts. Once you are up to speed on the fundamentals and components, the next step is to identify and dedicate needed resources to use the process on a project. This may mean convincing upper management of the need for such a system and gaining approval for the acquisition and deployment of the resources needed.

At the same time, you must be ready to empower employees with authority and responsibility as well as provide needed skills and knowledge through training. This, in turn, will remove barriers to quality performance. Even if there is no support from upper management, you can still use most concepts and tools presented in this module. In time, you'll be able to document enough positive results to make a convincing argument to upper management for providing a higher level of support in the future.

3.3.2 Identifying Barriers to Quality Construction

Determining barriers to quality construction begins by identifying jobs in which quality-related problems are occurring or have occurred in the past. These might include the following:

- Rework
- Falling behind schedule
- Running over budget
- Subject to special warranties
- Excessive waste

Usually, the management team is aware of such problems. If not, this information can be readily attained by asking company employees. One of the greatest opportunities for improvement is to reduce rework. The average cost of rework to the contractor is 12 percent of the project's total installed cost. These are dollars that are coming straight from the contractor's potential profit, so any reduction in rework will result in a higher profit margin—not to mention the probability of repeat business from the owner and a more competitive market position for the contractor.

Many problems on a construction project involve both employees and people from outside the organization. During the early stages of implementing project quality management, focus on problems inside the organization, like communications between the office and the field, or a work activity that's being performed only by company employees. Working on internal problems first gives you the chance to work out any process difficulties before extending your efforts outside the organization.

3.3.3 Employee Involvement

The next step is involving the project employees in the process. You will be encouraging them to take ownership for their work and to solve problems without waiting to be told. Here are some reasons why employees should be involved:

- Management does not have all the answers.
- Most people have ideas about how their tasks can be performed more effectively and efficiently.
- Those closest to the problem often have the best solution.
- An almost unlimited source of knowledge and creativity can be tapped through employee involvement.
- People are often willing and eager to share their thoughts and participate in developing solutions to business problems.

Have participants list resources that may be required to remove barriers to quality performance.

Explain the importance of the project manager's commitment and empowerment of employees.

Show Transparency 4, Barriers to Quality Construction. Discuss these examples of problems that can affect quality.

Have participants suggest areas of opportunity for improvement in a typical company.

Explain how problems need to be addressed both internally and externally.

Point out that company employees are a good source of information regarding issues.

Discuss other reasons to include employees in the quality management process.

Ask students if they think Samantha took advantage of employee feedback to resolve the situation.

MODULE 44111-08 ◆ CONTINUOUS IMPROVEMENT 11.5

Explain the role of problem-solving teams in quality improvement.

Review the steps for solving a problem. Use a job-related example to work through the steps.

To be effective in solving quality-related problems, employee involvement must be carefully planned. Since solving problems will be done by those affected by them, the first thing you should do is assign a team for each problem. The team should consist of between five and eight individuals who are part of the process related to the problem. For example, if the problem is in the process of installing electrical conduit in ceilings and it is being performed by three crews, the team could be made up of craftworker representatives from each of the various crews, one or two first-line supervisors, and the superintendent. The project manager might even serve on the team.

If only one crew is involved in installing the conduit, the entire crew could make up the team, along with the superintendent and possibly the project manager. In either case, the highest-level manager would serve as team leader. The manager has the greatest amount of authority to obtain the needed resources that the team may need to solve the problem. The ultimate objective of implementing a comprehensive project management plan is to have every person work with others (such as their assigned crew) to seek ways of making improvements in their work as part of their day-to-day activities.

One recommendation is that those serving on the quality team be volunteers—that is, when asked to serve, they agree to do so unconditionally. This will result in less resistance to project quality management and a greater chance that the problem will be solved in a relatively short amount of time. Also, using volunteers sends a more positive message to the other employees and they will likely volunteer to serve on future teams.

Once the team has been organized, it is important that it receives training. This can be done using qualified in-house personnel, such as the project manager or others. Training should take place during normal working hours and be completed within one to two months. It will be up to you to determine the extent of training required, and it may be advantageous to seek a consultant. However, it is beyond the scope of this module to provide further details on this issue. A cautionary note to keep in mind as you're setting up your program is that both time and resources will be wasted if the quality team does not have the skills and knowledge to solve problems.

3.3.4 Solving the Problem

It is critical that only one or two problems are tackled in the early stages of the quality program, and that the problems selected should have the potential to be solved and completed within four to six months.

Keep away from non-quantifiable problems or those that deal with personal feelings, such as company morale. Research has shown that if you solve day-to-day technical problems facing employees, the more complicated ones, such as personnel morale, will take care of themselves. Working on relevant problems enables the team to arrive at a successful solution as they go through the following steps:

Step 1 Develop a statement that describes the problem, when it occurs, and its extent.

Step 2 Develop a flowchart of the process that is creating the problem. Identify the steps (or tasks) in the activity, or the job in which the problem occurs. Process mapping is a technique that can be effective in solving problems.

Step 3 Identify both the suppliers and customers in the various steps or tasks in the process.

Step 4 Indicate all possible causes of the problem.

Step 5 Agree on the basic cause of the problem.

Step 6 Develop an effective solution and action plan, then implement them.

Step 7 Establish a process to monitor the effort.

Step 8 Assess the results of the implementation effort and make any needed corrections to deviations.

Step 9 Follow up in three to four months (or any realistic time frame) to ensure the problem has been completely solved.

For many problems, you will be able to lead the team through the steps in a structured and timely manner. For more complicated problems, you may need to call on others from either inside or outside the company for assistance. If you are not the team leader, then select a leader who has the necessary skills, knowledge, and experience to be effective.

Instructor's Notes:

11.6 PROJECT MANAGEMENT

When beginning the process of solving the problem, it's important to establish a time limit. Your chances of success are greatly increased if you use the four-to-six-month guideline for the first two or three problems. In addition, any needed resources must be identified. The team needs to be ready to deal with any limitations that may exist within the company.

The team should have the opportunity to meet for at least two hours once or twice a week during the normal workday. Although these meetings will have an immediate negative effect on productivity, in the long run the investment will reap positive results.

Once a few internal company problems have been solved and changes made, others outside the company can become involved in the problem-solving process. For example, if a company is experiencing problems with a specific material and decides to organize a quality team to solve it, a representative of the supplier, and even possibly the manufacturer, should be a part of it.

3.3.5 Mapping the Process

The best way to understand a process is to map the workflow in the form of a flowchart. A flowchart is simply a diagram that illustrates the workflow or operations in a process in a graphical form. Mapping clearly depicts the steps that are necessary to do the work and complete a process. The mapping process provides insight into how to improve the process. Mapping also provides a template for determining where standards should be set regarding accuracy and cycle time. Once standards are in place, defects can be counted and cycle time can be measured. Data from these measurements can be applied to quality improvement techniques and be effectively used to insure continuous improvement. Diagramming the workflow process can help you see how to reduce errors and defects, increase productivity, and satisfy stakeholders.

The people closest to the process should be involved in the mapping. There are a variety of diagramming techniques that can be applied to create a **process map**. First, determine the start and stop points of the process. Start at a high level to capture the overall picture and then break the steps in the process into smaller components that track the actual work or activity performed.

You can use standard flowcharting techniques and symbols to represent workflow, inputs, and procedures to diagram your map. The process can be laid out informally on a pad, flip chart, or white board, using sticky notes to move items around. Once the process is fully diagrammed, it can be formally documented. There are many software tools that make the flowcharting process easy to diagram.

The mapping process can be accomplished from different perspectives. The following three examples illustrate how flowcharting is an effective way to think through a process. The visual representation of the process map makes it easier to logically determine where improvements can be made and waste eliminated. Review *Figures 3, 4,* and *5* to see examples of how mapping can be used to improve day-to-day problems such as removing and replacing equipment. On larger issues, there may be an assigned group to think through the procedures and policies. This will take more time, but the concept of thinking end-to-end is essentially the same as the three examples.

In Example A (*Figure 3*), the worker realizes a piece of equipment is defective, tags it, takes it out of service, and then notifies the foreman. The foreman inspects the equipment, determines if it is possible for the equipment to be repaired, then finds a worker to make the repairs. If the worker is able to repair the equipment, it is put back into service. If the worker cannot repair the equipment, he notifies the foreman again, who then reports the defective equipment to the supervisor. The supervisor identifies the defective piece of equipment, submits paperwork for replacement, awaits approval, and replaces the equipment.

In Example B (*Figure 4*), the worker recognizes the equipment is defective, but attempts to repair the equipment first. If he is able to repair the equipment, the worker removes tags and puts it back into service. If the worker is unable to repair the equipment, he notifies the foreman, who then notifies the supervisor. The supervisor verifies the defective equipment, then begins paperwork, awaits approval, and replaces the equipment. In this map, the worker attempts to repair the tool before notifying the foreman, cutting out the foreman's decision of need for repair.

And, finally, in Example C (*Figure 5*), the worker recognizes defective equipment and attempts to repair it. If the equipment is unable to be repaired, he goes directly to the supervisor who verifies the defective equipment, starts the paperwork, awaits approval, and then replaces the equipment. In this example, the foreman is cut out and the problem is taken from the worker directly to the supervisor.

Explain how mapping a flowchart helps to get a clear picture of the steps of a process.

Describe other benefits of process mapping.

Demonstrate how to map a process. Describe different techniques.

Show Transparencies 5–7 (Figures 3–5). Walk through Examples A, B, and C.

See the Teaching Tip for Section 3.3.5 at the end of this module.

Assign reading of Sections 4.0.0–6.0.0.

MODULE 44111-08 ◆ CONTINUOUS IMPROVEMENT 11.7

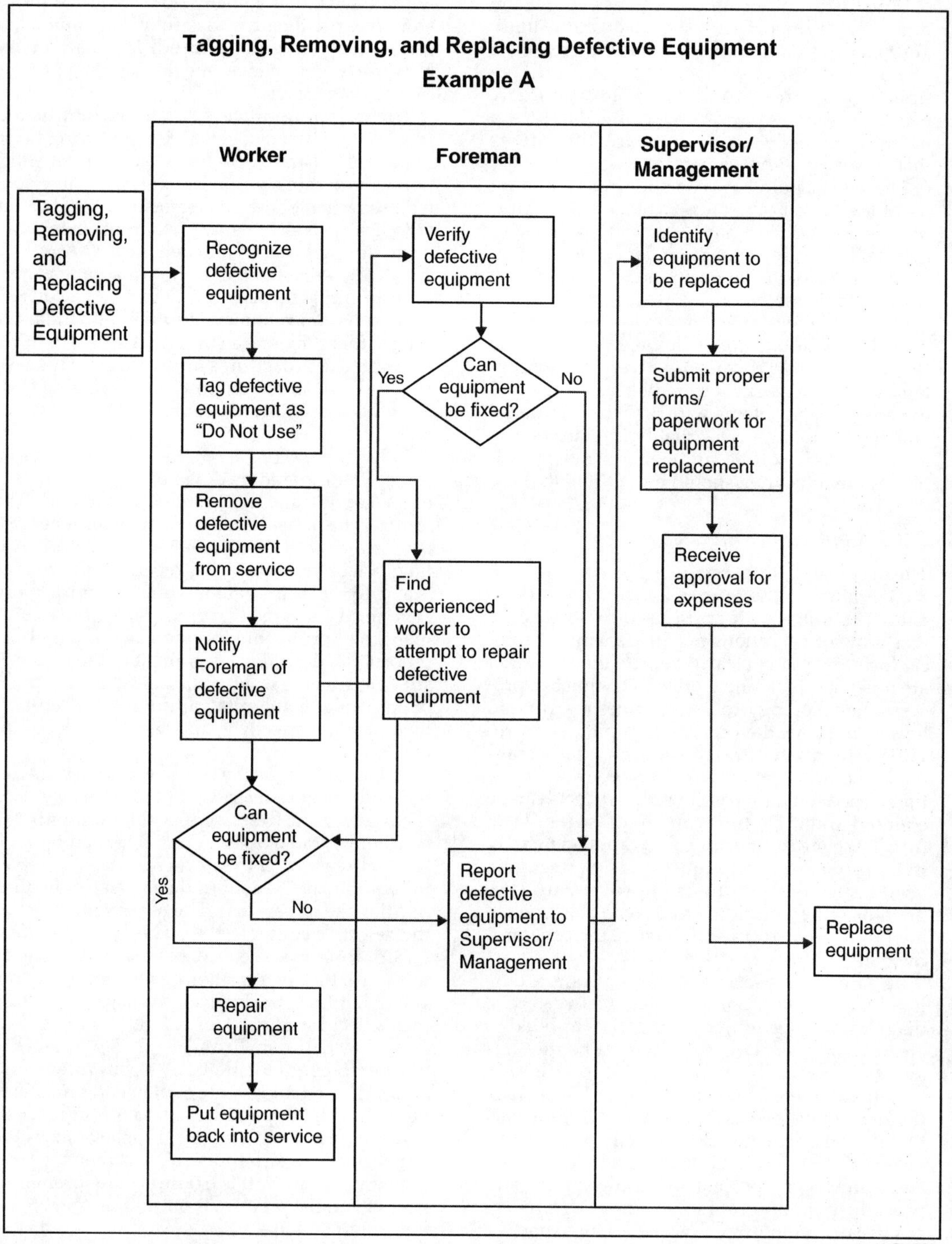

Figure 3 ◆ Example A, tagging, removing, and replacing defective equipment.

Instructor's Notes:

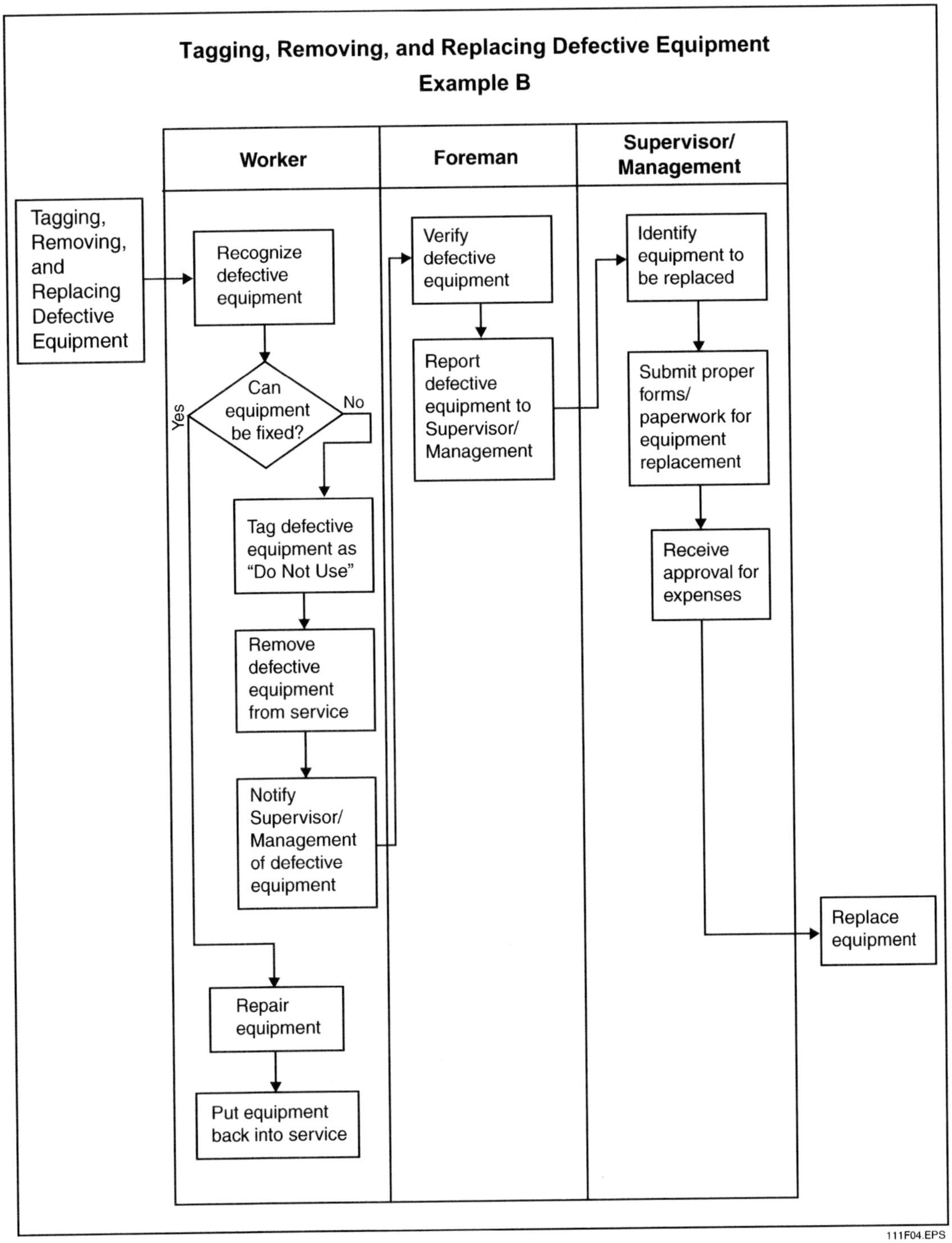

Figure 4 ◆ Example B, tagging, removing, and replacing defective equipment.

MODULE 44111-08 ◆ CONTINUOUS IMPROVEMENT 11.9

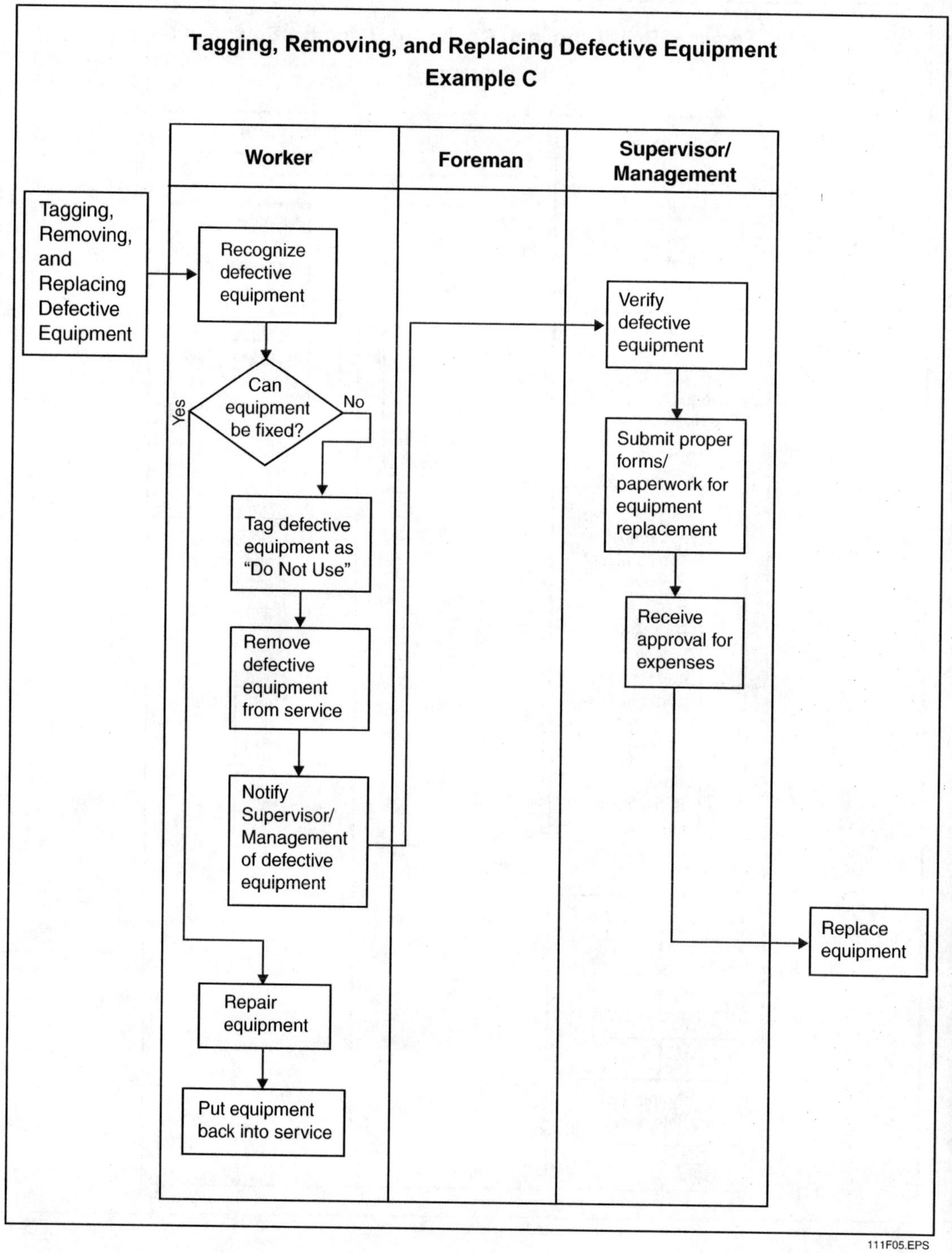

Figure 5 ◆ Example C, tagging, removing, and replacing defective equipment.

Instructor's Notes:

11.10 PROJECT MANAGEMENT

4.0.0 ◆ IMPLEMENTING CONTINUOUS IMPROVEMENT

A successful continuous improvement program permeates every aspect of the job to be performed and the processes that govern the work. For large firms and jobs, a continuous improvement program can be an elaborate business process that continually measures results. Smaller firms with smaller jobs may use less formal processes to gain improvements. No matter the size of the organization, continuous improvement will show results. Objectives of a continuous improvement program can be summarized as follows:

- Satisfying stakeholders
- Eliminating waste
- Maximizing productivity
- Preventing, not correcting

All paths in the continuous improvement process lead to stakeholder satisfaction.

A major objective of a continuous improvement program is to eliminate waste. That is accomplished through improving existing processes and seeking new methods to achieve the goal of quality delivery. Measurement and analysis are key components of a continuous improvement process. They affect the tasks, procedures, and policies associated with a specific job and with the company.

4.1.0 Applying Project Quality Management

Project quality management can be applied to the typical construction project by implementing crew meetings and constructive questions.

4.1.1 Crew Meetings

During crew meetings, make certain your supervisors explain to their workers what quality work is and how to achieve it. The workers should be asked their opinion concerning how the work is to be performed, and the time and other resources it will take. Supervisors should seriously consider any ideas that will improve the way the work is performed. As part of assigning work, the supervisor should emphasize the necessity to immediately cease any work that will result in unacceptable quality and not restart until problems have been resolved.

4.1.2 Questions To Ask To Improve Quality

The following questions can be used to help improve work quality. They should be asked in meetings of the project team members during the various phases of the construction project.

- What is preventing you from doing the job right the first time?
- What is being done to solve any quality-related problem? What help is needed?
- Are you satisfied with your recent work? Are your customers? Did you meet requirements without hassle?
- Do you know the contractual quality requirements for upcoming work? How will you determine compliance?
- For upcoming work, who are your customers? Identify both internal and external stakeholders when answering this question.
- What constitutes satisfaction for those customers? How will you determine customer satisfaction?
- What can we do to improve our work? Even if it's not "broken," there may be a better way.

4.2.0 Assessing Project Quality Management

The assessment process should occur during and upon completion of the project, and involve input from all parties. The input should be based on both qualitative and quantitative measurements. Keep in mind that assessment is a two-way process. All parties should be provided with the opportunity to formally assess each other. Parties performing work for the contractor should be formally evaluated by the contracting company and informed of the outcome of the rating. The contracting company should ask all parties with whom it had a relationship on the project to evaluate its performance. Finally, the company should conduct a post-project performance review with all stakeholders to assess its quality performance results. From these ratings, assessments, and reviews, lessons learned can be applied to the process of continuous improvement.

Methods of assessment include the following:

- Discussions at weekly project progress review meetings
- Individual discussion with parties to the project during construction
- Post-project review meetings
- Administration of satisfaction surveys and rating forms

Review the basic concept of continuous improvement and explain reasons to implement a continuous improvement program.

Show Transparency 8, Objectives of a Continuous Improvement Program. Review and discuss each objective.

Explain the role of crew meetings in the continous improvement process.

Review the list of questions that might be asked regarding quality improvement.

Describe the assessment process and various methods of assessment.

Show Transparency 9, Methods of Project Quality Management. Review and discuss each point.

Ask students what QC measures Samantha could employ to prevent serious problems from happening in the future.

MODULE 44111-08 ◆ CONTINUOUS IMPROVEMENT 11.11

Explain that quality issues should be addressed at weekly meetings and summarized at review meetings.

Discuss each question that should be asked at a post-project review meeting.

Explain the role of surveys and rating froms in the continuous improvement process.

Show Transparencies 10 and 11 (Figures 6 and 7). Review the sample rating forms.

Show Transparencies 12 and 13 (Figures 8 and 9). Review the sample satisfaction surveys.

Ask participants to list areas where a continuous improvement program might be effective.

Discuss typical applications for improvement.

The issue of quality needs to be placed on the agenda of every weekly project progress review meeting, and time allocated to discuss quality-related issues, not only problems. In addition, when quality problems arise on the project site, don't wait until the next weekly meeting to discuss them. Initiate immediate actions to correct the problem. This means talking with the affected individuals as problems are encountered.

Whether there were quality-related problems on the project or not, conduct a post-project review meeting. Include all parties involved in the project, including the architect/engineer, vendors, key employees that were involved in the project, and other stakeholders who can provide valuable feedback. Limit the review meeting to no more than a day and segment the sessions to include participants only in the review session that involves their input.

The review meeting should address both successes and problems that were encountered. Typical issues to discuss include the following:

- Areas that contributed to the success or failure of the project.
  - What worked well and what did not?
- Constraints that limited performance.
  - What did not work well, and how could it be prevented in the future?
- Problems that occurred and whether they could have been anticipated.
  - What kept us from recognizing these problem areas in advance?
- Innovations and breakthroughs that occurred and the outcomes they produced.
  - Can we adapt any of these to future projects?
- Improvements we might have made to this project and what effect they might have on our processes for future projects.
- Client and stakeholder satisfaction.
  - Did we achieve it? How do we know?
- Working relationship among all parties.
  - Did we have a good working relationship with all parties? Were there problems among the parties that we should know about? How do we know?

Finally, using satisfaction surveys and rating forms is an effective and efficient method of assessing quality. Rating forms are used to provide feedback to other contractors, vendors, and organizations the company has worked with on various quality-related items. If a company's ratings are generally good, most likely the group providing the rating will consider it for future projects. If the ratings are not generally good, it will be important for the representatives of the company receiving the unsatisfactory report to meet with rating representatives and to determine the specific problems to ensure future good relationships. *Figures 6* and *7* are examples of two rating forms.

Satisfaction surveys are for the purpose of having others rate the company's performance. They are completed in the same manner as the rating forms and used in the same way. *Figures 8* and *9* provide two examples of satisfaction surveys.

The company should go over the project at least a month before completion before warranties expire to assess any problems and report any warranty items to manufacturers. The company should attempt to have warranties extended until all problems are resolved. The company should also provide all commercial operating and maintenance manuals to the client.

4.3.0 Continuous Improvement Applications

Every process in the organization is a candidate for improvement. An organization that focuses on quality does not differentiate between the work-related processes and the workplace processes. A continuous improvement program is effective when starting a new project; developing a new or improved process, product or service; defining a repetitive work process; prioritizing problems or root causes; and implementing change.

Using lean management thinking and the principles discussed in the *Scheduling* module uncovers a number of opportunities for continuous improvement. Some of these will involve collaborating with your suppliers and partners.

Some examples of typical applications for improvement follow. Start the program with one key issue. Establish a regular schedule for introducing new issues to tackle.

Managing physical and human resources wisely and cost-effectively is an area that typically needs improvement. Recalling that eliminating waste is a major objective of quality and continuous improvement, actions can be taken to alter the status quo. Materials can be scheduled to arrive when needed, rather than sitting on the job site unused. The team at the work site can be motivated to consider changes and processes to perform the job better and faster. Have them rethink repetitive processes to determine how to eliminate waste and employ new techniques to improve performance. What measures can be taken to improve job safety to eliminate injuries and losses from accidents?

Instructor's Notes:

11.12 PROJECT MANAGEMENT

CONTRACTOR RATING FORM

Contractor Name ____________________ Project Name ____________________

Contractor Address ____

____

Contractor Project Manager ____________________ Date of Rating ____________________

In order to provide you with feedback to assist your organization in any assessment of our working relationship on the above noted project and to document this feedback for our internal purposes, our project management team has rated your performance on the below noted quality-related items that our organization feels are critical to a quality project.

ITEM	**RATING – WITH 5 BEING THE HIGHEST**				
Effectiveness of Communications	5	4	3	2	1
Overall Coordination	5	4	3	2	1
Overall Cooperation	5	4	3	2	1
Quality Workmanship	5	4	3	2	1
Adherence to Project Schedule	5	4	3	2	1
Adherence to Project Budget	5	4	3	2	1
Effectiveness of Contractor Management	5	4	3	2	1
Close-Out Documents	5	4	3	2	1
Punch List Completion	5	4	3	2	1
Safety Record	5	4	3	2	1
Overall Satisfaction	5	4	3	2	1

The following problems were noted as having a negative impact on the project:

____

____

____

We will consider your company on future projects. _______ Yes _______ No

111F06.EPS

Figure 6 ◆ Contractor rating form.

MODULE 44111-08 ◆ CONTINUOUS IMPROVEMENT 11.13

VENDOR RATING FORM

Vendor Name ______________________ Project Name ______________

Vendor Address ____

____

Vendor Contact ______________________ Date of Rating ______________

In order to provide you with feedback to assist your organization in any assessment of our working relationship on the above noted project and to document this feedback for our internal purposes, our project management team has rated your performance on the below noted quality-related items that our organization feels are critical to a quality project.

ITEM	RATING – WITH 5 BEING THE HIGHEST				
Timely Delivery	5	4	3	2	1
Quality of Product(s)	5	4	3	2	1
Adherence to Other Purchase Order Requirements	5	4	3	2	1
Submittal Procedures	5	4	3	2	1
Effectiveness of Communications	5	4	3	2	1
Overall Coordination	5	4	3	2	1
Responsiveness to Requests	5	4	3	2	1
Overall Satisfaction	5	4	3	2	1

The following problems were noted as having a negative impact on the project:

____

____

____

We will consider your company on future projects. ______ Yes ______ No

111F07.EPS

Figure 7 ◆ Vendor rating form.

Instructor's Notes:

____

____

____

11.14 PROJECT MANAGEMENT

CONTRACTOR SATISFACTION FORM

Project Name ____________________

Contractor Completing Form ____________________

Date Form Completed ____________________

At the completion of each project, our company conducts a post-project review for the purpose of identifying both positive and negative experiences which impacted on the quality of the project. It is important, as part of the review, to have input from the various parties to the project with which we worked. Therefore, please take a few minutes to complete this form and return it within the next ten (10) days in the enclosed stamped, self-addressed envelope. Thank you.

ITEM	RATING – WITH 5 BEING THE HIGHEST				
Relationship with our Project Management Team	5	4	3	2	1
Completeness/Correctness of Communications/Documents	5	4	3	2	1
Overall Coordination	5	4	3	2	1
Overall Cooperation	5	4	3	2	1
Safety Performance	5	4	3	2	1
Meeting Project Schedules	5	4	3	2	1
Achieving Project Budget	5	4	3	2	1
Close-Out Procedures	5	4	3	2	1
Punch List Completion	5	4	3	2	1
Responsiveness to Contractor's Needs	5	4	3	2	1
Overall Satisfaction	5	4	3	2	1

Comment on any specific problem areas you encountered with us on this project.

____

____

____

Would you want to work with our firm again? _______ Yes _______ No

111F08.EPS

Figure 8 ◆ Contractor satisfaction survey.

____

____

____

____

MODULE 44111-08 ◆ CONTINUOUS IMPROVEMENT 11.15

CLIENT SATISFACTION SURVEY

Project Name ______________________________

Date Form Completed ______________________________

Name of Client ______________________________

At the completion of each project, our company conducts a post-project review for the purpose of identifying both positive and negative experiences which impacted on the quality of the project. It is important, as part of the review, to have input from the various parties to the project with which we worked. Therefore, please take a few minutes to complete this form and return it within the next ten (10) days in the enclosed stamped, self-addressed envelope. Thank you.

Please circle the appropriate number:

	Unsatisfactory	Satisfactory	Excellent
I. Overall rating on how we performed on this project.	1 2	3 4	5 N/A
II. Key Function Rating			
A. Personnel (Overall)			
1. Principal-in-Charge	1 2	3 4	5 N/A
2. Project Manager	1 2	3 4	5 N/A
3. Project Supervisor	1 2	3 4	5 N/A
B. Project Documents			
1. Completeness of documents	1 2	3 4	5 N/A
2. Coordination of documents	1 2	3 4	5 N/A
3. Appearance of documents	1 2	3 4	5 N/A
C. Meeting Project Schedules	1 2	3 4	5 N/A
D. Achieving Project Budget	1 2	3 4	5 N/A
E. Responsiveness to Client Needs	1 2	3 4	5 N/A
F. Approach to Design			
1. Originality	1 2	3 4	5 N/A
2. Flexibility	1 2	3 4	5 N/A
3. Energy Awareness	1 2	3 4	5 N/A
G. Overall Professionalism	1 2	3 4	5 N/A
H. Overall Project Quality	1 2	3 4	5 N/A

III. Would you want to work with our firm again? ________ Yes ________ No

IV. Would you want to work with our project team again? ________ Yes ________ No

V. Please list any additional information/comments that may assist us in becoming more responsive to your needs.

____

____

____

111F09.EPS

Figure 9 ◆ Client satisfaction survey.

Instructor's Notes:

____

____

____

11.16 PROJECT MANAGEMENT

Focusing on quality suppliers rather than the low bidder, and reducing the number of suppliers has been shown to reduce costs and rework in the long run. Work with a select cadre of suppliers that subscribe to your company's quality and improvement standards. Enroll your suppliers in your own process of continuous improvement, so that each party measures and analyzes the processes and finds solutions for ongoing improvement. Team with them to develop a score card for all parties on the project and hold meetings to brainstorm and implement improvements.

Enlist other stakeholders in your continuous improvement program. The financing partner may be more creative with its terms when it recognizes that the improvement program is showing significant bottom line results. Build relationships with licensing agencies to reduce delays in permit processing. Ensuring stakeholder satisfaction is the ultimate payoff of a continuous improvement program. Check how you are doing on a regular basis. The company can measure stakeholder and partner satisfaction through regular communication that includes surveys, in-person visits and meetings, and inviting suggestions for improvement from these stakeholders.

The company's business processes and systems should be constantly improved. Eliminate costly mistakes in hiring. The hiring process should be fine-tuned so that new hires are the right fit for the company. Improve financial practices. Look at your accounts payable record and eliminate inaccurate contractor payments. When the continuous improvement program becomes an integral part of the company's culture, it should be integrated into the operational processes and the business systems. Obtain the tools to support the system of continuous improvement.

Create a learning organization. Provide opportunities for continued training and learning. Share experiences and company knowledge to eliminate redundancy and shorten the time to deliver the product or service. Provide systems and tools that enable collaboration and the collection of company-wide knowledge that can reduce waste.

Creating a culture of continuous improvement brings significant change to a company. It creates opportunities for every member of the team. At the same time, change makes some people uncomfortable and fearful. Communication is a critical component of a continuous improvement program. Management should build awareness of the need and opportunity for improvement. Part of the continuous improvement program is to involve all employees in the process and help them effect the change. Solicit input from the entire organization. Communicate results.

5.0.0 ◆ MEASURING IMPROVEMENT

You cannot know if a planned improvement succeeded unless you measure it. Evaluations are used to measure the success or failure of a change. Some plans do fail and they should be considered learning experiences. This section describes a few tools of measurement that allow you to solve problems and make decisions.

Measuring and evaluating improvement occurs throughout the duration of the process or program selected for improvement. In most cases, the change will be realized through incremental steps, making measurement an ongoing task.

Before we discuss methods of gathering data and performing measurement and analysis, it is important to determine which metrics are essential to the outcome. Thomas Pyzdek, a quality management consultant and author of *The Six Sigma Handbook*, says that managers and leaders are often overwhelmed with the amount of metrics that are presented. He provides the following guidelines for understanding and assessing the properties of the metrics to be used:

- The measurement must be valid and measure what it is supposed to measure. This is not always an easy task.
- The measurement must be reliable and give the same result, even when assessed by different people at different times. Some measurements are subjective, such as a stakeholder experience; however, subjective measurements can usually be crafted to evoke the same type of response from all respondents.
- A critical few metrics can be used to determine an outcome. Called critical-to-quality metrics (CTQs), they can be identified using subject matter experts, logic, common sense, or observation. Their importance can be assessed using statistical methods.
- If it is not clear which metrics are critical, they can be pared down by examining the correlation between them. Pyzdek recommends tracking the one that is easiest or cheapest to obtain.

Pyzdek suggests that the properties of critical metrics that yield a solid model of cause and effect can be summarized as follows:

- They are an important outcome.
- They are an input that affects an important outcome.
- They measure something you can control or something that can be compensated for by something you can control.

List opportunities and suggestions for implementing a continuous improvement program.

Discuss the challenges of creating a company culture of continuous improvement.

Explain that evaluations are used to measure the success or failure of planned improvement.

Review the metric guidelines for gathering data and performing measurements.

MODULE 44111-08 ◆ CONTINUOUS IMPROVEMENT 11.17

Review the metrics that a construction company would typically want to measure.

Explain how statistical methods are used to measure improvements.

Show Transparency 14 (Figure 10). Demonstrate how to construct a Pareto Chart.

Typical metrics that a construction company would measure include the following:

- Stakeholder satisfaction
- Schedules and on-time delivery history
- Number of defects and rework
- Cost of rework
- Safety
- Business management processes

Measuring improvement requires using statistical methods. These can be simple forms such as **Pareto analysis** or more complex methods that include multivariate analysis.

Pareto analysis is a good place to start to become familiar with measurement and analysis. More complex measurements and systems can be implemented as your continuous improvement process advances.

The **Pareto Chart** (*Figure 10*) is a graphical tool for ranking most to least significant causes. It is based on the principles of Vilfredo Pareto, a 19th-century Italian economist, and first defined by J. M. Juran. Widely known as the 80/20 Rule, it is based on the premise that a small percentage of a group accounts for the largest fraction of a value or impact. That is, 80 percent of the effects come from 20 percent of the possible causes. Juran labeled these 20 percent of causes the "vital few."

Some sample 80/20 Rule applications can include:

- 80 percent of process defects arise from 20 percent of the process issues.
- 80 percent of delays in schedule arise from 20 percent of the possible causes of the delays.
- 80 percent of stakeholder complaints arise from 20 percent of your product or services.

The goal of continuous improvement is to go after that 80 percent first by solving a few problems.

The Pareto Chart converts data into visual information, displaying in graphic form the relative importance of the differences between groups of data. It is one of the seven tools of quality management.

To construct a Pareto chart, the data are separated into segments or categories. After perform-

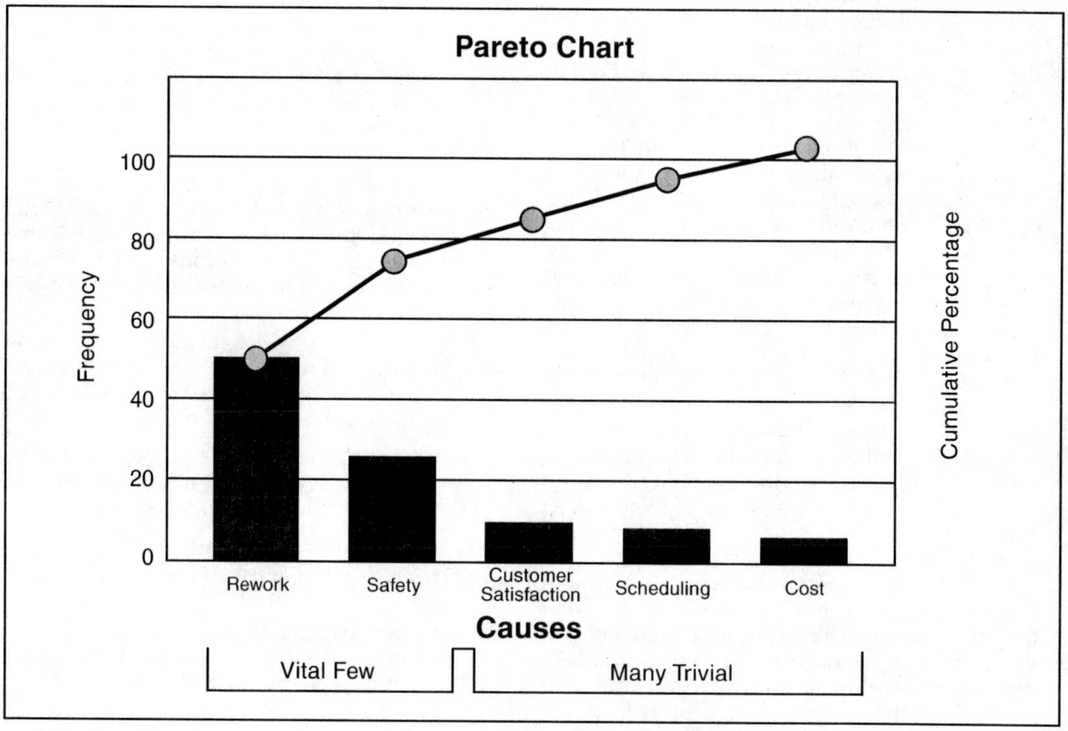

111F10.EPS

Figure 10 ◆ Pareto Chart.

Instructor's Notes:

11.18 PROJECT MANAGEMENT

ing a cause-and-effect analysis, gather the data on the frequency of the causes, ranked from most to least important. Plot the cumulative percentage of each group. The left vertical axis is the frequency (the counts for each category), the right vertical axis is the cumulative percentage, and the horizontal axis is labeled with the category names of the variables from most to least frequent. Determine the data points within each group and then construct the chart.

Typical questions the Pareto chart answers include the following:

- What are the largest issues facing our team or business?
- What 20 percent of sources are causing 80 percent of the problems (80/20 Rule)?
- Where should we focus our efforts to achieve the greatest improvements?

While Pareto is widely used, other graphical techniques that can be used to display measurements include bar charts and histograms.

6.0.0 ◆ EMPLOYEE RECOGNITION

Employee recognition is all about thanking people for doing a good job. It is also an effective motivational tool that reinforces and rewards the most important outcomes people create for any business. Each of us wants and needs to be recognized when we have worked hard and done an extra special job. When people feel they are being recognized, it reinforces the actions and behaviors of the company culture. However, recognition must be honest and specific and it should not be trivial. The recognition most important to employees is performance feedback.

Rewards should be flexible and tailored to be as diverse as your workforce. Keep in mind that not all types of recognition will work for all employees, especially across the generations. For example, Boomers like material rewards, Gen Xers like work/life balance, and Millennials want challenging job assignments. When it comes to recognition, one size does not fit all.

Susan Heathfield, a management and organizational development consultant stresses five important issues a company must consider for implementing a successful employee recognition program:

- The company needs to establish performance or contribution criteria that constitute rewardable behavior or actions.
- All employees must be eligible for recognition.
- The recognition must supply the employer and employee with specific information about which behaviors or actions are being rewarded and recognized.
- Anyone who then performs at the level or standard stated in the criteria receives the reward.
- The recognition should occur as close to the performance of the actions as possible, to reinforce behaviors the employer wants to encourage.

It is key to take the time to recognize those who helped make the process successful, specifically individuals who served on the teams. The following are some forms of recognition:

- Involving employees in discussions, analysis, and development of recommendations where they have shown an interest
- Mentioning the outstanding work or idea at appropriate meetings
- Having lunch with the employees
- Thanking employees when they have done something well or have presented a useful suggestion
- Presenting a cash incentive or material award for adopted suggestions
- Presenting meaningful, non-financial awards such as a parking space close to the entrance gate to the project

An example of a program implemented by a leading construction industry firm, MACTEC, follows a tiered approach to employee recognition:

- *Tier 1* – CEO level can include bonus program along with CEO recognition for persons who provide outstanding client service and project management.
- *Tier 2* – Most Valuable Performer (MVP) Club. Recognition for exceptional performance in one or more of the following:
  - Health and safety
  - Technical competence, professional knowledge, and self-development
  - Commitment, initiative, and productivity
- *Tier 3* – Division or Department Director Award. Cash or bonus for top-performing office in each division/company to emphasize the importance of teamwork in achieving goals.
- *Tier 4* – Bravo Cards. Formal acknowledgement for employees who go above and beyond for clients and co-workers.
- *Tier 5* – Service Anniversary Awards. CEO acknowledgement and gift selection for employees at milestone service anniversaries.

Explain how employee recognition can be used as a motivational tool.

Have participants suggest ways to reward employees who help to make the continuous improvement process successful.

Discuss the five criteria for implementing a successful employee recognition program.

Review the tiered approach to employee recognition.

MODULE 44111-08 ◆ CONTINUOUS IMPROVEMENT 11.19

Discuss the advantages of implementing an employee recognition program.

Emphasize that recognition is something that a project manger should do every day to ensure positive employee motivation.

Recognition can have big payoffs. The findings of a Facilities Operations Climate Assessment Survey conducted by Washington State University included the following:

- Recognition shows employees that they are appreciated and that their contributions count.
- When an employee goes out of his way and does something outstanding, management should signal, "Good job." This builds a culture of excellence and encourages giving that extra effort.
- Recognition improves interaction between supervisors and staff and also among staff when they recognize each other's notable achievements.
- Employee recognition decreases stress, absenteeism, turnover and related costs—essential to the firm in these times of heavier workloads and deflated morale.
- Recognition nurtures pride and enthusiasm.

Finally, recognition is something that a project manager can do, and should do, every day—not just on special occasions. A simple and honest thanks is a very positive motivator. A study of 1,500 employees conducted by management consultant Bob Nelson, author of the bestseller *1001 Ways to Reward Employees*, reported that when an employee received a personal message from a manager, that action ranked first of 67 potential sought-after incentives.

Instructor's Notes:

11.20 PROJECT MANAGEMENT

Review Questions

1. Describe continuous improvement.

____

____

2. Continuous improvement means making things better by focusing on _____.
   a. using scheduling software
   b. conducting daily inspections
   c. preventing problems
   d. filling out daily logs

3. Continuous improvement requires _____.
   a. trend analysis
   b. monitoring employee performance
   c. analyzing process and events
   d. a quality improvement focus

4. Pinpointing root causes of a process failure helps in what three major ways?
   1. ______________________________
   2. ______________________________
   3. ______________________________

5. Projects meet quality standards when _____.
   a. supervision and control is in place
   b. responsibility and evaluation is implemented
   c. technical conformance to specifications is followed
   d. barriers are identified and corrected

6. List the four steps to attain project quality management.
   1. ______________________________
   2. ______________________________
   3. ______________________________
   4. ______________________________

7. List the four core principles for continuous improvement in a company.
   1. ______________________________
   2. ______________________________
   3. ______________________________
   4. ______________________________

8. Which of the following is *not* an application for continuous improvement?
   a. focus on low bid suppliers
   b. enlist stakeholders in continuous improvement
   c. reduce the number of suppliers
   d. enroll suppliers in continuous improvement

9. List three metrics that Pyzdek suggests identify a model for cause and effect.
   1. ______________________________
   2. ______________________________
   3. ______________________________

10. A company's culture is reinforced by _____.
    a. revenue
    b. retaining people and quality consultants
    c. people's actions and behaviors
    d. a clear business strategy

Have participants complete the Review Questions, and go over the answers prior to administering the Module Examination.

Summarize the major concepts presented in the module.

Summary

In this module, you learned the fundamentals of continuous improvement and your role as a project manager in the continuous improvement process. The module covered the importance of measurement in determining the problems to tackle in a continuous improvement program, making changes incrementally to see if they work, and continuing the cycle of improvement. The basics of implementing a continuous improvement program to the job and the workplace were discussed, and examples of applications that are typical candidates for improvement were summarized. You learned that employee involvement and recognition are critical components of continuous improvement.

Notes

Instructor's Notes:

11.22 PROJECT MANAGEMENT

Trade Terms Quiz

1. A statistical technique in decision-making used for selection of a limited number of tasks that produce significant overall effects is known as a ____________.
2. The constant effort to eliminate waste, reduce response time, simplify the design of both products and processes, and improve quality and stakeholder service is called ____________.
3. The ____________ is a chart that ranks common occurrences so that improvement efforts can be focused on the "vital few," as opposed to the "trivial many."
4. A diagram that shows the flow of a process and its linkages among inputs, outputs, and tasks is called a ____________.

Trade Term List

Continuous improvement
Pareto analysis
Pareto Chart
Process map

Have participants complete the Trade Terms Quiz, and go over the answers.

Administer the Module Examination. Be sure to record the results of the testing on Craft Training Report Form 200, and submit the results to the Training Program Sponsor.

MODULE 44111-08 ◆ CONTINUOUS IMPROVEMENT 11.23

Trade Terms Introduced in This Module

Continuous improvement: The constant effort to eliminate waste, reduce response time, simplify the design of both products and processes, and improve quality and stakeholder service.

Pareto analysis: A statistical technique in decision-making used for selection of a limited of number of tasks that produce significant overall effect. It uses the Pareto principle—the idea that by doing 20 percent of the work you can generate 80 percent of the advantage of doing the entire job.

Pareto chart: A chart that ranks common occurrences so that improvement efforts can be focused on the "vital few," as opposed to the "trivial many."

Process map: A flowchart or diagram of a workflow process that shows linkages between inputs, outputs, and tasks.

Instructor's Notes:

11.24 PROJECT MANAGEMENT

Resources & Acknowledgments

References and Resources

1001 Ways to Reward Employees, Bob Nelson

Heathfield, Susan, available at http://humanresources.about.com/od/rewardrecognition/a/recognition_tip.htm

MACTEC, Atlanta, GA

Simon, Kerri, "80/20 Rule," accessed at www.isixsigma.com/library/content/c010527d.asp

Six Sigma Handbook, The, Thomas Pyzdek

Washington State University Pullman, WA 99164

Acknowledgment

Liska, Roger, Ph.D., Department Chair and Professor, Department of Construction Science and Management, Clemson University, Clemson, South Carolina.

MODULE 44111-08 ◆ CONTINUOUS IMPROVEMENT 11.25

MODULE 44111-08 — TEACHING TIPS

The following are suggested activities or instructional methods to help you teach the material in this module.

General

When you call on someone to answer a question, the rest of the class relaxes or even tunes out because they expect that the question and answer will take place only between you and the participant you called on. Instead, use this technique to involve more participants in answering questions and to keep them on their toes.

1. Ask participants to define a term or explain a concept.
2. After one participant has answered, ask a participant seated nearby if the answer is right. Then ask whether a participant in the back of the room agrees.
3. Ask participants to explain why they think an answer is right or wrong.
4. Use the session to clear up incorrect ideas and encourage participants to learn from their mistakes.

Section 3.3.5 *Process Mapping*

This exercise will give participants the opportunity to map a process. Participants will need their Participant Guides and pencil and paper to takes notes for discussion. The Instructor will need a chalkboard and chalk, dry-erase board and marker, or overhead projector and screen, blank transparency sheets, and marker. Allow 60 minutes for this exercise.

1. Divide participants into teams of 3 or 4. Explain that each team should brainstorm ideas and map the process for a common job-related problem. Allow 40 minutes for process mapping.
2. Have teams display their diagrams on the chalkboard, dry-erase board, or overhead projector.
3. Review and discuss each team's process map. Ask the following questions for each:
   - What could have been done differently?
   - Are there steps that should be added or could be removed?

MODULE 44111-08 — ANSWERS TO REVIEW QUESTIONS

	Answer	Section
1.	A process that continues to achieve acceptable and exceptional quality during a construction project, and commitment by the company to provide a skilled and responsible workforce.	(2.0.0)
2.	c.	(2.0.0)
3.	c.	(2.0.0)
4.	1. reducing overlap and redundancy 2. revising or eliminating activities 3. improving stakeholder relations and satisfaction	(2.0.0)
5.	d.	(3.2.0)
6.	1. project manager commitment 2. identifying barriers to quality construction 3. involving employees 4. solving problems	(3.3.0)
7.	1. meet customer needs 2. provide ways for employees to participate 3. develop programs based on reliable data 4. create an environment that encourages it	(3.0.0)
8.	a.	(4.3.0)
9.	1. identify important outcomes 2. identify inputs that affect important outcomes. 3. measure something controllable or something that can be compensated by controlling it.	(5.0.0)
10.	c.	(6.0.0)

MODULE 44111-08 — ANSWERS TO TRADE TERMS QUIZ

1. Pareto analysis
2. continuous improvement
3. Pareto chart
4. Process map

CONTREN® LEARNING SERIES — USER UPDATE

NCCER makes every effort to keep these textbooks up-to-date and free of technical errors. We appreciate your help in this process. If you have an idea for improving this textbook, or if you find an error, a typographical mistake, or an inaccuracy in NCCER's Contren® textbooks, please write us, using this form or a photocopy. Be sure to include the exact module number, page number, a detailed description, and the correction, if applicable. Your input will be brought to the attention of the Technical Review Committee. Thank you for your assistance.

Instructors – If you found that additional materials were necessary in order to teach this module effectively, please let us know so that we may include them in the Equipment/Materials list in the Annotated Instructor's Guide.

Write: Product Development and Revision
National Center for Construction Education and Research
3600 NW 43rd St, Bldg G, Gainesville, FL 32606

Fax: 352-334-0932

E-mail: curriculum@nccer.org

Craft ______ Module Name ______

Copyright Date ______ Module Number ______ Page Number(s) ______

Description

______

______

______

______

(Optional) Correction

______

______

______

(Optional) Your Name and Address

______

______

______

CONCLUDING EXERCISES FOR THE VIDEO

1. Conduct an open class discussion to summarize the problems and resolutions that have been portrayed in the video and that have been discussed in the related module.

2. Ask the students if they would agree with the way Samantha solved the problems on her site. Ask them to place themselves in Samantha's position and review again any situations they would have handled differently. How do they think these would have affected the outcome?

3. Describe any changes that Samantha was able to implement, and have class discussion about whether or not they will have a lasting effect. Would the class recommend different or other changes that might have a better outcome?

4. Finally, ask the students if they think that Samantha was even-handed in the final resolution with Keith. Do they think Keith should have been disciplined in this manner? Or was Keith out of line with Samantha? Should she have held anyone else accountable and implemented any other disciplinary actions?

Glossary of Trade Terms

Accident: An unplanned event that may or may not result in personal injury or property damage that interrupts the normal progress of an activity. Accidents are categorized by their severity and impact and invariably preceded by an unsafe act, unsafe condition, or both.

Accountable: Subject to giving an account; answerable.

Action level: The level of exposure to a substance or physical agent (usually one half of the permissible exposure limit) that requires medical surveillance under a particular OSHA standard.

Active listening: Giving undivided attention to a speaker in a genuine effort to understand the speaker's point of view.

Activity: The process of creating a project, or part of a project, by the employment of construction resources.

Actual cost: Final cost of the project.

Addenda: Documents used to modify the contract prior to receiving the bid.

Agency construction management (CM): A firm that provides various services to owners in the course of a construction project. These services can vary from program management to design and construction.

Arbitration: A form of alternative dispute resolution; a legal alternative to litigation in which the parties to a dispute agree to submit their respective positions to a neutral third party for resolution.

Behavioral interview techniques: Interviewing based on discovering how the interviewee acted in specific employment-related situations.

Bid bond: A guarantee that the contractor will enter into a contract, if it is awarded.

Bid: An offer to perform the work described in contract documents at a specified cost.

Bidding: Making an offer to perform work described in contract documents at a specified cost.

Bill of material: A listing of all of the components and other materials used in the assembly of a project or system.

Brainstorming: A group problem-solving technique that involves the spontaneous contribution of ideas from all members of the group.

Budget: The dollar and time amount allocated by the owner for a project.

Budgeted cost: Projected cost of the project, based on estimates.

Change order: A document that modifies the original contract documents due to changes that occur after the award of the contract.

Change: An event that occurs when something passes from one state or phase to another.

Close-out: Performing final inspections and documents to prepare for handing the project over for the owners' acceptance and possession.

Conciliation: An alternative dispute resolution process in which the parties to a dispute agree to use the services of a conciliator, who then meets with the parties separately in an attempt to resolve their differences.

Construction management (CM) at risk: The form of construction management and delivery in which the CM firm assumes the burdens and risk for the project; the CM is involved in the design phase and performs the construction delivery at a firm agreed-upon price.

Construction manager: The supervisor responsible to the project manager for day-to-day activities on the construction site.

Construction Specifications Institute (*CSI*): An organization that maintains and advances the standardization of construction language as pertains to building specifications.

Construction: A building project that is typically divided into three phases: development, design, and construction.

Continuous improvement: The constant effort to eliminate waste, reduce response time, simplify the design of both products and processes, and improve quality and stakeholder service.

Cost analysis: Breaking down the costs of some operation and reporting on each factor separately.

Critical path method (CPM): A technique for planning the most efficient way to achieve a given objective by determining the activities and events required and showing how they relate to each other in time.

Delegation: The act of empowering to act for another; assignment of duties to others.

Design-build: A construction delivery system whereby the contractor provides both design and construction services.

Glossary of Trade Terms

Engineering-Procurement-Construction (EPC): A variation of the design-build approach wherein there is a single point of contact with the owner throughout the project.

Ethics: A system of moral principles, rules and standards of conduct.

Expediting: The follow-up of purchase orders which are overdue, or are required by a prescribed deadline. The process or system by which suppliers are encouraged to meet the due date for delivery of outstanding purchase orders or to effect immediate delivery of overdue orders.

Experience modification rate (EMR): A rate computation to determine surcharge or credit to workers' compensation premium based on a company's previous accident experience. Accident experience from three previous years, not to include last year, is used to determine the rate.

Fact-finding: (1) A process designed to find information or ascertain facts, e.g., "a fact-finding committee." (2) An investigation of a dispute by an impartial third person who examines the issues and facts in the case, and may issue a report and recommended settlement.

Fishbone diagram: A graphic technique for identifying cause-and-effect relationships among factors in a given situation or problem; also called Ishikawa Diagramming.

Five-minute rating: The project manager observes the task area for five minutes and records whether each worker is working or not working.

Focused inspection program: OSHA's partial inspection process focusing on the hazards that cause the most deaths in the construction industry: falls, electrical, caught in/between, and struck by.

Free on board (FOB): A pricing term indicating that the quoted price includes the cost of loading the goods into transport at the specified place.

Grapevine: An informal person-to-person means of circulating information or gossip.

Incidence rate: The number of injuries and/or illnesses or lost workdays per 100 full-time employees per year or 200,000 hours of exposure.

Lags: Activities in schedules having successors that are scheduled later than activities that have no successors.

Lean scheduling: Planning construction in such a way that leads to the continuous process of eliminating waste and promoting efficiency through techniques that include team-based practices, efficient scheduling of resources (including workforce and just-in-time materials inventory and deliverables), and adjusting schedules in real time, with the goal of meeting or exceeding all customer requirements.

Litigation: A legal dispute between parties argued in a court.

Management style: Approach adopted by managers in exercising authority, encouraging participation in decision-making, motivating staff, delegating authority, communicating information and maintaining control.

Med-arb: A dispute resolution procedure where the mediator is armed with the power to settle unresolved issues by binding arbitration in the event they are not settled through mediation.

Mediation: A voluntary process in which the parties involved in a dispute work with a neutral third party to find a mutually acceptable solution.

Mini-trial: An articulated form of mediation that includes each party to the dispute presenting a summary of its case to the other side, with a neutral party taking the role of the judge but not making any decisions.

Nonverbal communication cues: Communication that occurs as a result of appearance, posture, gesture, eye contact, facial expressions, and other nonlinguistic factors.

Occupational Safety and Health Administration (OSHA): An agency of the United States Department of Labor created by Congress under the Occupational Safety and Health Act. Its mission is to prevent work-related injuries, illnesses, and deaths by issuing and enforcing rules (called standards) for workplace safety and health.

OSHA recordable incidence rate: A computation of the total number of OSHA-defined recordable injuries and illnesses times 200,000 divided by the hours worked for the period in question. It is one unit of measure used to evaluate a company's safety performance.

Glossary of Trade Terms

Pareto analysis: A statistical technique in decision-making used for selection of a limited of number of tasks that produce significant overall effect. It uses the Pareto principle—the idea that by doing 20 percent of the work you can generate 80 percent of the advantage of doing the entire job.

Pareto chart: A chart that ranks common occurrences so that improvement efforts can be focused on the "vital few," as opposed to the "trivial many."

Passive listening: Listening to another person's message without verbally responding; the obvious limitation is the speakers do not know whether they have been understood, only that they have been heard.

Personal protective equipment (PPE): The protective clothing, helmets, goggles, or other gear designed to protect the wearer's body or clothing from injury by electrical hazards, heat, chemicals, and infection, for job-related occupational safety and health purposes.

Planning: The process of anticipating future occurrences and problems, exploring their probable impact, and detailing policies, goals, objectives, and strategies to solve the problems.

Precedence diagramming method (PDM): A schedule network diagramming technique in which schedule activities are represented by boxes; schedule activities are graphically linked by one or more logical relationships to show the sequence in which the activities are to be performed.

Pre-construction: The planning phase that takes place before on-site work begins, involving issuing contracts, schedule development, identifying labor and equipment requirements, and other preparatory work.

Problem solving: The thought processes and techniques involved in finding a solution to an issue.

Process map: A flowchart or diagram of a workflow process that shows linkages between inputs, outputs, and tasks.

Production quality standard: Set by the specifications and must be followed by all construction managers.

Production resource standard: Work accomplished within the estimate.

Production time standard: Set by the owner, agreed to by the contractor, and incorporated into the job plan and schedule.

Production: The total quantity of work accomplished.

Productivity: Output per unit of input: a measure of efficiency.

Professional development plan (***PDP***): An individualized document used to record an employee's current training needs or desires and short- and long-term career goals; a written plan for developing knowledge, skills, and competencies that support both the employer's objectives and the employee's needs and goals.

Project manager: The person responsible for the planning, coordination, and controlling of a project from inception to completion, meeting the project's requirements and ensuring completion on time, within cost, and to quality standards specified in contract documents.

Project quality management (PQM): Controlling the day-to-day use of job resources to ensure that a project is completed in accordance with drawings and specifications.

Projected cost: Anticipated cost of a project based on estimates.

Punch list: An itemized list documenting incomplete or unsatisfactory items after the contractor has notified the owner that the tenant space is substantially complete.

Purchase order: Information sent to a vendor to request a product or service; typically includes item, quantity, price, discounts, vendor information, and ship-to information.

Quality assurance: All actions taken to ensure that standards and procedures are adhered to and that delivered products or services meet performance requirements.

Quality control: The control of variation of workmanship, processes, and materials in order to produce a consistent, uniform product.

Quality: The degree of excellence of a product or service; the degree to which a product or service satisfies the needs of a specific customer; or the degree to which a product or service conforms with a given requirement.

Quantity survey: The estimate of material quantity needed for a project.

Glossary of Trade Terms

Reporting system: A system for collecting data and capable of showing details and comparisons.

Risk assessment code: A numerical rating of the risk associated with a hazard based on probability, consequence(s), and exposure.

Root cause diagram: A diagram used to illustrate the relationships between a problem and its potential root causes.

Scope: The work that must be done to deliver a product with the specified features and functions.

Shop drawings: Detailed drawings showing how building elements will be fabricated

Span of control: The maximum number of employees that a person can control in a given situation; also the supervisory ratio of from three-to-seven individuals, with five-to-one being established as optimum.

Specifications: A part of the construction documents contained in the project manual consisting of written requirements for materials, equipment, construction systems, standards and workmanship.

Spreading: The faulty practice of reporting expenses or labor hours in a different budget area from that originally assigned; generally done to avoid reporting expenses or hours that bring that segment over budget.

Stakeholder: Specific people or groups who have a stake in the outcome of the project, including internal clients, management, employees, administrators, and external stakeholders, including suppliers, investors, community groups, and government organizations.

Traditional contract delivery: The standard approach to construction development in which the owner contracts separately with the architect/engineer and the contractor.

Two-way communication: A variety of methods for obtaining feedback from others, ranging from granting opportunities for presenting viewpoints to inviting those concerned to participate actively in the decision-making process.

Verbal communication cues: Communication that goes beyond the words spoken to include intonation, stress, rate of speech, and pauses or hesitations.

Work analysis: A study of each job, aimed at effectively and efficiently integrating all tasks and activities.

Work breakdown structure (WBS): A hierarchical decomposition of deliverables which breaks down the scope of work into unique and distinguishable activities and tasks.

Workers' compensation: Provides insurance to cover medical care and compensation for employees who are injured in the course of employment, in exchange for mandatory relinquishment of the employee's right to sue their employer for negligence.

Worker exposure: The number of persons who will be regularly exposed to a toxic substance or harmful physical agent in the course of employment through any route of entry (inhalation, ingestion, skin contact, absorption, etc.).

Index

Index

D

N

O

P

Q

R

S

T

U